VLSI System Design

VLSI SYSTEM DESIGN
When and How to Design
Very-Large-Scale Integrated Circuits

SABURO MUROGA

Departments of Computer Science and Electrical Engineering
University of Illinois, Urbana

1807 1982

175 YEARS OF PUBLISHING

A WILEY-INTERSCIENCE PUBLICATION
JOHN WILEY & SONS
New York / Chichester / Brisbane / Toronto / Singapore

Library of Congress Cataloging in Publication Data:

Muroga, Saburo.
  VLSI system design.

  Includes bibliographical references and indexes.
  1. Integrated circuits—Very large scale integration.
I. Title. II. Title: V.L.S.I. system design.

TK7874.M87 1982	621.381'73	82-8598
ISBN 0-471-86090-5		AACR2

Printed in the United States of America

10 9 8 7 6 5 4 3 2 1

Preface

In recent years, LSI and VLSI chips have been incorporated into an increasingly wide spectrum of products ranging from watches, toys, video games, home appliances and personal computers to large computers. Furthermore, software and systems have been incorporated into these chips. For these reasons LSI/VLSI system design is becoming of major interest to a broad range of people in industry, from chip designers and fabrication engineers to chip users, software designers, and managers. This book presents an overview of LSI/VLSI design and its key points.

This book is the result of my lecture notes. I have been enchanted by the progress of LSI/VLSI technology and wanted to convey my excitement to students. I started to teach LSI/VLSI system design in 1974 after a few years of preparation, and since then these notes have been repeatedly revised and expanded. This was not an easy task for the following reasons: integrated circuit design covers an extremely broad spectrum of areas, each area is very complex, and the interrelationship among these areas is even more complex. But the students' enthusiastic response has pressed me to continue.

The fact of the matter is that integrated circuits not only involve the integration of gates, connections, and memories in a single chip, but they also require the students' integrated knowledge of diverse areas. It has become extremely difficult for students to obtain integrated knowledge after taking specialized courses in diverse areas, because of the extremely intricate relationships among different areas and tradeoff considerations of different design approaches. Thus a course which "pulls together" the material of specialized courses is itself now needed. I believe that such a course will be increasingly important in the future.

Although all related areas such as architecture, logic design, electronic circuits, and layout are discussed, the discussions in this book are mainly centered on logic design, for the following reasons. First, a severe shortage of logic designers is expected because each manufacturer needs to produce a number of custom-designed chips, and a certain number of logic designers are tied down to each chip throughout its design. (As shown in Table 4.7.1*b*, layout is the most time-consuming task, although mostly done by draftpersons, but logic designers who have a knowledge of architecture and electronic circuit design must oversee chip design from the initial stage to the

v

last layout stage.) Second, tradeoff considerations among different methodologies in logic design are essential for LSI/VLSI system design. Methodologies are a new important problem in logic design because the selection of an appropriate methodology in logic design shortens the chip design time, in particular in layout and testing. For example, the logic design with PLAs greatly reduces the design time since algebraic minimization, which can be implemented in CAD, can be used, and layout is simple due to the PLA's regular structure. In contrast, logic design with random-logic networks is much more time-consuming, but the performance of chips so designed is much higher, with smaller chip sizes. In each case we have to compare different methodologies in logic design to find out what we gain or lose. Third, students should know the basic theory of some design aspects because LSI/VLSI system design is by no means an elementary subject. LSI/VLSI system design is based on the integration of many different areas, such as fabrication, electronic circuitry, solid-state physics, and architecture. The discussions in this book, however, would be too shallow if no previous knowledge in any area were assumed. In this sense, some knowledge of logic design and also elementary electronics is assumed. Logic design is assumed because it is relatively easy for students to learn (even for those who have no electronics background). At the University of Illinois, students take the course with this book, after having taken the logic design course with my previous book *Logic Design and Switching Theory* published in 1979. [That book provides a theoretical background on LSI/VLSI logic design and efficiency improvement techniques of CAD for logic design, such as PLA minimization, MOS network analysis, NOR (or NAND) network design, diagnosis, and simulation.] But students can start directly with this book by reading pertinent pages of the other book by themselves, or by supplementing them by classroom instruction. Also if readers are not interested in the details of logic design, they can probably get the flavor of LSI/VLSI system design without using the other book. Next, some knowledge of elementary electronics is assumed because LSI/VLSI technology is most directly related to electronics. But such knowledge is minimized, and readers need to know only the terminal behavior of transistors and diodes. Reading a few chapters of the book *Electronics: circuits and devices*, 2nd ed., by R. J. Smith, for example, would be helpful.

In Chapter 1, the advantages and disadvantages of the use of LSI/VLSI chips are outlined, with many examples of new products based on LSI/VLSI systems. Then, after a review of fabrication technology, cost analysis of LSI/VLSI chips, which is affected by many complex factors, such as yield, chip size, design man-power, and production volume, is discussed in Chapter 2. Such economic considerations are essential for chip design methodology in later chapters. In some cases the use of MSI or SSI packages, instead of LSI or VLSI chips, may be more appropriate in designing digital systems. In Chapters 3 and 4, all the key features of the major logic families are discussed, along with logic design procedures and layout design. In Chapter

5, all logic families are compared, and the technical trends are outlined. In Chapter 6, all types of memories which are currently available for LSI/VLSI chip design are summarized. The use of RAMs as part of logic networks is discussed. In Chapter 7, the use of ROMs as part of logic networks is extensively discussed. Excessively long design time, mostly spent in testing and layout, is becoming a serious bottleneck as the integration size increases. In order to reduce design time, regular-structured networks, such as ROMs and PLAs, are being used more and more. Advantages and disadvantages of ROMs as substitutes for logic gates are discussed. The use of ROMs is not straightforward. Designers must develop appropriate algorithms in using ROMs, tailored to specific computational tasks. Such algorithms are discussed as well as minimization procedures of ROM size. In Chapter 8, the use of CAD for every aspect of LSI/VLSI system design is outlined. In Chapter 9, full-custom and semi-custom designs are discussed. The full-custom design of LSI/VLSI chips is best in terms of performance improvement and cost reduction, but its design time is the longest. In order to find appropriate tradeoffs among design time, cost, performance, and possibly other factors, a gamut of semi-custom design approaches has been developed. For example, the use of microcomputers is one approach to reduce design time. All these problems are discussed from the viewpoint of tradeoffs. Problems in designing microprocessor chips are also outlined. In Chapter 10, the problems of system design as well as hardware–software tradeoffs are discussed, and then future problems in designing VLSI systems are predicted.

To my pleasant surprise, a large number of students who did not show any previous interest in related areas, such as electronics and fabrication, started to take related specialized courses, after having been stimulated by the chip design course in which this book was used. A fair number of these students have become LSI/VLSI system designers or have joined related CAD professions.

Because of the complexity of LSI/VLSI design and its popularity as a research topic, certain problems arise in writing a book which attempts to provide an overview of the technology. For example, to avoid excessive complexity, related areas have not been discussed in detail and, consequently, references which present tutorials or surveys have been, for the most part, cited. Also, one of the main objectives of this book is to compare different technologies. Yet a completely fair comparison is not really possible, because new results in one technology are always compared with old results in other technologies, and because technologies are in continual change. Similarly, a staggering number of papers have been published on every aspect of chip design, and I have exerted every effort not to overlook any key papers which are appropriate to the context. More particularly, in the enormously broad spectrum of areas related to LSI/VLSI system design, I have made every effort not to misrepresent any topics and not to overlook key points. Ambiguities and mistakes in original articles can sometimes aggravate this situation. In particular, many layout figures in publications contain mistakes. With regard to this, I am grateful to those authors who have provided clarifications or who have corrected me, and to the manufacturers who have provided die photos along with pertinent information.

Architecture design of general-purpose microprocessor chips is not discussed in detail, though the differences between chip architecture and conventional computer architecture have been discussed in Chapters 9 and 10. This is because at the present time, chip architecture is not drastically different from conventional computer architecture of mainframes or minicomputers, and readers can learn the subject from the many textbooks which discuss conventional computer architecture in full detail. LSI/VLSI technology more directly influences low-level design stages, namely, layout, circuit, and logic design stages, and the differences of chip architecture from mainframe or minicomputer architecture are a result of these influences which are discussed in all early chapters.

Research results by my group which are mentioned throughout this book have over many years been produced under grants by the National Science Foundation through Computer Systems Design Program of Dr. J. R. Lehman. I am also extremely grateful to Mrs. Z. Arbatsky, who painstakingly typed this manuscript.

The progress of the LSI/VLSI technology is, in itself, beautiful and exciting to learn. I hope that some of this beauty and excitement is conveyed to the readers.

SABURO MUROGA

Urbana, Illinois
Summer 1982

About Sections Labeled with ▲ and Remarks

These sections (or subsections) and Remarks throughout this book present advanced information, and skipping them will not impede the understanding of later material. It is advised that in college teaching, the remarks be completely skipped.

Contents

Conversion Table

1 mil	$\begin{cases} 10^{-3} \\ 25.4 \ \mu m \end{cases}$
1 cubic foot	28 liters
1 liter	1 cubic decimeter
1Å	$\begin{cases} 10^{-10} \ \text{meter (m)} \\ 10^{-8} \ \text{centimeter (cm)} \\ 10^{-4} \ \text{micrometer} \ (\mu m) \\ 10^{-1} \ \text{nanometer (nm)} \end{cases}$
1,000 Å	0.1 μm

Multiplier–Prefix–Abbreviation Table

Multiplier	Prefix	Abbreviation
10^{12}	tera	
10^{9}	giga	G
10^{6}	mega	M
10^{3}	kilo	k
10^{2}	hecto	
10	deka	
10^{-1}	deci	d
10^{-2}	centi	c
10^{-3}	milli	m
10^{-6}	micro	μ
10^{-9}	nano	n
10^{-12}	pico	p
10^{-15}	femto	f
10^{-18}	atto	

CHAPTER 1

Introduction

In this chapter, a review of the progress of integrated circuit technology is followed by a discussion of the social impacts of large-scale integration (LSI) or very-large-scale integration (VLSI) technology. Then the basic definitions and features of LSI/VLSI are described.

1.1 Progress and Impacts of Integrated Circuit Technology

The enormous progress of integrated circuit (IC) technology in recent years has been changing many things in our daily lives because digital systems can be manufactured with much lower costs, lower power consumption, higher speeds and in smaller sizes. A comparison of the ENIAC, the first electronic computer, and the HP-67, the pocket calculator of Hewlett-Packard, in Table 1.1.1 gives us some idea of the progress. The latter is essentially an ENIAC one can hold in one's hand. Also, the ENIAC had 18,000 vacuum tubes on a floor space of 300 m^2, whereas the microprocessor chip Z8000 introduced in 1978 has 17,500 transistors on a space of 3.9 mm^2. In 30 years the changes have been dramatic.

The TI-59, the pocket calculator of Texas Instruments with a price tag of $300, has a data memory of up to 100 registers and a program memory of up to 960 instructions. The computing power of the TI-59, according to Texas Instruments Corporation, is equal to that of the IBM 1401, a room-size computer introduced in 1959 for about $70,000. The IBM 5100 introduced in 1975 is a desk-top computer whose memory size is comparable to the IBM 360/30 introduced 10 years earlier, yet its outright sale price is roughly equal to the monthly rent of a 360/30. (The IBM 5110 introduced in 1978 is 3 to 30 times faster and priced at $9,900.) While 1 megabyte of main memory was rare in the large computers of the early 1960s, that much memory is in the pedestal of the Hewlett-Packard Amigo-300 small-business system, introduced in 1978 at a price of $32,000. The number of IC chips used inside is declining. In the mid-1960s a typical calculator contained 90 to 150 bipolar IC chips. In 1969 four MOS IC chips could do the same job, and in 1971 only one MOS chip is sufficient. The price dwindled to about 1/50. Now performance of the single chip is improving. In the next 10 or 15 years comparable progress in size, cost, and performance is expected.

1

Table 1.1.1 Comparison of Old Computer and Recent Calculator

	ENIAC (1946)	HP-67 (Introduced in 1977)	APPROXIMATE RATIO
Word size	10 decimal digits	10 decimal digits	1 : 1
Data memory capacity	20 registers	26 registers	1 : 1.3
Program memory capacity	750 instructions	224 instructions	3.3 : 1
Input/output	punched cards	magnetic cards	
Read rate	200 digits/sec	50 digits/sec	4 : 1
Manual switches	5,000	35	140 : 1
Number of multiplications	360/sec	15/sec	24 : 1
Cost	$480,000	$375 (1980)	1,300 : 1
Power requirements	50,000 W	$\frac{1}{2}$ W	100,000 : 1
Size	10 × 100 × 4 ft (4,000 ft^3)	3.2 × 1.4 × 6 in. (27 in.^{3})	270,000 : 1
Weight	30 tons	298 g	100,000 : 1

This book is intended to trace the progress and provide an overview of IC chip design—every aspect from IC fabrication to digital system design, but without going into the details of electronics, solid-state physics, or chemistry. It is becoming extremely important for computer engineers (or computer scientists) and company executives to learn when and how to use what kinds of ICs, and what are the advantages and disadvantages. Here are some of the reasons.

1. IC chips will incorporate increasingly more complete systems, including logic networks, memories, and software. The border line between logic designers and programmers is becoming blurred, and often designers must have the ability of not only designing and programming but also developing algorithms for realizing specific tasks in ICs. In order to design good digital systems in IC chips, designers must know what can be done with IC chips. Whether or not designers are able to design and use IC chips properly makes a drastic difference in the performance, cost, and other characteristics of a system, as we will see in later chapters.

2. Due to lower cost, smaller size, and improved performance of digital systems using modern ICs, designers increasingly often have to consider tradeoffs between software and hardware (i.e., ICs) in realizing tasks. If a task is realized by hardware, performance is usually much higher than when realized by software, but the task is much more time-consuming and expensive to realize and change. For example, significant parts of software, particularly operating systems, can be realized in hardware for improved speed and programming ease. The knowledge of LSI/VLSI is essential for hardware realization of software (Falk, 1974; *Computerworld,* Apr. 1975, p. 47; Goetz, 1975; Wagner, 1976, p. 91). Another example is floating-point operations which can be faster in a hardware realization by at least one or two orders of magnitude compared to a software realization. Also, dedicated processor chips have been developed for processing programs written in some high-level languages such as Pascal and Ada (Bursky, May 10, 1980; Schindler, Jan. 8, 1981).

   3.   Many new products have been developed because of low cost, small size, and/or low power consumption. Some of these became feasible for the first time because of the technological breakthrough in LSI or VLSI. Conventional products augmented by LSI/VLSI with new functions often greatly improve profitability for manufacturers. Also, completely new products open new markets. Some of the many examples follow:

Video Games.   Many video games, such as those by Atari and Mattel, are introduced and are becoming more sophisticated (*Fortune*, July 27, 1981, pp. 40–46).

Toys.   Electronic games by Mattel, Speak-and-Spell by Texas Instruments (Wiggins and Brantigham, 1978), radio-controlled toys, and other toys are being introduced with increasingly sophisticated functions (Giles et al., 1980; Mokhoff, Nov. 1980).

Electric Home Appliances with Added Functions.   Microwave ovens, refrigerators, and TV sets are controlled by microprocessors and provided with talking capabilities (Walker, 1977; *Electronics,* Sept. 27, 1973, p. 35; Fischer, 1981).

Hi-Fi and Electronic Musical Instruments.   The performance of hi-fi is greatly improving. Super-hi-fi by digital recording, FM stereo-tuners, and cordless infrared-transmitted headphones from Sennheiser (*Popular Science,* Jan. 1976, p. 91) are such examples. Also many electronic musical instruments with synthesized sounds are available with LSI chips (Burstein et al., 1979; *Electronics,* Mar. 24, 1981, p. 84).

Cameras.   Automatic focusing cameras by Konica (*Business Week,* Oct. 31, 1977, p. 108B) and Honeywell (Electronics, April 27, 1978, pp. 139–143), the Polaroid SX-70, many other cameras (*Electronics*, Aug. 21, 1975, pp. 74–81), and Kodak's movie (*Popular Science,* Dec. 1973) contain LSI chips inside. For example, the Canon AE-1 35-mm camera with 4,000 I^2L gate LSI in microcomputers and programmable logic array, eliminating 300 mechanical parts, is easy to use because exposure and shutter speed are automatically set (*Electronics,* Mar. 30, 1978, pp. 60, 62).

Electronic Watches.   The accuracy of wrist watches is 1 to 2 min per week for mechanical watches, 1 to 2 min per month for electronic watches with a tuning fork, and 1 to 2 min per year for quartz watches. Calculator watches, possibly with pager and heart rate, and watches with melody-playing capability are available (*Electronics,* Jan. 3, 1980, pp. 196, 198).

Electronic Locks.   A sophisticated electronic lock can be used with the control box inside the wall (*Popular Science,* Dec. 1973). Also room keys at some hotels are replaced by electronic locks, eliminating theft possibilities (*Business Week,* Oct. 20, 1980, pp. 540P).

Applications to Automobiles.   Electronic fuel injection, antiskid braking, seat-belt alarm, instant diagnosis by computer (e.g., VW and Honeywell), car-driving guidance, and tire-pressure sensor while driving (*Electronics,* Jan. 8, 1976, p. 58) are now feasible with extensive use of microcomputers and dedicated processor LSI chips. (Rivard, 1980; Bassak, 1980; Zeskind, 1981; Jurgen, 1981).

Sewing Machines.   The Singer Athena 2000 sewing machine contains an n-MOS LSI chip. Eliminating some 350 mechanical parts, the chip controls the linear motor movement and gives the user push-button control of 24 stitch patterns like dogs and tulips. It contains more than 800 transistors and can make 36,000 operations during one hour. It optimizes stitch width, length, and density (Hatfield et al., 1979; Shackil, Feb. 1981).

Health-Care Products.   Physical fitness and health-care products are being developed as well as many medical electronics products (McDermott, Sep. 20, 1980; Mokhoff, June 1980; Shackil, Apr. 1981).

Industrial Production and Energy Saving.   Industrial robots and numerically controlled machines are extensively used in industry in order to improve productivity, reliability, and design flexibility (Nevins and Whitney, 1978). Also, microcomputers are extensively applied to the operation of electric motors (Pearman, 1980) and air conditioners to save energy.

Printers.   Printers are quieter and more reliable when mechanical parts are replaced by IC packages, and they are becoming intelligent through the use of microcomputers (Williams, May 14, 1981). For example, in Hypertype I about 100 mechanical parts for control are reduced to a dozen due to the use of LSI.

Communications.   The current analog transmission of telephone signals will be replaced by digital signals in the future (Falk, Feb. 1977; Melving, 1980; Falconer and Skrzypczak, 1981). Private branch exchanges are becoming intelligent (Burck, 1980). Also, voice messages are stored in telephone exchanges and will be delivered to addressed persons whenever they are available. (*Business Week,* June 9, 1980; *Computerworld,* June 23, 1980, p. 65; Teja, 1979.) Mobile phones and pagers will be extensively used (Donlan 1981).

Personal Computers, Home Computers, and Intelligent Terminals.   In addition to increasingly sophisticated calculators (McDermott, June 20, 1980), personal computers, such as IBM, Apple Computer, and TRS-80 of Radio Shack, and home computers, such as Intellivision of Mattel, are available (Weisbecker, 1974; Raskin and Whitney, 1981; Marshall, 1980). Intelligent terminals are becoming more intelligent. In the future intelligent terminals, possibly with electronic files, electronic mails, and others, may be seen in nearly every office (Mokhoff, Oct. 1979). LSI/VLSI chips have been extensively used not only in these computers and terminals but also in **mainframes** in order to improve performance and reduce costs. (Large computers are usually called mainframes to differentiate them from minicomputers or microcomputers.)

As seen in the above examples, a large number of manufacturers who had never used electronics are now incorporating LSI/VLSI chips in their products, with new functions which could not be imagined in the past (e.g., dolls with talking capability).

Because of the continuous introduction of products with new functions, the marketing of products and the ways of our daily lives are changing. Even old products such as food blenders can command much higher prices when microcomputers are incorporated. It is expected that every home will have a video room with a projection television, a video disk player, a video cassette player, video games, home computers, digital hi-fi, and others. TV programs can be automatically recorded by video cassette players at a number of preset dates. Extremely accurate watches can be made so cheaply and compactly that a large number of products incorporate watches, such as calculators with clocks, speaking watches, desk-top clocks with memo notes, melody-playing watches, and cameras with exposure date recording. All of a sudden a large number of watch manufacturers who failed to keep up with this technological change faced financial disaster (Galling and Ball, 1980; *Business Week,* June 7, 1976, p. 20E). Retail stores of watches could no longer make a living out of watch repairs because watches with LSI have no mechanical parts to wear except batteries. The watch industry and related aspects of our daily lives have changed literally overnight. There are many other examples. LSI/VLSI is indeed causing significant technological and social changes and consequently will be the basis of a second industrial revolution.

4.   Users increasingly often have to custom design LSI/VLSI chips. Yet semiconductor manufacturers are usually not interested in custom design because of low profits. (There are many approaches in custom design, as discussed in Chapter 9. By some approaches custom design can be done in a short time, but the performance of designed products will not be good. Users can reduce the development time of custom chips by in-house design. [Digital Equipment Corporation estimates savings of 20% (*Business Week,* Dec. 3, 1979).] Also, users must protect their trade secrets (see Remark 1.1.1). Users know their own products best; so if they design custom chips by themselves after having acquired LSI/VLSI chip design knowledge, they can obtain products that are superior to those designed by designers of semiconductor manufacturers who do not have the time to learn the nature of the customers' products. For these reasons users have been hiring chip

designers for in-house custom design. Some of them start the in-house production of custom-designed or off-the-shelf LSI/VLSI chips in order to secure second sources, because assuring a constant supply of these chips is often difficult. (As a matter of fact, there have been numerous cases where user firms could not assemble their products because of the interruption of IC delivery from semiconductor firms, and went bankrupt.) In some cases semiconductor manufacturers give proprietary know-how to customer firms in exchange for high-volume orders. Historically only semiconductor manufacturers and very few computer or electronic manufacturers (such as IBM and RCA) had IC production facilities. But for the above reasons, an increasingly number of firms, including those that had no previous relation to electronics or computers (e.g., watch manufacturers and toy manufacturers), now have IC production facilities and are hiring hundreds of workers (Marion, 1980).

As more components, systems, and software go into single chips of LSI or VLSI, the manufacturing processes of electronic products have been changing drastically. Workers on assembly lines in conventional manufacturing methods are being replaced by LSI/VLSI chip designers. Each different product requires many chip designers. So if a firm has many products in its product line, it needs a large number of chip designers, which will create an enormous chip designer shortage.

When logic gates are to be realized in electronic circuits on IC chips, there are many different types of electronic circuits, the **IC logic families**. Currently a large number of IC logic families are available, such as ECL, TTL, IIL, n-MOS, p-MOS, CMOS, and VMOS. All have different features, and there are a number of different ways to use them (e.g., PLAs, gate arrays, and microcomputers). Unless designers understand manufacturing processes very well (e.g., mask preparation), they cannot design good IC chips, unlike the case of designing networks with discrete components, where engineers can determine manufacturing processes almost independently of design due to the great flexibility in assembling discrete components. Also, the cost structure is very complex. There are many other factors which greatly influence the cost, performance, and design time.

Therefore if we want to develop the best products (or best computer systems), it is not possible to do so without a knowledge of different IC logic families and the related complex problems of LSI/VLSI design. Since many products, whether electronic or not, incorporate IC packages, the knowledge of these complex situations is indispensable for new product development, whether in hardware or in software. (For example, when we purchase such new products from vendors, we will lose crucial key points in the negotiation of the purchase without such knowledge.)

This book provides the reader with such knowledge, namely, when and how to use which IC logic families with what advantages and disadvantages. We will attempt this without going into the details of electronics, solid-state physics, or architecture, because the space is limited and the author wants to avoid overly taxing the reader with jargons.* (Which of these subjects is important depends on where one works in a wide spectrum of IC fabrication and design. But no matter where one works, one must know of all other subjects to some extent.)

> ***Remark 1.1.1:***   Custom-designed chips can hide trade secrets far better than the use of off-the-shelf packages, because competitors are familiar with the off-the-shelf packages and can quickly understand new products designed with them. The protection, however, is not perfect because of **reverse engineering**, that is, stripping down a competitor's chip to analyze how it works and possibly to learn a few tricks which can help improve performance in future products. Reverse engineering is relatively easy for competitors: all that is needed is a microscope, a few small pots of acid to etch away the circuit layer by layer, and a camera to record the successive steps.
>
> Reverse engineering is very annoying for semiconductor manufacturers who pioneer new LSI chips, often spending several million dollars, because competitors can produce copies without paying royalties. In 1979 some manufacturers tried to amend the federal copyright law to protect chip design. To the surprise of observers, many semiconductor manufacturers strongly oppose it on the ground that they lose more than they gain by the copyright protection, and reverse engineering is a very common practice (*Business Week,* Apr. 21, 1980). Reverse engineering, however, is becoming extremely difficult or practically impossible, except for simple LSI chips, as the number of transistors on a chip and the complexity of processing increase.   □

1.2   Implementation of Components

IC technology is essentially a new way of realizing components of electronic circuits, such as transistors, diodes, resistors, and capacitors. In this section

---

* Without a discussion of electronics, solid-state physics, architecture, and chemistry even the reader who is not familiar with those subjects but is familiar with logic design can learn an overview of IC usage. It is, however, not possible to discuss IC technology without any knowledge of electronics. Readers who are not familiar with electroncs are advised to read, for example, Smith's readable textbook (Smith, 1980, Chaps. 1–6; or, 1976, pp. 3–65, 158–163, 287–354).

For those who want to specialize in chip design, a knowledge of logic design, electronics, solid-state physics, architecture, chemistry, and, in particular, of electronic circuit and logic design is indispensable. The author wishes this book to be an appetizer to stimulate the reader to learn of these related subjects, after having grasped an overview of LSI/VLSI design practice. Stimulation of interest is perhaps one of the most important aspects in education. Also many students have difficulty in integrating the knowledge gained by taking courses specialized in related subjects. This book tries to provide them with an integrated picture of their possible future profession in LSI/VLSI chip design.

let us discuss how to realize these components. For simplicity, each component is explained in its simplest structure, although that structure is not necessarily widely used now. More updated but complex structures for some components are discussed in later chapters.

Silicon is a widely used material of these components, though other materials such as gallium arsenide are sometimes used, as discussed in Chapter 5. But silicon, which contains a trace of impurity, such as n-type silicon or p-type silicon, is more useful than pure silicon. The n-type silicon, or **n-type region** (or simply n-region), inside a thin substrate of p-type silicon, is made by adding an impurity in column V of the periodic table (which is called a donor), such as P (phosphorus), As, or Sb. The p-type silicon, or **p-type region** (or simply p-region), inside a thin substrate of n-type silicon, is made by adding an impurity in column III of the periodic table (which is called an acceptor), such as B (boron), Ga, Al, or In. The **n$^+$-type region** (or simply n^+-region) is made by adding a greater quantity of impurity than n. The **p$^+$-region** is similarly defined. The electrical conductivity of silicon is increased by adding a greater amount of impurity.

Diode

A diode is made by a **junction** of p-type silicon and n-type silicon, as illustrated in Fig. 1.2.1a. A metal lead wire is attached to each side. When a voltage is applied between the p- and n-types through the lead wires, a current flows through the junction if the voltage at the p-type is higher than at the other side, that is, if the diode is **forward biased**. The current increases as the voltage increases, as shown in Fig. 1.2.1b. If the voltage at the p-type is lower, that is, if the diode is **reverse biased**, or **back biased**, no current (actually a very small current) flows, as shown in Fig. 1.2.1b.

Transistors

There are different types of transistors, but the two currently most important ones for IC are the bipolar transistor and the MOSFET.

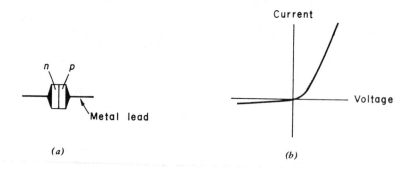

(a) (b)

Fig. 1.2.1   Diode.

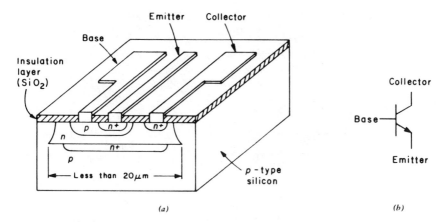

Fig. 1.2.2 *n-p-n* transistor. (*a*) *n-p-n* transistor. (*b*) Symbol for *n-p-n* transistor.

1. Bipolar transistor. An implementation example of an *n–p–n* bipolar transistor, which has an emitter, a base, and a collector, is shown along with its symbol in Fig. 1.2.2.

An *n–p–n* transistor essentially consists of two junctions, a *p–n* junction where *n* is the emitter and *p* is the base, and a *p–n* junction where *p* is the base and *n* is the collector. Thus an *n–p–n* transistor is a pair of diodes whose common *p*-type region is very thin and used as a base. In Chapter 3 we discuss how an *n–p–n* transistor works. If *n* and *p* are interchanged, a *p–n–p* bipolar transistor is obtained.

2. MOSFET. MOSFET is the abbreviation for **metal oxide semiconductor field effect transistor** and is often simply called **MOS**. An implementation example of an **n-MOS** (or an **n-channel MOS**) is shown in Fig. 1.2.3. The gate is realized with metal, whereas the drain and source are realized with n^+-regions connected to metal electrodes. (For the sake of simplicity, the simplest structure of a MOSFET is shown here, although this is not widely used at present. More complex structures, such as silicon-gate MOSFET, which are widely used, are discussed in later chapters.) An *n*-MOS is denoted by the symbol shown in Fig. 1.2.3*c*. Because the gate is made of metal, the MOSFET in Fig. 1.2.3*a* is called a **metal-gate MOSFET** (in contrast to the silicon-gate MOSFET described in later chapters). Also, since the two n^+-regions are separated simply by the *p*-region of the substrate, the *n*-MOS in Fig. 1.2.3*a* is called an **enhancement-mode n-MOS**, in contrast to a depletion-mode *n*-MOS where the two n^+-regions are connected by a thin *n*-type layer, as discussed in Chapter 4. **The "gate" in Fig. 1.2.3a should not be confused with "logic gates," and whenever it is necessary to differentiate the two, a logic gate, which consists of MOSFETs, will be called an MOS cell.**

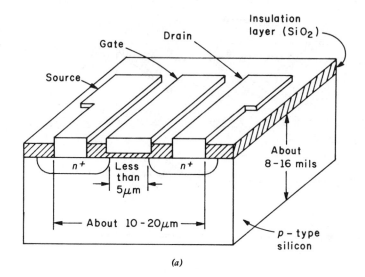

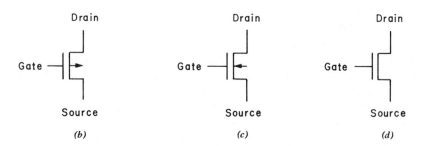

Fig. 1.2.3 MOSFET. (*a*) **n-channel MOSFET.** (*b*) **Symbol for p-MOS.** (*c*) **Symbol for n-MOS in** (*a*). (*d*) **Simplified symbol for MOS.**

If n and p are interchanged in Fig. 1.2.3*a*, a **p-MOS** is obtained, and its symbol is shown in Fig. 1.2.3*b*. Since an n-MOS in **positive logic*** expresses the same logic operation as a p-MOS in **negative logic**,† both the n-MOS and the p-MOS are denoted by the symbol shown in Fig. 1.2.3*d* instead of those in Fig. 1.2.3*b* or *c* when we are concerned with logic operations only (see Muroga, 1979, Sec. 2.3).

The two n^+-regions in the n-MOS in Fig. 1.2.3*a* or the corresponding p^+-regions in the p-MOS are called **diffusion regions**, or **diffusion areas**, since they are usually manufactured by the diffusion process, as explained in Chapter 2.

---

* If logic values 0 and 1 are represented by low and high voltages, respectively, this is called positive logic.
† If logic values 0 and 1 are represented by high and low voltages, respectively, this is called negative logic.

Discrete Components

When components are put into separate containers, one component per container, they are called **discrete components**, or **discrete devices**. A transistor or a diode prepared as explained above is put into a container. A resistor in the form of a discrete component is made of a carbon compound or metal glaze. A capacitor in the form of a discrete component is made of waxpaper sheets sandwiched between aluminum foils, mica sheets sandwiched between aluminum foils, or others.

When we want to make networks with logic gates at home for hobby work or in laboratories for experiments, we connect these discrete components with conductor wires and solder the connecting points. If a fixed pattern of connections is to be manufactured in large quantities, manufacturers put it on a **printed circuit (pc) board**. (The pc board is a thin insulator sheet on which flat metal conductors are placed, as if being printed.) Users mount discrete components on pc boards and solder the connecting points. Since the size of a discrete component varies between an inch and its fractions, a pc board of 10×10 in. may have several hundred discrete components on it.

Integrated Circuits

In the case of an IC, all components are placed on a single small thin semiconductor sheet measuring a few to a few hundred square mils, which is called a **chip**. A chip is sometimes also called a **die** (plural: dice). This is called "integrated" because the components and wiring are all an "integrated" part of the chip and cannot be separated from each other. For example, a MOS cell (i.e., logic gate) as shown in Fig. 1.2.4*a* is actually assembled as shown in Fig. 1.2.4*b*, which is the top view of an IC implementation, i.e., a **layout**. Both source and drain electrodes are shown in a single MOSFET in Fig. 1.2.3*a*. But when the drain and source electrodes of MOSFETs are not connected to any external terminals, they are not always necessary, and adjacent MOSFETS are connected only with the diffusion regions, as shown in Fig. 1.2.4*c*. Only the windows (the so-called **metal contacts**) of the insulation layer connected to the ground, the output, and the power source are shown as black areas. (Wavy ends mean possible extensions of the areas.)

Notice that the connections in Fig. 1.2.4*a* can be implemented by the metal connections shown by the nonshaded areas in Fig. 1.2.4*b* or by the diffusion regions (i.e., nonmetal connections) shown by the shaded areas in Fig. 1.2.4*b*. Also polysilicon and others are used as connections, as is discussed in later chapters.

If the MOS cell in Fig. 1.2.4*a* is to be implemented with discrete components, it is implemented as shown in Fig. 1.2.5 by encasing MOSFETS into separate containers and connecting them by wires. Of course, this implementation is very bulky.

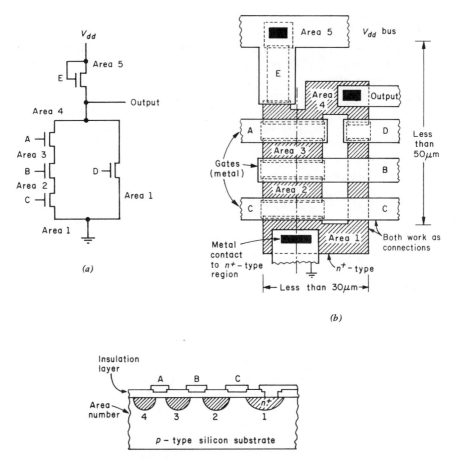

Fig. 1.2.4   Implementation of a MOS cell. (*a*) **A MOS cell to be implemented.** (*b*) **IC implementation.** (*c*) **Cutout view along the chained line in** (*b*).

When one connection needs to cross over another connection, this is realized as shown in Fig. 1.2.6 by means of a diffusion region (i.e., a short *n*-type region in a *p*-type substrate). Since a diffusion region has some resistance, if it is used as a connection, as shown by the shaded areas in Fig. 1.2.4*b*, it must be generally short; otherwise if it is long, it must be wide. When resistors are needed for a MOS IC, a resistor is usually realized by a diffusion region, as shown in Fig. 1.2.7, or by polysilicon,* as explained in a later chapter. (In other words, a diffusion region can be used for either a resistor or a short connection.) The resistance can be changed by changing the length and width of the diffusion region. Resistors often occupy more

---

*A polysilicon strip on the insulation layer can be used as a resistor (*Electronics*, Sep. 13, 1979, p. 112).

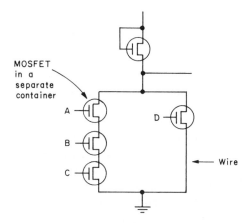

MOSFET
in a
separate
container

A

B

C

D

← Wire

Fig. 1.2.5 Implementation of MOS cell of Fig. 1.2.4a with discrete components. Circles denote discrete components. The entire cell occupies about 1 in.^{2} on a pc board.

area than transistors (particularly MOSFETs), and in this sense resistors are more costly than transistors.

When electronic circuits (or logic gates) which contain bipolar transistors are to be realized in an IC, we need to realize other components such as resistors, diodes, and sometimes capacitors on a silicon substrate. Bipolar transistors are realized as shown in Fig. 1.2.2. Resistors are realized in the same manner as shown in Fig. 1.2.7. This is discussed later in detail. A capacitor is realized by a capacitance between the large aluminum electrode and the n^+-type diffusion area, as shown in Fig. 1.2.8.

In contrast to the discrete components described above, resistors and capacitors are implemented in ICs in different ways, as shown in Figs. 1.2.7 and 1.2.8. [For some new realizations, see *Electronic Design* (Apr. 30, 1981, p. 53).] A 100-pF (i.e., picofarads) capacitor takes up as much area on the chip as 100 transistors. As for resistors, their practical range of values is limited in standard IC processes. Values of less than 10 Ω (i.e., ohms) or more than 100,000 Ω are difficult and sometimes costly to realize. Furthermore, the absolute values of resistors and capacitors are rarely controllable to better than ±20% of nominal values (*Spectrum,* Feb. 1979, p. 24), although ratios between two resistors can be maintained to ±2–5%. Thus relative costs of different components implemented in ICs are different from those of discrete devices. Capacitors cost more than resistors (in terms of covering areas), and resistors cost more than transistors. In other words,

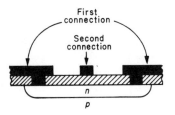

First
connection

Second
connection

n

p

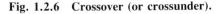

Fig. 1.2.6 Crossover (or crossunder).

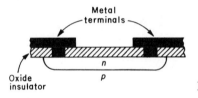

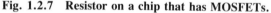

Fig. 1.2.7 **Resistor on a chip that has MOSFETs.**

the order of costs of different discrete components is reversed in ICs (Suran 1970). Consequently, in the case of IC design we avoid the use of components other than transistors or diodes, unless these components are really necessary, unlike in the case of circuit design with discrete components.

A diode is realized as shown in Fig. 1.2.9. Sometimes it is also realized by connecting the base of a bipolar transistor to the collector (Fig. 1.2.10*a*) or to the emitter (Fig. 1.2.10*b*). As an example, an electronic circuit for a logic gate which contains a bipolar transistor, is realized as shown in Fig. 1.2.11. (We discuss in detail later how to prepare the layout.) In an IC realization bipolar transistors, diodes, and resistors must be surrounded by the isolation regions (to be explained in a later chapter), as shown in bold solid lines in Fig. 1.2.11*c*, because without the isolation regions, current from one component may leak out into other components, thus creating mutual interference (Hibberd 1969).

Packaging and Assembling

An IC chip is mounted in a container. The terminals on the chip (i.e., inputs and outputs of networks, the ground, and the power supply on the chip), which are shown as small squares in Fig. 1.2.12 and are called **pads**, are connected to the terminals of the container, usually called **pins**, as illustrated in Fig. 1.2.12. When a lid is sealed to the container, or plastic is molded over it, we have an **IC package**. For high-speed computers, pins of a con-

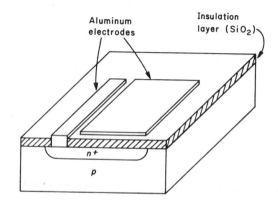

Fig. 1.2.8 **Capacitor.**

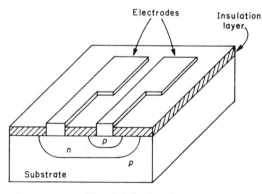

Fig. 1.2.9 Diode.

tainer are not desirable because they add a time delay. Accordingly a chip is attached to a ceramic module of a larger size, and the module is directly soldered to a pc board.

When a gate, gates, or networks are implemented on a single chip, as discussed so far, and encased into a single package, it is called a **monolithic integrated circuit**. When a package contains more than one chip or a mixture of chips and discrete components, all of which are put on a large substrate (usually a ceramic substrate), it is called a **hybrid integrated circuit**. Thus there are two types of ICs:

1. Monolithic IC.

2. Hybrid IC.

Henceforth IC will mean monolithic IC unless otherwise noted.

IC packages are usually placed on a PC board. The pins of IC packages and the holes on the pc board are soldered. Then pc boards are inserted into mother boards which are placed in cabinets along with backplanes.

Another approach to assemble IC chips is to place chips on a **ceramic carrier** and then the ceramic carriers on a **ceramic mother carrier**. The ceramic mother carriers are placed on a pc board. This approach has the advantage of smaller space than when IC packages are assembled on a pc board and of lower cost than hybrid ICs, as compared in Table 1.2.1.

(a)   *(b)*   **Fig. 1.2.10 Diodes realized with transistors.**

Table 1.2.1   Comparison of Different Assembling Methods

	WEIGHT (grams)	AREA (cm²)	YIELD	COST
PC board	52	81	High	Low
Chip carrier/mother carrier	12	12	High	Medium
Hybrid IC	10	6	Low	High

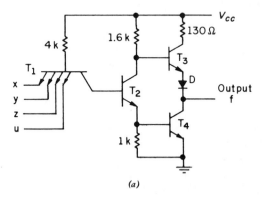

(a)

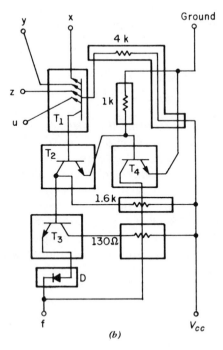

(b)

Fig. 1.2.11 (*a*) **TTL gate.** (*b*) **Schematic configuration of** (*a*) **to be laid out in the left half of** (*c*). (*c*) **Layout of a pair of 4-input NAND gates in TTL (SN7420N). (Courtesy of Texas Instruments, Inc.)**

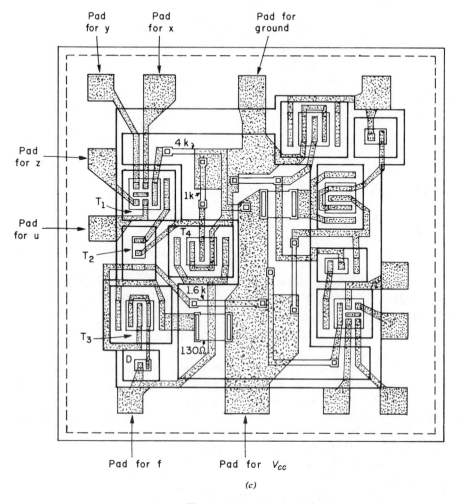

Fig. 1.2.11 (*Continued*)

References: Lyman, Nov. 24, 1977, Sept. 28, 1978; Balde and Amey, 1977, 1978; Tsantes, 1981; *Electronics,* June 30, 1981, pp. 39–40.

Integration Size

The different sizes of integration of IC chips depend on the applications and are defined as follows:

SSI (Small-Scale Integration). The number of logic gates on a chip is less than 10. (Sometimes this is defined as less than 20 gates.)

MSI (Medium-Scale Integration). The number of logic gates on a chip is 10 to 100. (Sometimes this is defined as 20 to 100.)

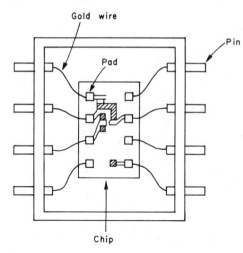

Gold wire

Pin

Pad

Chip

Fig. 1.2.12 IC chip mounted in a container.

LSI (Large-Scale Integration).   The number of logic gates on a chip is 100 to 100,000. (Sometimes this is defined as 100 to 1,000.)

VLSI (Very-Large-Scale Integration).   The number of gates on a chip is more than 100,000. (Sometimes this is defined as more than 1,000.)

The definition is not definite, as indicated by the parenthetical remarks. In particular, the definition of VLSI is very confused. VLSI often means more than 1,000 gates on a chip. But more precisely speaking, it appears to mean more than 100,000 gates on a chip, because a chip with up to 100,000 gates can be realized with the same technology, while the new submicron technology is applied when more than 100,000 gates are to be placed on a chip. Some authors (e.g., Barna and Porat, 1973, p. 361) define the integration size differently for MOS and bipolar; for example, LSI is defined as 100 gates or more in the case of bipolar transistors and 300 MOSFETS or more in the case of MOS.

In the preceding the integration size is defined in terms of the number of gates. The integration size in terms of the number of devices (i.e., transistors, resistors, and others) is defined by multiplying the number of gates by 3 to 5 in each case. For example, LSI contains about 300 to 300,000 devices on a chip.

The integration size is limited by production yield or heat generation. In other words, if chips are too large, the yield gets too low, and the cost of the chips is too high. If too many transistors (in particular, bipolar transistors) are put in a chip, the chip gets too hot. The maximum power consumption on a chip is usually limited to 4 to 5 W.

The decision of whether to use SSI, MSI, or LSI is usually made by considering many factors. First, if the production volume is very high, it is more advantageous to use ICs, as discussed in Section 1.3, as the larger integration size is used. However, LSI/VLSI is less flexible than MSI, and MSI is less flexible than SSI. In other words, when new networks are needed, any LSI prepared for previous design purposes must be abandoned. But if the previous networks are assembled with SSI packages, the new networks can be assembled by replacing some of these packages. [Because of flexibility, or unique properties such as those discussed in Nelson (Jan. 5, 1980), SSI and discrete devices will not disappear from the market.] Second, the complexity of required networks may not need large integration sizes. Third, as is explained in later chapters, the initial design and the manufacturing process of ICs are complicated, time-consuming, and costly. As the integration size increases, this is increasingly true, and the yield in manufacturing decreases. Thus if the integration size is too large, the cost per gate becomes excessively high, losing the cost advantage of the large integration size.

1.3   Advantages of ICs over Discrete Components

What are the advantages of IC gates over discrete devices?

1.   Digital networks with IC gates are inexpensive under certain conditions. For cost comparisons let us assume that we want to manufacture a network of 1,000 gates. Table 1.3.1 shows costs for five different realizations of the network: discrete components, SSI of TTL gates, a mixture of MSI and SSI, six MOS/LSI chips, and a single MOS/LSI chip. Also let us assume that the network of 1,000 gates requires 4,500 discrete components, and that discrete components or IC chips must be assembled with pc boards. As shown in Table 1.3.1, the total cost (i.e., assembly plus component cost) decreases as the integration size increases. Initial investment, such as design cost and special tooling costs for the custom design (last two rows), is not included in this total cost. If the network is manufactured in quantities of 10,000, for example, an initial investment of $60,000 would break down to $6 a piece, resulting in a cost of $14.20 per IC package. (The figures in Table 1.3.1 show only rough estimations.) However, sometimes manufacturers do not want to charge the full cost to this network, since part of the initial investment will be useful for the future manufacture of other networks. When we consider the costs for pc boards, tests, connectors, power supply, fans, cabinets, and others required for a complete system, the cost difference is far greater, as is explained in detail in later chapters. In other words, as we use fewer IC packages of greater integration sizes, the pc boards, power supply, cabinets, and other items can be smaller and less costly. The cost saving is greater for greater integration sizes of IC packages, although cost comparison of the entire systems is complicated.

As an example of cost comparison of entire systems, let us consider an intelligent terminal, Bantam, by Perkin Elmer, though this contains a cath-

Table 1.3.1 Cost Comparison of IC and Discrete Devices

COMPONENT TYPE	NUMBER OF COMPONENTS	NUMBER OF CONNECTIONS	ASSEMBLY COST		COMPONENT COST ($)	ASSEMBLY PLUS COMPONENT COST ($)	INITIAL INVESTMENT[a] ($)
			PER HOLE (¢)	TOTAL ($)			
Discrete	4,500	11,000	1	110.00	22.50	132.50	0
SSI (TTL)	300	4,000	1	40.00	40.00	80.00	0
MSI and SSI[b]	100	2,500	1	25.00	30.00	55.00	0
6-chip MOS	6	180	3[c]	5.00	30.00	35.00	120,000
1-chip MOS	1	40 or less	3[c]	1.20	7.00	8.20	60,000

[a] May or may not be borne entirely by customer.

[b] Off-the-shelf packages.

[c] Because of lower board density.

ode-ray tube (CRT) and a keyboard whose costs are not affected by the integration size of the IC packages (*Digital Design*, Nov. 1978, pp. 66–71; *Mini-Micro Systems*, Jan. 1979, pp. 104–106; *Electronic Design*, Nov. 8, 1978, p. 49). The prices decreased as the integration size increased, as follows:

Prior to 1974   First generation with TTL discrete components and IC packages, in 150 IC packages: $1,600–2,000

1976   Second generation with TTL discrete components and IC packages and a microprocessor (M6800), in 100 IC packages: $970–1,600

Third generation with TTL discrete components and IC packages and a microprocessor–CRT controller LSI chip, in 60 IC packages: $971

1978   Fourth generation with TTL discrete components and IC packages, and a custom LSI chip (for character decoding and controller), in 19 IC packages: $600

The fourth-generation terminal was introduced in 1978, using custom-designed LSI instead of the off-the-shelf microprocessor chip. (This was designed in-house, but if designed by an outside vendor, it would cost $200,000.) Custom-designed LSI reduced the cost to $600 with improved performance. Power consumption of the terminal was reduced to almost half by LSI. (The second-generation terminal also reduced the power consumption by using the microprocessor chip, eliminating cooling fans.) Since the power supply of the previous model costs 10 to 15% of the terminal cost, this is also a significant improvement.

In general, when production volume is high, a system in LSI or VLSI packages costs much less than one in discrete components.

2. The design of digital networks can be done more quickly by assembling off-the-shelf MSI or LSI/VLSI packages than by the use of discrete components. As will be seen later, many convenient networks are available as IC packages. When we can use them, the design time of networks can be shortened. **If we have to custom-design LSI/VLSI packages, the design is very time-consuming, but assembly is quicker and cheaper.** When the production volume of custom-designed LSI/VLSI packages is high, the initial investment for design, mask preparation, and others can be spread over a large number of packages, so each package costs very little. But when the production volume is very low, the cost of each package is extremely high. Different cases are compared in Table 1.3.2. (This is a rather simplified comparison. As we will see later, custom design has many variations.)

3. IC improves speed. As will be explained later, the signal propagation delay is longer on metal connections on a pc board than on connections

Table 1.3.2   Comparison of Different Implementation Approaches

	DESIGN		ASSEMBLY		COST PER SYSTEM WHEN PRODUCTION VOLUME IS	
	TIME	COST	TIME	COST	HIGH	LOW
Discrete components	Medium	Medium	Long	High	Almost unchanged	
Off-the-shelf packages	Short	Low	Medium	Medium	Almost unchanged	
Custom-designed IC	Very long	Very high	Very short	Very low	Very low	Very high

inside the chip. Also, by making the physical size of the entire digital system small, the speed is improved. The propagation time of an electrical signal along a 1-ft wire is about 1 nsec (i.e., nanoseconds), but parasitic capacitance* on the pc board further slows down the speed. As the switching speed of electronic gates improves (some gates have subnanosecond speed), the propagation time along wires and also connections on pc boards cannot be ignored. If the entire digital system is small, the length of the longest connections also becomes short, and the parasitic capacitance is reduced. Consequently the overall speed of the system improves.

4.   Power consumption is reduced for the following reasons. The reduction in size of a digital system and also the realization of the majority of connections on chips result in a reduction of parasitic capacitance throughout the system, reducing the electric power to drive signals over parasitic capacitance. Gates inside chips have lower noise disturbances, allowing the use of lower signal voltage levels and consequently reducing power consumption. The reduction of power consumption results in the increase in integration size of each chip if the chip size is limited by high power consumption, and this further improves speed, as we will see later. Also since the power supply is expensive, the power consumption reduction reduces the system cost, as seen in the above example of the cost of an intelligent terminal.

5.   Electronic circuits which realize logic gates can often be simplified. For example, some metal connections are eliminated in Fig. 1.2.4b (i.e., two MOSFETs with *A* and *B* can be connected by a common diffusion region without metal connection) when the entire network is placed on the same chip. As will be explained later, some components can be eliminated because large output power is not necessary inside a chip. (For example, if a gate must deliver a signal to a gate in another chip, totem-pole transistors are

---

* A parasitic capacitance is a capacitance embodied without a capacitor. It is usually embodied by terminals or connections against the ground and has a small capacitance value.

necessary in order to deliver large output power. Totem-pole transistors in a TTL gate may be eliminated if the gate delivers signals to another gate in the same chip. Also input diodes for the suppression of ringing on a transmission line can be removed.)

6.   The reliability of ICs is superior to that of assembled discrete components. When a large number of discrete components are assembled on pc boards, there is a chance of loose contacts or bad soldering of wires, and this is often difficult to detect after assembly. The reliability of assembled networks tends to be inversely proportional to the number of connections among components. But in the case of ICs, because the connections of IC gates are part of a package, there are few external connections to solder, and the reliability is high. (Connections inside IC chips could be faulty, but the chance of encountering such faulty connections is much lower than for external faulty connections.) It is reported that the failure rate of certain MSI chips with 70 to 100 gates each is only one seventh that of the SSI (*Computer,* Apr. 1979, p. 102).

7.   Since IC gates are so tiny, digital systems and computers are small and light. Because of this very important feature, many new products are being introduced, as described in Section 1.1.

The above are only direct advantages, but their combinations have deep technological and economical impacts on numerous new products, as we will see in later chapters.

If we want to gain the above advantages of ICs to the full extent, we need to learn key points in the very complex situation of LSI/VLSI design and fabrication. The situation is certainly very complex. For example, there are many IC logic families and variations, and even custom design has many variations. Also, when going from LSI to VLSI, many things, such as design and test, become much more complex than sheer increases in integration size. This book gives an overview of this complex situation and its key points.

CHAPTER **2**

Fabrication and Cost Analysis

A knowledge of the fabrication of IC chips and of cost analysis is very basic to LSI/VLSI system design. In this chapter, it is discussed as a prerequisite to the succeeding chapters.

2.1 Fabrication of IC Packages

Thin, round sheets of silicon about 4 or 5 in. in diameter, which are called **wafers**, are sliced from a single crystal grown by a method called the Czochralski method (or, probably less frequently, by the zone-refining method). On a wafer tens to several hundreds of IC chips are to be produced simultaneously by the process described in the following paragraphs. The many chips on a wafer look like an exquisitely detailed inlaid floor.

As an example let us now describe the process, using a p-channel MOS, that is, we want to produce the p-channel MOSFET* shown in Fig. 2.1.1. (Although only one MOSFET is shown, a large number of them are made simultaneously.) This is produced from the wafer by many steps, as shown in Fig. 2.1.2. An n-type silicon wafer sliced from a single crystal is oxidized to form an insulation layer (shaded), such as a silicon-oxide layer, at its top surface (Fig. 2.1.2a). A **photoresist process**, which is described later, is to form windows (Fig. 2.1.2b). (The distance between two windows is usually less than 15 μm.) A thin surface layer of boron concentration is formed in a diffusion furnace of gas (Fig. 2.1.2c). Diffusion–oxidation is not only to further diffuse boron into the silicon substrate, but also to grow a thin protective layer of oxide over the diffused regions (Fig. 2.1.2d). Windows for source, drain, and gate are opened by a second photoresist process (Fig. 2.1.2e). Since overlap between the gate and the p^+-regions is a source of parasitic capacitance (but no overlap is bad too, as will be explained in a later chapter), designed overlap is kept to a minimum. A new thin oxide layer is grown in the windows (Fig. 2.1.2f), and the oxide layer over the source and drain is removed to form **contact windows** (contact between a

---

* For the sake of simplicity, a metal-gate MOSFET is shown here. A silicon-gate MOSFET, which has a more complex structure but is widely used, is discussed in Chapter 4.

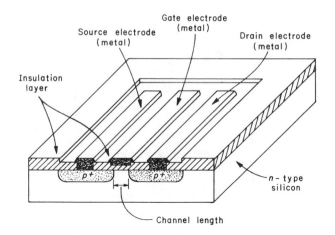

Fig. 2.1.1 *p*-channel MOSFET.

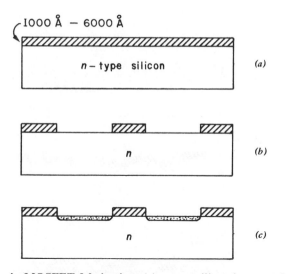

Fig. 2.1.2 Steps in MOSFET fabrication. (*a*) *n*-type silicon is covered with insulation layer (e.g., silicon oxide). (*b*) Windows are cut in insulation layer by first photoresist process shown in Fig. 2.1.3. (*c*) Predeposition is done to make the thin surface layer of boron concentration. (*d*) Diffusion–oxidation is to diffuse boron into source and drain, and a protective layer of dioxide is formed over the entire surface. (*e*) Windows for source, drain, and gate are cut by second photoresist process. Alignment of the mask is important in relation to the first mask. (*f*) Thin oxide layer is formed. Thickness of the layer over the gate is important (since it determines the threshold voltage). (*g*) Oxide layer of p^+-diffused regions is removed by third photoresist process. If it were not completely removed, it would be a cause for bad contacts. (*h*) Metalization is done over the entire surface. (*i*) Metal is partially removed by fourth photoresist process. (Possibly one more mask for another connection layer.) (*j*) Passivation, except on pads, is made to protect the surface from scratch or chemical erosion.

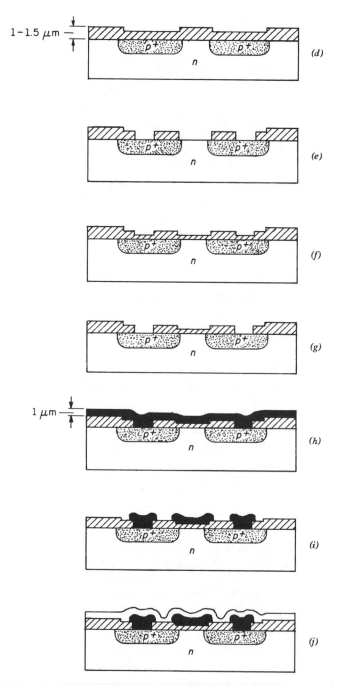

Fig. 2.1.2   *(Continued)*

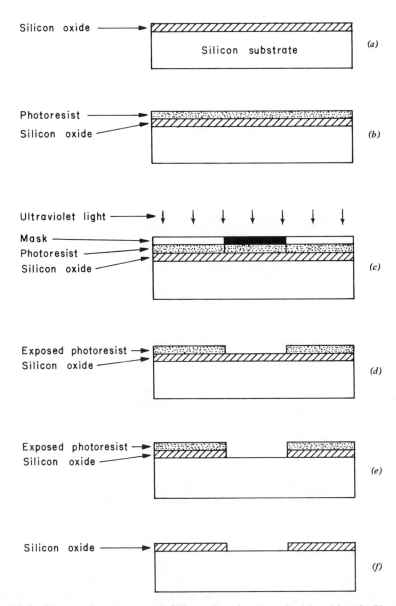

Fig. 2.1.3 Photoresist process. (*a*) Silicon sheet is covered with oxide. (*b*) Photoresist lacquer is applied to surface. (*c*) Photoresist is exposed to ultraviolet light through photomask. (*d*) Unexposed photoresist is removed with solvent. (*e*) Silicon oxide is removed by etching. (*f*) Photoresist is removed to leave "window" in silicon oxide.

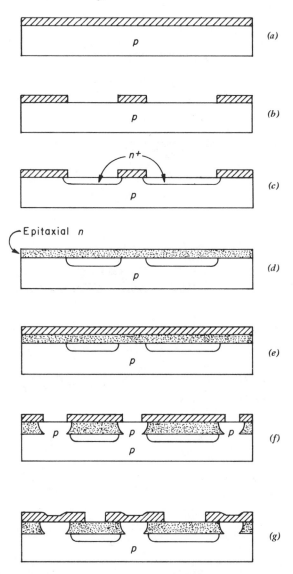

Fig. 2.1.4 Steps in bipolar fabrication. (*a*) *p*-type silicon covered with insulation layer (e.g., silicon oxide) is made. (*b*) Windows are cut in insulation layer by the first photoresist process (with the first mask for buried layers). (*c*) n^+-type dopant is diffused through the windows. (*d*) After washing away the oxide insulation layer, a thin layer of *n*-type silicon is grown on the silicon in an epitaxial reactior. (*e*) A new oxide insulation layer is grown. (*f*) Windows are cut for *p*-type dopant by the second photoresist process (with the second mask for isolation). The diffused regions connect with the underlying *p*-region, forming isolation between epitaxial regions. (*g*) After adding a thin insulation layer, windows are cut by the third photoresist process (with the third mask for base and resistor). (*h*) *p*-type dopant is diffused through the wondows. Then another oxide layer is added. (*i*) Windows are cut by the fourth photoresist (with the fourth mask for collector and emitter). (*j*) A shallow layer of n^+-type dopant is diffused.

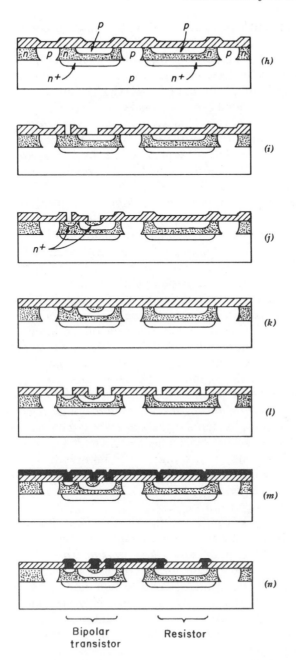

(k) **Another oxide layer is grown.** (l) **Windows are cut by the fifth photoresist (with the fifth mask for contacts).** (m) **The entire wafer is covered with a thin metal layer.** (n) **Undesired portions of the metal layer are etched away after the sixth photoresist process (with the sixth mask for metal connections). Another metal layer after growing an oxide layer is usually made, forming metal connections by the seventh photoresist process. Then passivation, except on pads, is made.**

diffusion region and a metal electrode) by a third photoresist process (Fig. 2.1.2g). Finally metal (e.g., aluminum) is evaporated over the entire surface (Fig. 2.1.2h), and then removed everywhere except at the source and drain contacts and the gate electrode region by a fourth photoresist process (Fig. 2.1.2i). (When connections among MOSFETs and also connections for power supply and ground are to be made, metal portions for these connections must not be removed. If more than one layer of connections is to be made, the formation of an insulation layer and a metal connection layer along with contact windows is repeated.) Lastly, a layer to protect against scratch or chemical erosion is formed as an option, and this is called **passivation** (Fig. 2.1.2j) (AMI staff, 1972).

During the above process, the selective removal of silicon dioxide or metal is done by the photoresist process, which is a photolithographic process using photoresist material. Making the window in Fig. 2.1.2b is illustrated in Fig. 2.1.3. The oxidized surface is coated with a thin layer of photoresist lacquer, as shown in Fig. 2.1.3b. This is an organic substance which polymerizes when exposed to ultraviolet light, and then, in that form, it resists attack by acids and solvents. A photographic **mask**, with opaque areas located where it is required to remove the silicon oxide, is placed over the wafer (simply a mechanical contact to the silicon-oxide surface of the wafer) and illuminated with ultraviolet light (Fig. 2.1.3c). (In this case of mechanical contact, if there is a gap between mask and photoresist, the edges of opaque areas may be blurred due to multiple light reflection.) The photoresist under the opaque areas of the photomask is unaffected and can be removed with a solvent; the exposed photoresist remains in the other regions (Fig. 2.1.3d). The wafer is baked to harden the photoresist and then soaked in a chemical solution to etch away the silicon oxide where it is not protected by the polymerized photoresist (Fig. 2.1.3e). The photoresist is removed from the surface by soaking in a solution of a mixture of inorganic acids (Fig. 2.1.3f). Then it is ready for the step of Fig. 2.1.2c. The same photoresist process is repeated in other steps in Fig. 2.1.2.

The above process is probably the simplest and thus the easiest to understand. [p-MOS is still used, for example, by Texas Instruments (*Electronics*, Dec. 20, 1979, p. 34) and Fairchild (*Electronic News*, Jan. 7, 1980, p. 12) although it is generally being replaced by n-MOS.]

The entire fabrication process used to take about 4 weeks and require more than 100 steps. (Notice that the steps shown in Fig. 2.1.2 are greatly simplified.) More steps are involved in current processes.

As described above, MOSFET fabrication requires at least 4 masks, namely, diffusion region mask, gate mask, contact mask, and metal mask. Bipolar transistor fabrication requires nearly three times the steps for MOSFET with at least 6 masks, as shown in Fig. 2.1.4, where an **epitaxial layer** is a single-crystal silicon layer grown on the substrate.

The current trend is use of more masks, that is, about 10 to 12 masks for MOS and 12 to 16 masks for bipolar transistors (*Electronic Design*, Mar.

29, 1979, p. 77; *Electro 80*, 22/3, p. 1). The exact alignment of masks is difficult. A deviation of each mask position by 20 to 30% of the minimum line width is usually expected. Thus as more masks are used, we will have an increasing number of faulty chips (*Spectrum*, Mar. 1979, p. 44).

References: Fairchild, 1979; Glaser and Subak-Sharpe, 1977; Richman, 1973; AMI staff, 1972; Hibberd, 1969; Carr and Mize, 1972.

In recent years ion implantation has been used to add impurities, replacing diffusion in some steps. In the case of diffusion, the surface of a silicon substrate is exposed to an impurity vapor at high temperature (around 1000 °C) to diffuse the impurity into a silicon substrate, so a high concentration of impurity results near the surface, as illustrated in Fig. 2.1.5. The **ion implantation** injects impurity ions into a silicon substrate, directly or through a thin oxide layer, by accelerating them by electric field. The depth of penetration is controlled by the acceleration energy (i.e., by the strength of the electric field). As shown in Fig. 2.1.5, the impurity peak position and the impurity amount are more accurately controlled. Thus by using ion implantation in appropriate steps, the quality of an IC chip can be significantly improved (Moorehead and Crowder, 1973; Dobriner, 1974).

The processing of wafers, or fabrication, explained with Figs. 2.1.2 through 2.1.4, has been done increasingly often with dry methods, such as plasma etching, reactive ion etching, molecular beam epitaxy, ion implantation, and possibly others, because **dry processing** by these methods has many advantages over the **wet processing** by liquid chemicals shown in Figs. 2.1.2 and 2.1.4. Processing parameters can be controlled better by dry processing. Also wet etching undercuts the photoresist at the same rate it etches downward, as shown in Fig. 2.1.6a. Dry etching reduces this undercutting, and also it etches cleanly downward, as shown in Fig. 2.1.6b. Thus dry processing is indispensable for VLSI.

References: Lyman, June 19, 1980; *Solid State Techn.*, Mar., Apr., Aug., Nov., 1980; Feb., Apr., May 1981; Panish and Cho, 1980; Seliger and Sullivan, 1980; Luscher et al., 1980.

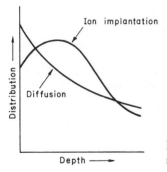

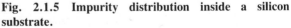

Fig. 2.1.5 Impurity distribution inside a silicon substrate.

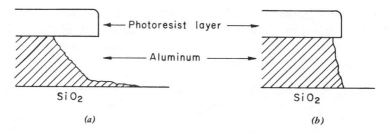

(a)   *(b)*

Fig. 2.1.6   Wet etching versus dry etching. *(a)* **Wet etching.** *(b)* **Dry etching.**

Notice that in the case of bipolar fabrication, n-type, p-type, and n-type layers (or p-type, n-type, and p-type layers) are stacked up. At each stacking a greater amount of impurity of the opposite type is added to the upper layers than to the lower layers. Thus after the completion of fabrication, the upper layers contain much greater amounts of impurities than do lower layers, and consequently the unit resistance of the upper layers is much lower.

The air in a fabrication room must be very clean because the components on a chip are smaller than dust particles. (Actually, local air cleanness around the wafer, which may get dust from human operators, machines, containers, and many others, is much more important than the cleanness of the entire room.) The cleanness of different places is compared in Table 2.1.1. The sizes of dust particles and others are compared in Fig. 2.1.7 (Gossen, 1979).

As seen in the above fabrication, all information about logic networks is contained in the masks used during the photoresist processes, that is, the

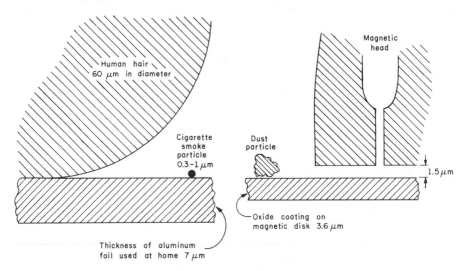

Fig. 2.1.7   Relative size.

Table 2.1.1 Cleanness

	NUMBER OF DUST PARTICLES OF 0.5 μm OR GREATER DIAMETER IN AIR PER LITER	NUMBER OF DUST PARTICLES OF 5 μm OR GREATER DIAMETER PER LITER
Fabrication room (for LSI)	3.57	
Stratosphere		
40 km	7	
10 km	20	
Ocean	2,500	28
City		
Clean	177,000	
Dirty	350,000	69

opaque areas in these masks. These masks are prepared in the following manner.

The architecture of a digital system to be realized on a chip, such as the connection configuration of registers and networks and the size of memories, is designed, and then the logic design of all the networks with gates such as adders and a control logic network is done. Subsequently the logic networks are converted into electronic circuits, and the layouts of electronic circuits are designed. They are drawn on mylar sheets which can be as large as a few feet to several feet on each side (depending on chip size), or on a CRT screen by which layouts are stored digitally in a magnetic tape or disk. In this case we need to prepare many layouts since each layout represents a pattern on each mask to be used during fabrication. As the integration size of a chip increases, the layout becomes more time-consuming. In particular, layout design, electronic circuit design, and logic design are highly interactive and not separable. The entire sequence of all designs is complex and very long, requiring as many as tens of man-months or possibly more. Each step must be examined for possible mistakes and corrected. When there are no more easily detectable mistakes, the layouts are complete.

After the layouts of the entire chip have been drawn on mylar sheets, they are converted into digital data by a **digitizer** with the aid of a computer, and these data are stored digitally on a magnetic tape or disk. Then different portions of each layout are displayed on a CRT one by one and inspected visually for further mistakes. After all corrections have been made (although more mistakes may be discovered in later stages), a **reticle**, which is a small photographic plate of the layout image, is prepared from each layout stored on the magnetic tape or disk. (At the same time each layout stored on the magnetic tape is sometimes prepared mechanically by a plotter on a mylar sheet, a few feet long on each side, in order to check further for mistakes or to find possibilities of improvement in the reduction of the layout area. Such large mylar sheets are convenient for visual inspection of the global layout as compared with the inspection on the CRT screen.) The layout image on the reticle is then reproduced on a glass plate covered with chromium or emulsion photographic film by means of optical reduction equipment called a **photorepeater**. In this case the original layout image is repeated on the glass plate as many times as there are chips on the wafer, with a photographic reduction of about 10:1 (see Fig. 2.1.8*f*). The glass plate thus obtained is called a **master mask** and looks like a tile floor where each rectangular tile has the same layout image of the chip. From this master a large number of **working masks** are reproduced which will be actually used for fabrication.

In the photoresist process of Fig. 2.1.3 a mask is directly contacted to a wafer. This is called **contact printing**. For high-volume fabrication of wafers, each mask will be used repeatedly, but actually it can only be used a limited number of times since its surface gets scratched by the wafers when an operator adjusts the registration of the mask. An emulsion-type mask can be used roughly 10 times and a chromium-type mask about 100 times. (The skill of the operator and the required defect density determine how many times a mask can be used.) In the case of **projection printing**, the layout image on a mask is projected optically onto a wafer without contacting the mask to the wafer. Then mask can be used a large number of times, substantially reducing the mask costs per wafer—several times reduction (*Electronics*, Aug. 28, 1980, p. 48). [Theoretically a mask can be used an unlimited number of times, but actually usage is limited to about 100,000 times because the mask must be cleaned due to dust accumulation, and it is scratched at each cleaning (Hamilton and Lyman, 1980).] Because of this advantage, projection lithography is becoming widely used (King, 1979, 1980; Markle, 1974; Hamilton and Lyman, 1980). For a comparison of contact and projection printing, see, for example, Lyman (July 21, 1977, p. 86).

The photographic preparation of a master mask (excluding its design) costs about $2,000. Since fabrication requires at least several masks, $10,000 is needed easily for an entire fabrication process. Each working mask costs less, but notice that a large number of them is needed for manufacturing.

Figure 2.1.8 illustrates the entire sequence of design and fabrication. The fabrication process shown in Figs. 2.1.2 and 2.1.4 treats simultaneously a number of wafers, each of which has tens to several hundreds of chips. The chips are tested by probes (Fig. 2.1.8*l*) because usually there are many faulty chips after such highly complicated fabrication steps. A **probe test** is automatically carried out by contacting the pads of every chip with microelectrode probes (registration of each chip with respect to the probes is done automatically by fine mechanical adjustment) and electric signals from some pads are analyzed by a computer. (These signals are from the outputs of the networks in response to electrical signals fed to other terminals.) If faulty chips are found, they are marked and will be thrown away later. Then the entire wafer is broken into individual chips by a laser scriber (or a diamond saw). Since this process is similar to glass cutting, it is called

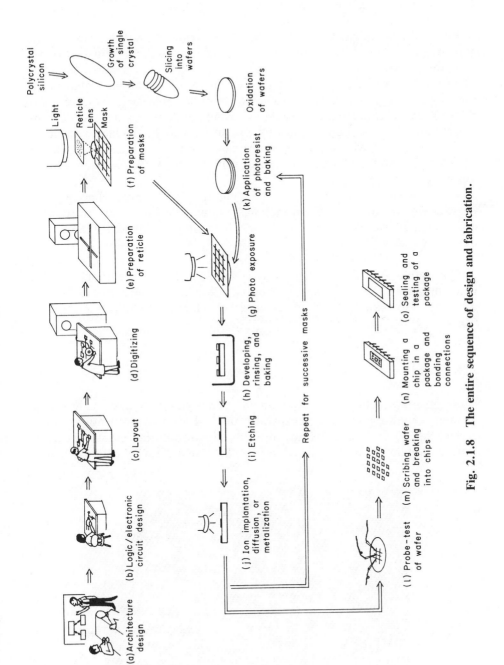

Fig. 2.1.8 The entire sequence of design and fabrication.

(a) Architecture design

(b) Logic/electronic circuit design

(c) Layout

(d) Digitizing

(e) Preparation of reticle

(f) Preparation of masks

Polycrystal silicon

Growth of single crystal

Slicing into wafers

Oxidation of wafers

(k) Application of photoresist and baking

(g) Photo exposure

(h) Developing, rinsing, and baking

(i) Etching

(j) Ion implantation, diffusion, or metalization

Repeat for successive masks

(l) Probe-test of wafer

(m) Scribing wafer and breaking into chips

(n) Mounting a chip in a package and bonding connections

(o) Sealing and testing of a package

Light

Reticle
Lens
Mask

scribing and **breaking**. Only good chips are mounted in containers (often ceramic), and then the pads of chips are connected to the terminals, that is, pins, of the container with gold or aluminum wires. This is called **bonding**. Then the containers are sealed with lids, or, more commonly, chips with their pins bonded are molded in plastic. In either approach we have completed IC packages. (Sometimes chips are mounted on ceramic sheets without containers. ICs prepared in this way are called IC **modules**.) Each IC package is tested with its external terminals by feeding electric signals to its input pins and analyzing those at its output pins by a computer. This is a **package test**.

When the networks on a chip are complicated, the probe test and the package test would require astronomical amounts of time if they were to be done completely. Thus these tests are done only within reasonable time limits.

For References to tests, see, for example, Williams and Parker, 1979; Kitamura et al., 1978; *Computer*, Sept. 1979; and also Muroga, 1979, Sec. 9.6. See also Akers, 1980; Harrison et al., 1980; Muehldorf, 1980; and Chrones, 1981.

2.2   Yield

During the fabrication of IC chips, many things can go wrong. Wafers may be faulty (e.g., impurities other than the desired ones, or dislocations in the crystal due to an uneven temperature distribution during crystal growth). During the fabrication illustrated in Figs. 2.1.2 and 2.1.4 (mostly during the

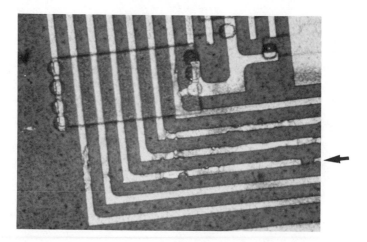

Fig. 2.2.1   Open circuit. (Reproduced by permission of Zilog, Inc. This material shall not be reproduced without the written consent of Zilog, Inc.)

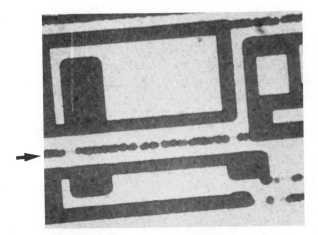

Fig. 2.2.2   Short-circuit mechanism caused by bridged metal connection. (Reproduced by permission of Zilog, Inc. This material shall not be reproduced without the written consent of Zilog, Inc.)

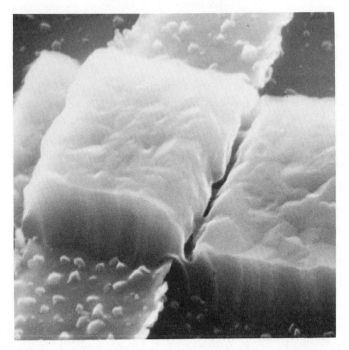

Fig. 2.2.3   Basic connection failure mode. (Reproduced by permission of Zilog, Inc. This material shall not be reproduced without the written consent of Zilog, Inc.)

photoresist process), many faulty chips may result due to dust particles, insufficient etching, misalignment of masks, scratches on masks, chemical contamination, image blurring, and others. Figures 2.2.1 and 2.2.2 show an open circuit and a short circuit, respectively. Figure 2.2.3 shows a partly broken metal connection caused by a sharp discontinuity in metal thickness, which later may result in an open circuit due to burning of the thin portion of the connection during operation. Figure 2.2.4 shows a good connection which has a reasonably uniform thickness.

Even during the manufacturing processes between the probe test and the package test many things can go wrong. Some chips may be damaged by the scribing–breaking process. Bonding may not be done perfectly. Lids may not be completely sealed (so moisture will leak into the package, causing chemical erosion later).

Even if each step during fabrication is almost faultless, the yield of IC packages becomes very low after a large number of fabrication steps, as the theoretical calculation in Fig. 2.2.5 shows. Here **yield** is defined as the ratio of the number of faultless products to the number of all products, that is,

$$(\text{yield}) = \frac{(\text{number of faultless products})}{(\text{number of all products})} \times 100\%$$

Although the yield is greatly influenced by many factors, it usually decreases as the chip size increases. The actually observed yield decreases slightly more slowly than its exponential decrease:

$$k_1 \, e^{-k_2 D}$$

where D is the area of the chip, and k_1 and k_2 are constants. This is illustrated in Fig. 2.2.6 for loosely and densely packed chips. The yield varies greatly from about 90% for SSI packages to 10% for LSI packages. Often chips with the most advanced processing technology, such as microprocessor chips, start with yields of 1% or less, but after many years' improvement the yield can become as high as 50%.

Usually a yield for a bipolar chip is much lower than that for a MOS chip because of the inherent complexity of its fabrication (Fig. 2.1.4 compared with Fig. 2.1.2). We can now manufacture chips of increasingly larger integration sizes without lowering their yields because of technological progress. (The figures of yield are hard to obtain since they are proprietary information of the manufacturers. Also, they fluctuate during manufacturing and differ from company to company. The figures in this book are quoted only to give rough ideas.)

References: Murphy, 1964; Seeds, 1967; Noyce, 1968; Moore, 1970; Gupta and Lathrop, 1972; Warner, 1974; Glaser and Subak-Sharpe, 1977; Phister, Oct. 1979. For the causes of low yields, see, e.g., Hochman, 1969 and Phelps, 1971.

If we can increase the wafer size, assuming that chip size and yield are the same, then we can obtain more faultless chips. Since this can be done

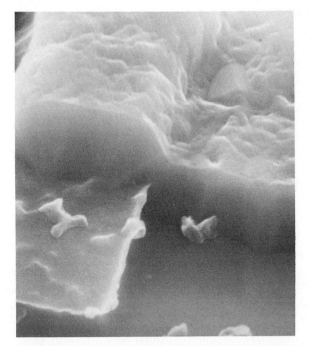

Fig. 2.2.4 **Satisfactory metal connection. (Reproduced by permission of Zilog, Inc. This material shall not be reproduced without the written consent of Zilog, Inc.)**

without any significant cost increase for processing wafers, we can proportionately reduce the chip cost. But the wafer size is currently limited by the following factors:

1. Larger wafers develop more dislocations and nonuniform distribution of impurities due to nonuniform temperature distribution during single crystal growth, although even larger wafers have been introduced by technology

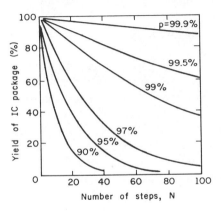

Fig. 2.2.5 **Yield of IC package p^N, where p is the yield for each step.**

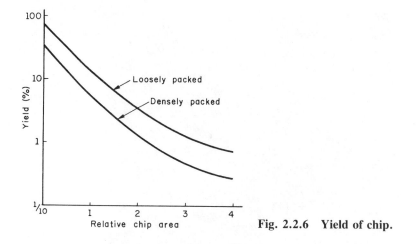

Fig. 2.2.6   Yield of chip.

progress. [The zone-refining method is better in this regard than the Czochralski method, but it is more costly. It has been proposed to grow a crystal under almost no gravity in outer space, since a 10 times bigger crystal can be grown without dislocation (Bylinsky, 1979).]

2.   Semiconductor manufacturers must invest heavily in new equipment to handle larger wafers.

3.   There are many other problems which lower the yield, such as the following, though some of them are becoming less pressing: During fabrication, heat is not evenly distributed and wafer warping or distortion results at high temperature. Higher accuracy of photorepeaters or other mask preparation equipment is desired. Heat expansion of mask glass plates must be controlled. Higher accuracy of mask alignment is required.

Notice that the chip yield is influenced by both the chip size and the packing density, as seen in Fig. 2.2.6. When the packing density increases with the chip size fixed, the integration size increases with decreasing yield. Also, when the number of masks used during fabrication increases, the yield decreases because mask misalignment and imperfect etching tend to accumulate.

Another factor which limits the chip size and also the packing density (i.e., the integration size, without proportionately increasing the chip size) is heat dissipation. The maximum allowable heat dissipation on a chip is generally 4 to 5 W. Biopolar transistor logic families, ECL and TTL (except IIL), cannot attain as large an integration size as MOS logic families because of the large power consumption of each logic gate. But the chip size and the integration size (without proportionately increasing the chip size) of MOS logic families are also limited by heat dissipation as a very large number of logic gates are placed on a chip (*Computer*, Sept. 1978, p. 14).

Improvement of the yield is vitally important for the manufacturer if the cost is to be lowered and the profit increased. For example, Intel Corporation increased its unit output nearly four times between the first quarter of 1973 and the first quarter of 1974. Nearly half of that increased output came from yield improvement. (Note that the manufacturing costs remain the same for this yield improvement.) The conversion from 2-in. to 3-in. wafers was a distant second. (This conversion requires the purchase of new equipment.) Far down the line in the third place as a factor in the increased output, was new facilities (*Electronic News*, Apr. 15, 1974, p. 50).

Since the manufacture of IC packages is very complicated, yields are improved and costs are lowered as manufacturers gain experience with the same products over the years. This improvement is called the **learning curve**. It was recognized by Henry Ford in the beginning of the century and is particularly important in the case of LSI/VLSI fabrication. This is due to gradually finding an optimum set of processing parameters, such as impurity levels, temperature, and time, combined with the proper handling of wafers by human operators and, possibly, design modifications. Although it depends on individual manufacturers and products, and accordingly is proprietary how improvements are made, the learning curve principle roughly asserts that as the production volume doubles, the manufacturing cost drops by 30% (*Business Week*, Mar. 1, 1976, p. 43). Thus the first company to deliver a new product in high volume can drop its price fastest and retain the largest market share (Cunningham, 1980).

The management of operators is also important for yield improvement. Vitally important is a teamwork for **quality control**, which was first introduced to Japan in 1950 by W. E. Deming in its theoretical aspects, and sometime later by J. Juran in its practical aspect, and has since been refined in practice in Japan. This team effort is often called the **quality control circle**. Combined with the learning curve it sometimes easily doubles the yield that otherwise can be attained by the learning curve alone, without the quality control circle (Goto and Manabe, 1980; *New York Times*, May 10, 1981, Business Sec.; Juran, 1974; Lyman and Rosenblatt, 1981; *Electronics*, Apr. 10, 1980, pp. 24, 81-82; May 8, 1980, p. 8; Nov. 6, 1980, p. 46; December 4, 1980, pp. 6, 24, 95–100; Dec. 18, 1980, pp. 41-42; Mar. 24, 1981, pp. 95-96; *Spectrum*, July 1980, p. 21; Ishikawa, 1976; *Business Week*, July 20, 1981, pp. 18–44).

The delivery of IC packages in sufficient quantities by semiconductor manufacturers is vitally important for the user. In the early 1970s many firms that assembled calculators went bankrupt, and some computer firms had lower revenues because ordered IC packages were not delivered. This has been repeated with many manufacturers in later years. Thus **second sourcing** is very important. Users have to find other semiconductor manufacturers or in-house IC production as second sources before making a firm commitment to one semiconductor manufacturer for specific IC packages (*Electronics*, May 15, 1975, p. 74).

The purchase of faultless or reliable IC packages is also vitally important for users. For example, Hamond Electronic Organ uses 10 million transistors a year. If 1% of the transistors are faulty, then 70% of the assembled electronic organs are faulty at final inspection (*Electronics*, Apr. 26, 1973, p. 75). In the beginning of 1975 some firm continued to delay the introduction of a new version of minicomputers with a 4k RAM because the 4k RAMs delivered were too unreliable.

Semiconductor manufacturers test finished IC packages to some extent before shipment. But a complete test is often too expensive, or impossible, for complex LSI such as microprocessor chips. Even with partial tests, test costs occupy a significant portion of IC manufacturing costs, as we will see later.

Semiconductor manufacturers, when required, and users generally test IC packages more carefully to screen out those that do not meet the specifications. These extra tests are usually thermal shock, mechanical shock, and **burn-in** (EDN, Dec. 16, 1981, pp. 330–336). It was not rare to reject 2 to 5%, or more, of IC packages by these tests during the 1970s, until quality control prevailed.

References: Muehldorf, 1980; Turner, 1981; Koo, 1980; Mattera, 1975.

2.3   Cost Analysis

Unlike when digital systems with discrete components are designed, the cost analysis is essential when designing digital systems or computers in LSI/VLSI, because the pricing of LSI/VLSI has a very complex relationship with the design approaches. As we will see in later chapters, a large number of different and complex design approaches have been devised, and designers must single out one of them, considering tradeoffs between performance, cost, design time, production volume, design flexibility, and many other factors, in each of a large number of different development situations discussed in Section 1.1. In other words, the engineering problem of the design cannot be separated from the manufacturing problem; the two are highly interactive. In contrast, in the case of designing and manufacturing digital systems with discrete components (also SSI or MSI packages), designers have very few design approaches, each of these approaches is simple and straightforward, and design and manufacturing can be done almost independently of each other. Indeed, "integrated circuit" in the case of LSI or VLSI means not only the integration of all the components on a single chip, but also the designers' integrated knowledge of all the aspects, including cost analysis, design approaches, design flexibility, ease in correcting design mistakes, and fabrication, as will be discussed in detail in later chapters. Thus **the cost analysis is essential for choosing the appropriate integration size and design approaches discussed in the succeeding chapters.**

The cost of an IC package, roughly speaking, is determined by the following two items:

1. Initial investment. When designers decide to develop an LSI/VLSI chip realizing a certain digital system or logic network, they have to spend a long time (months or years) in architecture design, logic design, electronic circuit design, and layout design. Then they have to test their design by making a prototype chip before the final version of the layout is prepared for the production phase. The expenses for this design phase are mainly the salaries of the designers and draftpersons plus the computer time and material cost of the prototype masks, wafers, and chips.

When the integration size is small, the above design phase is simple, and this initial investment (in particular, salaries) is small.

2. Manufacturing cost. This is the cost for fabricating each chip, assembling it in a package, and testing the package. If the fabrication cost of a wafer is divided by the number of faultless chips sliced from the wafer, the approximate fabrication cost of each chip is obtained. Then by adding the assembling and testing costs of each package, the manufacturing cost of each package is obtained.

Then, considering the above two items, the total cost of each IC package is derived by the following formula:

$$(\text{cost per IC package}) = \frac{A}{V} + B \qquad (2.3.1)$$

where A is the initial investment cost
        B is the manufacturing cost per package
        V is the production volume

A becomes large as the integration size of a chip increases. A is extremely large for LSI/VLSI, whereas A is very small for MSI or SSI. But if the production volume V is large enough, the first term in Eq. (2.3.1) is very small compared with the second term B, and the cost per IC package is almost equal to B. Thus even if A costs a million dollars for some LSI chip (such as an off-the-shelf microprocessor chip), the cost per package is only a few dollars when V is over a few millions and the production yield is high. But if the integration size increases, V generally decreases, unless designed for very general purposes, since usage of the chips become limited. The other extreme is discrete components or SSI whose usage is so wide that V can be extremely large, so its cost is very low. But a large number of discrete components or SSI packages must be assembled on pc boards; thus the unit cost of a digital system [to be calculated by Eq. (2.3.2) in the following] is far higher than that of the system with LSI/VLSI packages. In the case of LSI/VLSI, if V is very low, the cost per package becomes prohibitively expensive, and we need to find an appropriate LSI/VLSI design approach, as we will discuss in later chapters. Tradeoffs among integration size, initial investment cost A, and market size (i.e., V) are vitally important considerations in LSI/VLSI package development plans.

Notice that if the yield is low, B becomes high. Since the fabrication cost of each wafer is relatively fixed, an increase of yield immediately leads to a reduction of B.

The selling price of a package is the cost given by Eq. (2.3.1) plus indirect cost and profit, where the indirect cost is for sales support, and administrative personnel and clerical supports. Each amount of the indirect cost and profit is usually comparable to or less than the cost given by Eq. (2.3.1) (less for products in markets with tough competition).

In the case of designing and manufacturing a digital system with discrete components, SSI or MSI packages, or their mixture, we can have a formula similar to Eq. (2.3.1):

$$\text{(cost per unit of a digital system)} = \frac{a}{v} + b \qquad (2.3.2)$$

where a is the initial investment cost
        b is the manufacturing cost per unit
        v is the production volume

When a digital system is to be designed and manufactured with a large number of discrete components and SSI or MSI packages, b is large, no matter whether or not a/v in Eq. (2.3.2) is very small. Actually b is much greater than B in Eq. (2.3.1) (easily 100 times, 1000 times, or more), because a large number of components and packages are assembled on many pc boards, pc boards are assembled in cabinets, possibly with cooling fans, and pc boards and cabinets are interconnected by wires and cables. Of course, when a digital system to be designed is too large to fit into a single LSI/VLSI package, the unit cost of the system can be calculated by Eq. (2.3.2) using the package cost Eq. (2.3.1) in calculating the manufacturing cost b in Eq. (2.3.2).

In the following we analyze in detail the initial investment A and the manufacturing cost B, which appear in Eq. (2.3.1).

Initial Investment A

The development of a microprocessor LSI chip of the most advanced technology in the late 1970s took typically five engineers (or designers) and three draftpersons for 2 years. Thus their total salary was estimated at about a half million dollars or more. Depending on the chip complexity, we can easily spend 10 times more. They specify the architecture of a digital system to be placed on a chip and design logic networks. The networks are tested by logic simulation and often by hardware prototype. Then they design electronic circuits with the help of CAD programs. The layout of the electronic circuits, which is the most time-consuming, is done by draftpersons under the supervision of engineers. Usually the layout is tested by repeating the realization of chips at least a few times, until major design mistakes are eliminated and the final version of layout is reached. If the chip realization is repeated too many times, the cost of the preparation of layouts, masks, and wafers is very significant. Special CAD programs and testing programs usually have to be developed to reduce design and check time, and accordingly the design expenses throughout the above design phase.

All expenses incurred during the above design phase, including salaries and the amortization of the equipment needed, such as CAD terminals, constitute the initial investment A. The capabilities of the people involved and also the management can make a difference in keeping all expenses under control.

The fabrication cost of a chip decreases as the chip area D decreases, and this relationship can be approximated by the formula

$$(\text{fabrication cost per chip}) = K(10^{0.0243D})$$

where K is a constant determined by the fabrication technology and design parameters (Cragon, 1980). Thus when a semiconductor manufacturer wants to sell a chip in high-volume production (say, on the order of millions), the chip with the smallest size is designed painstakingly, taking a long period of time. This is called the **full-custom design approach**.

When the production volume V is low, the first term in Eq. (2.3.1) is much higher than the second term. So it is meaningless to minimize the fabrication cost, and consequently cost B, with packaging, testing, and others added. In this case it is better to use a quick design approach, based on logic design methodology and CAD, which is called a **semi-custom design approach** such as the following. (Many other semi-custom design approaches will be discussed in Chapter 9.) Once layouts of standard networks or their subnetworks are compactly designed by people with new technology and after having spent long periods of time, we can later design new chips by simply assembling these layouts, which are called **cells**. In this case, since cells are assembled on a chip by interconnecting them without changing the insides of the cells, the entire layout cannot be most compact, but the design time can be greatly reduced. Simulation is often not needed. For example, a single engineer can design a chip in a few months, while the chip would require 8 people for 2 years in the case of a completely new design with new technology. This design approach is called the **cell-library approach**. The initial investment A is substantially reduced, but the best performance and lowest manufacturing cost B cannot be obtained due to the larger chip size than in the full-custom approach. In other words, if we use a difficult and time-consuming design approach, A increases and B decreases, and if we use an easy and quick design approach, A decreases but B increases.

Table 2.3.1   Cost Breakdown of an IC Package (Excluding Test Cost)

ITEM	PERCENTAGE OF PACKAGE COST	
	LSI	SSI
Finished chip cost after fabrication (this decreases as yields improve)	50%	10%
Labor for assembling (such as bonding) and packaging	30%	40%
A container (plastic) and material for assembly (such as gold bonding wire and solder)	20%	50%

Thus A and B are highly dependent on each other. In contrast, a and b are almost independent in Eq. (2.3.2).

Manufacturing Cost B

A silicon wafer, which costs about \$6 (3-in. wafer) to \$15 (4-in. wafer), costs about \$40 to \$150 after fabrication. Thus if a wafer contains 10 faultless chips, each chip costs approximately \$4 to \$15 before assembly and testing. If a wafer contains 1000 faultless chips, each chip costs 4 to 15¢.

Table 2.3.1 shows the cost breakdown of a finished IC package gathered from different sources, though the percentages of the different items vary depending on manufacturers, logic families, integration size, yield, container material (plastic or ceramic), and wages. This table gives only a rough idea of the cost breakdown. Notice that labor is an important factor, though it greatly varies depending on where the package is manufactured (United States or offshore). Also, the combined cost of the package container and material exceeds that of the chip in the case of SSI. The major cost goes to the container and labor, as in the case of a lipstick, where the container is more expensive than the lipstick inside. The package test cost is not included in Table 2.3.1. It is becoming very expensive and occupies a significant portion of the total chip cost as the integration size increases.

Table 2.3.2 shows some cost analysis examples compiled by Integrated Circuit Engineering (*Electronic News*, Apr. 15, 1975). (The original explanation is slightly adjusted in this table. Integrated Circuit Engineering Corporation is a consulting company which provides microelectronics industry survey and technology know-how.) Detailed actual costs are rarely published.* Cost analysis examples including test costs are shown in Table 2.3.3 based on a pamphlet distributed by Advanced Micro Devices, Inc., in 1978.

When the integration size increases, the yield gets progressively lower (other conditions such as packing density being the same), and accordingly,

---

* For price trends of commercial packages, see *Electronic News* (in particular the "patterns in pricing" column), for example.

Table 2.3.2 Cost Analysis of Typical Commercial Packages[a,b]

	TTL SSI 7400 (bipolar)	TTL MSI 7442 (bipolar)	1K RAM 1103 (pMOS with Si gate[c])	CALCULATOR LSI (p MOS with Si gate)	CMOS SSI 4000
Cost of fabricated wafer	$35	$35	$55	$40	$40
Number of faultless chips per wafer	3,000	1,000	163	63	2,025
Fabrication cost per chip	1.2¢	3.5¢	34¢	63¢	2¢
Assembly cost per chip	7.5¢	10.1¢	38¢	58¢	8.2¢
Cost for package test and finishing[d] per chip	2¢	3.6¢	47¢	52¢	4.1¢
Total manufacturing cost per chip	10.7¢	17.2¢	$1.19	$1.73	14.3¢
Average selling price per chip	14¢	50¢	$3.05	$3.30	35¢
Manufacturing margin per chip	3.3¢	32.8¢	$1.86	$1.57	20.7¢
Manufacturing margin in percent	24%	66%	61%	48%	59%

[a] Reproduced with permission of Integrated Circuit Engineering Corporation.
[b] All packages are in plastic.
[c] Si gate is a MOSFET gate realized in silicon, instead of in aluminum, as discussed in Chapter 4.
[d] Finishing cost is for marking, sorting, packaging, and shipping.

Table 2.3.3   Cost Analysis Including Test Costs[a]

	SSI (50 mil², 14 pins)	MSI (80 mil², 16 pins)	LSI (130 mil², 24 pins)
Chip fabrication cost	20%	45%	70%
Packaging cost	75%	45%	20%
Test cost	5%	10%	10%

[a] Copyright © 1978, Advanced Micro Devices, Inc. Reproduced with permission of copyright owner.

the chip fabrication cost increases because many faulty IC packages must be thrown away. On the other hand, the cost per gate decreases as the integration size increases if the yield is 100%, since the cost of an IC package and test increases only slightly even if the integration size increases. Combining these two contradicting tendencies, we get the cost per gate in terms of the integration size as illustrated in Fig. 2.3.1. This curve shows that if the integration size exceeds a certain threshold, the cost per gate increases sharply (almost exponentially due to a yield decrease), and the cost advantage of VLSI is lost. In other words, there is an optimum chip size which minimizes the cost per gate. Beyond this optimum size the chip cost is determined primarily by yield, and below this optimum size the chip cost is determined primarily by the cost of packaging and testing. The optimum size of the chip, coupled with an increase in packing density, is gradually becoming larger every year, as illustrated in Fig. 2.3.2 (Phister, Nov. 1979).

The wafer size increases every few years, and the manufacturable chip size increases also, as shown in Fig. 2.3.3. It is important to notice that the economically manufacturable chip size is considerably smaller than the maximum manufacturable chip size because of the above optimum chip size. **Even if the maximum manufacturable chip size is increased by new technology in laboratories, of which we can see many examples in later chapters, the**

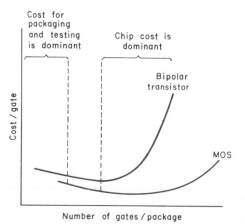

Fig. 2.3.1   Cost per gate versus integration size.

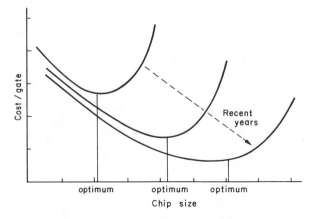

Fig. 2.3.2   Progress in optimal chip size.

actual manufacure of that size from an economical point of view is usually years away.

Many factors that influence yields and costs are summarized in Table 2.3.4. In the bottom row the importance of the manufacturers' learning curve is stressed. As the number of masks used increases and processing gets complicated, the yield decreases, as shown in the row "complexity."

Of course, by redesigning the layout of a digital system and by incorporating fabrication technology progress, the chip size can be reduced with speed improvement and cost reduction. The improvement of the 4-bit slice microprocessor chip Am2901 by Advanced Micro Devices, Inc., is shown in Fig. 2.3.4 as an example. (Am2901 is one of the most popular bipolar microprocessor chips.)

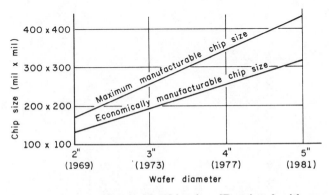

Fig. 2.3.3   Progress in manufacturable chip size. [Reprinted with permission from *Electronic Design* **(Mar. 29, 1978, p. 46); Copyright © by Hayden Publishing Company, Inc., 1978.]**

Table 2.3.4 Factors that Influence Yields and Costs[a]

IF THIS PARAMETER INCREASES	THEN THIS PARAMETER				COST		
	NUMBER OF FAULTLESS CHIPS PER WAFER	CHIP SIZE	PROBABILITY OF DEFECT	PROBE YIELD	PER CHIP	DEVELOPMENT	PER UNIT OF FUNCTIONAL AREA
Production volume	↑	↑	↑	↔	→	→	→
Complexity	→	←	←	→	←	←	←
Chip size	→	←	←	→	←	↑	↕
Number of gates on a wafer	→	←	←	→	←	←	↕
Input/output connections	→	←	←	→	←	←	←
Vendor experience in manufacturing chips	←	↳	←	←	←	→	→

[a] *Source*: Fairchild Camera booklet "OPTIMOS," 1972 (with permission of Fairchild Camera and Instrument Corporation).

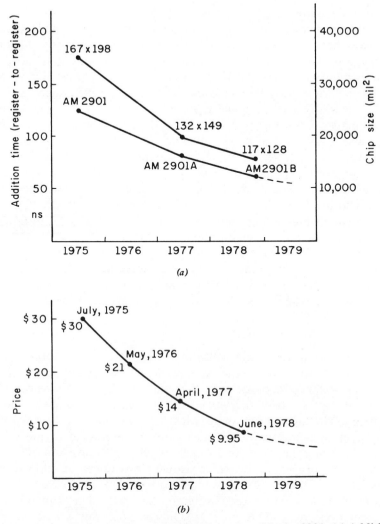

Fig. 2.3.4 Improvement of 4-bit slice microprocessor chip Am2901. (*a*) Addition time (register to register). (*b*) Price. (Copyright © 1978 by Advanced Micro Devices, Inc. Reproduced with permission of copyright owner.)

2.4 More About Fabrication

In this section we review current trends in chip fabrication after a discussion of protective circuits against static voltage.

Protective Circuits on a Chip

During the fabrication or handling of IC packages for testing or assembly on a pc board, a very high static voltage (a few hundred to a few thousand volts) often develops due to friction of worker clothes or plastic material. When such a high voltage is accidentally applied to the pins of an IC package, the insulation layer between a MOSFET gate and the substrate inside breaks down instantly. (A silicon-oxide insulation layer of about 1,000 Å or less, which is currently used in MOSFETs, breaks down at 100 to 150 volts.) The MOSFET is permanently damaged. So protective circuits, such as the one illustrated in Fig. 2.4.1, are usually provided right after the pads on an IC chip (Watanabe, 1979). Figure 2.4.1*a* shows the layout of a protective circuit together with a pad and an MOS cell, Fig. 2.4.1*b* shows the corresponding circuit, and Fig. 2.4.1*c* shows the cross section along *X–X'* of Fig. 2.4.1*a*. When a high voltage is applied at the pad, the diffusion region (A in Fig. 2.4.1*a*), which otherwise works as a resistor, works as a diode between the substrate and itself, discharging the current through the diode, and consequently reducing the voltage applied to the gate of MOSFET B in Fig. 2.4.1*b*.* There are many other protective circuits (Yenni and Huntsman, 1980; Kirk et al., 1976; *EDN*, Oct. 20, 1980, pp. 50-51; AMI staff, 1972).

In the case of bipolar transistors there is no need for special protective circuits because the resistance between base and emitter is so low that an input resistance connected to the base can protect the transistor.

Direct-Step-on-Wafer Projection

When the line width (e.g., the width of the metal connection on a chip) is reduced in order to increase the packing density of devices on a chip, or when larger wafers are processed, the contact or projection printing described in Section 2.1 does not work well. If a wafer is distorted or warped by the heat cycle during processing (Figs. 2.1.2 through 2.1.4), it will throw layer-to-layer registration of the masks on the wafer out of the required accuracy, lowering the yield. In order to alleviate this problem, **direct-step-on-wafer (DSW) projectors**, or **wafer steppers**, have been developed, which work as follows. The layout image on a reticle is directly projected onto the surface of a photoresist-covered wafer by an optical lens equipment with image reduction (10:1, 5:1, or 1:1). Only a small portion of the wafer, consisting of one chip or at most several chips (using a reticle that contains

---

* Although readers do not need to understand precisely how the protective circuit in Fig. 2.4.1 works, the following is provided for those interested. When a positive voltage is applied at the pad, a current flows from the diffusion region to the substrate since the positive voltage is applied to the forward direction of the diode constituted by diffusion region *A* and the substrate. When a negative voltage is applied at the pad, a small current still flows through the backward direction of the diode. In particular, a greater current flows under the aluminum electrode where a strong electric field develops, as shown by *C* in Fig. 2.4.1*c*, and this causes voltage drop across the large resistance of the diffusion region *A*, reducing the voltage applied to the gate of MOSFET *B*.

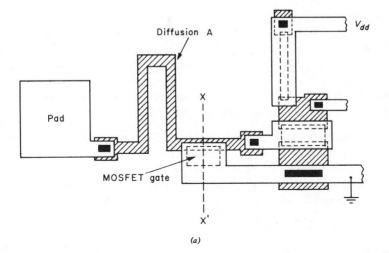

(a)

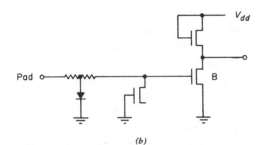

(b)

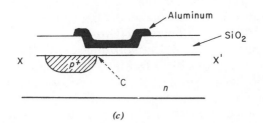

(c)

Fig. 2.4.1   Protective circuit.

many layouts), is exposed. Then the wafer moves one step further to a new position for the next exposure, being positioned accurately by a laser interferometer. (In this case the master and the working masks are not used at all.) The registration of the image can be done at each exposure, and also, by using a smaller lens aperture, the resolution is higher than for the printing methods described in Section 2.1, which project a large entire mask onto the entire wafer. On the other hand, repetition of the exposures over the entire wafer takes more time than the methods in Section 2.1 with one exposure (Lyman, Apr. 12, 1979; Lowe, 1981; *Solid State Tech.*, May 1981).

Line Width Reduction and Speed Improvement by Deep-Ultraviolet, Electron-Beam, or X-Ray Lithography

When the line width is reduced, the chip size is reduced by the improved packing density. Furthermore, the channel length in a MOSFET (i.e., the distance between two p^+-regions in Fig. 2.1.1) can be made shorter, and the parasitic capacitance of connections reduced. Consequently the speed of the transistors is improved and power is reduced.

However, as long as ultraviolet lithography of 3,000 Å wavelength or longer is used in IC fabrication (as discussed in Section 2.1), the line width is limited to 1 μm, because if we try to make a fine line, the line becomes blurred due to optical phenomena. If we want to reduce the line width, we need to use a shorter wavelength to improve the resolution. By using deep-ultraviolet lithography of 2,000 to 3,000 Å, the line width can be reduced to about 0.7 μm. Thus deep-ultraviolet lithography is becoming popular (Iwamatsu and Asanami, 1980; Tobey, 1979; Doane, 1980; Grossman, Mar. 29, 1979; *Electronics*, Sept. 13, 1979, pp. 44, 46).

By directly writing network patterns on a wafer by electron beam without masks, the line width can be reduced to about 0.5 μm. The on–off operation of the electron beam is controlled by an electric field, which is in turn controlled by a computer program. The entire wafer must be scanned by electron beam, so the exposure is time-consuming, although many techniques to reduce time have been devised. When the channel length of MOSFETs is 1 to 2 μm or shorter, masks used in Fig. 2.1.3 or reticles for deep-ultraviolet lithography described in the previous paragraph are usually prepared by electron-beam lithography (Grossman, Mar. 29, 1979, June 7, 1980; Weber and Yourke, 1977; McDermott, June 21, 1978; Lyman, Apr. 12, 1979; Reynolds, 1979; Eidson, 1981; *Hewlett Packard J.*, May 1981).

By using X-ray lithography of 5 to 50 Å, the line width can be further reduced to about 0.02 μm. Since it is difficult to form a very sharp X-ray beam and to deflect it, masks are prepared with electron-beam lithography,* and then each mask pattern is transferred to a wafer by the flood exposure of X rays, like the photolithography explained in Section 2.1 (Hughes and Fink, 1978; Queisser, 1977; Grossman, Mar. 29, 1979; Lepselter, 1981).

Any new technology is usually not immediately applicable in production,

---

* Although direct writing on a wafer by electron beam is very time-consuming, electron-beam lithography is better for the preparation of masks to be used for the fabrication by photolithography as described in Section 2.1 than are optical methods. The preparation of a set of masks takes 15 days by the conventional optical method but only 3 days by electron beam (*Business Week*, Oct. 24, 1977, pp. 94B–94L).

even if it is successful in laboratories, for it requires further refinement before production use. (Usually there is a delay of many years.) Although electron-beam lithography with low resolution is used in the production of some IC packages, electron-beam lithography with high resolution, and X-ray lithography require further refinement before actual use in production (*EDN*, May 20, 1978, pp. 22–26). Also, the production equipment for these new lithography techniques is much more expensive than for conventional techniques (Lyman, Apr. 12, 1979). The investment cost for electron-beam and X-ray equipment, in addition to that for dry processing mentioned in Section 2.1, is becoming very high. Capital equipment for the fabrication was about one tenth the annual sales of a semiconductor manufacturer in the early 1970s and about one third in the late 1970s. It is still rising.

Ion-beam lithography (IBL) is also in exploration. Ion beams do not scatter as much as electron beams, and unlike X rays, collimated sources are available. IBL has shown submicron (0.5 μm) resolution and is capable of a large throughput. Potentially IBL can handle many different IC processing steps, such as ion implantation, milling, and mask and wafer lithography (*Electronics*, Oct. 25, 1979, p. 217; Bollinger and Fink, 1980; Brown et al., Aug. 1981).

With the new techniques discussed above, an improvement in packing density and also in speed of at least a few times to 10 times or more is expected.

References: Capece, 1978; *Business Week*, Oct. 24, 1977.

Automation of IC Production

As IC production (particularly packaging) is highly labor intensive, production automation is desirable. But it is extremely hard because of the intricate nature of processing. The production is only partially automated, such as the photorepeater in preparing a master mask, the automatic mask aligner (actually interactive), and the automatic probe test. Some companies such as IBM are trying to automate at least fabrication (Brunner et al., 1981).

Currently most important is the automation of packaging. The **film carrier technique**, which is often called **tape automated bonding (TAB)**, slashed packaging costs and boosted packaging rates more than tenfold:

Manual bonding	60 chips/hour (about 16¢/chip)
High-speed automated wire bonder	600 chips/hour
Film carrier technique (TAB)	1000–2000 chips/hour

The film carrier technique uses a film that consists of copper and polyimide layers (and possibly another layer of adhesive). The copper layer is etched into interconnection leads in a spidery pattern, as shown in Fig.

2.4.2. Chips (shown as the square in Fig. 2.4.2) are automatically attached to the inner ends of these copper leads. Then the chip and the connected leads are automatically cut from each frame of the film and attached to the pins of a package. The high cost of the polyimide layer is a disadvantage of the film carrier technique (Lyman, Oct. 16, 1975, Dec. 25, 1975, Dec. 18, 1980; Ryan, 1980; *Solid State Tech.*, March 1980; Dais et al., 1980; Devitt and George, 1978).

High-speed automated wire bonders simply bond pads and pins one by one automatically and are widely used.

The **flip-chip soldering** technique was adopted by IBM around 1979, although this was an LSI extension of the solder reflow technology developed in 1964. A large number of small solder bumps (132 bumps for the Schottky TTL gate chip and 17 × 17 bumps for the ECL gate chip) are placed in matrix form, as illustrated in Fig. 2.4.3, where even under these bumps there are logic networks. Then the chip is inverted onto a ceramic carrier, which has many pins in the corresponding positions, and the two are heated until the solder reflows and bonds them together. This is done automatically. Thus conventional pads on a chip and bonding wires are replaced by the solder bumps (Pomeranz et al., 1979; Blumberg and Brenner, 1979).

Automation of the insertion of IC packages and other components into testers and also into holes of pc boards has been widely used for many years.

Production automation usually requires heavy initial investment, so it can be economically justified only with very high volume production.

References: Lyman, July 21, 1977; Clemens and Castleman, 1980; Paivanas and Hassan, 1980; Freese, Sept. 5, 1974; *Electronics*, Apr. 17, 1975, pp. 95–106.

Fig. 2.4.2 Top view of a film. The white square in the center of each film frame is a chip.

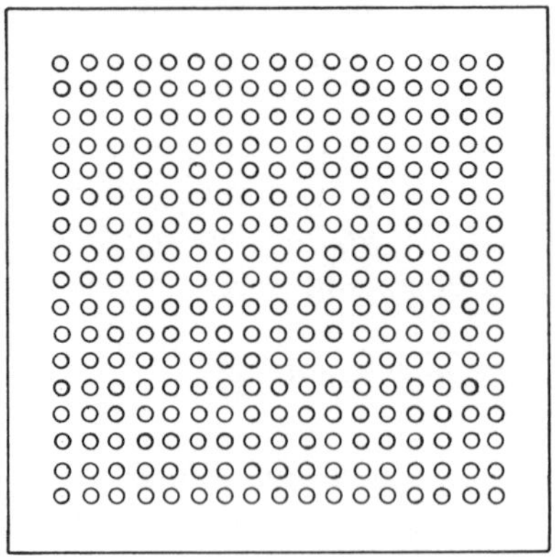

Fig. 2.4.3 Chip with a matrix of solder bumps.

2.5   Selection of an IC Logic Family and Logic Design Approach

Until MOSFETs started to be widely available for designers of new products in the late 1960s, logic gates based on bipolar transistors such as ECL and TTL were the designers' only choice. Tradeoffs between ECL and TTL (or its variations) in terms of speed and costs were straightforward. If the designers wanted high speed, they chose ECL. Otherwise they chose TTL or a variation.

Now there are many choices for logic designers, as follows. When logic gates or networks are to be realized on IC chips, there are many different ways to realize them in electronic circuits. They are called **IC logic families**. They can be divided into two major classes of families and many other minor ones which are currently in the development stage and are not as widely used as the major ones, as follows:

Logic Families Based on Bipolar Transistors.   RTL, DTL, HTL, TTL, ECL, TRL, EFL, IIL

Logic Families Based on MOSFET.   *p*-MOS, *n*-MOS, CMOS, CMOS/ SOS, DMOS, VMOS, and their static and dynamic versions

Other Minor Logic Families.   GaAs MESFETs, bubble logic, CCD logic, Josephson junctions

Some IC logic families have their variations. For example, TTL has variations of low-power TTL, Schottky TTL, tristate TTL, and others. Each MOS logic family has two different electronic realizations, static and dynamic MOS circuits.

There are so many logic families available that designers have difficulties in finding the most appropriate ones. They are all different in terms of **electronic performance** (throughout the rest of this book the word "performance" is used to mean mostly speed, although it can have many other meanings, such as power consumption, noise immunity, power supply voltage, stability, signal amplification, and possibly others), **logic capability** (logic networks with a certain logic family require fewer gates or levels than with other logic families to realize given functions), cost, integration size, design procedures, ease in correcting design mistakes, ease of design changes, availability, and possibly other aspects, though they do have overlaps in some aspects. Since there are now vastly diversified motivations in choosing IC packages, as discussed in Section 1.1, designers must compare every aspect of these logic families. It compounds their selection difficulties that most of these logic families are still in progress technologically, and designers must anticipate the technological status of families a few years later.

Major logic families have the following features. ECL is the fastest, though it has the largest power consumption and it is the most expensive. For mainframes which require the highest speed, ECL is preferred. TTL is somewhat slower and less expensive, but it is generally still faster than other logic families. Thus TTL is usually used in minicomputers. MOS logic families such as *p*-MOS, *n*-MOS, and CMOS consume much less power, though their speeds and costs are lower. Thus MOS logic families are more appropriate for LSI. CMOS has the smallest power consumption and the maximum noise immunity, so it is the most appropriate for automobile controls and watches. For some products the selection of the appropriate logic families is not simple. When we need higher speed, will ECL with future improvements be preferred to MOSFETs or Josephson junctions? For products that require low power consumption, such as electronic toys, CMOS and IIL are probably preferred. But IIL can be realized easily with TTL on the same chip. When power-consuming devices such as digital displays need to be connected to a logic network, IIL may be preferrable to CMOS in this aspect since TTL can derive sufficiently large currents for digital displays. But can CMOS also derive sufficiently large currents by future improvement? In many cases, answers to the future technological status are not clear. But designers have to decide one way or another, often with great risks. The features of the different logic families, logic design, and mask design are discussed in Chapters 3 through 5.

The appropriate use of memories is becoming vitally important in LSI/VLSI design because they have unique features compared with logic gate networks. Memories, particularly read-only memories (ROM), can be used not only for storing software or firmware, but also for replacing some logic gates in networks. By using ROMs we can reduce the cost of logic networks and enhance flexibility in design changes, though speed tends to be sacrificed. This is because the layout time of ROMs is much shorter than that for with logic gate networks due to the regular structure of the ROMs. The features and usage of memories are discussed in Chapters 6 and 7.

Even after designers have chosen a logic family somehow, they have to face another question: whether to use off-the-shelf IC packages or custom-designed IC packages. If they want to introduce new commercial products quickly into the market, they can finish development of the products quicker with off-the-shelf packages than with custom-designed IC packages. But if the production volume is high, their new products in off-the-shelf IC packages will be more expensive with lower performance than their competitors' if the latter use custom-designed IC packages. Thus designers have to make tough decisions after having carefully analyzed contradicting factors such as these.

The custom design of IC packages has many different approaches, such as random-logic networks, gate arrays, cell library approaches and programmable logic arrays (PLA), as is discussed in Chapters 7 (PLA) and 9. Each of these examples has variations. Generally, approaches with the advantage of short design time, partly due to the proper use of computer-aided design (CAD) programs discussed in Chapter 8, have the disadvantage of higher manufacturing costs or lower performance than other approaches. Tradeoff consideration is not simple.

Finally Chapter 10 discusses system design aspects and the problems with VLSI. In particular VLSI is not just LSI with increased integration size, but has completely different problems due to its enormous complexity. VLSI will bring about not only technical but also social changes.

Remark 2.5.1: As shown in the text, there are many different logic families available now, and probably more families or variations are coming. An interesting situation is that almost every family would be used for years to come, though some of them are more popular than others. This is understandable if we consider that as logic networks are being applied to many commercial products as well as computers, there are many different situations corresponding to different combinations of cost, speed, power consumption, voltage (different size batteries), size, weight, convenience (mechanical logic devices or cryogenics may be avoided in many cases), noise immunity, and volatility (in case of memories), so that each family can find its own niche. Despite the popularity of IC logic families, even electromechanical relays [for comparison with solid-state relays, see Grossman Dec. 20, 1978] and the fluidic logic devices [e.g., *Electronics* (Oct. 27, 1967, p. 94); Rohner (1979)] are still used.☐

Problems

2.1 Table 2.3.1 shows that the finished chip cost is only a small part of the cost of an entire IC package. It appears to indicate that the increase of integration size reduces the cost per gate, considering other costs in Table 2.3.1. Discuss whether this is possible. What are the limiting factors?

2.2 When we manufacture an LSI or VLSI package which realizes given logic functions, discuss the following:

(a) What steps require computer programs?

(b) What steps in the manufacture of the package must be considered in the logic design stage for best results?

2.3 Suppose that we want to manufacture chips where each side of the chip is L units in length. Assume that a processed wafer (i.e., gates are fabricated on this wafer) costs K dollars, the probability of no fault in a unit area is p (so the probability that a chip of L in each side is faultless is p^{L^2}), the packaging cost (excluding the finished chip cost) is T, and each gate has the same shape with the same size.

Find the optimal value of L which minimizes the cost per gate. Introduce appropriate assumptions whenever necessary. Since it is difficult to derive an exact formula, draw a curve which loosely shows the relationship between L and the cost per gate.

2.4 Discuss the advantages and disadvantages of the following two approaches to manufacture IC chips:

(a) During fabrication, each device on a chip is tested by probes. Then a connection mask (used in Fig. 2.1.2*i*) is prepared, using only faultless devices, while all other masks are prepared regardless of faulty devices. (This type of approach, called "discretionary wiring," was at one time used by Texas Instruments during the days when the industry could not achieve high yields.)

(b)   Each LSI chip is designed with extra subcircuits. After processing a wafer with such chips, each subcircuit is tested by probes, and then some connections are burned off by a laser beam or a heavy current such that only faultless subcircuits are connected. (This type of approach was used by IBM in a 64-k bit RAM, announced in 1978, and also by Texas Instruments in magnetic bubble memories.)

2.5   Suppose that the fabrication of some chips consists of 100 steps with equal yield at each step, for simplicity. Using Fig. 2.2.5, calculate how much yield each step should have when the yield of the IC package is 90, 60, 30, and 10%, respectively.

2.6   Given a wafer of diameter d, how many chips can we obtain from this wafer? Assume that each chip is a square with sides length a.

(a)   Draw approximate curves for $d = 3$, 4, and 6 in. only for $a = 10$ to 300 mil.

(b)   Find the cost of a chip with $a = 300$ mil, assuming that a processed wafer costs \$150. (Simply divide \$150 by the number of chips, ignoring other factors.)

2.7.   Find the cost per IC package, $A/V + B$, for an initial investment cost A = \$10,000, \$100,000, and \$300,000 when the manufacturing cost per package B is \$10. [This is Eq. (2.3.1) in Section 2.3.] Draw curves for the cost per package for these three values of A over the range of production volume $V = 5,000$ to 1,000,000.

2.8   Suppose that a manufacturer wants to develop a new product with a microprocessor. If designers develop the product with an off-the-shelf programmable microprocessor chip (designers can store programs in ROMs inside the chip since it is programmable), they can purchase the chip at \$10 each [i.e., $B = \$10$ in Eq. (2.3.1)], and the initial investment cost A is \$10,000. As an alternate approach, designers can develop their own custom-made processor chip (or a dedicated processor LSI chip). In this case they can make the chip size smaller by eliminating unnecessary instructions (and the corresponding networks), since it is tailored to the specific functional needs of the new product. (Note that off-the-shelf microprocessor chips have a wide spectrum of instructions as general-purpose processors.) Consequently B can be reduced to \$5 when fabrication is mature. But A will be \$500,000 because of the long design time with talented designers. The manufacturer expects the production volume to be at least 5,000 but does not know how many can be sold eventually. In order to help in the selection of an appropriate approach, draw curves for the cost per package [Eq. (2.3.1)] versus production volume V for these two approaches (off-the-shelf and custom-made packages). Then determine which approach is more cost-effective for what range of V. Also what cost per package do we get at the crosspoint volume V_0? When the production volume is $2V_0$, determine what cost per package we have and how much we can save in total, compared with the off-the-shelf package approach.

CHAPTER **3**

Logic and Mask Designs for Bipolar Logic Families

In this chapter we discuss important IC logic families based on bipolar transistors, namely, TTL, ECL, IIL, and their variations. Although TTL and ECL are historically older logic families than IIL and MOS, they are still extensively used. In particular, ECL is most important for high-speed applications, TTL is useful for high output power with reasonably high speed, though slower than ECL, and IIL is much slower but useful for high packing density. Improvements of these logic families are still going on.

3.1 Bipolar Transistors

In this section we outline* the basic mechanism of bipolar transistors and bipolar transistor circuits.

Carriers

When voltages of appropriate magnitudes are applied to the collector, emitter, and base electrodes of an *n-p-n* transistor, as shown in Fig. 3.1.1, electrons, which are denoted by $\ominus$, are emitted from the emitter electrode. The majority of them reach the collector because the base is so thin that electrons can penetrate through the base easily, while some of them recombine with holes, which are denoted by $\oplus$. (A small amount of holes are injected from the base to the emitter.) The stream of **holes** is actually the stream of electrons in the direction opposite to the arrows of $\oplus$. Because of the difference of the movement mechanism of these electrons from that of the electrons denoted by $\ominus$, the stream of these electrons is regarded as flow of positive electric charge, i.e., holes in the opposite direction. The stream of electrons is mainly concentrated underneath the emitter region. Since the **buried layer** has good conductivity due to high doping (n^+), the stream flows more inside the buried layer than inside the *n*-region.

When the voltage at the base against the emitter increases, the streams

---

* For details, see textbooks of electronics, such as Smith (1980, 1976).

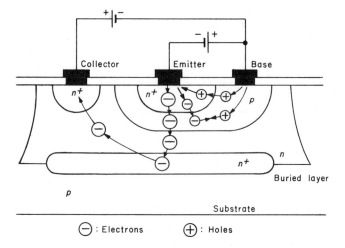

Fig. 3.1.1   Carrier movement in an *n-p-n* transistor.

of electrons and holes increase. When the polarity of this voltage is reversed (i.e., the minus voltage is applied at the base against the emitter), there will be no stream because the diode, which consists of the base (*p*) and the emitter (n^+), is reverse biased.

The electrons and holes are called **carriers**. In the case of a bipolar transistor, electrons which are negatively charged and holes which are positively charged, in other words, both types of carriers, constitute the current through this transistor, while in the case of MOSFETs only one type of carrier constitutes an electric current, as discussed in Chapter 4. Thus the transistor here is called a "bi"-polar transistor because of both types of carriers involved.

Suppose p, p^+, n, and n^+ in Fig. 3.1.1 are replaced by n, n^+, p, and p^+, respectively, and the polarity of two power supply voltages is reversed. We then have a *p-n-p* transistor and the corresponding carrier movement where holes and electrons are interchanged.

Inverter Circuits

Suppose that an *n-p-n* transistor is connected to a power supply with a resistor, and to the ground, as shown in Fig. 3.1.2*a*. When the input voltage v_i increases, the collector current i_c gradually increases, as shown in Fig. 3.1.2*b*. (Actually i_c is 0 until v_i reaches 0.5 to 0.6 volts. Then i_c gradually increases and starts to saturate.) As i_c increases, the output voltage v_o decreases from 5 to 0.3 volts or less, as shown in Fig. 3.1.2*c*. Therefore, when the input v_i is a high voltage (about 5 volts), the output v_o is a low voltage (about 0.3 volts), and when v_i is a low voltage (about 0.3 volts), v_o is a high voltage (about 5 volts). (When v_i is 5 volts, i_c is about 5 mA (i.e.,

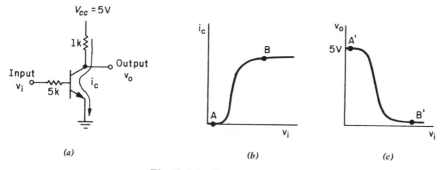

Fig. 3.1.2   Inverter circuit.

milliamperes), and the base current is about 0.8 mA.) This is illustrated in Table 3.1.1a. Thus if binary logic values 0 and 1 are represented by low and high voltages, respectively, we have the truth table in Table 3.1.1b. This means that the circuit in Fig. 3.1.2a works as an inverter. In other words, if v_i represents a switching variable x, then v_0 represents the switching function $\bar{x}$.

Since throughout this book we are concerned with binary logic values in designing logic networks, we will from now on consider only the on–off states of currents or the corresponding voltages in electronic circuits (e.g., A and B in Fig. 3.1.2b, or A' and B' in Fig. 3.1.2c), without considering their magnitudes.

As we will see later, the transistor circuit in Fig. 3.1.2a is often used as part of other transistor circuits. Here notice that if resistor R is added between the emitter and the ground, and the output terminal v_0 is connected to the emitter, as shown in Fig. 3.1.3, then the new circuit does not work as an inverter. When v_i is a high voltage, v_0 is also a high voltage, because the current that flows through the transistor produces the voltage difference across resistor R. When v_i is a low voltage, v_0 is also a low voltage, because no current flows and consequently no voltage difference develops across R. So if v_i represents a switching variable x, v_0 represents the switching function x, and no logic operation is performed. The transistor circuit in Fig. 3.1.3 is also used often as part of other circuits to supply a large output current. Thus the circuit works as a current amplifier.

Table 3.1.1   Input/Output Relation of the Inverter in Fig. 3.1.2a

(a) VOLTAGE RELATION		(b) TRUTH TABLE	
INPUT v_i	OUTPUT v_0	v_i	v_0
Low voltage	High voltage	0	1
High voltage	Low voltage	1	0

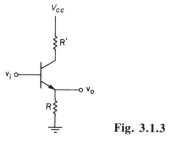

Fig. 3.1.3   Transistor circuit.

Speed Limits of Bipolar Transistors

There are many factors which limit the speed of a bipolar transistor.

The bipolar transistor shown in Fig. 3.1.4a, under closer examination, contains **parasitic capacitances*** and **parasitic resistances** (or **internal resistances**), as illustrated in Fig. 3.1.4b. For example, the collector region has the parasitic resistance R_c (between the collector electrode and the border surface with the base region), the parasitic capacitance C_{bc} against the base region, and the parasitic capacitance C_{cs} against the substrate (which is usually grounded). Thus the transistor in Fig. 3.1.4a has the equivalent circuit shown in Fig. 3.1.4c. These parasitic capacitances and resistances, which are undesirable, limit the speed of the transistor as follows. When a voltage as shown in Fig. 3.1.5a is applied to the base terminal in Fig. 3.1.4 (i.e., between the base and the emitter electrodes), C_{be} must be charged up through R_b and R_e, while capacitances C_{bc} and C_{cs} are discharged. Thus the voltage waveform at the collector electrode falls slowly, as shown in Fig. 3.1.5b, and the transistor becomes conductive more slowly than the dotted line, which corresponds to the sharp rise in Fig. 3.1.5a. When the voltage falls in Fig. 3.1.5a, C_{be} is discharged. As the transistor becomes nonconductive, C_{bc} and C_{cs} are charged up. So the waveform at the collector electrode rises slowly, as shown in Fig. 3.1.5b.

If we want to estimate the speed of transistor circuits, we need to have a detailed analysis of solid-state circuits to find parasitic capacitances and resistances and a detailed analysis of equivalent circuits to find exact magnitudes of currents and voltages (instead of simply considering the on–off states of currents and voltages). As we will see later, means to reduce parasitic capacitances and resistances for the speed-up of solid-state circuits have been devised.

---

* Parasitic capacitances C_{bc} and C_{be} actually consist of the junction (or space-charge-layer) capacitance and the diffusion capacitance. The junction capacitance represents the redistribution of the electrical charge on either side of a junction when the junction voltage is changed. The diffusion capacitance represents the change in excess carriers associated with a change in current through the junction. Also C_{cs} is the space-charge-layer capacitance of the collector–substrate junction. For details on modeling equivalent circuits, see textbooks on solid-state circuits.

Since the transit time of carriers to cross the base affects the transistor speed, the speed can be improved by making the base very thin, though it tends to increase the resistance R_b. In the case of an *n-p-n* transistor, the carriers from the emitter to the collector are electrons, as illustrated with large circles ⊖ in Fig. 3.1.1, while in the case of a *p-n-p* transistor, the carriers from the emitter to the collector are holes. Since the mobility of electrons is about 2.5 times of that of holes, an *n-p-n* transistor is faster than a *p-n-p* transistor if the dimensions are the same.

If we reduce (**scale-down**) the size of a bipolar transistor, along with

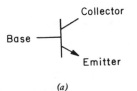

(*a*)

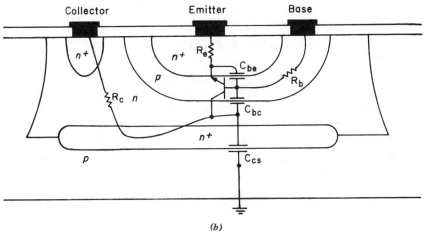

(*b*)

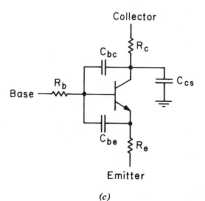

(*c*)

Fig. 3.1.4 **Analaysis of bipolar transistor.** (*a*) **Bipolar transistor.** (*b*) **Cross section of bipolar transistor.** (*c*) **Equivalent circuit of (*a*).**

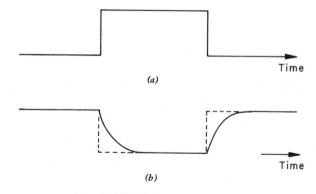

Fig. 3.1.5 Voltage waveforms.

appropriate adjustments of the parameter values, all speed-limiting factors discussed above are lessened, and the transistor speed is improved, though complex physical problems arise as we approach the scale-down limit. (The thickness limit of the base is regarded to be 0.03 to 0.05 μm).

3.2 TTL (Transistor–Transistor Logic)

TTL, or T^2L, has been widely used since the early 1960s. A simple TTL circuit is shown in Fig. 3.2.1, which is used when medium output power is sufficient. (When lower power is sufficient, this gate can be further simplified, as will be shown later.) The operation of this gate may be analyzed as follows. [For an electronic analysis of current and voltage magnitudes, see electronics textbooks, such as Taub and Schilling (1977); Ilardi (1976).]

When we want to see how a logic gate performs its logic operation, it is sufficient to consider only on–off states of currents, that is, no currents (actually very small currents flow) or sufficiently large currents throughout

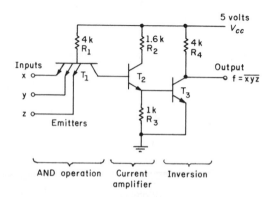

Fig. 3.2.1 TTL for medium output power.

the electronic circuit, ignoring the magnitudes of these currents. When a low voltage, which represents logic value 0, is applied at any one input, say at x, in Fig. 3.2.2a (which is identical to Fig. 3.2.1), that is, when $x = 0$, a current flows from power supply V_{cc} (about 5 volts) to x because the base of transistor T_1 has much higher voltage than its emitter connected to x, thus making T_1 conductive. Because of the voltage drop across R_1 due to this current, the base of T_1, and accordingly the base of tansistor T_2, have low voltages, making T_2 nonconductive. (There is almost no voltage difference, that is, about 0.1 volts, between x and the collector of T_1.) Thus no current flows through T_2. Accordingly no voltage develops across resistor R_3, and the base of transistor T_3 has low voltage, making T_3 nonconductive. Consequently no current flows through T_3. Thus we have a high voltage at output f, which represents logic value 1. Even when a low voltage is applied at input y or z, the above situation does not change.

When high voltages are applied at all inputs x, y, and z, as shown in Fig. 3.2.2b, that is, when $x = y = z = 1$, no current flows through the emitters of transistor T_1 since the voltage at the base of T_1 is not sufficiently higher than that at its emitters. (In this case a small current flows through R_1, T_1,

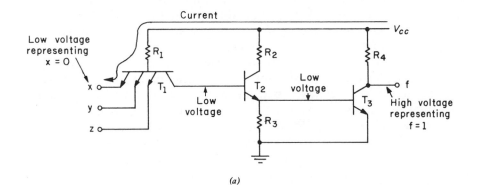

(a)

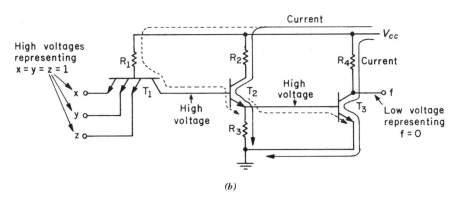

(b)

Fig. 3.2.2 Logic operation of TTL gate in Fig. 3.2.1.

T_2, and R_3, as shown by a dotted line.) Thus the base of transistor T_2 also has a high voltage (about 1.2 volts). Then a current flows through T_2, since its base has a sufficiently higher voltage than its emitter. Consequently a voltage develops across resistor R_3, and the base of transistor T_3 has a high voltage (about 0.6 volts, which is higher than a low voltage of almost 0 V in Fig. 3.2.2a). Then a current flows through T_3, and the voltage at output f is lowered. (In this case the small current through R_2, T_2, and T_3, shown by a dotted line, is essential to make T_2 and T_3 conductive.)

From the above analysis with Fig. 3.2.2 we obtain the truth table in Table 3.2.1, since low and high voltages represent logic values 0 and 1, respectively. Thus output f represents the function NAND, that is,

$$f = \overline{xyz}$$

The TTL gate in Fig. 3.2.1 is consequently denoted by the symbol given in Fig. 3.2.3a. Notice that multi-emitter transistor T_1 is to perform the AND operation because only when at least one of the inputs x, y, and z has a low voltage, its collector (accordingly the base of T_2) has low voltage. Transistor T_2 is to supply T_3 with a current which is sufficiently large to let T_3 work fast, even if the output f has many fan-outs. This subcircuit consisting of R_2, T_2, and R_3 is essentially the circuit shown in Fig. 3.1.3. Thus T_2 has no direct relation with the logic operation of the entire gate in Fig. 3.2.1. (When the output f does not require high output power, we can connect the collector of T_1 directly to the base of T_3, eliminating the subcircuit which consists of R_2, T_2, and R_3. This is essentially the circuit in Fig. 3.2.13b for two inputs x and y, though a diode is added for performance improvement.) Transistor T_3 is to invert its base voltage and to deliver it to its collector. This subcircuit, consisting of R_4 and T_3, is essentially the inverter shown in Fig. 3.1.2a (though the role of 5 kΩ in Fig. 3.1.2a is replaced by the preceding subcircuit of T_2 which performs the AND operation as well).

More as well as fewer inputs can be connected to transistor T_1. For example, the connection of inputs x, y, z, and u to T_1 in Fig. 3.2.1 (by providing one more emitter for u) results in the output function $f = \overline{xyzu}$.

Table 3.2.1 Truth Table for Fig. 3.2.2

INPUTS			OUTPUT
x	y	z	f
0	0	0	1
0	0	1	1
0	1	0	1
0	1	1	1
1	0	0	1
1	0	1	1
1	1	0	1
1	1	1	0

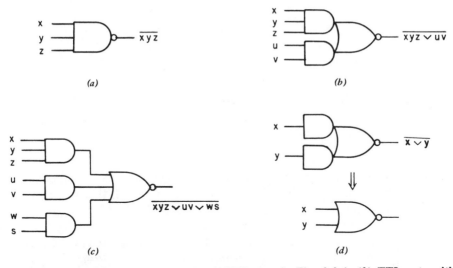

(a)

(b)

(c)

(d)

Fig. 3.2.3 Symbols for TTL gates. (a) TTL gate in Fig. 3.2.1. (b) TTL gate with expander, as in Fig. 3.2.4. (c) TTL gate with three expanders ($\overline{xyz \lor uv \lor ws}$). (d) TTL gate in Fig. 3.2.5 ($\overline{x \lor y}$).

The maximum fan-out of a TTL gate is usually about 10, though it may be fewer or more, depending on the output power which the gate can deliver.

A TTL gate can express a more complex function by connecting additional multi-emitter transistors. For example, Fig. 3.2.4 represents $\overline{xyz \lor uv}$ by adding one more multi-emitter transistor along with a resistor, as shown inside the dotted lines. (Here, "$\lor$" denotes the OR operation.) This configuration shown inside the dotted line is called an **expander** and is commercially

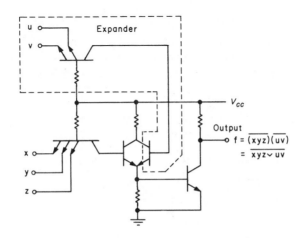

Fig. 3.2.4 TTL gate with an expander.

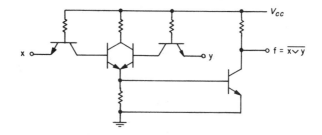

Fig. 3.2.5   TTL NOR gate.

available as an independent IC package (e.g., SN7460 of Texas Instruments). A TTL gate package which has a pair of extra pins such that an expander can be connected is called an **expandable gate** (e.g., SN5450 of Texas Instruments).

The addition of an expander to the TTL gate in Fig. 3.2.1, as shown in Fig. 3.2.4, is essentially an AND operation of output functions, that is,

$$\overline{(xyz)\cdot(uv)} = \overline{xyz} \vee \overline{uv}$$

This complex function represented by a single TTL gate is usually denoted by a combination of gate symbols, as shown in Fig. 3.2.3*b*, and is called **AND-OR-INVERT**. The number of inputs can easily be enlarged within a certain limit by adding many expanders, as shown in Fig. 3.2.3*c*.

A TTL gate can realize a NOR gate also. By providing two inputs in Fig. 3.2.3*c*, as shown in Fig. 3.2.3*d*, a NOR gate is realized. The corresponding electronic circuit is shown in Fig. 3.2.5. [Notice that the electronic circuit for NAND gate (Fig. 3.2.3*a*) is simpler than that for NOR gate (Fig. 3.2.3*d*), even though both are single gates. This example shows that a simpler logic operation does not necessarily require a simpler electronic circuit. In the case of TTL, NAND is simpler and faster than NOR.]

An AND gate can be implemented by cascading the NAND TTL gate of Fig. 3.2.3*a* and a NOT TTL gate (which is Fig. 3.2.3*a* with a single input). But as shown in Fig. 3.2.6, a special TTL gate which modifies the gate in Fig. 3.2.1 is usually preferred because of its simplicity.

The AND operation can also be realized by the following means. When the outputs of gates shown in Fig. 3.2.1 are connected as shown in Fig.

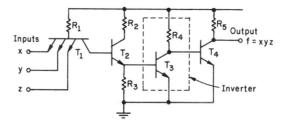

Fig. 3.2.6   AND gate obtained by modifcation of TTL gate in Fig. 3.2.1.

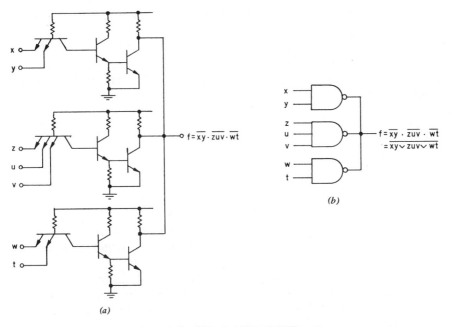

Fig. 3.2.7 Wired-AND of TTL gates.

3.2.7, the connected point realizes the conjunction of the three output functions. (Readers should confirm this by forming a truth table based on voltage levels.) In other words, the AND operation is implemented simply by connection, without any other components or gate, though we can connect, at most, several outputs (*Electronic Design*, Jan. 18, 1973, pp. 7, 11). This is called **Wired-AND**. Wired-AND is very useful since it implements a function which otherwise would require more than two levels of gates. One important limitation is that if an output is combined with other outputs to form Wired-AND, the output does not represent the original output function. For example, the top gate in Fig. 3.2.7 no longer represents the original function $\overline{xy}$. Notice that when more sophisticated versions of TTL for high output power are used, we may not be allowed to used Wired-AND, as will be explained later.

In summary, each TTL gate can realize a negative function or AND operation. Here a **negative function** is the complement of a disjunctive form* with noncomplemented literals, such as

$$\overline{x_1 x_2 \vee x_3 x_4 x_5 \vee x_6 x_7}$$

* A **conjunction** (i.e., a **logic product**) of literals where a literal for each variable appears at most once is called a **term** (or a **product**). A **disjunction** (i.e., a **logic sum**) of terms is called a **disjunctive form** (or a **sum-of-product form**). In this example, $x_1 x_2 \vee x_3 x_4 x_5 \vee x_6 x_7$ is a disjunctive form consisting of terms $x_1 x_2$, $x_3 x_4 x_5$, and $x_6 x_7$. [See, for example, Muroga (1979).]

Procedure 3.2.1:   Design of TTL Networks in Double-Rail Input Logic

Suppose that we want to design a TTL gate network for a given switching function f, assuming that both noncomplemented and complemented variables are available as network inputs, that is, that the network is in **double-rail input logic**.

1. Rewrite f as

$$f = \overline{g}$$

Then derive a minimum sum for g by a minimization procedure known in switching theory [e.g., see Muroga (1979)].

2. Then f can be realized by one of the following two approaches:

    (a) f can be realized by a single TTL gate in the form of Fig. 3.2.3c. The number of literals in the minimum sum is equal to the number of inputs, and the number of terms is equal to the total number of expanders plus the original multi-emitter transistor. (For example, for $f = \overline{xyz \vee uv \vee ws}$ in Fig. 3.2.3c we have $g = xyz \vee uv \vee ws$ as its minimum sum. Since the number of literals is 7, this TTL gate has seven inputs: x, y, z, u, v, w, and s. The number of terms in g is 3, so we need two expanders and the original multi-emitter transistor for terms xyz, uv, and ws.)

    If g contains complemented variables, say $\overline{w}$, connect $\overline{w}$ as input to this gate, since $\overline{w}$ is available as network input.

    If the number of expanders, or the number of inputs, in some multi-emitter transistors exceeds a permissible limit, there is no simple design procedure. Thus we need to design networks on a trial-and-error basis by finding appropriate switching expressions with parentheses and then by connecting AND gates (each by the TTL gate of Fig. 3.2.6) to the inputs of some multi-emitter transistors.

    (b)   The complement of each term in the minimum sum for g can be realized by a single TTL gate. Then f is realized by the Wired-AND of the outputs of these gates, as illustrated in Fig. 3.2.7. (For example, $f = \overline{xyz \vee uv \vee ws}$ in Fig. 3.2.3c can be realized by the Wired-AND of TTL gates for $\overline{xyz}$, $\overline{uv}$, and $\overline{ws}$.) If the number of inputs to the Wired-AND exceeds a permissible limit, we need to design on a trial-and-error basis.

    When more than one function is to be realized, a network with many TTL gates can be similarly designed.   □

The design of TTL networks in double-rail input logic is probably easier than with other logic families. If only noncomplemented variables were available as network inputs, the design of TTL gates would be much harder.

Procedure 3.2.2 Design of TTL Networks in Single-Rail Input Logic

Suppose that we want to design a TTL network for a given switching function f, assuming that only noncomplemented variables are available as network inputs, that is, that the network is in **single-rail input logic**.

1. Rewrite f as

    $$f = \overline{g}$$

    Then derive a minimum sum for g by a minimization procedure.

2. If the minimum sum for g does not contain complemented literals, then f is realized by one of the following approaches:

    (a) f can be realized by a single TTL gate in the form of Fig. 3.2.3c.

    (b) The complement of each term in the minimum sum for g can be realized by a single TTL gate. Then f is realized by the Wired-AND of the outputs of these gates, as illustrated in Fig. 3.2.7.

    If the number of expanders or the number of inputs in some multi-emitter transistors in case (a), or the number of inputs to the Wired-AND in case (b), exceeds a permissible limit, we need to design on a trial-and-error basis by finding appropriate switching expressions with parentheses.

3. If the minimum sum for g contains complemented literals, we can realize these complemented literals with TTL gates. In this case if the complemented literals in this switching expression can be combined by appropriately using parentheses, we can reduce the number of gates. Then, using the outputs of these gates and the noncomplemented literals in g as inputs, we realize a TTL gate for f, based on the form of Fig. 3.2.3c. For example, when $f = \overline{xy \vee \overline{u}vw \vee \overline{s}vw}$ is given, Fig. 3.2.8a results from $g = xy \vee \overline{u}vw \vee \overline{s}vw$ in its minimum sum. But g can be rewritten as follows: $f = \overline{xy \vee (\overline{u} \vee \overline{s})vw} = \overline{xy \vee (\overline{us})vw}$. Based on this, the TTL network in Fig. 3.2.8b is obtained. This network has not only fewer gates than that in Fig. 3.2.8a, but also the output gate has a much simpler structure than the corresponding gate in Fig. 3.2.8a.

    Instead of the TTL gate with expanders shown in Fig. 3.2.3c, we can design a network, on a trial-and-error basis, with the Wired-AND in Fig. 3.2.7.

    Instead of using g, we can realize a network, on a trial-and-error basis, by rewriting the given f into an appropriate switching expression with parentheses where fewer complements appear. Also when we want to reduce the number of gates (rather than the number of levels) as much as possible, this approach may be appropriate.

4. We may be able to derive many different TTL networks by the approaches discussed. The complexity of these networks is not obvious from the switching expressions on which these networks are based.

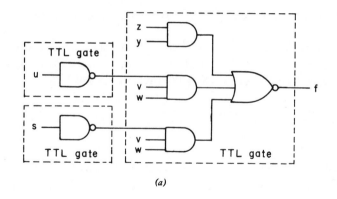

(a)

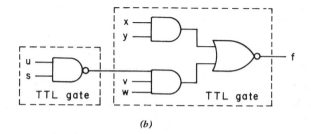

(b)

Fig. 3.2.8   Simplification of TTL network.

Thus we probably need to design different networks and compare them to choose the best one, though this may not be a simple task. Networks with many output functions can be similarly designed.   □

Another way to design TTL networks in single-rail input logic is Procedure 4.2.1 described in Chapter 4. (In this case each negative gate can easily be converted into an AND-OR-INVERT, i.e., a TTL gate.) By this procedure we can always minimize the number of TTL gates (i.e., AND-OR-IN-VERTs). If some gates become excessively complicated, we need to convert them into simpler ones, using additional gates.

Since TTL gates have been used for many years, and many good TTL gates were designed in the past, we can often derive good TTL networks by modifying those designed in the past. We can often improve the network performance by modifying electronic circuits. This requires a knowledge of electronics.

References: Morris and Miller, 1971; Lancaster, 1974; Maley, 1970; Burke and van Bosse, 1965; and many publications by major semiconductor manufacturers. For wired logic, see Kambayash and Muroga, and Kawasaki et al. (both to be published).

High-Power TTL Gates and Their Problems

The basic circuit for the TTL gate shown in Fig. 3.2.1 can be modified in many different ways, depending on the situation (Priel, 1971). Fig. 3.2.9a shows a TTL gate for high-speed and high-power output. The so-called **Darlington connection** of two transistors (shown within a dashed oval) has a high gain, generating a large output current through these transistors quickly for a small signal change. The **totem-pole** connection of two transistors (shown also within a dashed oval) is for fast switching of an output current. The subcircuit in the center (shown in a dashed rectangle), which is called a **phase splitter**, is a variation of the current amplifier in Fig. 3.2.1 and is used for the totem-pole connection. When a gate has an output that goes out to a connection on a pc board, this TTL gate is used since it can generate high output power. When this gate receives signals through connections on a pc board, the connections work as transmission lines, and input signals tend to oscillate. The diodes in the input transistor are useful for the elimination of this signal ringing.

An expander (or expanders) can be added as shown in Fig. 3.2.9b.

Suppose that the output of the TTL gate in Fig. 3.2.9 is connected to another TTL gate, as shown in Fig. 3.2.10. When the output of the left-hand TTL gate represents logic value 1, that is, T_2 is off while T_1 is on, no current* flows in the multi-emitter transistor T_m in the next TTL gate and the output transistors T_1 and T_2. as shown in Fig. 3.2.10a. When the output of the left-hand TTL gate represents logic value 0, that is, T_2 is on while T_1 is off, a current flows through T_m and T_2, as shown in Fig. 3.2.10b.

Suppose that two TTL gates are Wired-ANDed, as shown in Fig. 3.2.11. When T_1 and T_4 are on, currents I and I_m flow through T_4, possibly damaging transistor T_4 by excessive magnitude. (In this case I is much greater than I_m since R and the internal resistance of T_1 combined are much smaller than R_m.) This possibility of damage is aggravated if we have Wired-AND of more gates and greater fan-out.

Therefore TTL gates which contain totem-pole transistors are usually not allowed to connect Wired-AND. So TTL gates in SSI or MSI packages do not get much benefit from Wired-AND. However, in the case of an LSI package, all TTL gates except those that supply output power to the outside of the package may be realized by the circuit shown in Fig. 3.2.1 (or even simpler versions, as will be shown later), and Wired-AND may be used freely, possibly simplifying the network. This is one advantage of using LSI.

---

* If the connection from output f to one emitter of transistor T_m is placed on a pc board and consequently has a large parasitic capacitance, a large current flows through T_1 to charge up the capacitance, when T_1 turns on.

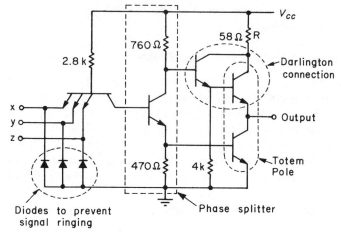

(a)

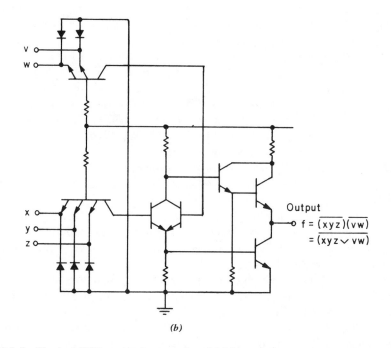

(b)

Fig. 3.2.9   Typical TTL gates for speed and high output power. (Courtesy of Texas Instruments, Inc.) (a) SN54H00 with 6-nsec delay time. (b) Addition of an expander to (a).

When we need to connect the outputs of TTL gates in Fig. 3.2.9, we use the modification of TTL, shown in Fig. 3.2.12, which is called **tri-state TTL**.

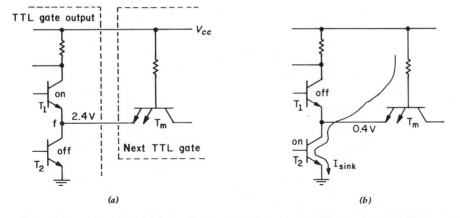

Fig. 3.2.10 **Output current of a TTL gate.** (*a*) **Current for logic value 1 at TTL gate output.** (*b*) **Current for logic value 0 at TTL gate output.**

A tri-state TTL still has totem-pole transistors, but it has a control input. When the control input has a low voltage, both totem-pole transistors are nonconductive and the impedance looked back from the output terminal is high. In other words, the control input adds the high-impedance state in addition to the ordinary TTL output states 1 and 0. Thus tri-state TTL permits tying down the outputs of many TTL gates to a common line. (Notice

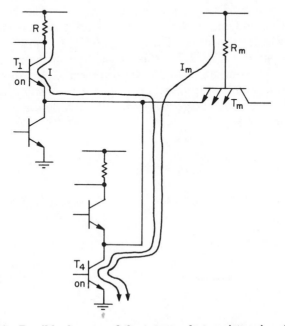

Fig. 3.2.11 **Possible damage of the totem-pole transistors in a TTL gate.**

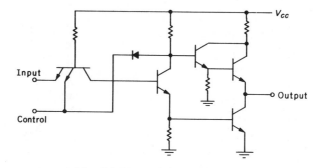

Fig. 3.2.12   Tri-state TTL gate.

that this does not realize Wired-AND.) Tri-state TTL gates are used for bus line connection in minicomputers (National Semiconductor Publications AN43, AN68, AN73; Femling, 1971; Mick, 1974; Calebotta, 1972).

Variants of TTL

Since gates inside a chip (i.e., those not used for the inputs or outputs of the chip) have lower noise disturbances, **voltage swings** and consequently voltage levels (i.e., magnitudes of voltages) can be low. (If logic values 1 and 0 are represented by 3.7 volts and 0.3 volts, respectively, the difference $3.7 - 0.3 = 3.4$ volts is the voltage swing in this case.) Also, as discussed above, the connections within a chip have very low parasitic capacitance. So the output power of a gate can be low. Thus the TTL gate in Fig. 3.2.1 can be further simplified, as shown in Fig. 3.2.13. (In this case diodes can be added using little extra area.) The gate in Fig. 3.2.13*a* is used also without the diode. The gate in Fig. 3.2.13*b* is often used in the input section of a complex chip. Appropriate use of these different TTL gates requires consideration of the relationship between voltage and current magnitudes and possibly other electronics aspects (Priel, 1971; Xylander, 1972; DeFalco, 1972; Armstrong, 1981).

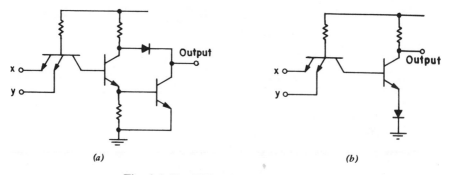

(a)   (b)

Fig. 3.2.13   TTL gates inside a chip.

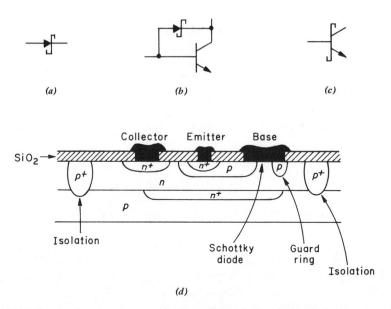

(d)

Fig. 3.2.14 Schottky diode and Schottky transistor. (*a*) Schottky diode. (*b*) Combination of Schottky diode and transistor. (*c*) Schottky transistor. (*d*) Cross section of Schottky transistor.

Schottky diodes are often used to improve the performance of electronic circuits (by preventing saturation of bipolar transistors). This is a variation of an ordinary diode and denoted by the symbol shown in Fig. 3.2.14*a*. A Schottky diode is formed by bonding a metal, such as aluminum or platinum (instead of *p*-type silicon in the case of the ordinary diode in Fig. 1.2.1*a*) to *n*-type silicon (Schilling and Belove, 1979; Glaser and Subak-Sharpe, 1977). The combination of a Schottky diode and a transistor shown in Fig. 3.2.14*b* is called a **Schottky transistor**. It is expressed by the symbol in Fig. 3.2.14*c* and realized as shown in Fig. 3.2.14*d*, taking little extra area.

Using Schottky transistors, we can get a TTL gate with high speed, as shown in Fig. 3.2.15. This is called **Schottky TTL** (Tarui et al., 1969). It has a delay time of 3 nsec versus 6 nsec of the TTL gate shown in Fig. 3.2.9*a*.

The **low-power Schottky TTL** (introduced by Texas Instruments in 1973), which is a low-power version of the Schottky TTL, is widely used. Electronic circuits are shown in Fig. 3.2.16.* (Notice that the multi-emitter transistor at the input of Fig. 3.2.15 is replaced in Fig. 3.2.16 by the combination of a resistor and Schottky diodes.) The low-power Schottky TTL of Fig. 3.2.16*a* has totem-pole transistors for high output power, but the low-power Schottky TTL of Fig. 3.2.16*b* is used inside a chip, being greatly simplified. A low-power Schottky TTL gate consumes the least power among all TTL gates

* These look like diode-transistor logic (DTL), but are called Schottky TTL.

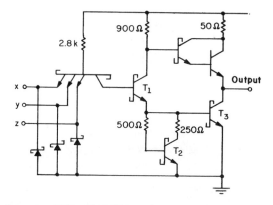

Fig. 3.2.15   Schottky TTL (SN54S00). (Courtesy of Texas Instruments, Inc.)

discussed so far (about one tenth the power of a high-power TTL, such as SN54H00, and one fifth the power of an ordinary TTL, such as the SN54 series). Since the low-power Schottky TTL is still fairly fast despite low power consumption [the low-power Schottky TTL of Texas Instruments, for example, has 8- to 10-nsec delay time, and the Fairchild version (9LS) has 5-nsec delay time], the low-power Schottky TTL has been used extensively in LSI (Cavinaugh, 1973; Alfke and Alford, 1975).

Faster TTL Gates

In 1978 further performance improvements of the Schottky TTL were made by a few companies (commanding also higher prices), as compared in Fig. 3.2.17 (*Electronics*, Feb. 1, 1979, pp. 14, 88-89; Oct. 25, 1979, p. 110). For example, Fairchild's new FAST TTL has achieved a delay time of 2 nsec at a power dissipation of 4 mW/gate (much less than a conventional Schottky TTL) by using the isolation technique (discussed in the following), ion implantation, and size scale-down (*EDN*, Dec. 15, 1978, p. 56). Texas Instruments has two series, AS (advanced Schottky) and ALS (advanced low-power Schottky). ASTTL has 2.3 nsec at 14 mW/gate, and a typical gate delay for MSI packages is 1 nsec at 12 mW. ALSTTL has 4 nsec at 1 mW/gate (Ranada, 1979; Hoffpauir, 1980; Stehlin, 1980; *Electronics*, May 5, 1981, pp. 104, 106, 107).

Isoplanar Process

The performance and costs of digital systems or computers are influenced by many factors, such as operating system software, architecture, logic networks, electronic circuits, and packaging. If even one of these factors is not properly designed, good performance or low cost cannot be attained, and in the worst case the digital system may not work at all. In the following we discuss how an improvement in another factor, namely, solid-state circuit realization of an electronic circuit, influences the performance of digital systems, using the FAST TTL gate (Fig. 3.2.17) as an example.

By combining a Schottky diode with a bipolar transistor, as shown in Fig. 3.2.14*b* or *c*, we can improve the speed of the gate (i.e., the waveform in Fig. 3.1.5*b* falls and rises more sharply than the solid line). Further speed-up can be obtained by reducing parasitic capacitances and resistances. This

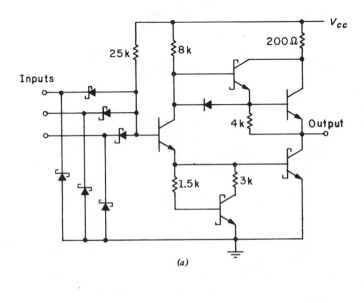

(a)

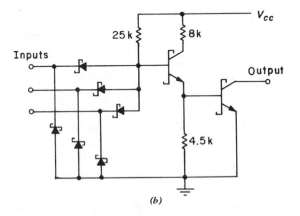

(b)

Fig. 3.2.16 **Low-power Schottky TTL. (Courtesy of Texas Instruments, Inc.)** (*a*) **With high output power (SN54LS00).** (*b*) **Inside chip (SN54LS01).**

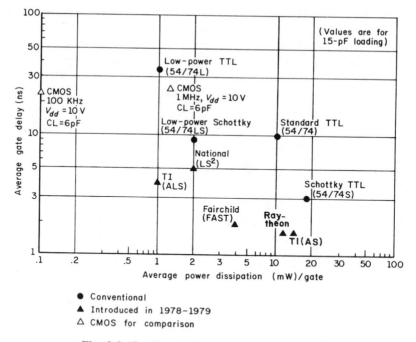

- ● Conventional
- ▲ Introduced in 1978–1979
- △ CMOS for comparison

Fig. 3.2.17   Comparison of different types of TTL.

can be achieved by scaling down the size of the transistor and improving insulation, as discussed in the following.

Figure 3.2.18*a* shows the lateral geometry and cross section of a Schottky transistor which is realized in **planar transistor** structure, well established, and used conventionally. In this case the area underneath the emitter contributes mainly to the operation of the transistor. The entire area outside this adds only parasitic resistances and capacitances. The minimum geometry is largely determined by the difficulties in the alignment of masks and wafer processing. **Isoplanar technology** is very different from conventional planar processes in that the *p*-type isolation diffusion (see Fig. 2.1.4*f*) is replaced by an oxide sidewall (more precisely, a silicon-dioxide sidewall). An isoplanar transistor is shown in Fig. 3.2.18*b*. Since the oxide isolation is nonconductive, unlike the *p*-type isolation diffusion, the base region need not be surrounded by the collector region, without short-circuiting to the *p*-type substrate. (Notice that the *p*-type substrate is not an insulator.) This eliminates the problem of misalignment of masks to keep the base and the *p*-type isolation diffusion separated. Even if the base mask is larger than the collector area, the base region is determined by the coinciding area of the base mask and the collector area within the oxide isolation. Since the collector area is greatly reduced, parasitic capacitances C_{bc} and C_{cs} are significantly reduced.

In the **Isoplanar II** transistor in Fig. 3.2.18*c* the emitter region extends to the wall of the oxide isolation, reducing parasitic capacitance C_{be}. Since it is easier to print a long, narrow rectangle on masks and wafers than to print shorter rectangles of the same width, the emitter in Isoplanar II can be made smaller than that in Isoplanar. The Isoplanar II process reduces the collector area by more than 70% compared to the conventional planar transistor, and by more than 40% compared to an Isoplanar transistor. Even more important for high speed, the parasitic capacitance C_{bc} of an Isoplanar II transistor is reduced by 60%, compared to planar and Isoplanar transistors. These area reductions, combined with the use of shallower junctions and enhanced processing, have resulted in Schottky transistors with very high speed.

Using the Schottky transistors discussed above, the 54F/74F00 in the FAST TTL gate shown in Fig. 3.2.19*a* is realized for driving other TTL gates outside the package. For gates which are to be used inside a chip but are to have a large fan-out, this circuit can be simplified with a gate delay time of 1.7 nsec. For gates for a small fan-out inside a chip, the TTL gate in Fig. 3.2.19*b* can be used with subnanosecond delay time, where the current to the output transistor T_2 is to be supplied from outside the circuit through a resistor (Alford et al., 1979; Bechdolt et al., 1979).

The chip which is adopted in IBM processors 4331 and 4341, announced in 1979, contains 704 Schottky TTL gates fabricated with oxide isolation, called **recessed-oxide isolation**, using three metal layers of connections. The gate delay time is 3 nsec (*Electronic Design*, June 7, 1979, pp. 26–28; Dec. 20, 1979, pp. 25-26; *Electronics*, Nov. 9, 1978, p. 81; Pomeranz et al., 1979).

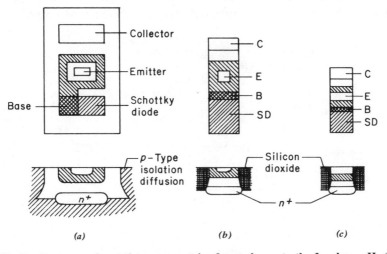

Fig. 3.2.18 **Progress of transistor geometries from planar to the Isoplanar II. (Courtesy of Fairchild Camera and Instrument Corporation.)** (*a*) **Planar.** (*b*) **Isoplanar.** (*c*) **Isoplanar II.**

Despite an increasing number of IC logic families, TTL will continue to be used where power outputs and reasonably high speed are required. The speed of MOS is improving and is sometimes faster than that of TTL. But high-speed MOS gates cannot deliver power outputs. TTL gates are cheaper than ECL gates, which are discussed in Section 3.4.

Speed–Power Product

The **speed–power product** or, more accurately, the **delay–power product** of a gate is defined as (delay time) × (power dissipation). This is a basic concept which is convenient in comparing different logic families and variants in each family. The smaller the speed–power product, the better. Every effort has been exerted toward the reduction of the speed–power product. The curve with a constant speed–power product is shown by a straight line in a figure with log–log scale (e.g., dashed lines in Fig. 3.2.20). Examples of such straight lines are illustrated in Fig. 3.2.20 for TTL and CMOS (to be discussed later) as solid lines.

Roughly speaking, each logic family or its variant can stay on the same straight line. But it cannot stay over the entire ranges of delay time and power consumption. When we increase power consumption beyond a certain value, the delay time does not decrease proportionately but starts to saturate, as seen at the ends of the solid lines in Fig. 3.2.20.

The delay–power product is usually measured by forming a ring counter (i.e., a loop of logic gates where each gate has only one fan-out connection). In other words, it is measured under the best conditions. However, if we form a network where some gates have many fan-out connections, a logic family which can deliver only low output power, like MOS families, shows a much greater delay–power product. So the influence of output load (or

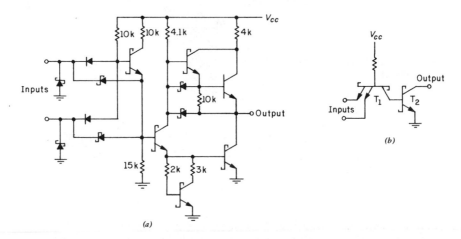

Fig. 3.2.19   FAST TTL gate. (*a*) 54F/74F00. (Courtesy of Fairchild Camera and Instrument Corporation.) (*b*) Internal gate for small fan-out.

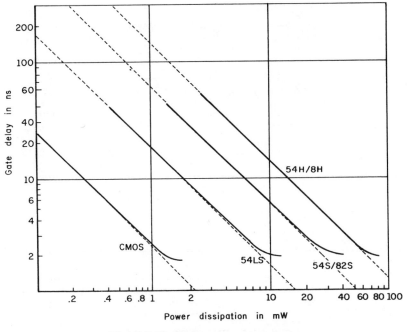

Fig. 3.2.20 Delay–power products.

the number of fan-out connections) cannot be properly represented by the delay–power product concept, unless it is used as an additional parameter. In this sense the delay–power product is not a precise but only a reasonably good measure to compare logic families and variants in each family.

3.3   Layout Design of Bipolar Transistor Circuits

Since all the information of logic networks is expressed in masks, let us discuss the preparation of masks for bipolar transistor circuits.

A typical geometry of a bipolar transistor is illustrated in Fig. 3.3.1. The buried layer is to provide a low-resistance layer for uniform distribution of current. (In this figure there is an n^+-region between the buried layer and the collector metal in order to improve the performance. But this requires an extra mask. Thus unless necessary, this n^+-region is formed at the same time as the emitter is diffused, as shown in Fig. 2.1.4j.)

Layout Rules

There are many rules for the layout of bipolar transistor circuits. Since all the layout rules must be met simultaneously, the layout is a very complex and time-consuming task. In the following, only typical rules are presented.

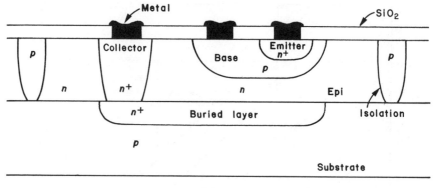

(a)

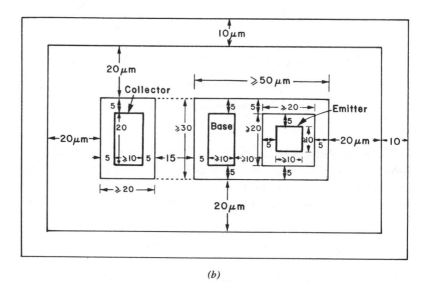

(b)

Fig. 3.3.1 Bipolar transistor geometry. Dimensions are not scaled. (a) Cross section. (b) Top view (25.4 μm = 1 mil).

1. Bipolar transistors can have various configurations. If a heavy current is expected, a uniform distribution of the current is attempted by different configurations, as shown in Fig. 3.3.2. In each case a collector, base, or emitter may be divided into many electrodes, but they are connected by a metal connection. Each case has only one common buried layer. Note that these transistors have different device parameters because of different configurations.

2. Isolation walls of wells can be made common. (Transistors, resistors, and diodes are placed inside isolation wells, as illustrated in Fig. 1.2.11c.)

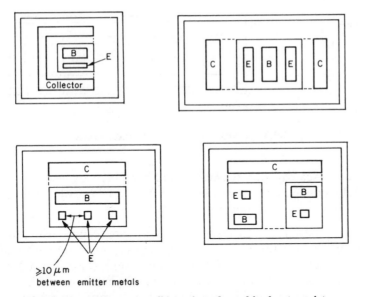

Fig. 3.3.2 Different configurations for a bipolar transistor.

If two isolation walls are close, as shown in Fig. 3.3.3*a*, they can be merged as shown in Fig. 3.3.3*b*. The wall width is 10 μm or more. All isolation walls are connected to the most negative voltage throughout the chip.

3. Transistors whose collectors are directly connected (as shown in Fig. 3.3.4*a*) can be fabricated using one large buried layer as the common collector. These transistors can be placed in the same isolation well, as shown in Fig. 3.3.4*b*. Transistors that do not have connections among them are rarely placed in a single common isolation well. The reason is that if they are placed in a single common isolation well, these transistors are connected by the *n*-type epitaxial layer.

4. The space between the base diffusion and the collector contact can be increased to permit metal connections (one or two metal connections) to be run between collector contact and base contact. (This can be done because the current between base and collector flows mostly between the base diffusion and the buried layer.) However, the greater base-to-collector

(a)   *(b)*

Fig. 3.3.3 Isolation walls.

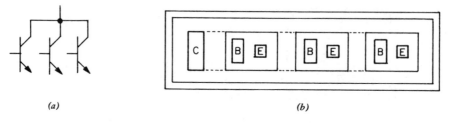

(a)   (b)

Fig. 3.3.4 Transistors with collectors connected.

distance increases parasitic resistance and parasitic capacitance of the collector. Metal connections cannot be run between base and emitter contacts by making base diffusion longer.

    5.   All transistors (as well as diodes if diodes are realized with transistors, as shown in Fig. 1.2.10) which are to be closely matched with respect to device parameters (e.g., parasitic resistance and amplification) must be oriented in the same direction so that any misalignment of masks affects all transitors in the same way. In other words, if an emitter is east of a base in one transistor, an emitter in another transistor must also be east of its base, as illustrated in Fig. 3.3.5.

    6.   A resistor is most frequently realized by the diffused base, as shown in Fig. 3.3.6. This diffused base resistor is formed simultaneously with the transistor bases. [Other possibilities are a polysilicon strip on top of the silicon-dioxide layer (e.g., see *Electronics* (September 13, 1979, p. 112) and Okada et al. (1980)), buried base, collector–resistor, buried collector, emitter–resistor, and thin film.]

Metal connections can run across resistors.

Notice that three layers in the diffused base resistor in Fig. 3.3.6 actually form a parasitic *p-n-p* transistor, as shown in Fig. 3.3.7, whose collector is the *p*-substrate in Fig. 3.3.6. In order to prevent the parasitic *p-n-p* transistor from working as a transistor (i.e., in order to make it always nonconductive), V_{cc}, which is connected to the middle *n*-layer, must be connected to a voltage

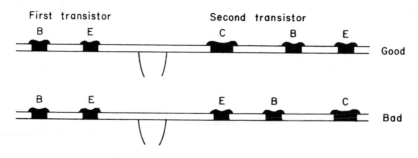

Fig. 3.3.5 Orientation of transistors.

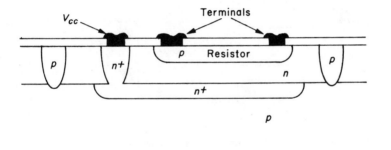

Fig. 3.3.6   Resistor (base diffusion).

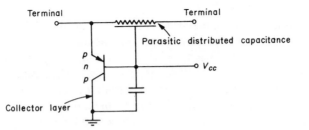

Fig. 3.3.7   Parasitic transistor in the resistor of Fig. 3.3.6.

higher than the *p*-regions so that two diodes are reverse biased. The upper two layers in Fig. 3.3.6, which form a reverse-biased diode, work as a parasitic capacitance also.

7.   In order to minimize any resistor value deviation due to a misalignment of masks, two terminals of a resistor should be on opposite sides with respect to the resistor, as illustrated in Fig. 3.3.8, if resistance accuracy is required.

8.   All resistors can be placed in the same isolation well using one large buried layer. Or resistors can be grouped separately, and each group can be placed in the same isolation well.

9.   If parasitic capacitance is not a problem, resistors can be placed in the same isolation well as a transistor, sharing one common buried layer, no matter whether they have connections among each other. Resistors must have at least 10-μm spacing among them. The transistor collector and a resistor must have at least 12-μm spacing.

10.   If an isolation wall is very wide, a metal connection running on it may have a large parasitic capacitance. (Usually we do not need to worry about this.)

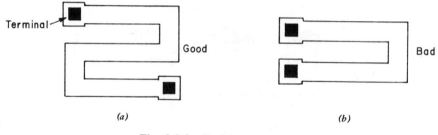

Terminal

Good

Bad

(a)

(b)

Fig. 3.3.8   Resistor geometry.

11.   A diode is usually implemented by connecting the collector and the base of a transistor as one electrode, using the emitter as the other electrode, as illustrated in Fig. 1.2.10*a*.

12.   A crossunder is realized as shown in Fig. 3.3.9. Since the n^+ diffusion can be done at the same time as the emitter diffusion, no extra processing steps are necessary. Also a crossunder can be realized in the same manner as the resistor illustrated in Fig. 3.3.6, placing the second connection (perpendicular to this page) between the two electrodes. (A buried layer can be skipped if the parasitic transistor effect is small.) When a resistor is necessary somewhere on a connection, Fig. 3.3.6 is a convenient means.

In either case, a crossunder adds some resistance, so crossunders should be avoided for power supply and ground connections. Since crossunders are not desirable for other connections also, designers should reduce them as much as possible by rearranging electronic circuits.

13.   A large buried layer (in the transistor or diode) yields a large parasitic capacitance. (For fast gates, such as ECL gates to be discussed later, the minimization of parasitic capacitance is very important for the improvement of speed.)

Layout Dimensions

In the following we show an example (see Fig. 3.3.1) of layout dimensions. (This is not the most advanced, but it still may be used.)

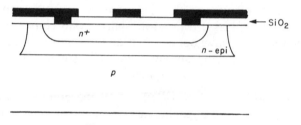

SiO_2

n^+

n – epi

p

Fig. 3.3.9   Crossunder.

1. The isolation wall is at least 10 μm wide.

2. The isolation wall must be 20 μm away from any buried layer of transistor or resistor.

3. The emitter diffusion is at least 20 × 20 μm.

4. The spacing between multiple emitter diffusions is 10 μm.

5. The base diffusion is at least 30 × 50 μm.

6. The collector n^+ diffusion is at least 30 × 20 μm.

7. The minimum distance between collector contact and base diffusions is 15 μm.

8. The minimum distance between multiple base diffusions is 10 μm.

9. The emitter contact window is at least 10 × 10 μm.

10. The base contact window is at least 10 × 20 μm.

11. The collector contact window is at least 10 × 20 μm.

12. The base contact window and emitter diffusion region must be at least 10 μm apart.

13. The metal connection is 15 μm wide or wider, with a spacing of 10 μm or more between connections.

14. The resistors can be realized based on diffusion squares. A diffusion in the form of a square on a chip has 160 Ω between two opposing edges, regardless of its size, as illustrated in Fig. 3.3.10. This is usually denoted by 160 Ω/□.

    In general, a resistor is 10 μm wide or wider. Thus a resistor of 480 Ω, for example, is formed by three squares in cascade, each 20 μm wide, as shown in Fig. 3.3.11a. A resistor of 80 Ω is formed by a half square, as shown in Fig. 3.3.11b.

    A diffusion square has 96 Ω (i.e., 60% of 160 Ω/□) between two adjacent edges, as illustrated in Fig. 3.3.12a. When a resistor is realized by bending its shape, as shown in Fig. 3.3.12b, the resistance of the square at the bending corner is calculated based on this, the total resistance being 160 × 5 + 160 × 0.60 Ω.

    The contact window size of a resistor is at least 20 × 20 μm.

    For simplicity we calculate the resistance from the distance between the inward edges of two resistor windows.

    Designers should use resistors which are as wide as possible whenever such shapes fit the space under consideration, since this improves the accuracy of the resistance.

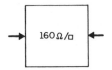

Fig. 3.3.10 Square resistor.

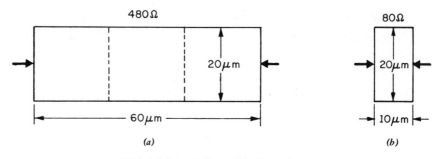

Fig. 3.3.11   Different resistor shapes.

15. A crossunder is realized based on the diffusion of 4 $\Omega/\square$. Its width is 10 μm or more.

16. Any metal must overlap a contact window by 2.5 to 5 μm on each side, and the distance between any contact window and the edge of a diffusion region (emitter, base, collector n^+, or resistor) must be 5 μm or more, as illustrated in Fig. 3.3.13.

17. Pads are usually placed on the chip periphery.* Each side of a pad must be at least 100 μm. The spacing between pads must be 50 μm or more. Each pad must be at least 25 μm away from any metal.

Electronic Circuit Design

Since electric performance such as speed and power consumption is the primary design objective, the electronic circuit design is actually done at the same time as the configuration is laid out and the dimension are de-

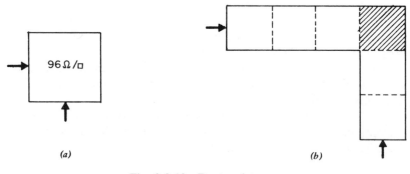

Fig. 3.3.12   Bent resistor.

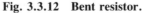

* On some chips, pads are also placed inside a chip, as illustrated in Fig. 2.4.3.

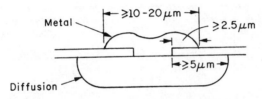

Fig. 3.3.13 Contact window dimensions.

termined. More precisely speaking, appropriate electronic circuits must be chosen with a detailed estimation of speed, power consumption, and waveforms (although so far many electronic circuits were shown without detailed analysis), and then the configurations and dimensions of all the components (i.e., transistors, diodes, resistors, and possibly others) are determined by precisely analyzing parasitic resistances, parasitic capacitances, waveforms, power consumption, and others with the aid of computer-aided design programs (or circuit modeling programs), such as SPICE2 (Nagel and Pederson, 1973; Nagel, 1975). For example, if a transistor is to deliver high output power, the sizes of the buried layer, emitter, and base are designed larger than the dimensions in the above example of layout dimensions. Also when a long connection is used, we have to check with SPICE2 whether the speed is impaired. However, since this book is not intended for a detailed analysis of electronic circuit design (Chua and Lin, 1975), this discussion is omitted here.

With the technological progress of fabrication, layout rules (including dimensions) and electronic circuits continue to change. Whenever designers want to develop IC chips with new technology, they need to use new layout rules and electronic circuits. Layout rules (in particular, dimensions) are proprietary and are rarely published by semiconductor manufacturers.

Figures 1.2.11c and 3.3.14 show layout examples of TTL gate networks.

References: Glaser and Subak-Sharpe, 1977; Hamilton and Howard, 1975.

Problems

3.3.1 Explain concisely why the TTL gate with an expander shown in Fig. 3.2.4 realizes the output function $f = \overline{(xyz)} \cdot \overline{(uv)}$ by analyzing the on–off states of currents through transistors and drawing the corresponding truth table.

3.3.2 Design an *S–R* flip-flop with TTL gates. Show an electronic circuit.

3.3.3 Design a TTL network in single-rail input logic (only noncomplemented variables are available as network inputs), using as few TTL gates as possible, for the following functions, under the constraints that at most one expander can be

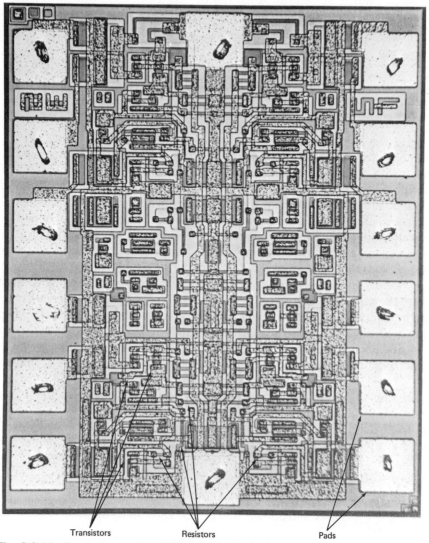

Transistors   Resistors   Pads

Fig. 3.3.14 **Layout example of Schottky TTL gate network: four NAND gates of 2-input Schmitt trigger (54/74LS132). (Courtesy of Texas Instruments, Inc.)**

added to a TTL gate and each multi-emitter transistor can have at most three emitters. Show an electronic circuit (not a logic network).

   (a) $f = \overline{xy \lor zuvw}$

   (b) $f = \overline{xy \lor uvw \lor uvz}$

3.3.4 Design a TTL network in double-rail input logic, using as few TTL gates as possible, for the following function, under the constraints that at most one expander can be added to a TTL gate and each multi-emitter transistor can have at most three emitters. Show an electronic circuit.

$$f = \overline{xy \vee \overline{u}z \vee \overline{v}z \vee \overline{w}z}$$

3.3.5 Discuss whether the derivation of an appropriate switching expression, using parentheses, in Procedure 3.2.1 can yield a better TTL network than those without parentheses obtainable by Procedure 3.2.1. Show an example.

3.3.6

(a) Discuss the advantages and disadvantages of the Wired-AND in Fig. 3.2.7 and the expander addition in Fig. 3.2.4 with respect to electronic circuit problems (if any) and ease of logic design.

(b) Design a TTL network in single-rail input logic for the following function by the two approaches illustrated in Figs. 3.2.4 and 3.2.7. Use as few TTL gates as possible, under the constraints that at most one expander can be added to a TTL gate and each multi-emitter transistor can have at most three emitters. Show electronic circuits.

$$f = \overline{xy \vee \overline{u}z \vee \overline{v}z \vee \overline{w}z}$$

3.3.7 Design a TTL network in single-rail input logic for $x \oplus y \oplus z$, using as few TTL gates as possible. Show an electronic circuit.

3.3.8 Design an adder with 2-bit positions in single-rail input logic, using TTL gates. The adder has an end-around carry input. Show an electronic circuit.

3.3.9 Suppose that we have designed a network in single-rail input logic with a minimum number of NAND gates for function f as well as for its complement $\overline{f}$. When we remove the output NAND gate in the network for $\overline{f}$ and tie together the output connections of all NAND gates that feed to this output gate, we obtain a new network, which realizes the function f. For example, from the NAND network for $\overline{f}$ in Fig. P3.3.9a we obtain the network with Wired-AND for f in Fig. P3.3.9b. Often this network has fewer gates than the network of NAND gates only (without Wired-AND) which is obtained for f.

(a) Only when the network for $\overline{f}$ has a certain type of connection configuration, can we derive the new network for f by the above procedure. State what the connection configuration is.

(b) Prove that the new network has at most two gates less than the network of only NAND gates for f, when the network for $\overline{f}$ has the type of connection configuration found in (a).

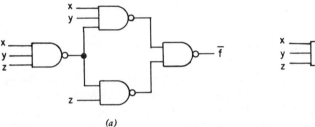

(a)

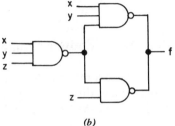

(b)

Fig. P3.3.9

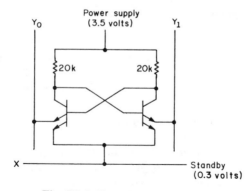

Fig. P3.3.10 TTL memory cell.

3.3.10

(a) Draw a layout for the TTL gate in either *A*, *B*, or *C* in the following. Try to make the chip area as small as possible, using a shape as close to a square as possible. Pads including those for power supply and ground must be provided.

To reduce parasitic capacitance for high speed, minimize the buried-layer areas.

(b) Draw different masks for the layout made in (a).

(c) State the total area in square micrometer.

Do not use oversize paper because drawing will be too time-consuming. Use letter-size paper, $8\frac{1}{2} \times 11$ in., sacrificing accuracy. For conversion of a transistor circuit into layout, see Figs. 1.2.11 and 3.3.14 as examples.

    A. TTL gate in Fig. 3.2.1, with diodes added like in Fig. 3.2.9*a*.

    B. TTL gate in Fig. 3.2.9*a*, with input *z* (as well as its emitter and diode) deleted.

    C. TTL memory cell in Fig. P3.3.10. (Since this is a memory cell, the layout must be repeatable. In other words, when the layout is repeated vertically and horizontally, the *X*, Y_0, Y_1 connections and the power supply must be connected to those of the adjacent memory cells. Pads are not necessary.)

3.3.11 Discuss why all resistors can be placed in the same well as mentioned in item 8 of the layout rules, while this is not necessarily true for all transistors, as mentioned in item 3.

3.4 ECL (Emitter-Coupled Logic)

ECL is currently the logic family with the highest speed and a high output power capability, although power consumption is also the highest. A basic ECL gate is shown in Fig. 3.4.1. Since this is more complicated and covers more chip area than the gate of any other logic family (including the TTL gate in Fig. 3.2.1), ECL is more expensive than TTL or any other logic family.

The logic operation of the ECL gate shown in Fig. 3.4.1 is analyzed in Fig. 3.4.2; the input z in Fig. 3.4.1 is eliminated for simplicity. When input x has a high voltage representing logic value 1 and y has a low voltage representing logic value 0, transistor T_x becomes conductive and a current flows through T_x and resistor R_1, as illustrated in Fig. 3.4.2a. In this case the voltage at the emitter of transistor T_r is not low enough (higher than -4 volts due to the current shown) against its base to make T_r conductive, so there is no current through T_r and resistor R_2. Consequently transistor T_f has a high voltage at its base, which makes T_f conductive, and output f has a high voltage representing logic value 1. (Notice that transistor T_f and resistor R_3 work in the same manner as that in Fig. 3.1.3 with $R' = 0$.) On the other hand, transistor T_g is almost nonconductive (actually a small current flows, but let us ignore it for simplicity), since its base has a low voltage due to the voltage drop developed across resistor R_1 by the current shown. Thus output g has a low voltage representing logic value 0.

Even when y (instead of x), or both x and y, has a high voltage, the above situation is not changed, except for the current through T_y.

Next suppose that both inputs x and y have low voltages, as shown in Fig. 3.4.2b. Then there is no current through resistor R_1. Since the base of transistor T_r has a higher voltage than its emitter (about 0 volt at the base and -0.8 volts at the emitter), a current flows through R_2 and T_r, as illustrated in Fig. 3.4.2b. Thus T_f has a low voltage at its base and becomes almost nonconductive (more precisely speaking, less conductive). Output f has consequently a low voltage, representing logic value 0. Transistor T_g has a high voltage at its base and becomes conductive. Thus output g has a high voltage, representing logic value 1.

The above analysis of Fig. 3.4.2 leads to the truth table in Table 3.4.1.

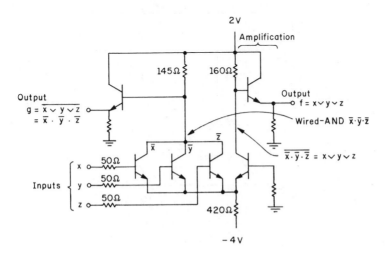

Fig. 3.4.1   Basic ECL gate (ECL 2500). (Courtesy of Texas Instruments, Inc.)

From this table the network in Fig. 3.4.2 has two outputs:

$$f = x \lor y \quad \text{and} \quad g = \overline{x \lor y}$$

[For a detailed electronic analysis, see electronics textbooks such as Barna and Porat (1973); Taub and Schilling (1977); Schilling and Belove (1979); and Ilardi (1976).]

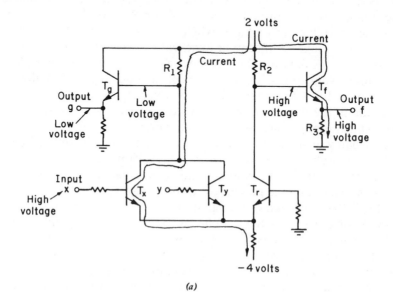

(a)

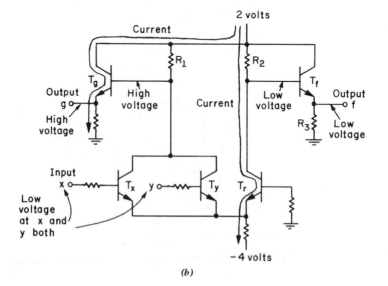

(b)

Fig. 3.4.2   Logic operation of ECL gate in Fig. 3.4.1.

Table 3.4.1   Truth Table for Fig. 3.4.2

INPUTS		OUTPUTS	
x	y	f	g
0	0	0	1
0	1	1	0
1	0	1	0
1	1	1	0

In a similar manner we can find that the ECL gate in Fig. 3.4.1 has two outputs:

$$f = x \lor y \lor z \quad \text{and} \quad g = \overline{x \lor y \lor z}$$

The gate is denoted by the symbol shown in Fig. 3.4.3. Because of the unique structure of an ECL gate, if either output $x \lor y \lor z$ or $\overline{x \lor y \lor z}$ is implemented by the gate, the other output can be provided easily, namely, by adding only one more pair of transistor and resistor. The simultaneous availability of OR and NOR as the double-rail output logic, with few extra components, is the unique feature of the ECL gate, making its logic capability powerful.

Another important feature of the ECL gate is Wired-OR. As shown in Fig. 3.4.4, the OR of the outputs can be realized without using an extra gate, simply by tying together these outputs (electronically speaking, this is possible because the outputs are those of emitter–followers). This is called **Wired-OR,*** **emitter-dotting,** or **Implied-OR.** It costs nothing except the area covered by the connections. This is very convenient in logic design. If one output is Wired-ORed with another ECL gate, it does not express the original function, and it cannot be connected to other gates. But if the same output is repeated by adding an extra pair of transistor and resistor inside the gate, as shown in Fig. 3.4.5, then the new output can be Wired-ORed with another gate output or connected to the succeeding gates. In the ECL gate at the top position in Fig. 3.4.5, for example, the first output $f = x \lor y \lor z$ is connected to the next level of gates, while the same f in the second output is used to produce the output $\overline{u \lor v} \lor x \lor y \lor z$ by Wired-ORing with the output of the second ECL gate. [For Wired-OR, see for example, Powers (1965) and Danielsson (1970).]

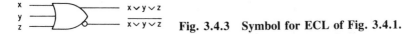

Fig. 3.4.3   Symbol for ECL of Fig. 3.4.1.

---

* Some authors call this Wired-AND, but this is a misnomer. It should be differentiated from the Wired-AND that is described in the following.

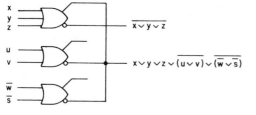

Fig. 3.4.4 Wired-OR of ECL gates.

The maximum fan-out of an ECL is about 10 to 25, depending on speed and other conditions.

A network of ECL gates may be designed starting with a network of only NOR gates for the following reason. Consider a network of ECL gates as shown in Fig. 3.4.6*a*, where Wired-OR is not shown for the sake of simplicity. This network may be converted into the network shown in Fig. 3.4.6*b* by eliminating the OR output of the left-hand gate in Fig. 3.4.6*a* and connecting inputs *x* and *y* of this gate directly to the right-hand gate. Thus the network in Fig. 3.4.6*b*, which expresses the same output as the network in Fig. 3.4.6*a*, consists of NOR gates only, at the sacrifice of the increase of fan-in of the right-hand gate (though signals on these connections for *x* and *y* propagate faster). Notice that the number of gates does not change in this conversion. When the network in Fig. 3.4.6*a* contains Wired-OR, it can be treated in the same manner. For example, the ECL network with Wired-OR in Fig. 3.4.6*c* can be converted to the network without Wired-OR shown in Fig. 3.4.6*d* at the sacrifice of the increase in fan-in. Notice that two NOR outputs in the upper left gate in Fig. 3.4.6*c* can be combined into one in Fig. 3.4.6*d*. Thus we can apply the above conversions to get a NOR network of the same number of gates, possibly with greater fan-in at some gates. Even if each gate has two or more NOR outputs or OR outputs, the situation does not change.

Any ECL network which is more complex than these simple examples can be converted in the same manner into a network with NOR gates only.

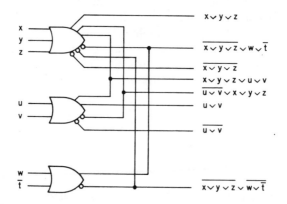

Fig. 3.4.5 Multiple-output ECL gates for Wired-OR.

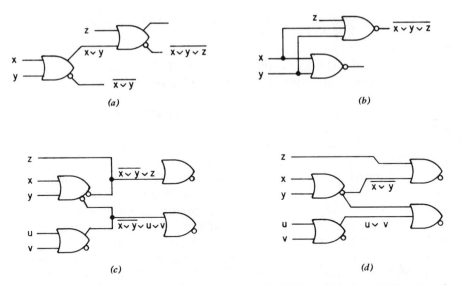

Fig. 3.4.6 **Conversion between ECL network and NOR network.** (*a*) **ECL network.** (*b*) **NOR network.** (*c*) **ECL network with Wired-OR.** (*d*) **ECL network without Wired-OR.**

When we want to design a network with ECL gates for given functions f and $\bar{f}$, it can be designed by reversing the above conversion, as follows.

Procedure 3.4.1 *Design of Logic Networks with ECL Gates*

1. Design a network for f and another network for $\bar{f}$ using NOR gates only without maximum fan-in or fan-out restriction on each gate. Use a minimum number of NOR gates in each case. [As discussed in Muroga (1979), the map-factoring method is usually convenient for designing networks with a minimum number of NOR gates, no matter whether it is in single- or double-rail input logic.]

2. Choose the network of fewer gates among the two networks obtained.
   Reduce the number of input connections to each gate by providing Wired-ORs, or by using OR outputs of other gates, if possible. In this case extra NOR or OR outputs at each ECL gate must be provided whenever necessary (such as Fig. 3.4.5). Thus if any gate violates the maximum fan-in restriction, we can avoid it by Wired-ORs or OR outputs.

3. This generally reduces the fan-out of gates also, but if any gate still violates a maximum fan-out, avoid it by using extra ECL gates. The output ECL gate of this network presents f and $\bar{f}$.

If no gate violates a maximum fan-out after step 2, the number of NOR gates in the original NOR network is equal to the number of ECL gates in the resultant ECL network. So if we originally have a network with a minimum number of NOR gates, the designed ECL network also has the minimum number of ECL gates, unless we need extra gates in step 3. □

Notice that the use of OR outputs and Wired-ORs generally reduces the number of connections or fan-in (i.e., input transistors). This also reduces the total sum of connection lengths, thus saving chip area. For example, the total length of the connections for x and y in Fig. 3.4.6b can be twice the connection length between two gates in Fig. 3.4.6a. Also, the total length of two connections in Fig. 3.4.6d can be twice the length in Fig. 3.4.6c. In procedure 3.4.1 NOR networks with a minimum number of gates are important initial networks. It is known that when the number of gates is minimized, the number of connections in the networks tends to be minimized (Muroga and Lai, 1976). The number of connections cannot be reduced independently of the number of gates (Alkhateeb et al., 1980). [For the properties of wired logic, see Kambayashi and Muroga, and Kawasaki et al. (both to be published).]

Remark 3.4.1: When we need to connect many gates by a single connection, the connection with a minimum total length is known as the minimum-cost spanning-tree problem (Aho et al., 1974; Deo, 1974). If we need to avoid branching of a connection into many in order to avoid transmission-line reflections of signal on a long connection on a pc board, the problem becomes the traveling salesman problem known in integer programming (Salkin 1975).□

Transistor-Level Logic Design with ECL

So far in this section we have discussed logic design with ECL gates. If we connect points inside one ECL gate to some points of another ECL gate, we can realize a complex logic function with a simpler electronic circuit configuration. In other words, we can realize logic functions by freely connecting transistors, resistors, and diodes, instead of regarding the fixed connection configuration of transistors, resistors, and diodes as logic gates whose structure cannot be changed. This approach could be called **transistor-level logic design**.

Consider two ECL gates as shown in Fig. 3.4.7a. Point A in gate 1 can be connected to point A' or B' in gate 2. Point B also can be connected to point A' or B'. The connection realizes **Wired-AND**. This is also called **collector-dotting**.* Notice that the voltage level at A represents the function $\overline{x \vee y}$ because the voltage is high, representing logic value 1, only when both x and y are low voltages. Also the voltage level at B represents $x \vee y$ because

---

* Connection of expanders in the case of TTL is also collector-dotting.

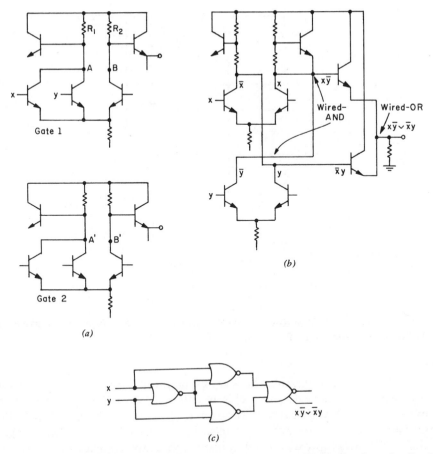

Fig. 3.4.7 **Example of Wired-AND.** (*a*) **Two ECL gates.** (*b*) **Realization of** $x \oplus y$ **by transistor-level logic design.** (*c*) **Realization of** $x \oplus y$ **with ECL gates.**

the voltage at B is high or low when the voltage at A is low or high, respectively.

For example, by connecting two ECL gates as shown in Fig. 3.4.7*b*, $x\bar{y}$ $\vee \bar{x}y$, which requires four ECL gates if designed with gates as shown in Fig. 3.4.7*c*, can be realized by the much simpler electronic circuit of Fig. 3.4.7*b*. In other words, $x\bar{y}$ and $\bar{x}y$ are realized by Wired-AND, and then these two products are Wired-ORed in order to realize $x\bar{y} \vee \bar{x}y$. In this case some resistors or transistors may be necessary for electronic performance improvement (since resistors R_1 and R_2 draw too much current in gate 1 in Fig. 3.4.7*a*, new resistors are added in Fig. 3.4.7*b* in order to clamp the currents), and unnecessary resistors or transistors may be deleted, although such an addition or elimination of resistors or transistors has nothing to do with logic operations (Fox and Nestork, 1971; *Computer Design*, Oct. 1972, p. 81). Another example of Wired-AND is shown in Fig. 3.4.8.

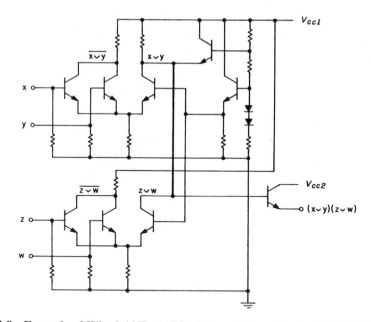

Fig. 3.4.8   Example of Wired-AND (*MECL System Design Handbook*, 1980. Courtesy of Motorola Semiconductor Products, Inc.)

Another approach in realizing a logic operation is the series connection of transistors in an ECL gate. This is called **series-gating**. Although it is sometimes called Wired-AND, series-gating actually realizes Wired-OR. Transistors T_1 and T_2 are connected in series in Fig. 3.4.9. Since the voltages at their collectors represent $\bar{y}$ and $\bar{x}$, respectively, the connection represents $\bar{x} \vee \bar{y}$, that is, $\overline{xy}$. So series-gating looks like the series connection of relay contacts.

A more complex example is the full adder in Fig. 3.4.10.* Usually at most 3 transistors are connected in series (the 2 transistors in the bottom level are part of the power supply). This is because too many transistors in series tend to slow down the speed of the gate (due to parasitic capacitances against the substrate). The transistor T_1 whose collector voltage represents $\bar{B}$ (since it assumes the logic value 0 when $B = 1$, B is complemented) is connected in series to the adjacent transistor T_2. The collector voltage of transistor T_2 represents function $\overline{AB}$ (AB is complemented because the voltage at the collector represents the logic value 0 only when $A = B = 1$). Then this is

---

* Figure 3.4.10 shows a full adder with ECL gates, modifying the ECL gate shown in Fig. 3.4.1 drastically. Full adders with ECL gates in single-rail logic can be designed with fewer transistors but probably with slower speed (T. Ueda, Japanese patent Sho 51-22779, 1976). C. R. Baugh and B. A. Wooley designed a full adder in double-rail logic (U.S. patent 3,978,329, 1976).

connected in series to transistor T_3 whose collector voltage represents $\overline{ABC}$. Wired-AND is implemented by parallel connections of transistors whose collector voltages represent $\overline{ABC}$, $\overline{A}B\overline{C}$, $A\overline{B}\overline{C}$, and $\overline{A}\overline{B}C$, as shown in the upper left portion of Fig. 3.4.10. Then the complement of the sum, that is, $\overline{S} = \overline{ABC} \cdot \overline{A}B\overline{C} \cdot A\overline{B}\overline{C} \cdot \overline{A}\overline{B}C = ABC \vee \overline{A}B\overline{C} \vee A\overline{B}\overline{C} \vee \overline{A}\overline{B}C$, is obtained by this Wired-AND. (A connection of this type is already used in the inputs in Fig. 3.4.1. In other words, the NOR output of Fig. 3.4.1 is Wired-AND of $\bar{x}$, $\bar{y}$, and $\bar{z}$, that is, $\overline{xyz} = \overline{x} \vee \overline{y} \vee \overline{z}$.) Usually at most 8 transistors can be connected in parallel.

The implementation of Wired-AND in this manner requires careful consideration of readjustments of voltages and currents. (Thus transistors or resistors may be added or changed in order to improve electronic performance, but this is not directly related to logic operations.)

Notice that collector-dotting and series-gating are possible only when an ECL network is implemented in a single chip. It is not possible, without impairing the speed advantage of ECL, when each ECL gate is implemented in a separate IC package with extra pins for internal connections. Wired-AND, along with Wired-OR, and OR and NOR outputs make logic design with ECL gates very flexible. Because of this, an ECL network designed without the concept of gates can be faster and simpler than a network realized with ECL gates when gates are treated as independent building

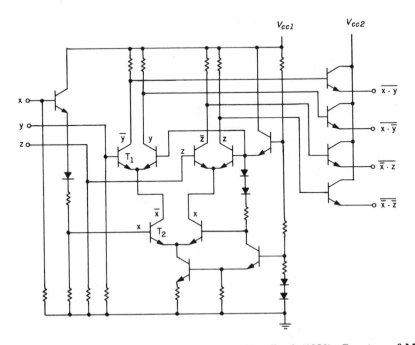

Fig. 3.4.9 Series-gating [*MECL System Design Handbook* (1980); Courtesy of Motorola Semiconductor Products, Inc.]

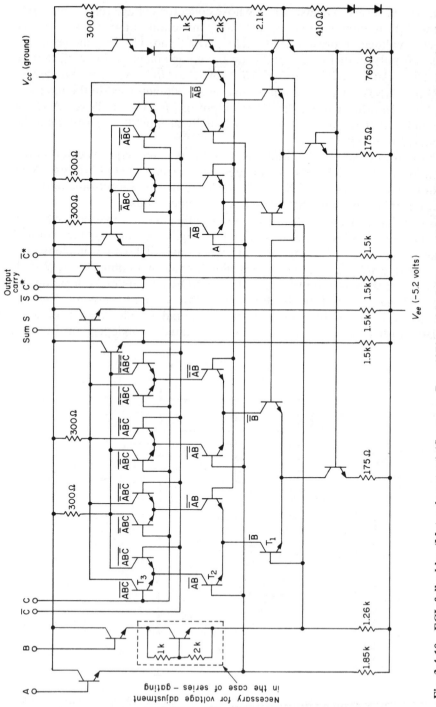

Fig. 3.4.10 ECL full adder (Motorola patent) (*Spectrum*, Dec. 1970, p. 34). Letters beside transistors represent switching functions at representative points. (Copyright © 1970 IEEE.)

blocks. Even a network implemented with ECL gates, treating gates as independent building blocks, usually needs fewer ECL gates and fewer levels than when it is synthesized with NOR gates only, because each gate has duo outputs, NOR and OR. This adds extra speed to the network in addition to the fast speed of the ECL gate itself. It is said that an ECL network on the average is roughly 50% faster than a network with NOR gates only. In this sense, the best logic design does not necessarily result from the concept of gates, but may result from freer connections of transistors, although we cannot do transistor-level logic design as often as gate-level logic design because adjustments of voltage and current magnitudes are cumbersome.

ECL permits using transistor-level logic design freely since ECL is a current switch device like a relay. In this sense, ECL may be regarded as a hybrid of electronic gate and relay. In the case of TTL, the transistor-level logic design is restricted because of voltage levels. TTL is essentially diode logic in its input (i.e., multi-emitters in its input transistor). Only connections of expanders and Wired-OR are permitted, in more restricted ways than ECL (the free connection of B to A' or B' in Fig. 3.4.7a for ECL is not permitted for TTL), and only outside TTL gate packages.

In summary, the logic design of ECL is accomplished as follows:

1. Gate-level logic design. Treat each ECL gate as an independent building block. Without changing the inside of each ECL gate, design a network with ECL gates which have duo outputs, NOR and OR (each NOR or OR may be available in more than one output with some ECL gate packages), and with Wired-OR.

2. Transistor-level logic design. By modifying ECL gates by the following approaches, an ECL network is designed:

(a) Wired-AND: Collector-dotting (at most 8 transistors in parallel).

(b) Wired-OR: (i) Emitter-dotting. (ii) Series-gating (at most 3 transistors in series).

Problems in Using ECL

ECL has several disadvantages. One is large power consumption. An ECL gate consumes 20 to 60 mW, compared to 1 to 20 mW for a TTL gate. Because of this, ECL is not suitable for as large LSI as that with MOS. At most a few hundred ECL gates (depending on the tradeoff of speed and power consumption) can be put on a single chip, though this is improved by the isoplanar process to be described later. (Thus networks are usually divided into many chips, and many ECL gates must deliver high output power over connections on pc boards.) For example, the LSI chip used in the Amdahl computer 470 V/6 contains 100 ECL gates of 0.6-nsec delay time (each chip has 80 pins). Since it generates 3.5 W, it has a finned heat sink installed on its stud (*Computer Design*, Jan. 1976, pp. 27, 28).

The second disadvantage is tricky wiring. Since connections on a pc board work as transmission lines, signal waveforms are usually excessively distorted without appropriate termination of the connections. [See, for example, Blood et al. (1972); Barna and Porat (1973, chap. 12); Davidson and Lane (1976); Ho (1971); Balph (1972); Bandler and Rizk (1979); Dworsky (1979).]

The third disadvantage is that two power supplies are required.

Although ECL gates require large chip areas and high power consumption, they are the fastest among all logic families which are widely used (excluding those that are still in the developmental stages and are discussed in later chapters). Also, they are able to operate over a wide range of temperatures and power supply voltages (*Electronics*, July 17, 1972, p. 68). For these reasons, ECL has been used for almost all high-speed models of mainframe computers. In particular, since Motorola introduced in 1971 the ECL 10,000 series (sometimes abbreviated as 10 K series) which has a typical 2-nsec gate delay, ECL has been widely used because the price difference between ECL and Schottky TTL became small (*Electronics*, Dec. 25, 1975, p. 36; Feb. 21, 1974, p. 85).

For design with ECL gates, see, for example, Vacca (1971) and Blood et al. (1972). There are many other references (Maul, 1973; Miles, 1972; Walker, 1973; Jaeger, 1974; Fairchild, 1974; Balph et al., 1975; Pierce et al., 1980).

Optimal networks with ECL gates for functions of three variables are covered in Baugh et al. (1972) and Kawasaki et al. (to be published).

Heat Dissipation

Since many components are packed into a tiny space and they generate heat when currents are on, cooling of these components is difficult because of the basic law of physics, "the larger a surface, the easier the cooling." Some components generate more heat than others. If such components are crowded somewhere, the temperature of this area may become excessively high, damaging the components. Thus placement of the components such that heat is generated uniformly over a chip may be important. Even so, the power density in a chip may be very high, though the absolute value of power is very small. (Bipolar transistors generate more heat than MOS.) The electric power density in a chip is approximately equal to that of an average-size 500-W lamp. Thus heat conduction between a chip and a package container must be carefully designed, or packing density on a chip must be lowered.

The heat design of ICs is vitally important for their reliability. There is a popular rule of thumb that for each 10-degree temperature rise, the failure rate doubles (*EDN*, Oct. 1, 1972, pp. 40–42). Because of heat, not many

ECL gates can be packed in a single chip, since a single chip can handle at most a few watts.

References: Mizell, 1969; Seely and Chu, 1972; *Electronics*, Nov. 8, 1973, pp. 98–104; Jan. 24, 1974, pp. 87–89; May 16, 1974, pp. 116–121; June 13, 1974, pp. 113–118; June 27, 1974, pp. 115–120; Oct. 31, 1974, pp. 87–90; Rodriguez, 1979.

ECL with Improved Isolation

The Isoplanar process discussed in Section 3.2, namely, replacement of *p*-type isolation diffusion by silicon-dixoide isolation, was actually applied to ECL before being applied to TTL.

As shown in Fig. 3.4.11, the planar bipolar transistor in Fig. 3.4.11*a* was improved to *b* by the original version of the Isoplanar process (*Electronics*, Mar. 1, 1971, pp. 54, 55), and then to *c* by Isoplanar II. Further reduction of the size is illustrated in Fig. 3.4.12. By Isoplanar II the area of a transistor is reduced by more than 70% compared to a conventional planar transistor and by over 40% compared to an Isoplanar transistor. In 1979 it is further reduced by Isoplanar S. The gate delay of ECL is reduced to 500 to 750 psec (i.e., picoseconds) by Isoplanar II and to 300 to 400 psec by Isoplanar S (Dhaka et al., 1973; Rice, 1979). Also because of the reduction of power consumption by the isoplanar technique, we can pack up to 1,500 ECL gates into a single chip (see Remark 3.4.2).

Fairchild developed the ECL 100K series (or F100K as trade name) based on Isoplanar II. The gate delay of the 100K ECL is 750 psec at 40 mW with a 50-Ω load in the case of SSI. In the case of internal gates in MSI, the gate delay is less than 500 psec at less than 10 mW (Fairchild, 1977). The Fairchild 100K series also includes ECL gates based on Isoplanar S with 400-psec gate delay.

Other semiconductor manufacturers developed similar isolation techniques (*Electronics*, Aug. 7, 1975, pp. 51, 53, 54; Feb. 19, 1976, p. 108; Apr. 28, 1977, p. 25). For example, Motorola reduced the typical gate delay of 2 nsec of the ECL 10K series to 1 nsec of the 10KH series without increasing power consumption, by the isoplanar process MOSAIC (*Electronic Design*, July 19, 1980, p. 38; *Electronics*, Dec. 4, 1980, p. 200). Based on MOSAIC, Motorola's MECL 10,000 Macrocell Array contains 750 equivalent gates on a chip (Prioste, 1978; Prioste et al., 1979; Reed and Ligget, 1980).

ECL in Isoplanar, 100 K, is compared with other logic families in Fig. 3.4.13.

The Isoplanar technology can also be applied to CMOS, in addition to TTL and ECL, for performance improvement. In particular, application to memories yields significant improvements.

Remark 3.4.2: ECL also has some variations, as shown in the following examples. The so-called buffered ECL, with extra components added, has good performance, which is independent of temperature and power supply voltage, unlike the ECL gates discussed in the text (Muller et al., 1973; *Electronics*, Feb. 21, 1974, p. 85). A low-power version of the ECL has been developed. It does not have the output emitter followers of the ECL, requiring a power supply of lower voltage than the ECL. Thus the power consumption is reduced (Akazawa et al., 1978; Yano et al., 1980; Sato et al., 1980). In the microcontroller of the series 4300 mainframes announced in 1979, IBM used a highly packed ECL gate chip, using the oxide isolation, with about 1,500 gates on a single chip. With collector-dotting the output power can be adjusted by con-

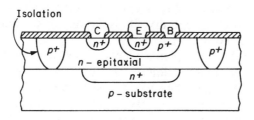

(a)

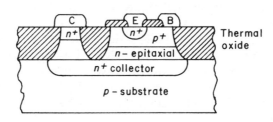

(b)

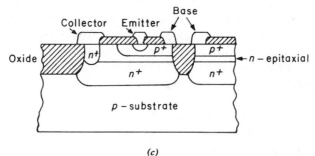

(c)

Fig. 3.4.11 Comparison of cross sections in Isoplanar improvements. (*a*) Planar. (*b*) Isoplanar. (*c*) Isoplanar II.

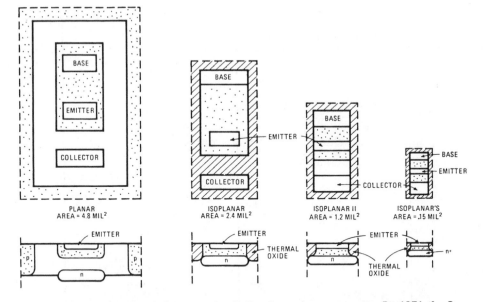

Fig. 3.4.12 Comparison of geometries in Isoplanar improvements. In 1971 the Iso-planar technique used oxide isolation. In 1975 Isoplanar II emitters extended to the walls of the oxide for a 50% reduction in area. In 1979 Isoplanar S has scaled those devices down. (Reprinted from *Electronics* Dec. 6, 1979, p. 139. Copyright © McGraw-Hill, Inc., 1979. All rights reserved.)

necting Schottky diodes. (These ECL circuits are shown in Chapter 9.) The gate has a delay of 1.5 nsec at 0.85 mW, or by changing the values of the resistors, 0.80 nsec at 1.7 mW. Three metal layers are used for the first time in a commercial product (two metal layers were the industry's common practice) (Blumberg and Brenner, 1979; Werbizky et al., 1979, 1980; *Electronics*, Feb. 15, 1979, pp. 46, 48; Apr. 7, 1981, pp. 149–152). Base-coupled logic (BCL) is a variant of the ECL with low power consumption, introduced by Siemens (*Electronic Design*, Mar. 1, 1976, p. 19).

ECL and its variations discussed so far are often called **current mode logic** (CML). But CML includes bipolar logic circuits where currents are turned on or off, unlike ECL and the above variations, where currents are switched between transistors (*Electronics*, Dec. 15, 1975, p. 36; Mar. 18, 1976, p. 69; Dec. 7, 1978, p. 65).

Emitter–emitter logic (E^2L) is a modification of ECL for the convenience of driving 50-Ω transmission lines (*Electronics*, Feb. 7, 1974, pp. 114–118; DiPietro, 1975). ☐

Problems

3.4.1 Convert the NOR network in Fig. P3.4.1 into an ECL network with as few ECL gates as possible, where each ECL gate can have many NOR and OR outputs. Wired-OR can be used.

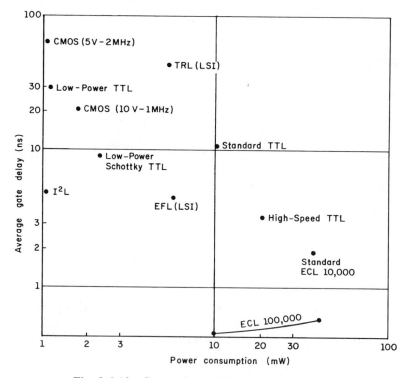

Fig. 3.4.13   Comparison of delay–power products.

3.4.2   Design an ECL network in double-rail input logic by Procedure 3.4.1 for the following function, with as few ECL gates as possible. Maximum fan-in is 2.

$$f = xy\bar{z} \vee \bar{x}z \vee \bar{y}z$$

3.4.3   Design an ECL network in single-rail input logic by Procedure 3.4.1 for one of the following functions, with as few ECL gates as possible. Maximum fan-in and fan-out is 3.

A. $\bar{x}z \vee \bar{y}\bar{z}$

B. $\bar{x}y \vee \bar{y}z$

3.4.4   Design a full adder using the ECL gates in Fig. 3.4.3 or 3.4.5, using as few gates as possible. Then compare the number of transistors with that in Fig. 3.4.10. Assume that the gate in Fig. 3.4.3 is implemented as shown in Fig. 3.4.1.

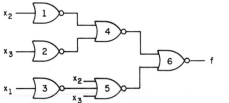

Fig. P3.4.1

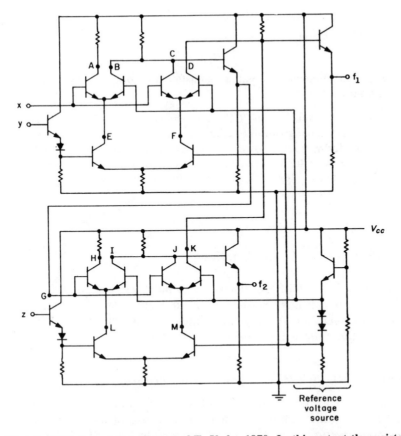

Fig. P3.4.5 ECL full adder. (Patent of T. Ueda, 1970. In this patent the resistors at *A* and *H* are set to resistance value 0, since they are not involved in the logic operation.)

3.4.5 The ECL network in Fig. P3.4.5 represents a full adder. Find what switching function is expressed by each of the points $A, B, C, D, E, F, G, H, I, J, K, L, M, f_1$, and f_2. Then compare the number of transistors with that in Fig. 3.4.10.

3.4.6 Design an ECL circuit to realize the output function f realized by the network shown in Fig. P3.4.6, by using collector-dotting. (Neither resistor values nor diodes which are often used for performance improvement or for voltage adjustment need to be shown.)

3.4.7 Design an ECL circuit to realize the output functions f and $\bar{f}$ realized by the gate shown in Fig. P3.4.7, by using series-gating. (Neither resistor values nor

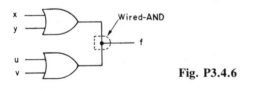

Fig. P3.4.6

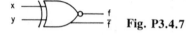

Fig. P3.4.7

diodes which are often used for voltage adjustment or for performance improvement need to be shown.)

3.4.8 Draw a layout for the ECL gate shown in Fig. P3.4.8, using the layout rules (with dimensions) given in Section 3.3. Try to make the chip area as small as possible. Pads for x, y r, f, power supply, and the ground must be provided. Assuming that we need high speed, make the buried-layer areas small. Do not use oversized paper because drawing will be too time-consuming. Use letter-size paper, $8\frac{1}{2} \times 11$ in., sacrificing accuracy.

3.5   Other Bipolar Logic Families

In addition to TTL and ECL discussed so far, there are many other bipolar logic families which have actually been in production or proposed, as discussed in this section. Among them integrated injection logic (IIL or I^2L) is probably the most widely used.

Schottky TRL (Schottky Transistor Resistor Logic)

TRL (or RTL) was popular before TTL was introduced. This old logic family came back with the Schottky transistor. Because of its simplicity and low power consumption (3 mW/gate), it is suitable for LSI. Several hundred gates can be put into a single chip. (Schottky TRL is used in Motorola's Megalogic, as discussed in Chapter 9.)

As shown in Fig. 3.5.1a, Schottky TRL realizes the NOR operation. Tying down collectors, we can realize Wired-AND as shown in Fig. 3.5.1b. The delay time is 25 nsec, and maximum fan-out is 5 inside a chip.

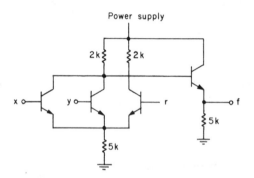

Fig. P3.4.8   ECL gate.

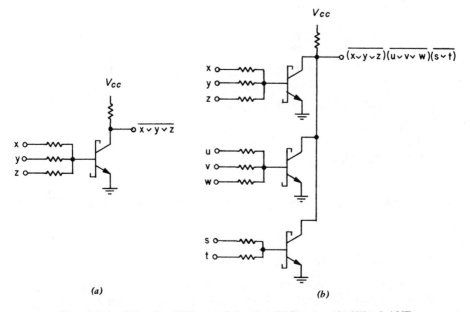

Fig. 3.5.1 Schottky TRL. (*a*) Schottky TRL gate. (*b*) Wired-AND.

EFL (Emitter Function Logic)*

EFL, shown in Fig. 3.5.2*a*, realizes only positive functions, as shown in Fig. 3.5.2*b*, and this is unusual for an IC logic family. Since NOT cannot be realized by EFL, NOT realized by another IC gate family must be used. The AND operation is realized by a multi-emitter transistor at the input of the gate. Wired-OR is permitted at the emitters of the output transistors. [Where multiple input currents can occur, the transistor shown with a dashed line (actually it works as a diode) is added to prevent saturation of the input transistor. (Skokan, 1973; Wong and Jackson, 1979).]

EFL is much slower than ECL, though its power consumption is lower. But combining EFL in a series-gated arrangement with multilevel ECL results in a logic structure which has advantages over both conventional EFL and series-gating ECL in speed, delay–power product, and area (Scavuzzo, 1980; Cooperman, 1980).

Remark 3.5.1 EFL (Emitter Follower Logic): Old complementary transistor logic (CTL) of Fairchild came back as EFL by TRW. According to TRW, more than 5,000 EFL gates can be put on a single chip. Triple diffused** EFL further improves the density (*Electronics*, Jan. 24, 1974, pp. 29, 30; Feb. 1, 1974, pp. 26, 95; Mar. 20, 1975, pp. 29, 30). □

---

* This EFL should not be confused with EFL in Remark 3.5.1.
** For "triple diffusion," see Buie and Swartzlander (1977) and Buie (1975).

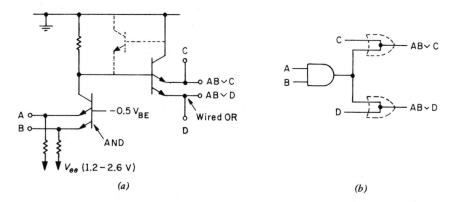

Fig. 3.5.2   **EFL gate and its function.** (*a*) **EFL gate.** (*b*) **Functions realized by an EFL gate.**

I²L (Integrated Injection Logic)

I²L, or IIL, is also called merged transistor logic (MTL). It was announced simultaneously by IBM and Philips in 1972 (Hart and Slob, 1972; Berger and Wiedmann, 1972). It is the first bipolar logic family that greatly reduces the chip area and power consumption per gate. An I²L gate requires only 1 to 2 mil², or the space of a single bipolar transistor.

IIL is illustrated in Fig. 3.5.3. Holes (denoted by ⊕) are injected from the p emitter (denoted as p_1) of a p-n-p transistor formed by p_1, n_1, and p_2 in Fig. 3.5.3a. The holes in the vicinity of the emitter–base junction (denoted by p_1 and n_1) work as currents in two other n-p-n bipolar transistors in Fig. 3.5.3a. (Each of the two transistors consists of n_2, p_2, and n_1.) In other words, currents are supplied through a small emitter, called an **injector**, whereas in the case of TTL or ECL, currents are supplied through resistors which occupy large chip areas. This makes an I²L gate occupy a very small area, which is unusual for a bipolar logic family. As we see in Fig. 3.5.3b, each I²L gate forms a rectangle. The electronic circuits which embody Fig. 3.5.3a and b are presented in Fig. 3.5.3c, where the injector shown in Fig. 3.5.3a (or b) is split into two p_1's in two p-n-p transistors. [Here notice that p_1 of the injector, n_1 of the substrate, and p_2 of the base constitute a p-n-p transistor, and n_1 of the substrate, p_2 of the base, and n_2 of the collector constitute an n-p-n transistor. These two sets of p-n-p and n-p-n transistors are shown in Fig. 3.5.3c. Since p-n-p and n-p-n transistors are merged well, I²L is also called merged transistor logic (MTL).] Each I²L gate is essentially an inverter with a single transistor. The logic operation NOR is realized by Wired-AND connection of more than one I²L gate, as illustrated in Fig. 3.5.3c. The electronic circuits shown in Fig. 3.5.3c are usually expressed by Fig. 3.5.3d, where each p-n-p transistor not involved in logic operations is expressed by a chain of two circles. The output of one I²L gate can be Wired-ANDed with those of many other gates, and more than one collector

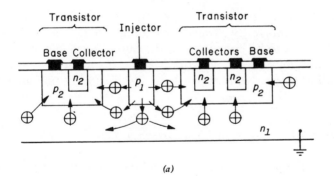

(a)

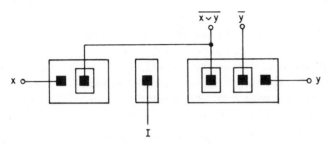

(b)

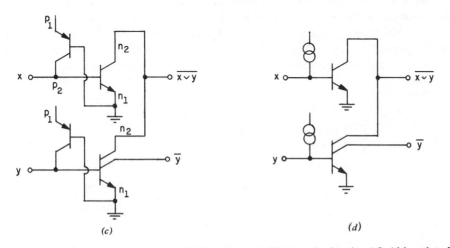

(c)   (d)

Fig. 3.5.3 I²L. (*a*) Cross section. (*b*) Top view. (*c*) Electronic circuits. (*d*) Abbreviated expression of (*c*).

can be provided at each gate, as shown by $\bar{y}$ in Fig. 3.5.3c and d. (Notice that collector currents for two transistors in Fig. 3.5.3c or d are to be supplied from the I²L gates to be connected to the output terminals.)

The logic design of an I²L network is easier if we interpret it as a NAND network (rather than a NOR network, as discussed in Problem 3.5.6).

Procedure 3.5.1:   Design of Logic Networks with I²L Gates

If we partition an I²L network in a way different from Fig. 3.5.3d, then each I²L may be regarded as a NAND gate, as shown in Fig. 3.5.4. Thus when an I²L network is given as shown in Fig. 3.5.5a, this may be interpreted as a NAND network, as illustrated in Fig. 3.5.5b. Based on this conversion, we can derive an I²L network of a minimum number of I²L gates as follows:

1.   Design a NAND gate network for a given function f, with a minimum number of NAND gates as the primary objective, and with a minimum number of connections as the secondary objective. [As discussed in Muroga (1979), the map-factoring method is usually convenient for designing networks with a minimum number of NAND gates, no matter whether it is in single- or in double-rail input logic.]

2.   Then convert each NAND gate to an I²L gate.

Notice that external inputs are usually Wired-ANDed in the inputs of some gates in the resultant I²L network. (For example, in Fig. 3.5.5 x is Wired-ANDed with y at gate 1, and also with the output of gate 1 at gate 2.) So, **if Wired-AND of external inputs is not permitted, extra I²L gates must be added as buffers of the external inputs.** If each fan-out of an external input is provided separately by a different collector of a multicollector transistor in the preceding IIL network (e.g., x at gate 1 in Fig. 3.5.5 is provided by a collector different from the collector for x at gate 2, where these collectors

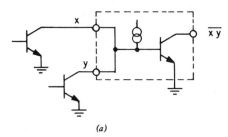

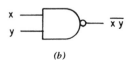

(a)   (b)

Fig. 3.5.4   NAND operation by I²L.

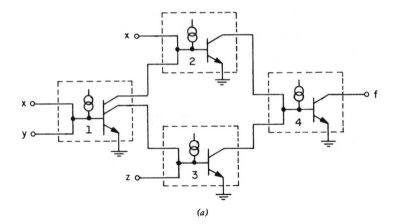

(a)

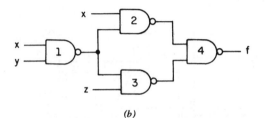

(b)

Fig. 3.5.5 **Conversion of an I²L network into a NAND network.**

are in the multicollector I²L gate not shown in Fig. 3.5.5), then the external inputs can be freely Wired-ANDed with others at any I²L gate. □

Fig. 3.5.6 shows the logic network of NAND gates, the electronic circuits of I²L gates, and the I²L layout for a D-type flip-flop. Notice that the two lines for clock C in Fig. 3.5.6b and c cannot be merged, since each line is Wired-ANDed with different gate outputs. Fig. 3.5.7 shows a layout for a decoder, where lines are abbreviated to simple forms.

As seen in Figs. 3.5.6 and 3.5.7, the layout configuration of injectors is flexible, although it determines the speed. As illustrated in Fig. 3.5.8a, when the base region is laid out perpendicular to the injector [shown as (1)], nonuniform current density, due to base resistance inside each n-p-n transistor, results in different delay times for different collectors (the speed is slower for C_3 than for C_1 because of a smaller current as illustrated in Fig. 3.5.8b). A layout with the base region parallel to the injector [shown as (3) in Fig. 3.5.8a] ensures the minimum delay time for all collectors, but reduces

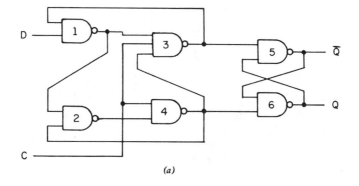

(a)

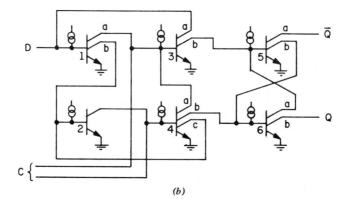

(b)

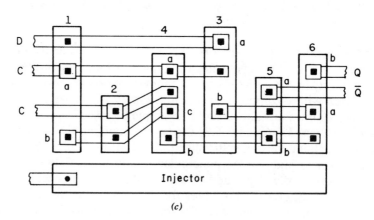

(c)

Fig. 3.5.6 *D*-type flip-flop. (*a*) Network of NAND gates. (*b*) Network of I²L gates. (*c*) I²L layout.

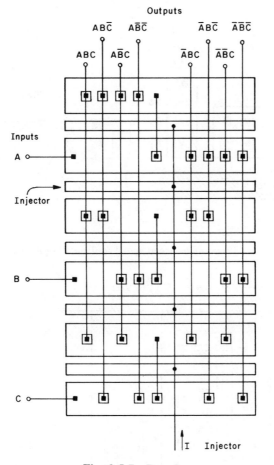

Fig. 3.5.7   Decoder.

the packing density because injectors must be repeated more often throughout the chip.

Since I^2L gates are laid out in rectangular form, compact layouts of I^2L gates are relatively easy, compared with layouts of other logic families (in particular MOS) which look like jig-saw puzzles of odd-shaped pieces. Consequently computer-aided layout can be more effectively done, compared with that for other logic families.

Gate areas of different logic families are compared in Fig. 3.5.9 (based on the technology prevailing around 1975). Although a small gate area is very desirable, this does not mean small chip areas for realizing given functions. Despite small gate areas, I^2L has low logic capability because it can realize only the NAND operation. This also tends to increase the connection areas on a chip. Thus depending on the functions to be realized, other logic

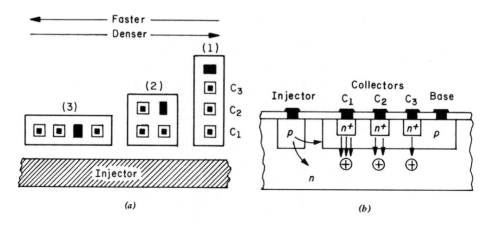

Fig. 3.5.8   Tradeoff between speed and packing density.

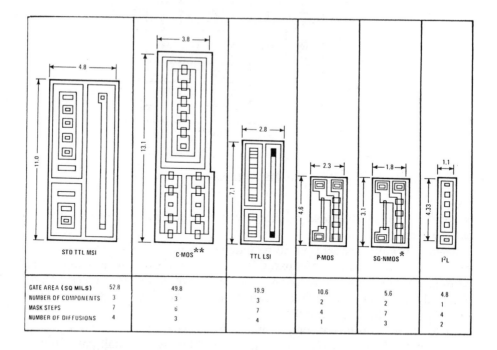

Fig. 3.5.9   Comparison of gate areas of different logic families. Each structure shows 4 inputs or outputs. All dimensions are shown in mils. *Silicon-gate *n*-MOS to be discussed in the next chapter. **Recent CMOS uses silicon gate instead of metal gate shown here. (Reprinted from *Electronics*, Feb. 6, 1975, p. 85. Copyright © McGraw-Hill, Inc., 1975. All rights reserved.).

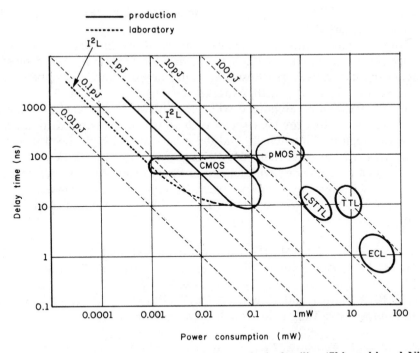

Fig. 3.5.10 **Delay–power products of different logic families (Shinozaki and Nishi, 1978).**

families which may have larger gate areas may yield a smaller chip area than I^2L because of greater logic capability.

Other Features of I^2L

An I^2L gate has a delay time of 10 nsec or more, dissipating 50 μW per gate. The delay–power product is 0.1 pJ (i.e., picojoules) for I^2L gates in the laboratory, whereas ECL and TTL have often 10 pJ or more. The delay–power products of different logic families are compared in Fig. 3.5.10. (Notice that logic families in production usually have lower performance than those in laboratories, and this figure does not accurately show the most advanced technology.) In Table 3.5.1 a typical I^2L is compared with a low-

Table 3.5.1 **Comparison of I^2L with Others**

	LSTTL	I^2L	ISL
Speed	9.5 nsec	15 nsec	3.5 nsec
Power	2 mW	250 μW	250 μW
Delay–power product	19 pJ	3.75 pJ	0.88 pJ

power Schottky TTL (LSTTL) and an integrated Schottky logic (ISL) described in Remark 3.5.2.

Another unique feature of I^2L is its compatibility with other bipolar logic families on the same chip. Although I^2L usually needs only 4 or 5 masks, ordinary bipolar transistors (those shown in Chapters 1 and 2) can be placed on the same chip as I^2L with 7 masks, as illustrated in Fig. 3.5.11. This means that I^2L can be combined with other bipolar logic families, such as ECL, TTL and EFL, for high speed or high output power. I^2L is for low speed with high packing density and the others are for high speed with low density (*Electronics*, February 21, 1974, p. 84; Wong and Jackson, 1979). Also I^2L can be combined with analog or linear electronic circuits, making I^2L useful for many applications, such as consumer electronic products (*Electronics*, Apr. 13, 1978, pp. 102, 103; O'Neil, 1979; Amazeen and Timko, 1979).

References on I^2L: Hart et al., 1974; Berger and Wiedmann, 1975; DeTroye, 1974; Pedersen, 1976; Warner, 1976; Garry, 1977; Bruederle and Smith, 1975; *Spectrum*, June 1977, pp. 29, 33–35; *IEEE JSSC*, Apr. 1977; Evans, 1979; Berger and Kelwig, 1979; Taub and Schilling, 1977; Halbo, 1980; Smith, 1980; Tang et al., 1980; Stewart-Warner, 1976.

Remark 3.5.2   *Variants of I^2L*: The performance of I^2L can be improved by the isoplanar process. Fairchild calls it Isoplanar I^2L, or I^3L (Crippen and Hingarh, 1980).

The performance of I^2L is also improved by use of the Schottky diode. The **integrated Schottky logic (ISL)**, shown in Fig. 3.5.12, is such an improvement by Philips. The gate delay is about 2 to 4 nsec, which is twice that of the low-power Schottky and a quarter that of the I^2L. So the ISL fills the gap between the I^2L (which has high packing density, low speed, and low power consumption) and the low-power Schottky TTL (which has high speed, low packing density, and high power consumption). The ISL is different from the **STL** proposed by Berger and Wiedmann of IBM (*ISSCC 1975*, pp. 172, 173) or the Schottky Di-istor logic proposed by Schuenemann and Wiedmann in 1973, in that the latter lacks p-n-p transistors T_2–T_3. Its complex structure makes the ISL occupy a 20% greater area than I^2L, and a current source must be supplied

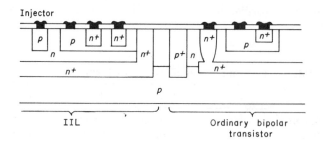

Fig. 3.5.11 I^2L and bipolar transistor on the same chip.

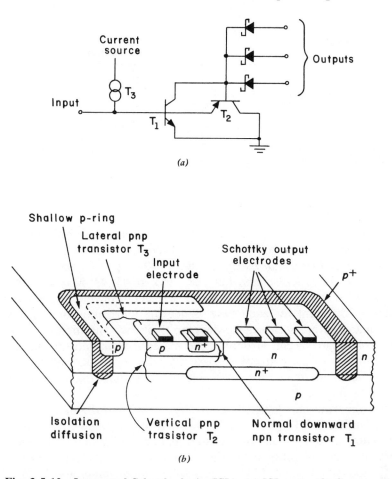

Fig. 3.5.12   Integrated Schottky logic (ISL). (*a*) ISL gate. (*b*) Cross section.

from the input electrode, unlike for I²L where the current supply is provided as injector (*Electronics*, June 8, 1978, pp. 41, 42; Sept. 13, 1979, pp. 111, 113; Lau, 1979; *Spectrum*, Jan. 1980, p. 47; *Electronic Design*, Sept. 1, 1979, p. 44; Hewlett, 1980).

The **Schottky transistor logic (STL)** pursued by Texas Instruments is reported to have a gate delay of less than 500 psec and a delay–power product of less than 50 fJ. The STL is a candidate for electron-beam lithography, since Texas Instruments has scaled the bipolar process down to 1 μm active regions (*Electronics*, Oct. 25, 1979, p. 110; Dec. 6, 1979, p. 126; July 14, 1981, p. 133; Evans et al., 1980).

The **static induction transistor logic (SITL)** is also a high-performance version with Schottky diodes, developed by Nishizawa (Nishizawa et al., 1979).

There are other variations of I²L (Peltier, 1975; *Electronics*, Dec. 12, 1974, pp. 36, 38; Hoffmann, 1976).   □

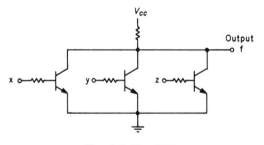

Fig. 3.5.13 RTL.

Other Logic Families

There are many variants and also other logic families. Examples are **complementary transistor logic (CTL)**, which is different from the CTL mentioned in Remark 3.5.1 (*Electronic Design*, Mar. 1, 1976, p. 18), and **low-voltage inverter logic (LVI)** (*Electronics*, Feb. 28, 1981, p. 41, 42; May 19, 1981, p. 8; Konian and Walsh, 1981). Logic design and electronic circuit design may be somewhat different for each case. Since the fabrication of ICs is very complex, manufacturers can easily develop their own variations by slightly changing solid-state circuit structure and material. They do this since it is not only difficult to copy structures exactly with the same material (fabrication still has black magic aspects), but also desirable to avoid patents of other manufacturers.

The **resistor-transistor logic** (RTL) in Fig. 3.5.13 and the **diode-transistor logic (DTL)** in Fig. 3.5.14 were widely used before integrated circuits became popular. They are not used in new design, but are still manufactured for the replacement of components in existing digital systems. RTL has poor noise immunity and small fan-out. [Let us explain **noise immunity** for a DTL inverter, that is, a DTL gate with input x only, though the explanation can be applied to other logic families also. When input voltages p (for logic value 0) and q (for logic value 1) change to p' and q', respectively, by noise, the corresponding output voltages change to a' and b' from a and b, respectively, for the slowly changing transfer curve A in Fig. 3.5.15. Thus for large noise

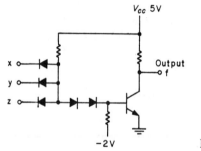

Fig. 3.5.14 DTL.

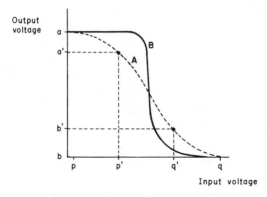

Fig. 3.5.15   Noise immunity improvement.

a' and b' may be confused, and the inverter may malfunction. A sharply changing transfer curve B greatly reduces this possibility, that is, B has better noise immunity.] DTL does not have these shortcomings but is slower than TTL (Millman and Halkias, 1972; Taub and Schilling, 1977; Garrett, 1970).

Problems

3.5.1
(a)   Develop a procedure to design a network in single-rail input logic for a given function f, using as few Schottky TRL gates as possible. Maximum fan-in is restricted.

(b)   Design a network by your procedure for (a) for the following function . Assume that maximum fan-in is 2.

$$f = \overline{x \vee y \vee uv}$$

3.5.2
(a)   Convert the J–K master–slave flip-flop in Fig. P3.5.2 into an I^2L network (i.e., in the format of Fig. 3.5.6b).

(b)   Then draw a layout of the network obtained in (a). Exact dimensions are not necessary.

3.5.3
(a)   Design an I^2L network in single-rail input logic by Procedure 3.5.1 for the following function, with as few gates as possible. Assume that external inputs can be Wired-ANDed with any other external inputs or gate outputs.

$$f = x(y\overline{z} \vee \overline{y}z)$$

(b)   Draw a layout for the I^2L network designed for (a). Exact dimensions are not necessary.

3.5.4   Solve Problem 3.5.3 assuming that Wired-AND cannot be used for external inputs. [*Hint:* design a NAND network with only complemented input variables.

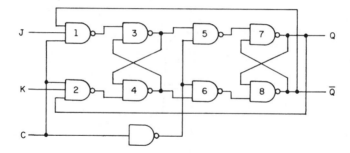

Fig. P3.5.2   J–K master–slave flip-flop.

[This can be done by modifying the map-factoring method (Muroga, 1979).] Then convert it into an I^2L network.]

3.5.5 Solve Problem 3.5.3 assuming that both complemented and noncomplemented variables are available as the network inputs.

3.5.6 An I^2L network can be interpreted as a network of NOR gates, as illustrated in Fig. P3.5.6. The I^2L network in Fig. P3.5.6a can be rewritten as Fig. P3.5.6b, splitting multicollector T_2 into single-collector transistors. Regarding each dotted rectangle in Fig. P3.5.6b as a NOR gate, we have the NOR network in Fig. P3.5.6c.

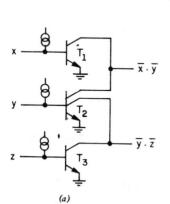

(a)

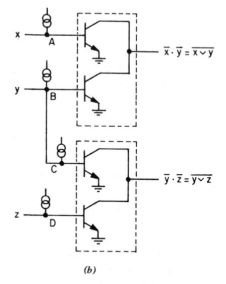

(b)

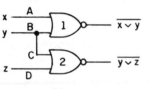

(c)

Fig. P3.5.6

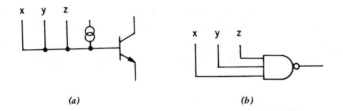

Fig. P3.5.7 (*a*) I²L gate. (*b*) Other logic-family gate.

(a) If we want to design an I²L network with a minimum number of gates as the primary objective and a minimum number of connections as the secondary objective, develop a procedure to design the corresponding NOR network such that reversal of the conversion in Fig. P3.5.6 leads to such an I²L network. What should be minimized in this NOR network?

(b) Design an I²L network for the following function by your procedure developed for (a).

$$f = x\overline{u}(\overline{y} \vee z) \vee \overline{x}u(y \vee \overline{z})$$

(c) Discuss the advantages and disadvantages of your procedure for (a), compared with Procedure 3.5.1.

3.5.7 In the case of a layout for I²L, inputs x, y, and z at distant places can be connected to a single trunk line (Fig. P3.5.7a), whereas in the case of a layout for other logic families, all these inputs must have long lines to the gate, as shown in Fig. P3.5.7b. In this sense I²L has the advantage of small areas for input connections, but it has the disadvantage of large areas for output connections (i.e., separate collector electrodes are required, as shown in Fig. 3.5.3).

(a) Does the entire I²L network have an advantage or disadvantage over the other logic families, counterbalancing the above advantage and disadvantage?

(b) Discuss what types of networks (or network configurations) are appropriate for I²L in this regard.

3.5.8 Given the layout for a network consisting of six I²L gates, as shown in Fig. P3.5.8, try to reduce the height by permuting the positions of these gates. It is assumed that in each rectangle of gate, collectors and a base must be placed at equal distances. Thus the height of the network layout in Fig. P3.5.8 is five.

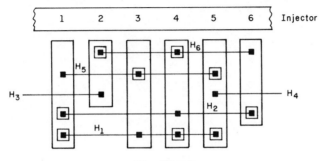

Fig. P3.5.8

3.5.9 Determine the output function f of the RTL gate in Fig. 3.5.13 by writing a truth table based on the on–off states of currents through the transistors.

3.5.10 Determine the output function f of the DTL gate in Fig. 3.5.14 by writing a truth table based on the on–off states of currents through the transistor and diodes.

3.6 Chapter Summary

In this chapter (and also in later chapters) it is seen that logic design, electronic circuit design, and layout design are highly interactive. For each logic family, logic networks must be modified because of different constraints due to electronic circuits, and electronic circuits must be modified further because of different constraints due to layouts. Yet we cannot make layouts directly, skipping logic design and electronic circuit design, because of the enormous complexity if a chip contains a few hundred gates or more. Thus after architecture design and the placement of networks in different areas on a chip, we have to go through logic design, electronic circuit design, and then layouts, in this order, and in each step previous designs must be modified.

Generally the integration size which can be achieved with bipolar logic families is not as large as that of MOS logic families to be discussed in the next chapter. The integration size decreases in this order: I^2L, TTL, ECL. The integration size of ECL is smallest because of heat dissipation. The integration size of TTL is greater because of less heat, but it is still restricted by its gate size. The integration size of I^2L is far greater, but still smaller than that of MOS logic families because the structure of I^2L is more complex than that of MOS, and consequently it has a lower yield. (A precise comparison is difficult since maximum integration sizes attempted have not been reported.)

CHAPTER 4

Logic and Mask Designs
for MOS Logic Families

The largest integration size can be obtained by MOS, because the TTL and ECL integration sizes are limited by the heat generated by their heavy power consumption and also by large gate sizes. (I^2L does not have these problems, but it is not used in chip sizes as large as those of MOS because of rather slow speeds and low yield.) So LSI or VLSI usually means MOS. Numerous new products, such as electronic calculators and games, could not exist if MOS had not been developed. In this chapter, logic and mask designs of MOS logic families are discussed.

4.1 MOSFET (Metal-Oxide Field Effect Transistor)

In this section we outline the basic mechanism of MOSFETs and their circuits.

Carriers

The p-channel MOSFET discussed in Section 2.1 is, more precisely speaking, a **p-channel enhancement-mode MOSFET** (or p-channel enhancement MOS). When voltages are applied to the drain, source,* and gate of a p-channel enhancement MOSFET, as shown in Fig. 4.1.1a, a thin channel for holes is formed underneath the gate, and a current which consists of holes moving from the source to the drain flows. When the voltage at the gate becomes more negative, the channel becomes denser, and a greater current flows. The channel has a thickness of less than 100 Å. (In contrast, the insulation layer between gate and substrate has a thickness of the order of 100 Å.) The stream of holes is restricted to a very thin layer close to the surface, unlike the bipolar transistor where the stream of holes and electrons flows in a much deeper place or, so to speak, in a three-dimensional manner. When the voltage at the gate increases toward a positive value, the channel

---

* Drain and source are interchangeable. Either p-diffusion region in Fig. 4.1.1a can be used as a drain and the other p-diffusion region as a source.

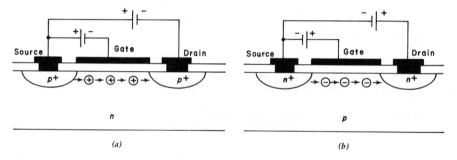

Fig. 4.1.1 Enhancement-mode MOSFETs. (*a*) *p*-Channel enhancement-mode MOS-FET. (*b*) *n*-Channel enhancement-mode MOSFET.

gets thinner, and beyond a certain value, which is called **threshold voltage**, the channel is not completely formed and no current flows.

When *p* and *n* in Fig. 4.1.1*a* are exchanged, we have an **n-channel enhancement-mode MOSFET** and the corresponding carrier movement, where holes are replaced by electrons, as shown in Fig. 4.1.1*b*. In other words, an *n*-channel is formed by the positive voltage at the gate which exceeds the threshold voltage, and a current flows in Fig. 4.1.1*b*. (Notice that the direction of the current is the same as that of the hole movement, but opposite to that of the electron movement.)

In each of the *p*-channel and *n*-channel enhancement MOSFETs only one type of carrier contributes to the current flow (i.e., holes in Fig. 4.1.1*a* and electrons in Fig. 4.1.1*b*), unlike with bipolar transistors where both holes and electrons contribute to the current flow.

References: Schilling and Belove, 1979; Richman, 1973; Streetman, 1980.

Comparison of p-MOS and n-MOS

n-Channel MOSFETs have some inherent performance advantages over *p*-channel MOSFETs. The mobility of electrons, which are carriers in the case of an *n*-channel MOSFET, is about 2.5 times greater than that of holes, which are the carriers in a *p*-channel MOSFET. Thus an *n*-channel MOSFET is faster than a *p*-channel MOSFET (though other limiting factors tend to dilute the 2.5 times factor) (Puckett and Lattin, 1973; Markkula, 1973).

Since *n*-channel enhancement MOSFETs are more often used than *p*-channel enhancement MOSFETs, an older technology, we will consider the former throughout the rest of this book, unless otherwise noted. [Since *p*-MOS technology is well established and less expensive than *n*-MOS, *p*-MOS is still used despite its lower performance (*Electronics,* Dec. 20, 1979, p. 34).]

Inverter Circuits

Suppose that the source of an *n*-channel enhancement-mode MOSFET is grounded and the drain is connected to the power supply of 5 volts through

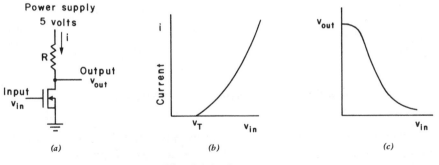

Fig. 4.1.2   Inverter.

resistor R, as illustrated in Fig. 4.1.2a. (The substrate is usually grounded or connected to a constant bias voltage, and this is denoted by connecting the end of the arrow in the symbol for MOS to the ground, as shown in Fig. 4.1.2a. Henceforth this substrate connection will not be shown in figures, since this is not directly involved in logic operations.) When the input voltage v_{in} increases from 0 volt, the current i, which flows from the power supply to the ground through R and the MOSFET, increases as shown in Fig. 4.1.2b, but for v_{in} smaller than the threshold voltage v_T, essentially no current flows. Then because of the voltage drop across R, the output voltage v_{out} decreases, as shown in Fig. 4.1.2c. Since we use binary logic, we need to use only two different voltage values, say 0.2 and 5 volts. If v_{in} is 0.2 volts, no current flows from the power supply to the ground through the MOSFET and v_{out} is 5 volts. If v_{in} is 5 volts, the MOSFET becomes conductive and a current flows. v_{out} is 0.2 volts because of the voltage drop across R. Thus if v_{in} is a low voltage (0.2 volts), v_{out} is a high voltage (5 volts); and if v_{in} is a high voltage, v_{out} is a low voltage, as shown in Table 4.1.1a. If low and high voltages represent logic values 0 and 1, respectively, in other words, if we use **positive logic**, Table 4.1.1a is converted to the truth table 4.1.1b. (When low and high voltages represent 1 and 0, respectively, this is said to be in **negative logic**.) Thus if v_{in} represents switching variable x, output v_{out} represents function $\bar{x}$. This means that the electronic circuit in Fig. 4.1.2a works as an **inverter**.

In this case, if we want to have a sufficiently low voltage at v_{out} (as low as 0.2 volts), representing logic value 0, R must be much greater than the resistance of the MOSFET when the MOSFET is conductive. (Otherwise

Table 4.1.1   Truth Table for the Circuit in Fig. 4.1.2

(*a*) VOLTAGE RELATION		(*b*) TRUTH TABLE	
INPUT v_{in}	OUTPUT v_{out}	v_{in}	v_{out}
Low voltage	High voltage	0	1
High voltage	Low voltage	1	0

Power supply

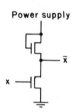

Fig. 4.1.3   Inverter circuit with MOSFET load.

the voltage swing can be easily masked by noise and the correct logic operation $\bar{x}$ is not performed.) By calculation, R must be on the order of 20 to 100 kΩ, requiring a large area on a chip, if it is realized by a diffusion area in the substrate (as shown in Fig. 1.2.7). Thus the resistor R is usually* replaced by an n-channel enhancement MOSFET whose gate is connected to its power supply, as shown in Fig. 4.1.3. This MOSFET occupies the chip area of only $\frac{1}{20}$ or less the area of resistor R realized by diffusion, so the circuit in Fig. 4.1.3 is much more compact than that in Fig. 4.1.2a.

By connecting many n-channel enhancement MOSFETs, we can realize any negative function. For example, if we connect three MOSFETs in series, including the one for resistor replacement, as shown in Fig. 4.1.4, the output f realizes the NAND function of variables x and y. Only when both inputs x and y have high voltages, two MOSFETs for x and y become conductive and a current flows through them. Then the output voltage is low. Otherwise at least one of them is nonconductive and no current flows. Then the output voltage is high. This relationship is shown in Table 4.1.2a. In positive logic this is converted to the truth table 4.1.2b, concluding that the circuit represents $\overline{xy}$. Other examples are shown in Figs. 4.1.5 through 4.1.7.

The MOSFET that is directly connected to the power supply as a replacement of resistor R in Fig. 4.1.2a is called **load** or **load MOSFET** in each of Figs. 4.1.4 through 4.1.7. Other MOSFETs that are directly involved in logic operations are called a **driver** or **driver MOSFETs** in each of these circuits.

The switching function f for the output of each of these circuits can be obtained as follows. Calculate the transmission of the driver, regarding each

---

* As we will see later, the resistor is often realized by a polysilicon strip atop the insulation layer in the case of memories, in order to reduce the area further. [See, for example, *Electronics* (June 21, 1979, p. 66).]

Power supply

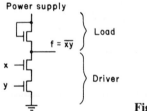

Fig. 4.1.4   NAND.

Table 4.1.2   Truth Table for the Circuit in Fig. 4.1.4

(*a*) VOLTAGE RELATION			(*b*) TRUTH TABLE		
x	*y*	*f*	*x*	*y*	*f*
Low	Low	High	0	0	1
Low	High	High	0	1	1
High	Low	High	1	0	1
High	High	Low	1	1	0

MOSFET as a make-contact of relay, and then complement it. [See Muroga (1979, Chapter 2).] For example, the transmission of the driver in Fig. 4.1.5 is $x \vee y$. Then by complementing it, we have the output function $f = \overline{x \vee y}$. Thus **a MOS circuit expresses a negative function of input variables connected to driver MOSFETs.** The circuit in Fig. 4.1.6 is constructed by repeating series or parallel connections of MOSFETs in its driver. The MOSFETs for z, u, and v are connected in parallel. This is connected in series with the MOSFET for y. Then this is connected in parallel with the MOSFET for x. Such a circuit is often called a **series–parallel circuit.** The circuit in Fig. 4.1.7 is not constructed by repeating series or parallel connections. Such a circuit is often called a **bridge circuit** or, more precisely, a **non-series–parallel circuit,** and can express a complex switching function. (A bridge circuit is used in some microprocessor chips such as Intel's 8080.)

A circuit that consists of one load MOSFET and driver MOSFETs connected to it, such as each of the circuits in Figs. 4.1.4 through 4.1.7, is called **a MOS cell, MOS gate** or **MOS logic gate.** (If we want to differentiate a MOS gate from MOSFET's gate (shown in Fig. 4.1.1), the MOS gate here should be called a MOS logic gate and the MOSFET's gate should be called a MOSFET gate. Usually the meaning is clear from the context.)

D-Load MOS Cell (D-MOS Cell or E/D-MOS Cell)

All the MOSFETs in the electronic circuits in Figs. 4.1.4 through 4.1.7 are enhancement mode. But if a load MOSFET is replaced by a depletion-mode MOSFET discussed in the following, still realizing all driver MOSFETs with

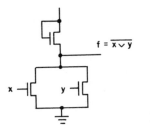

$f = \overline{x \vee y}$

Fig. 4.1.5   NOR.

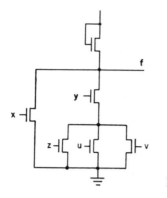

Fig. 4.1.6   Series–parallel circuit.

enhancement-mode MOSFETs, the chip area is further reduced, and the speed is improved also, though one extra mask is required and power consumption may increase.

The *n*-channel **depletion-mode MOSFET** is different from the *n*-channel enhancement-mode MOSFET in having a channel by forming a thin *n*-type layer embedded (usually by ion implantation through a thin silicon oxide layer, before forming the gate) underneath the gate, as illustrated in Fig. 4.1.8. When a positive voltage is applied at the drain (against the source), a current flows through this channel even if the voltage at the gate is 0 volt (against the source). As the gate voltage becomes more positive, the channel becomes denser, carrying a greater current. Or, as the gate voltage becomes more negative, the channel becomes thinner, carrying a smaller current. If the gate voltage decreases beyond threshold voltage v_T, no current flows. This relationship between the gate voltage v_{GS} (against the source) and the channel current i_D is shown in Fig. 4.1.9*a* as compared with that for the *n*-channel enhancement-mode MOSFET shown in Fig. 4.1.9*b*.

A D-load *n*-MOS cell is an electronic circuit where *n*-channel enhancement-mode MOSFETs are used in the driver and an *n*-channel depletion-mode MOSFET is used as a load. Fig. 4.1.10 shows an inverter realized with a D-load *n*-MOS cell. Henceforth let us denote the depletion-mode MOSFET by the MOSFET symbol with double lines. Notice that the MOS-

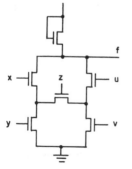

Fig. 4.1.7   Bridge circuit.

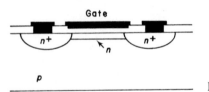

Fig. 4.1.8 *n*-Channel depletion-mode MOSFET.

FET gate of the load MOSFET is connected to the output terminal instead of the power supply.

When v_{in} changes from a high voltage to a low voltage in the circuit in Fig. 4.1.10, the output voltage v_{out}, labeled *D* in Fig. 4.1.11*a*, increases faster than in the circuit with a resistor load (Fig. 4.1.2*a*) labeled *R* or in the circuit with an enhancement-mode load (Fig. 4.1.3) labeled *E*, since it supplies a constant current i_D for a wide range of voltages, as shown by curve *D* in Fig. 4.1.11*b*. Thus the D-load MOS cell in Fig. 4.1.10 operates faster than the resistor load or the enhancement-mode load MOS cells. (The speed improves by several times for the same power consumption.) The voltage swing for the D-load MOS cell is greater than that for the enhancement-mode load as illustrated in Fig. 4.1.11*c*, because in the latter case there is a small voltage drop across the load MOSFET when the driver is nonconductive. The output voltage v_{out} is almost equal to the power supply in the case of a D-load MOS cell.

Furthermore the **transfer curve**, that is, the relationship between v_{in} and v_{out}, changes more sharply than for the other two cases, as shown in Fig. 4.1.11*c*, improving the noise immunity (*Electronics,* Feb. 15, 1971, p. 81).

The load realized by the depletion-mode MOSFET occupies only 40% of the chip area occupied by the enhancement-mode MOSFET. Also a D-load MOS cell works even if the power supply voltage fluctuates, because a constant current flows through the D-load MOSFET, as illustrated in Fig. 4.1.11*b*.

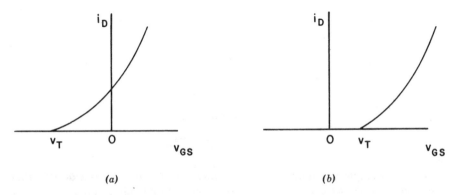

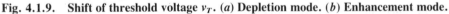

Fig. 4.1.9. Shift of threshold voltage v_T. (*a*) Depletion mode. (*b*) Enhancement mode.

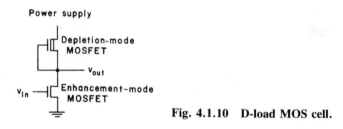

Fig. 4.1.10   D-load MOS cell.

For the reasons given above, D-load MOS cells are widely used.

References: Hoffman, 1971; McKenny, 1971; Cole, 1976; Knepper, 1978.

Speed Limit of MOSFETs

There are many factors that limit the speed of a MOSFET.

Parasitic capacitances and resistances of a MOSFET are illustrated in Fig. 4.1.12a. Because of these parasitics, channel current and consequently output voltage change slowly, as shown by the solid line in Fig. 4.1.11a, instead of the change indicated by the dashed line when an input voltage which changes instantly is applied at the MOSFET gate. In other words, when the input voltage at the MOSFET gate changes, parasitic capacitances

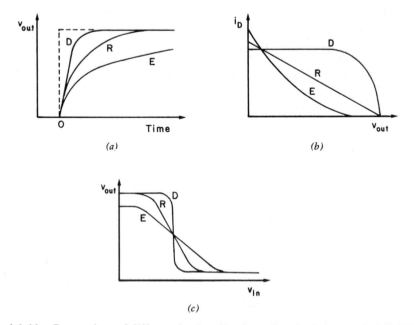

Fig. 4.1.11   Comparison of different load realizations. *D*—depletion-mode MOSFET; *R*—resistor; *E*—enhancement-mode MOSFET. (*a*) Output voltage rise. (*b*) Current. (*c*) Relationship between input and output voltages.

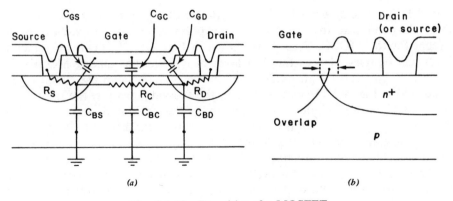

Fig. 4.1.12 Parasitics of a MOSFET.

(the subscript B in C_{BS}, for example, stands for bulk, which means substrate) must be charged, or discharged, through parasitic resistances. The greater these parasitic capacitances or resistances are, the slower the charging or discharging.

Gate electrode and n^+ diffusion regions of an n-channel MOSFET (or p^+ -diffusion regions in the case of a p-channel MOSFET) must have a small overlap, as illustrated in Fig. 4.2.12b, because otherwise a channel cannot be formed between the drain and the source throughout the entire area underneath the MOSFET gate, even when a sufficiently positive voltage is applied at the MOSFET gate. But the parasitic capacitances C_{GD} and C_{GS} due to the necessary overlaps can be large. They increase as the overlaps increase. Because of mask misalignment problems, the overlap is made with sufficient margin, but this adds more parasitic capacitance.

The parasitic capacitances of diffusion regions against the gate and the substrate, namely, C_{GD}, C_{GS}, C_{BD}, and C_{BS}, have significant values. (The parasitic capacitance C_{GD} is sometimes called **Miller capacitance**.)

These parasitics can be reduced by the self-aligned gate structure and the scale-down of the dimensions, as discussed in the following.

Self-Aligned Gate Structure

The overlap between the MOSFET gate and the diffusion region shown in Fig. 4.1.12b has a critical importance as follows. If the overlap is designed too small, no overlap may result by a misalignment of masks, and even if appropriate voltage is applied to the MOSFET gate, no channel is formed, making this MOSFET not workable. If the overlap is designed too large as a precaution against possible misalignment, parasitic capacitance C_{GD} becomes too large, slowing down the speed of the MOSFET.

A self-aligned gate structure solved this dilemma in the late 1960s, making the desired overlap accurately without using a mask. As illustrated in Fig. 4.1.13, a layer of polysilicon is deposited over the oxide insulation layer in

Fig. 4.1.13a. Using a mask which is opaque except for the dotted area in the right-hand side in Fig. 4.1.13b, the polysilicon layer is etched away except for the MOSFET gate, as shown in Fig. 4.1.13b. (This polysilicon layer can be used also as the first layer of connections among MOSFETs since it is covered by an oxide layer in Fig. 4.1.13c, and the metal layer deposited in Fig. 4.1.13d can be used as the second layer.) At the same time the polysilicon and oxide layer areas above the drain and source are also etched away. Next the drain and source regions are diffused. In this case the silicon gate (shown by the dotted area) itself acts as a mask, determining the overlap between the silicon gate and the diffusion regions accurately, as shown in Fig. 4.1.13b. Thus we have a self-aligned gate structure. Another layer of silicon dioxide is deposited and the contact windows are etched

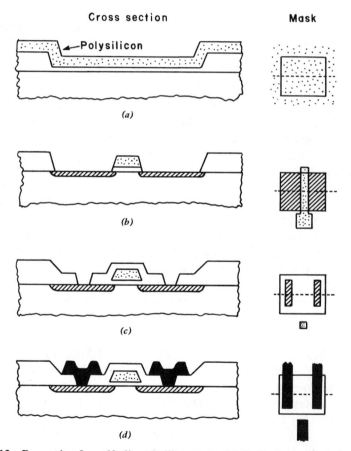

Fig. 4.1.13 **Processing for self-aligned silicon gate.** (a) Oxide is grown and polysilicon is deposited. (b) Boron is diffused, forming source and drain. (c) Sio$_2$ is deposited, and windows for contacts are masked and etched. (d) Aluminum is deposited, and second connection layer is defined.

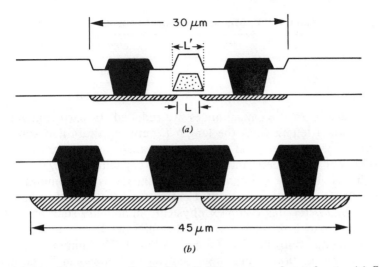

Fig. 4.1.14 Comparison of self-aligned silicon gate and metal gate. (*a*) Self-aligned silicon gate. (*b*) Metal gate.

away with a mask, as illustrated in Fig. 4.1.13*c*. Aluminum is then deposited as the second connection layer, as shown in Fig. 4.1.13*d*. Since a thin oxide layer underneath the gate is protected by the polysilicon gate, subsequent steps can be applied with higher yield during fabrication.

By reducing the parasitic capacitances C_{GD} and C_{GS} in this way, the speed is greatly improved. In addition the self-aligned silicon gate has the advantage of reducing the area for the MOSFET, as illustrated in Fig. 4.1.14. The area reduction is due to the elimination of the margin necessary for possible misalignment in the case of the metal gate structure discussed so far. Another advantage of the silicon gate is the ease of running metal connections over polysilicon connections which connect silicon gates. (These polysilicon connections and gates are formed simultaneously during fabrication.) Thus the self-aligned silicon gate structure is widely used, although it is more expensive to fabricate than the metal gate structure.

When the **silicon gate** is used, the name "metal-oxide semiconductor (MOS)," which means the arrangement of metal gate, oxide layer, and semiconductor, is no longer appropriate, although it is still in common use.

References: Faggin et al., 1968; Faggin and Klein, 1969, 1974; Clemens et al., 1975.

Scale-Down of a MOSFET

The scale-down of the dimensions of a MOSFET with a metal gate or a silicon gate (i.e., a self-aligned gate structure), along with the appropriate adjustment of the parameters, also improves the speed and power con-

Table 4.1.3   Minimum Channel Length

1973	1975	1977	1979
7.5 μm	6 μm	5 μm	3–4 μm

sumption because parasitic capacitances are reduced. In particular, reduction of the channel length (i.e., the length L between drain and source in Fig. 4.1.14a) increases the speed because the transit time of the carriers to cross the channel is reduced, and the parasitic capacitances C_{GC} and C_{BC} are also reduced. Usually the channel length chosen is the minimum with which a MOSFET works properly. If it is too short (estimated to be about 0.2 μm), the MOSFET has complex physical phenomena such as voltage breakdown, and **punch-through** (i.e., a current flows between source and drain without being controlled by the gate voltage). The minimum channel length used in production is becoming shorter, as shown in Table 4.1.3, improving the speed of the MOSFETs. It is important to notice that for the sake of convenience **the horizontal length L' of the silicon gate in Fig. 4.1.14a is usually called channel length by the manufacturers** because this length appears in a mask. In order to differentiate it from the real channel length L, this could be called a **mask channel length**. A scaled down n-MOS is usually called **high-performance MOS (HMOS)**, as announced in 1977 by Intel and improved in later years. Their features are shown in Table 4.1.4 (*Electronics,* July 7, 1977, p. 32; Aug. 18, 1977, pp. 92, 94–99; Oct. 27, 1977, p. 92; Feb. 16, 1978, p. 48; Sept. 13, 1979, pp. 124–128; Oct. 25, 1979, p.

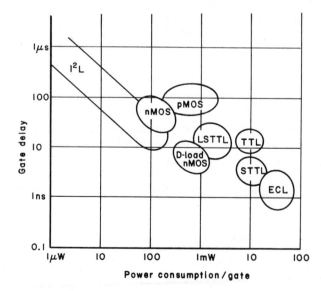

Fig. 4.1.15   Delay–power products.

Table 4.1.4 HMOS

PARAMETER	DEPLETION-MODE n-MOS (1976)	HMOS (1977)	HMOS II (1979)	HMOS III (1982)
Channel length (μm)	6	3	2	1.5
Gate delay (nsec)	4	1	0.4	0.2
Delay–power product (pJ)	4	1.0	0.5	0.25

105; Oct. 23, 1980, pp. 114, 116; *Intel Preview,* July/Aug., 1980, pp. 3, 4; Pashley et al., 1977; Liu et al., 1982).

In Fig. 4.1.15 the delay–power products of *p*-MOS and *n*-MOS with enhancement-mode loads and also of *n*-MOS with depletion-mode loads are compared with those of bipolar logic families.

4.2 Logic Design of MOS Networks

Since a MOS cell can express an arbitrary negative function and it is not directly associated with a simple switching expression such as a minimal sum, it is not a simple task to design a network with MOS cells so that the logic capability of each MOS cell to express a negative function is utilized to the fullest extent. Each MOS cell can contain as many as 40 MOSFETs if the speed is not important (and also if a large chip area is not a problem in the case of static MOS, as discussed in Section 4.3). In practice, for high speed a MOS cell that consists of at most 3 to 5 MOSFETs is used. For low speed more MOSFETs may be used, but not as many as 40. Let us consider the layout of a MOS cell. The layout of the NAND cell of Fig. 4.2.1*a* is shown in Fig. 4.2.1*b*, using a metal gate structure for simplicity. MOSFETs occupy only a small area of the entire layout (the dashed lines show thin

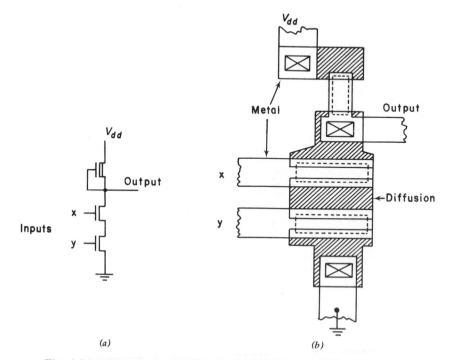

(a)   (b)

Fig. 4.2.1 Layout of a NAND cell. (*a*) NAND cell. (*b*) Layout of (*a*).

insulation layer areas above MOSFET channels), unlike with ECL and TTL gates where bipolar transistors cover significant areas in each gate layout (e.g., Fig. 1.2.11c). The major part of the layout is occupied by **intraconnections**, in other words, diffusion regions (shaded), metal connections, and contact windows. So in designing the layout of a MOS cell it is important to find an appropriate compact configuration by mainly considering intraconnection areas. When we consider the layout of the entire network of many MOS cells, **interconnections** among the cells occupy relatively greater chip areas than those among gates in TTL or ECL networks, since MOS cells are much smaller than ECL or TTL gates. MOSFETs alone probably occupy only 5 to 10% of the entire area.

Therefore if we consider the layout, the objectives in the logic design of MOS networks are not as simple as those for ECL or TTL networks where the minimization of the number of gates tends to minimize the layout size. However, it seems that minimization of the number of MOS cells tends to reduce the layout size only if an appropriate placement of the MOS cells, with appropriate shapes of MOS cells such that not many long interconnections appear, is found for a logic network. As the relationship between the compactness of the layout and a logic network is very complex, in addition to the complexity of logic design itself, there is probably no better guideline.

In this section a procedure is discussed to design a MOS network with each cell expressing a negative function. Then in Section 4.4 we discuss a procedure to design a MOS network with a mixture of NAND cells, NOR MOS cells, and slightly more complex cells, which is often used in industry. Then we compare it with the procedure described in this section.

Logic Design of MOS Networks, under the Assumption that Each MOS Cell Can Express a Negative Function

A logic gate whose output represents a negative function is called a **negative gate**. We now design a network with a minimum number of negative gates. A MOS cell is a negative gate. The **feed-forward network** shown in Fig. 4.2.2 (the output of each gate feeds forward to the gates in the next level) can express any loopless network. (For the reason, see Appendix.) Let us use Fig. 4.2.2 as the general model of a loopless network.

Procedure 4.2.1: Design of Logic Networks with Negative Gates in Single-Rail Input Logic

We want to design a MOS network with a minimum number of MOS cells (i.e., negative gates) for the given function $f(x_1, x_2, \ldots, x_n)$. (The number of interconnections among cells is not necessarily minimized.) It is assumed that only noncomplemented variables are available as network inputs. The

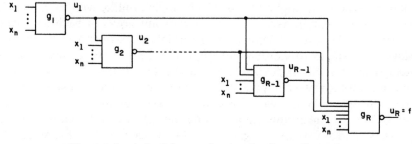

Fig. 4.2.2   A feed-forward network of negative gates.

network is supposed to consist of MOS cells g_i whose outputs are denoted by u_i, as shown in Fig. 4.2.2.

Phase 1

1.   Arrange all input vectors $V = (x_1, \ldots, x_n)$ in a lattice, as shown in Fig. 4.2.3, where the nodes denote the corresponding vectors in parentheses. White nodes, black nodes, and nodes with a cross ⊗ denote true vectors, false vectors, and don't-care vectors, respectively. The number of 1's con-

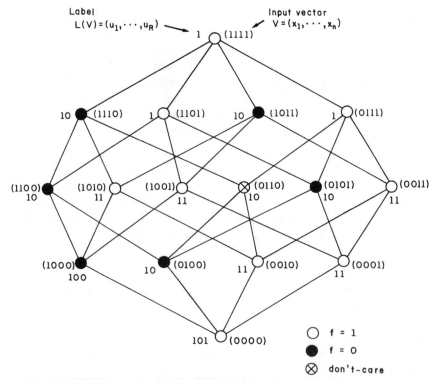

Fig. 4.2.3   Example for Procedure 4.2.1.

tained in each vector V is defined as the **weight** of the vector. All vectors with the same weight are on the same level, placing vectors with greater weights in higher levels, and every pair of vectors that differ only in one bit position is connected by a short line.

Fig. 4.2.3 is an example for a function of four variables.

2. We assign the label $L(V)$ to each vector $V = (x_1, \ldots, x_n)$ in the lattice in steps 2 and 3. Henceforth $L(V)$ is shown without parentheses in the lattice.

First assign the value of f to the vector $(11\ldots1)$ of weight n at the top node as $L(11\ldots1)$. If f for the top node is "don't-care," assign 0.

3. When we finish the assignment of $L(V)$ to each vector of weight w, where $0 < w \le n$, assign $L(V')$ to each vector V' of weight $w - 1$, the smallest binary number satisfying the following conditions: If $f(V')$ is not "don't-care,"

(a) The least significant bit of $L(V')$ is $f(V')$ (the least significant bit of $L(V')$ is 0 or 1, according to whether f is 0 or 1), and

(b) The other bits of $L(V')$ must be determined such that $L(V') \ge L(V)$ holds for every vector V of weight w which differs from V' in only one bit position.

If $f(V')$ is "don't-care," ignore (a), but consider (b) only. [Consequently the least significant bit of $L(V')$ is determined such that (b) is met.]

For the example we get $L(1110) = 10$ for vector $V' = (1110)$ because the last bit must be $0 = f(1110)$ by (a), and the number must be equal to or greater than 1 by (b). Also for vector $V' = (1000)$ we get $L(1000) = 100$ because the last bit must be $0 = f(1000)$ by (a), and the number must be equal to or greater than each of 10, 11, and 11 assigned by (b) to the three nodes (1100), (1010), and (1001), respectively.

4. Repeat step 3 until $L(00\ldots0)$ is assigned to the bottom vector $(00\ldots0)$. Then the bit length of $L(00\ldots0)$ is the minimum number of MOS cells required to realize f. Denote it by R. Make all previously obtained $L(V)$ into binary numbers of the same length as $L(00\ldots0)$ by adding 0's in front of them. For the example, $L(11\ldots1) = 1$ obtained for the top node is changed to 001 in Fig. 4.2.3.

Phase 2
Now let us derive MOS cells from the $L(V)$ found in Phase 1.

1. Set $i = 0$ (i is to be introduced in the following.)

2. Denote each $L(V)$ obtained in Phase 1 by $L(V) = (u_1, \ldots, u_R)$.
For each $L(V)$ which has $u_{i+1} = 0$ make a new vector $(V, u_1, \ldots, u_i)$.
For the example let us first explain the case of $i = 1$, instead of starting with the case of $i = 0$ which is somewhat special. For $u_2 = 0$ a new vector

(V, u_1), i.e., (11110), results from $L(1111) = 001$ because $V = 1111$ and $u_1 = 0$. Also a new vector (11010) results from $L(1101) = 001$, and so on.

3.   Find all the minimal vectors from the set of all the vectors found in step 2, where the minimal vectors are defined as follows. When $a_k \geq b_k$ holds for every k for a pair of distinct vectors $A = (a_1, \ldots, a_m)$ and $B = (b_1, \ldots, b_m)$, then the relation is denoted by

$$A > B$$

and B is said to be smaller than A. If no vector in the set is smaller than B, B is called a minimal vector of the set.

For the example, for $u_2 = 0$ we get (11110), (11010), (01110), (10001), and (00001) by step 2. Then the minimal vectors are (11010), (01110), and (00001).

For $u_1 = 0$, which is a somewhat specialized case, the minimal vectors are (0100), (0010), (0001) (i.e., new vectors consist of V's only).

4.   For every minimal vector make the product of the variables that have 1's in the coordinates of the vector, where the coordinates of the vector $(V, u_1, \ldots, u_i)$ denote variables $x_1, x_2, \ldots, x_n, g_1, \ldots, g_i$, in this order. For example, we form $x_1 x_2 x_4$ for vector (11010). Then make a disjunction of all these products and denote it by $\bar{g}_{i+1}$.

For the example we get

$$\bar{g}_2 = x_1 x_2 x_4 \vee x_2 x_3 x_4 \vee g_1$$

from the minimal vectors (11010), (01110), and (00001) for $u_2 = 0$.

5.   Repeat steps 2 through 4 for each of $i = 1, 2, \ldots, R - 1$. For the example, we get

$$\bar{g}_1 = x_2 \vee x_3 \vee x_4 \text{ and } \bar{g}_3 = x_1 x_3 x_4 g_2 \vee x_1 g_1 \vee x_2 g_2$$

6.   Arrange R MOS cells in a line and label them $g_1, \ldots, g_R$. Then construct each MOS cell according to the disjunctive forms obtained in steps 4 and 5, and make connections from other cells and input variables (e.g., MOS cell g_2 has connections from x_1, x_2, x_3, x_4, and g_1 to the corresponding MOSFETs of g_2, according to disjunctive form $\bar{g}_2 = x_1 x_2 x_4 \vee x_2 x_3 x_4 \vee g_1$). The network output realizes the given function f. For the example we get the network shown in Fig. 4.2.4.

Phase 3

The bit length R in label $L(00. . .0)$ for the bottom node shows the number of MOS cells in the network given at the end of Phase 2. Thus if we do not necessarily choose the smallest binary number in Phase 1, step 3, but choose a binary number still satisfying the other conditions in step 3, then we can still obtain a MOS network of the same minimum number of MOS cells as long as the bit length R of $L(00. . .0)$ is kept the same. (For the top node

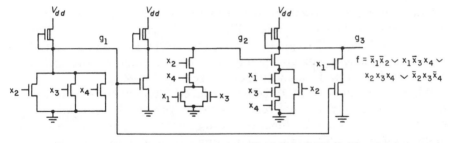

Fig. 4.2.4   MOS network based on the labels $L(V)$ in Fig. 4.2.3.

also we do not need to choose the smallest binary number as $L(11\ldots)$, no matter whether f for the node is don't-care.)

This freedom may change the structure of the network, though the number of cells is still the minimum. Among all the networks obtained, there is a network that has a minimum number of cells as its primary objective, and a minimum number of interconnections as its secondary objective. (Generally it is not easy to find such a network because there are too many possibilities.)   □

Although the number of MOS cells in the network designed by Procedure 4.2.1 is always minimized (see Appendix for the proof of why Procedure 4.2.1 works), the networks designed by Procedure 4.2.1 may have the following two problems: (1) The number of interconnections is not always minimized. (2) Some cells may become very complex so that these cells may not work properly with reasonable gate delay time. (As discussed in Section 4.3, the complexity of some cells is a serious problem for static MOS networks, though much less so for dynamic MOS networks as discussed in Section 4.8 or for CMOS discussed in Section 4.9.) If so, we need to split these cells into a greater number of reasonably simple cells, giving up the minimality of the number of cells. Also after designing several networks according to Phase 3, we may be able to find a satisfactory network. (See Remark 4.2.2 for other possibilities.)

Chip Area Minimization

The minimization of the number of MOS cells, which is the design objective in Procedure 4.2.1 (or further modification of designed networks considering constraints such as the simplicity of each cell and maximum fan-out), has the following justifications in the case of random-logic network. (This is much more so in the case of gate arrays discussed in Chapter 9 where enough routing channels are provided.) Here **random-logic networks** are logic networks of gates (or MOS cells) which are laid out most compactly by placing cells and connections freely, without necessarily using straight connections, after having designed logic networks suitable for such compact layouts. When we refer to its layout aspect, this is called **random-logic layout**. In contrast, in **regular-structured networks** (or **regular-structured layouts**) all connections are laid out straight like a mesh screen, such as memories discussed in Chapter 7 (e.g., Figs. 7.7.1 and 7.9.1). Regular structured networks are much less time-consuming to lay out than random-logic networks, at the sacrifice of performance or chip size.

When the number of MOS cells is reduced, the number of interconnections tends to be reduced, as verified at least with small networks. (For the majority of a few hundred functions tested, networks designed with a minimum number of gates as the primary objective and a minimum number of connections as the secondary objective, though the case is limited to NOR gates, are completely identical to those designed with a minimum number of connections as the primary objective and a minimum number of gates as the secondary objective, as discussed in Muroga and Lai (1976), along with theoretical considerations.) Since the majority of MOS cells are roughly equal in size, let us assume that every MOS cell occupies the same area. Then as the number of cells increases (and the number of interconnections tends to increase accordingly), the total area covered by these cells tends to increase linearly, and consequently interconnections get longer on the average because they need to span further. (In other words, when there are two networks which have the same number of connections but a different number of gates, the network with the greater number of gates tends to require a larger connection area than does the other network.) Therefore the minimization of the number of MOS cells tends to reduce the entire chip area, though there are exceptions (Muroga and Lai, 1976).

Even when a logic network with a minimum number of cells and a minimum number of interconnections is to be obtained, there are usually many logic networks with different configurations, realizing the given output functions with the same number of cells and interconnections. The area covered by the network greatly depends on the **network configuration**. (See Remark 4.2.2 for means to find different network configurations.) Furthermore even for a chosen network configuration, the area greatly depends on where on a chip we place MOS cells and their interconnections and also MOSFETs and their intraconnections in each MOS cell. This **placement problem** looks like a jig-saw puzzle, but its complexity is far greater. When a MOS network has a modular structure, its layout is easy because each module can be laid out without considering the rest of the network, and the layout of the module can be repeated without repeating the layout design effort if the module is laid out in repeatable shape (i.e., the module layout can be repeated without a gap between adjacent modules). Since each interconnection goes to adjacent modules only, each interconnection is short, keeping its parasitic capacitance at a minimum and consequently not increasing the gate delay of the MOS cell. For these reasons, networks designed in modular structure are preferred in order to reduce design time along with compact size, although the speed of the entire network is not necessarily faster than for more random-structured networks, as will be explained later.

When a number of networks are laid out on a chip, the placement of each network at an appropriate location is vitally important for the reduction of the entire chip area and the improvement of speed. If two networks which have a large number of connections between them are placed far distant, these connections become very long, covering a large area, with greater signal propagation time (and possibly larger MOS cells are required to deliver greater output power on these connections). If the networks which are inappropriately connected are designed with poor architecture, we may not be able to reduce the chip size or improve the speed, no matter how well the placement of the networks is done. In this sense the minimization of the areas of connections among the networks is highly dependent on the architecture design and is much more difficult than the minimization of connection areas inside each network or gate, which can be done locally. As the integration size of the chip increases, the areas of connections among the networks increase more sharply than those inside each network or gate, unless different architecture is chosen, possibly at the sacrifice of performance.

One rough estimate of different areas on an LSI chip of predominantly random-logic networks (designed in the late 1970's) is shown in Table 4.2.1, though the percentages may be different depending on the network types, designer skill, and integration size. The recent trend is toward a more frequent use of regular-structured layouts, consequently increasing the percentage of the connection areas, in order to save on layout time, though random-logic layouts are still used for key networks whose speed is vitally important for the entire chip performance.

Good experienced designers may be able to reduce the percentage of the interconnection area beyond the above figure, making the total area for the circuits really compact. The layout of Zilog's microprocessor chip Z8000, designed and highly packed by M. Shima, is shown in Fig. 4.2.5 as an example of a random-logic LSI chip.

In summary, although connections occupy large areas on a chip, the areas for intraconnections inside each MOS cell, interconnections among cells in each network, and interconnections among networks are each reduced by a different means. The area for intraconnections inside each cell can be reduced with almost no consideration of interconnections outside, once the output function for the cell to realize is determined. The area for intercon-

Table 4.2.1 Percentages of Areas on Random-Logic LSI Chip

ITEMS	PERCENTAGE OF ENTIRE CHIP AREA
MOS cells (intraconnections included)	30%
Interconnections among cells and among networks	40%
Power supply and ground lines	20%
Pads	10%

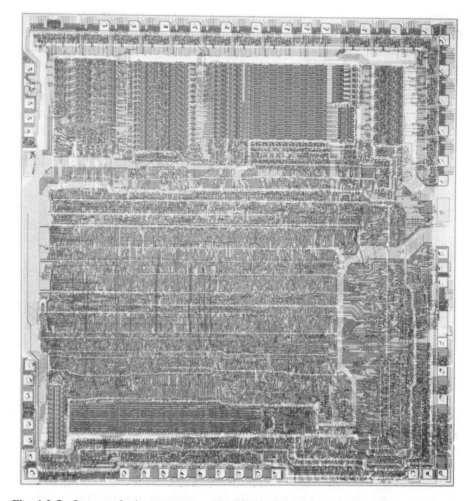

Fig. 4.2.5 Layout of microprocessor chip Z8000. (Reproduced by permission of Zilog, Inc. Z8000™ is a trademark of Zilog, Inc., with whom the author and publisher are not associated.)

nections among gates inside each network can be reduced by the minimization of the number of MOS cells (with the minimization of the number of connections as the secondary objective), considering an appropriate network configuration, though the layout shapes of individual networks are also important, as discussed in Section 4.6.5. Since there is no good procedure known to design a network with a minimum connection area, the minimization of the number of cells still appears to be a good design objective. In this case, after design, modification of the networks is usually necessary in order to incorporate constraints due to electronic circuits and layouts and also to simplify the layout, along with gaining improved performance, by using appropriate electronic circuits, such as transfer gates to be discussed later. [The integer programming logic design method can design minimal networks, by incorporating some of the constraints, but is not a simple procedure (Muroga, 1970, 1971, 1979).] Lastly, the area for interconnections among networks can be reduced by a good architecture design and an appropriate placement of networks.

In addition to area minimization, minimization of the number of MOS cells has the advantage of reducing power consumption and failure rate (Remark 4.2.1).

Remark 4.2.1

1. Minimization of the number of MOS cells results in the minimization of the power consumption in the entire chip, since each cell consumes approximately equal power on the average. As an increasingly large number of MOSFETs are packed into a single chip every year, this is very desirable. When the integration size is small, this has not been a serious problem because MOS cells consume much less power than TTL or ECL gates. But when 100,000 MOSFETs or more are packed into a single chip, the total power consumption presents a real bottleneck in LSI/VLSI design. Also, if the number of MOS cells is reduced, the number of power supply and ground lines are proportionally reduced, reducing the area for connections and further making the chip smaller. Also, the reduction of power consumption lowers the capacity of a power supply which is usually expensive.

2. Since the number of cells and the number of connections tend to decrease simultaneously, the chip has a better yield in fabrication because, generally speaking, the fewer components a MOS circuit has, the higher the yield.

3. The reliability of a chip during use is improved because the fewer components the chip has, the lower the chances of failure. □

Remark 4.2.2: Although Procedure 4.2.1, for the sake of simplicity, is described with a minor modification based on Nakamura et al. (1972) and partly on Liu (1972, 1977) and Lai (1976), the design of MOS networks with negative gates was first discussed by Ibaraki and Muroga (1969, 1971). Though it is more complicated, it has the advantage of being easily extended to the minimization of the number of interconnections among gates after the number of gates has been minimized in the case of two level networks, as discussed in Ibaraki (1971). Liu developed a procedure to design a network where MOSFETs are distributed in all cells as uniformly as possible. This procedure was implemented in a computer program by Shinozaki (1972). Lai developed a powerful procedure to design a MOS network such that the number of cells is minimized and every interconnection is irredundant (i.e., if any interconnection is removed, the network output function will be changed) (Lai, 1976; Lai and Muroga, to be published). This procedure was implemented in a computer program by Yamamoto (1976) and Yeh (1977).

Methods to transform a MOS network into others which possibly have more desirable cell complexities were discussed by Culliney (1977). (Problem 4.5.9 is such an example.) There are heuristic procedures to interactively simplify complex cells in the networks designed by Procedure 4.2.1 (Shimizu and Muroga, to be published; Fiduccia and Muroga, to be published).

Design procedures of MOS networks in two levels and their properties are discussed in detail by Hu (1978) and Nakamura (1979).

MOS network design was also discussed in El-Ziq and Su (1979) and El-Ziq (1978).

Heuristic design procedures of NOR networks, which are called **transduction methods**, were developed to design large networks in reasonably short computation time, though the number of gates is not necessarily minimized (Muroga, Kambayashi, Lai, Culliney and Hu, to be published). Also transformations to convert given networks into others with different configurations were developed (Kambayashi, Lai and Muroga, to be published). Although these methods were developed for networks of NOR gates, they can be extended to the design of large MOS networks.   □

Remark 4.2.3:   Since aluminum has a much lower melting point than polysilicon, an aluminum layer is formed usually at the last stage during fabrication. Suppose that aluminum is used for more than one connection layer. If the second aluminum layer is to be made after covering the first aluminum layer by an insulation layer, that insulation layer must be made with low temperature, but this can be done only with lower yield than an insulation layer formed under high temperature. But when polysilicon or silicide is used for a connection layer, no such problem occurs.

In the case of ECL and TTL, only two metal layers (usually one for power supply and ground and the other for interconnections among logic gates) are used for connections, except for short crossunders. This is because diffusions and polysilicons, which have high resistance, are not appropriate as connections where ECL and TTL require heavy currents, and consequently we have to use metal layers as connections despite the above yield problem. [IBM's ECL gate array with three metal layers announced in 1979 is probably an exception (Blumberg and Brenner, 1979).]

In the case of MOS circuits, diffusion and polysilicon layers can be used as connections because heavy currents are not used in MOS circuits. Then a metal layer can be formed as the third layer at the last stage of fabrication, without lowering the yield. The availability of three layers contributes to the compactness of MOS circuits in general (*Electronic Design,* Mar. 29, 1979, pp. 76, 77, 79). In the experimental VLSI chip, Hewlett-Packard Corporation increased the number of connection layers to 4: two layers by tungsten, one layer by polysilicon, and one layer by diffusion (Mikkelson et al., 1981). More connection layers are desirable for the compact layout, but this is not easy because as the number of layers increases the surface becomes excessively rugged for the next layer.   □

4.3   Problems in Electronic Circuit Design

Logic design, electronic circuit design, and layout design are highly interactive because problems in one of them are imposed on the others as con-

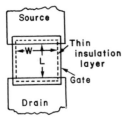

Fig. 4.3.1   MOSFET dimensions.

straints. So let us discuss key problems in electronic circuit design, now that we have some idea of how to design logic networks.

Size of a MOS Cell

High and low voltages at the output terminal of a MOS cell, which represent logic values 1 and 0, respectively, in positive logic (or 0 and 1, respectively, in negative logic), must have a sufficiently large difference in magnitude. This difference is called the **voltage swing.** If the voltage swing is too small, the next MOS cell may not be able to differentiate the two logic values 0 and 1 because of interference by noise, and the MOS network may malfunction. The size of MOSFETs and consequently the size of the MOS cell are determined such that we have sufficiently large voltage swing, as explained in the following.

The resistance value of a MOSFET, when it is conductive, is proportional to L/W, where the **channel length*** L is the distance between the source and drain diffusion regions, underneath the MOSFET gate, and the **channel width** W is the width of the thin insulation layer underneath the MOSFET gate, as illustrated in Fig. 4.3.1. The larger L, the larger the resistance; and the larger W, the smaller the resistance. (When L becomes too long with narrow W and the MOSFET does not fit a given space, the MOSFET can be bent like in Fig. 3.3.12. Then the resistance value at the corner should be estimated as about 60% of the value between the two opposing sides.)

Suppose that we want to design a layout for the MOS cell in Fig. 4.3.2. For some combinations of values of inputs x_1, x_2, x_3, and x_4 (e.g., for the combinations of $x_2 = x_4 = 1$ and $x_1 = x_3 = 0$) the driver becomes conductive, that is, a current flows from the power supply to the ground through the driver, providing the output a low voltage which represents logic value 0. **In each case where the driver is conductive for a different combination of input values of x_1 through x_4 the resistance value R_L of the load MOSFET M must be much greater than the resistance value of the driver.** This is because

---

* In the case of the silicon-gate MOSFET in Fig. 4.1.14, the silicon-gate length L' is often called channel length for the sake of convenience because it appears on a mask. Notice that L' is slightly longer than the real channel length L. Also in the case of metal-gate MOSFET, the gate length is longer than the real channel length L.

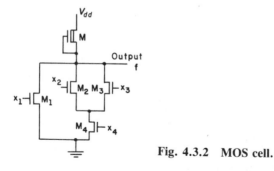

Fig. 4.3.2   MOS cell.

otherwise the output voltage at f becomes too high when it is supposed to be very low, and the voltage swing becomes too small. In the case of the inverter in Fig. 4.3.4, **the ratio of L/W for the load to L/W for any driver MOSFET is usually chosen to be 4 to 6 times. In this book let us use 6 times.** When the driver consists of many MOSFETs, the L/W ratio for each driver MOSFET must be proportionately lowered, as described in the following.

Suppose that we have $x_1 = 1$, and among the remaining inputs, at least $x_4 = 0$. Then we have a conductive path from the output terminal f to the ground, only through driver MOSFET M_1. Then the resistance R of M_1 must be much smaller than R_L from the above requirement. Next suppose that we have $x_2 = x_4 = 1$ and $x_1 = x_3 = 0$. Then we have a conductive path through M_2 and M_4 only. The total resistance value of M_2 and M_4 must be much smaller than R_L to meet the above requirement. Similarly, the total resistance value of M_3 and M_4 (corresponding to the combination $x_3 = x_4 = 1$ and $x_1 = x_2 = 0$) must be much smaller than R_L. If the resistance value of each of M_2, M_3, and M_4 is chosen to be $R/2$, where R is the resistance value of M_1 chosen sufficiently low compared with R_L, then the above requirement is satisfied. This means that if all M_1, M_2, M_3, and M_4 are to have the same minimum channel length L (we usually do this to keep the maximum speed with every MOSFET), then M_2, M_3, and M_4 must have W twice as wide as M_1, as illustrated in Fig. 4.3.3. If this makes W too wide, L of M_2, M_3, and M_4 must be shortened to reduce the channel resistance value. But the technically feasible minimum value is chosen as L (as discussed with Table 4.1.3), so a reduction of L is not possible, forcing us to limit the number of MOSFETs connected in series which become conductive for each possible combination of input variable values.

Notice that paths that do not become conductive for any combinations of input values, if any, need not be considered.

The above problem of low resistance requirement for driver MOSFETs is called the **high ratio problem** (or **requirement**). The high ratio problem is different for a network consisting of only resistors (not MOSFETs) where currents can go through all resistors simultaneously.

Suppose that the inverter in Fig. 4.1.10 is laid out as shown in Fig. 4.3.4:

Load MOSFET	10 μm wide, 10 μm long
Driver MOSFET	30 μm wide, 5 μm long

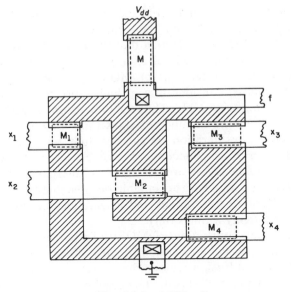

Fig. 4.3.3   MOS cell.

although many different dimensions are used in different cases of speed, power consumption, and shape. (For example, for low speed the channel width of these MOSFETs can be smaller, reducing the size of the entire circuit, as shown in the example of Fig. 4.4.3. The dimensions in Fig. 4.3.4 are for high speed.) The minimum channel length which can be feasible with current technology is usually chosen as L (e.g., 5 μm in the above example, which was the shortest feasible value some years ago), because the MOSFET can work fastest. The ratio L/W must be maintained constant if we want to

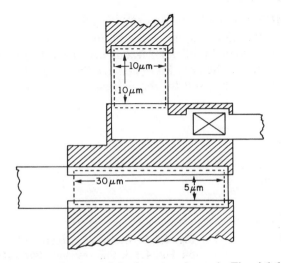

Fig. 4.3.4   Layout of the inverter shown in Fig. 4.1.10.

attain the same resistance value. (For example, if the length is doubled to $2L$, the width must be $2W$ to have the same resistance value, considerably increasing the MOSFET size. Notice that the larger MOSFET can handle greater power consumption without excessive temperature rise.) So when we determine the size of each driver MOSFET in each conductive case according to the above consideration of high ratio, this minimum channel length is used whenever possible. Thus the dimensions of MOSFETs for the MOS cell in Fig. 4.3.3 are obtained as shown in Fig. 4.3.5.

From the above consideration we can conclude that if many MOSFETs are connected in series, they become excessively large. If k MOSFETs are connected in series, the layout area of the driver increases roughly k^2 times. Many MOSFETs in parallel are acceptable because they do not present this problem. If k MOSFETs are connected in parallel, the layout area of the driver increases roughly k times only.

Also in the series connection of MOSFETs parasitic capacitances are greater than in the parallel connection because of a wider overlap between MOSFET gates and diffusion regions due to large channel width W. Thus even from the viewpoint of speed, the series connection in any path in a MOS cell is less desirable, as discussed in the following. Of course even in the case of parallel connection, if we connect too many MOSFETs in parallel, the parasitic capacitance at the output terminal becomes too large.

Usually the number of MOSFETs in series is limited to at most 3 for medium speed, and at most 2 in the case of high speed. The number of MOSFETs in parallel is limited to 8. (Because of this, the example network designed by Procedure 4.2.1 in Fig. 4.2.4 cannot be used for medium speed, and we need to convert the last MOS cell into one that contains fewer MOSFETs in series.)

Speed of MOS Circuits

The speed of MOS circuits is strongly affected by the parasitic capacitance, as illustrated in Fig. 4.3.6. Suppose that an inverter which consists of load MOSFET M_L and driver MOSFET M_D has its output line connected to the gate of MOSFET M_N in the next MOS cell, as illustrated in Fig. 4.3.6a. This MOS circuit can be expressed by the equivalent electronic circuit in Fig.

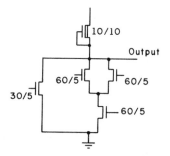

Fig. 4.3.5   Electronic circuit with dimensions for Fig. 4.3.3. Figures show W/L ratios.

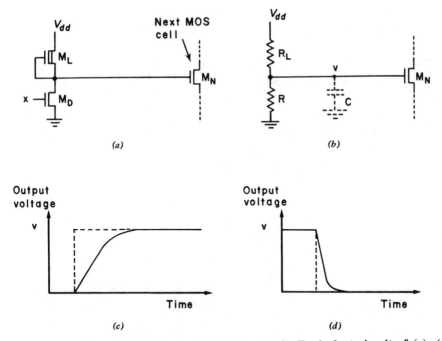

Fig. 4.3.6   Speed of MOS circuit. (*a*) **MOS cell.** (*b*) **Equivalent circuit of** (*a*). (*c*)
Charging. (*d*) **Discharging.**

4.3.6*b* with load resistance R_L, driver resistance R, and parasitic capacitance
C. (The connection has resistance, but this can be ignored unless the connection is long and narrow, made of polysilicon.) The parasitic capacitance
C is actually the sum of the following three parasitic capacitances:

C_D Parasitic capacitance* of diffusion regions (usually one common region, as seen in Fig. 4.3.4) of load MOSFET M_L and driver MOSFET
M_D against the substrate (or the ground). In other words, this is C_{BS}
of M_L and C_{BD} of M_D shown in Fig. 4.1.12. The smaller these diffusion
regions, the smaller C_D.

C_C Parasitic capacitance** of the connection line (metal layer or polysilicon layer) against the substrate (or the ground). This is smaller
than C_D or C_G, unless the connection line is long.

---

* The value of C_D is typically 0.05 pF (for a low-power MOS cell) to 0.5 pF (for a
high-power MOS cell), though it varies depending on technology. The value can be
5 pF or more for a very-high-power MOS cell which sends a signal off the chip.
** The value of C_C is typically 0.005 pF (for a very short metal connection) to 30
pF (for a very long connection), though it varies depending on technology. The value
is about 50% greater in the case of a polysilicon connection. (In contrast, a connection
line on a pc board is 50 pF to a few hundred pF.)

C_G Parasitic capacitance* of the gate of MOSFET M_N against the substrate** (or the ground), as shown in Fig. 4.1.12.

When the driver MOSFET M_D is conductive for $x = 1$, R is relatively small, and consequently the output voltage v is also almost 0 volt. When x changes from 1 to 0, M_D become nonconductive, and R becomes almost infinitely large (essentially open-circuited). Then the parasitic capacitance C is charged to voltage v by the power supply of voltage V_{dd} through load resistance R_L. Thus v changes gradually as shown by the solid line in Fig. 4.3.6c instead of the ideal response shown by the dashed line. The smaller R_L or C, the faster the response (i.e., the closer to the dashed line). Next when x changes from 0 to 1, M_D becomes conductive, and R becomes small (but not zero). Then the parasitic capacitance C is discharged through R. Thus v changes gradually as shown by the solid line in Fig. 4.3.6d, instead of the ideal response shown by the dashed line. The smaller R or C, the faster the response. Notice that the change is much faster than that in Fig. 4.3.6c, because R is much smaller than R_L.

When the MOS cell in Fig. 4.3.6a has output connection lines to m MOSFETs in the next MOS cells, C_G multiples by m and C_C increases in proportion to the total length. For a MOS cell which has short connections, C_D is dominant, unless the fan-out is large, and for a MOS cell which has a long connection, C_C is dominant. For a medium-speed MOS cell the values of C_D, C_C, and C_G are roughly comparable. For a high-speed MOS cell C_G is usually dominant, unless connection lines are long, and is typically 0.1 pF to 0.3 pF per connection.

When the parasitic capacitance is large (due to many fan-outs or very long connections), the response shown in Fig. 4.3.6c and d becomes slow, so the load resistance R_L must be reduced in order to keep the same response time. This means that the channel width of load MOSFET M_L must be increased, and then the channel width of driver MOSFET M_D must also be increased because of the high ratio requirement. This results in an increase of the size of the circuit, as can be seen from Fig. 4.3.4. (For example, if we want to improve the speed two times, the channel width of all the MOSFETs must be doubled, roughly doubling the size of the entire circuit.) But this increase of channel width results in an increase in the parasitic capacitances C_{BS} and C_{BD} (and others) shown in Fig. 4.1.12, offsetting the speed increase effort. If a significant improvement cannot be obtained, we need to redesign the logic network to overcome this difficulty. So in difficult cases we need to repeat the design of logic network, electronic circuit, and layout.

The determination of the parasitic capacitance C and the size of R_L and

---

* The value of C_G is typically 0.01 pF (for a small MOS cell) to a few pF (for a large MOS cell for high power), though it varies depending on technology.
** The input resistance of the MOSFET gate is very large and typically 10^{14} Ω or more.

R_D is usually done by repeatedly adjusting predetermined values, observing the waveforms generated by electronic circuit modeling programs such as SPICE2. But the details are omitted here, because they are beyond the scope of this book.

Crossover of Connections

When the crossover of two connections is realized by a diffusion region as shown in Fig. 1.2.6, a resistance of low value is essentially added. Also the parasitic capacitance increases, compared with that of a metal connection on the insulation layer. Thus an excessive number of crossovers is not desirable [though it does not matter if the total sum of resistances in one connection is small compared with the output resistance of the MOS cell (*Electronic Design,* June 22, 1972, p. 30)].

When two diffusions of two connections run in parallel and then a third metal connection runs perpendicular to them on the surface of the insulator layer, as shown in Fig. 4.3.7, an undesirable parasitic transistor, which is often called a **field transistor**, may be created. The conducting channel is formed by the voltage on the third metal connection. Thus two n^+-type diffusion regions and the channel form a parasitic MOSFET; or under certain conditions they work as a bipolar *n-p-n* transistor where the diffusion regions act as the emitter and the collector and the substrate acts as the base. Consideration of the property of these parasitic transistors is important in connection layout; it is included as sufficiently large spacing in the layout rules mentioned in Section 4.6, along with local doping and other precautions (AMI staff, 1972).

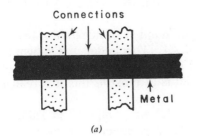

(a)

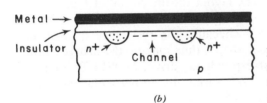

(b)

Fig. 4.3.7 **Parasitic MOSFET which may be induced by crossover of a connection over two other connections.** (*a*) **Top view.** (*b*) **Cross section.**

Distribution of Heat Generation

Power dissipation must not exceed about 10 μW per 100 μm^2 at any place within an LSI chip because of thermal resistance and the maximum temperature limit of about 150°C for reliable operation.

4.4   Logic Design of MOS Networks with a Mixture of NOR and NAND Gates

When the number of MOSFETs in series is 2 or 3 and the number of MOS-FETs in parallel is at most 8, the types of negative functions are very limited. Thus MOS networks are often first designed using only NOR and NAND gates, probably using NOR gates more often than NAND gates. This is also partly because designers are familiar with NOR and NAND gates, and checking is easier for them. After networks have been designed with a mixture of NOR and NAND gates, some cells (which realize NOR or NAND gates) are merged into fewer cells, which represent negative functions slightly more complex than NOR or NAND. Of course we may not be able to obtain the most compact networks by this approach.

In order to give some idea of logic design with negative gates (i.e., Procedure 4.2.1) and logic design with a mixture of NOR and NAND gates, let us design a full adder.

First a full adder with a minimum number of NOR and NAND gates as the primary objective and a minimum number of interconnections as the secondary objective is designed in Fig. 4.4.1a. [The minimality is proved by the integer programming logic design method (Liu et al., 1974).] Then the corresponding MOS network is obtained in Fig. 4.4.1b by converting each gate into a MOS cell.

Next a full adder with a minimum number of negative gates is designed in Fig. 4.4.2, based on Procedure 4.2.1. [This adder was shown in Spencer (1969).] Since the number of MOSFETs connected in series is at most 3 and the number of MOSFETs in parallel is at most 4, this is a realistic circuit for medium speed.

The circuits of Figs. 4.4.1b and 4.4.2 are compared in Table 4.4.1. Obviously the circuit shown in Fig. 4.4.2 is superior in every aspect. The number of levels is reduced from 5 to 3, but the circuit in Fig. 4.4.2 may not be that much faster because the switching times of some MOS cells may be slower than those in Fig. 4.4.1b. The layout for the circuit of Fig. 4.4.1b is shown in Fig. 4.4.3 (prepared by a student) and is for low speed, compared with those shown in later sections, because the resistance of each load MOSFET is high due to its slender shape, thus making the charge-up time of parasitic capacitance long. Correspondingly driver MOSFETs are small. (Full adder networks have many other variations, in particular those combined with transfer gates in Section 4.5.)

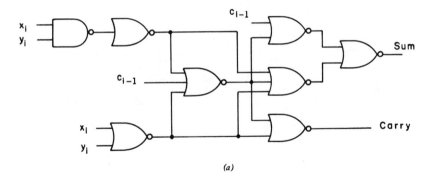

(a)

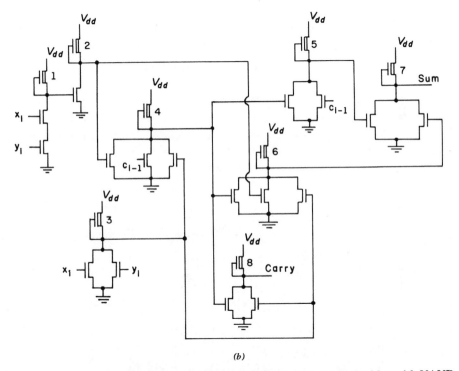

(b)

Fig. 4.4.1 Full adder designed with NOR and NAND gates. (*a*) Full adder with NAND and NOR gates. The number of gates is minimized as the primary objective and the number of interconnections is minimized as the secondary objective. (*b*) Full adder of MOSFETs converted from (*a*).

Table 4.4.1 Comparison of the Two Full Adder Circuits of Figs. 4.4.1*b* and 4.4.2

IMPLEMENTATION	NUMBER OF MOSFETs	NUMBER OF CELLS	NUMBER OF LEVELS	NUMBER OF CONNECTIONS	POWER CONSUMPTION
NOR and NAND	25	8	5	19	2
Negative gates	18	4	3	11	1

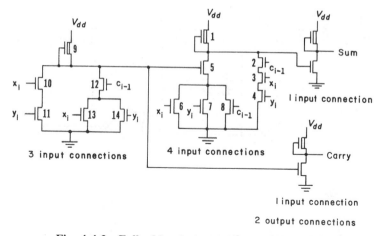

Fig. 4.4.2   Full adder designed with negative gates.

Since each MOS cell generally represents any negative function which is a much more complex function than those represented by NOR, NAND, OR, and AND gates, designers cannot quickly identify the functions which networks consisting of these MOS cells represent. Thus switching expressions are usually written along with the drawings of MOS cells and networks

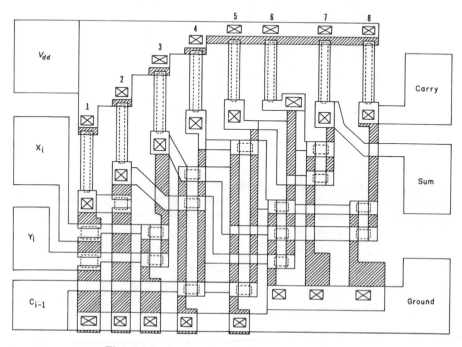

Fig. 4.4.3   Layout for the adder of Fig. 4.4.1b.

for convenience. Sometimes switching expressions that have an explicit correspondence to connection configurations are preferred rather than simplified switching expressions such as minimum sums.

4.5   Examples of Useful and Unconventional MOS Networks

Next we show some useful and unconventional MOS networks. An S–R latch, which is often used in sequential networks, is shown in Fig. 4.5.1. A D-type master–slave flip-flop is shown in Fig. 4.5.2; it is used often in clocked sequential networks.

In the case of TTL or ECL, complex functions can be realized by free connections of different points in electronic circuits, ignoring the gate concept. In the case of MOS we can also realize complex functions by simple circuits, if connections are freely made, ignoring the gate concept. (These functions require many more MOSFETs if they are designed based on gates.) They are useful in some applications, so let us show some examples. Notice that the following examples are not in the format of ordinary MOS cells as illustrated in Figs. 4.1.4 through 4.1.7 and Fig. 4.3.5.

The circuit in Fig. 4.5.3 realizes the even-parity function $\overline{x \oplus y}$. (Frei et al., 1980.) Notice that inputs x and y are connected to the sources of MOS-FETs 1 and 2, unlike all the previous circuits where inputs are connected to MOSFET gates only. Signal x at the source of MOSFET 1 is either sent to the drain or not, according to whether or not its MOSFET gate has a high positive voltage. Such a MOSFET (1 and also 2) is called a **transfer gate. The MOS networks which are designed by the procedures discussed so far can sometimes be simplified by the appropriate use of transfer gates, possibly along with improved performance.** Using transfer gates, a latch (more precisely speaking, a D latch to store data) can be constructed as shown in Fig. 4.5.4. When control input c is a high positive voltage, the feedback loop by a transfer gate with $\bar{c}$ is cut, and the input value is fed into the cascade

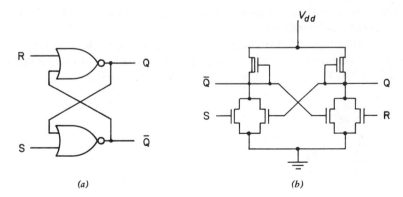

Fig. 4.5.1 S–R latch. (*a*) S–R latch. (*b*) MOS realization of (*a*).

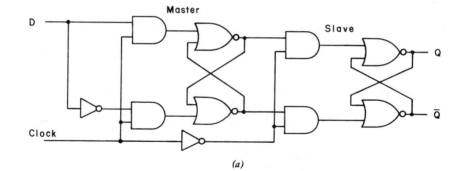

Master

Slave

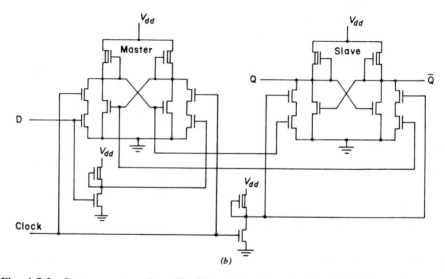

(a)

Fig. 4.5.2   *D*-type master–slave flip-flop. (*a*) Logic network. (*b*) MOS realization of (*a*).

of two inverters. When *c* becomes low voltage, the input is cut off and the loop retains the information.

The circuit in Fig. 4.5.5 realizes the parity function of three variables and its complement simultaneously. Notice that unlike all the previous circuits, some driver MOSFETs are shared by two load MOSFETs.

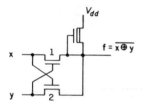

Fig. 4.5.3   **Parity circuit.**

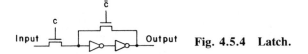

Fig. 4.5.4 Latch.

There are many other unconventional MOS networks. [For another example, see *Electronics* (Sept. 11, 1980, p. 121).] All these networks are useful for simplification of electronic realization of logic networks or for improvement of performance, though complex adjustments of voltages or currents are often required.

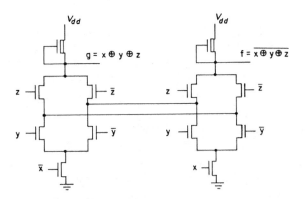

Fig. 4.5.5 **Circuit with complementary parity function outputs.**

Problems

4.5.1 By Procedure 4.2.1, design a MOS network with a minimum number of cells in single-rail input logic, for one of the following functions.

A. $f(x_1, x_2, x_3, x_4) = \Sigma(0, 3, 4, 7, 8, 9, 12, 13, 14)$

B. $f(x_1, x_2, x_3, x_4) = \Sigma(9, 10, 12, 15)$

4.5.2 Derive a MOS network different from that of Fig. 4.2.4 by deriving labels $L(V)$ different from those in Fig. 4.2.3 for the same function, according to Phase 3 of Procedure 4.2.1, without the restriction that $L(V')$ be the smallest binary number.

4.5.3 Derive an upper bound on the number of MOS cells required for a function of n variables by Procedure 4.2.1. Is it a tight bound? If there is a function that requires exactly this bound, show that function.

4.5.4 (a) Design a network with MOS cells for

$$f = x_1 \oplus x_2 \oplus x_3 \oplus x_4$$

using as few MOSFETs as possible.

(b) When designed by Procedure 4.2.1, examine whether some of $g_1, g_2, \ldots, g_{i-1}$ contained in the expression for some $\bar{g}_i$ can be deleted without

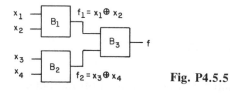

Fig. P4.5.5

changing the value of f. Then if some expressions are simplified, draw the corresponding network.

4.5.5 If we design a MOS network for the function

$$f = x_1 \oplus x_2 \oplus x_3 \oplus x_4$$

in three blocks, as shown in Fig. P4.5.5, where block B_1 has output $x_1 \oplus x_2$, block B_2 has output $x_3 \oplus x_4$, and block B_3 has inputs from B_1 and B_2, then we might be able to reduce the total number of MOSFETs, though the number of cells may increase. Discuss whether or not this is true.

4.5.6 Design a full adder network in single-rail input logic, using Procedure 4.2.1.

4.5.7 Explain why the circuits in Figs. 4.5.3 and 4.5.5 realize the functions shown.

4.5.8 Find what function the output of the circuit in Fig. P4.5.8 expresses. Explain why. Discuss whether the circuit works reliably.

4.5.9 When MOS cells $g_1, \ldots, g_r$ with driver configurations $T_1, \ldots, T_r$, respectively, each feeds to only a single MOSFET, and these r MOSFETs are connected in parallel in some place in the driver of a MOS cell h, as illustrated in Fig. P4.5.9a, then this parallel connection of MOSFETs can be replaced by a single MOSFET fed by a new MOS cell g_w whose driver is formed by a series connection of the configurations $T_1, \ldots, T_r$, as shown in Fig. P4.5.9b. For example, the MOS network in Fig. P4.5.9c can be converted to that in Fig. P4.5.9d without changing output function f. Prove that the network in Fig. P4.5.9a can be converted into that of Fig. P4.5.9b and vice versa.

Remark: When either Fig. P4.5.9a or Fig. p4.5.9b is simpler than the other, we can obtain a more desirable network by this transformation. For other transformations, see Culliney (1977). This transformation would be more useful for dynamic MOS networks since there the number of MOSFETs in series is less restricted than in static MOS networks.   □

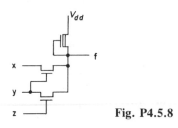

Fig. P4.5.8

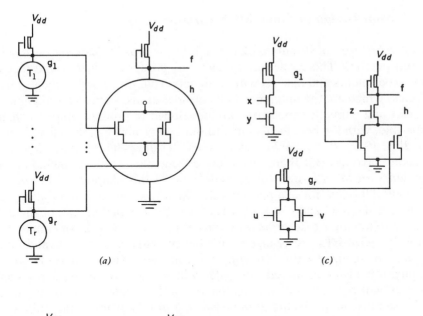

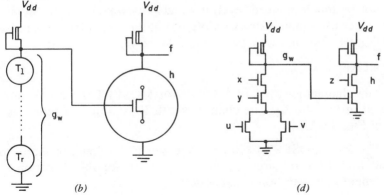

Fig. P4.5.9   Transformation

4.5.10 Suppose that MOS networks for high speed can be designed with MOS cells, where in each cell the number of MOSFETs in series, in each path in the driver, is at most 2 and the number of MOSFETs in parallel is at most 8, assuming that each cell is constructed by repeating series and parallel connection of MOSFETs.

(a) Discuss concisely what switching functions such a MOS cell can realize.

(b) Design a MOS network with a minimum number of cells for the following function f under the above constraint.

$$f = \overline{xyz \lor xyu \lor xzu \lor yzu}$$

4.6   Layout Design of Static MOS Circuits

Layout rules are constraints under which the layout is made by designers and draftpersons. Those rules are prepared by process engineers who repeat many experiments and discuss with designers what they want to have. Layout rules, which are the outcome of all considerations, such as misalignment of masks and parasitic transistors, are developed to produce chips as small as possible, with the best yields but without compromising reliability (Rung, 1981; Ipri, 1979). If we employ fewer rules, designers have greater freedom in making layouts. But there are a few dozen rules or more (including all constraints on MOS circuits discussed so far in this chapter). When there are about 100 rules, for example, it will take a few months until a designer remembers them all and can start a layout without violating any of them.

In the following we describe only typical rules for the sake of simplicity. (Here all MOSFETs are realized with the metal-gate structure, and connections are realized with diffusion or metal only, without polysilicon. If polysilicon is also considered, our rules will be more complex, and layout problems will be excessively time-consuming for students to solve. Also, for the small-scale problems to be presented in this chapter, the chip area would not be much reduced, even if we also used polysilicon. Metal gates are no longer used in industry very often, but the simplified rules presented here will give students some ideas about layout.)

Layout Rules

1.   The channel length L of every driver MOSFET is 5 μm or more, as illustrated in Fig. 4.6.1a. The channel width W is 10 μm or more, as illustrated in Fig. 4.6.1b.

2.   The load MOSFET is 10 μm wide and 10 μm long, or more. (Since the load MOSFET requires an additional mask for the depletion mode, misalignment consideration is necessary.)

3.   The metal overlap over the thin oxide layer (shown by the dashed line) must be 2 μm in all four edges, as illustrated by A in Fig. 4.6.1b.

4.   The width and spacing of the diffusion regions are shown as $B \geq 5$ μm and $C \geq 5$ μm, respectively, in Fig. 4.6.1b.

5.   The width of the metal connection line must be at least 5 μm, as shown by D in Fig. 4.6.1b. For a power supply connection the width must be greater, depending on the current magnitude. The spacing between metal connections must be at least 5 μm, as shown by E.

6.   The contact window shown by the dashed line in Fig. 4.6.1b must have a square shape of 5×5 μm, a rectangle of 4×6 μm, or a larger area.

7.   The metal layer margin around the contact window must be 2 μm in all directions, as shown by F in Fig. 4.6.1b. (This margin is necessary to

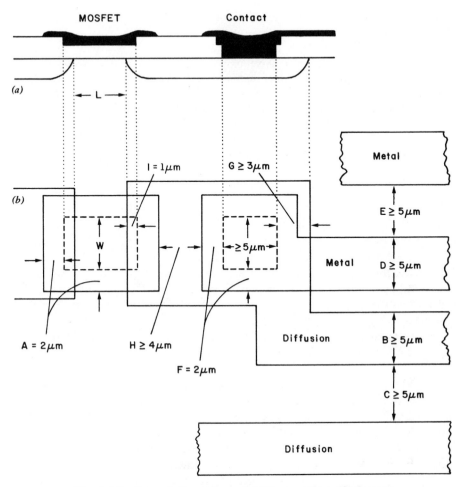

Fig. 4.6.1   Layout dimensions. (*a*) Cross section. (*b*) Layout.

have a good contact between the diffusion region and the metal layer which may be impaired by misalignment.)

8. The diffusion region must extend beyond the contact window by at least 3 μm, as shown by *G*. (This margin is necessary to prevent short-circuiting of the metal layer to the substrate by mask misalignment.)

9. The spacing between MOSFET gate and contact metal must be at least 4 μm, as shown by *H*.

10. Each pad must be at least 100 μm × 100 μm and must be at least 40 μm apart from any metal connection. Spacing between pads must be at least 90 μm.

Whenever a new processing technique is developed, layout rules need to be updated (Remark 4.6.1).

Remark 4.6.1:   In addition to shortening the channel length for the improvement of speed, efforts to reduce areas for connections have continued. As mentioned in Remark 4.2.3, the number of layers has been gradually increased. Also realization of contacts without peripheries (*G* and *F* in Fig. 4.6.1) have been tried, such as so-called **external contacts** (Mikkelson et al., 1981) and **self-aligning contacts** (Hosoya et al., 1981). This will reduce areas required for the peripheries and consequently the entire chip area.   □

General Principles in Layout Design

The entire chip is divided into many areas such that the total chip area is minimized, keeping the highest speed, where a different network such as a register, adder, or control logic is assigned to each area. (For example, if two networks which have a number of interconnections between them are assigned to distant areas, the chip area increases, and the speed is lowered.) This is called a **chip plan** or **floor plan**. Estimation of the location, area size, and shape of each area as accurate as possible is vitally important, because if some intended networks do not fit snugly in the allocated areas, the designers must repeat time-consuming layout with a new chip plan. Then we design each areas as follows.

1.   Whenever possible, run diffusion or polysilicon connections vertically and run metal connections horizontally, though not necessarily straight. Power supply and ground connections are usually realized in metal and run in parallel with other metal connections.

In this way, MOS cells, each of which has the shape of a vertically long rectangle (shown by the dashed line) are arranged horizontally, as illustrated in a schematic diagram in Fig. 4.6.2. Actually, metal layer and diffusion regions can be arranged in many different ways, as detailed layout examples in Figs. 4.6.3 through 4.6.5 show. Here, (*a*) and (*b*) in each figure show a

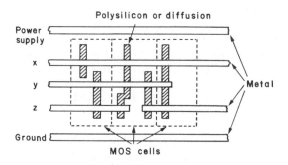

Fig. 4.6.2   General layout principle.

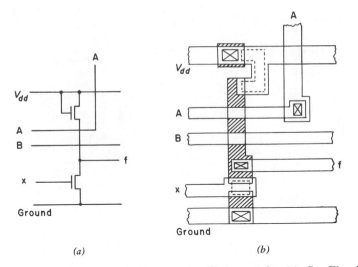

(a)   (b)

Fig. 4.6.3 **Layout example. (*a*) MOS circuit. (*b*) Layout for (*a*). See Fig. 4.6.5*c* for legend.**

circuit and its layout, respectively, and the legend for all these figures is shown in Fig. 4.6.5*c*.

Long connections which span many MOS cells are realized by metal connections which have essentially no resistance. Since polysilicon and diffusion connections have higher resistance, they are generally used for short distances only (vertically) so that their resistances will not be a problem. Connection dimensions, as well as transistor dimensions, which can be used are determined with the aid of electronic circuit modeling programs.

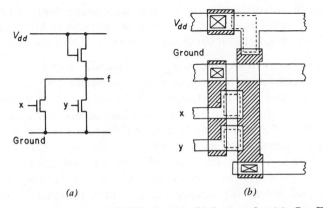

(a)   (b)

Fig. 4.6.4 **Layout example. (*a*) MOS circuit. (*b*) Layout for (*a*). See Fig. 4.6.5*c* for legend.**

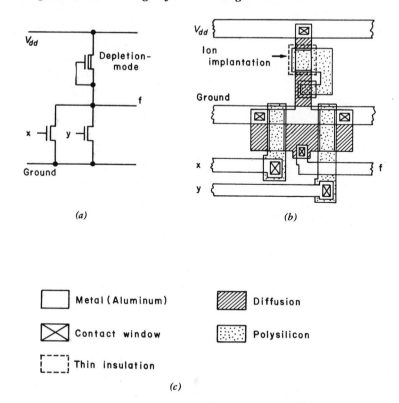

(a)

(b)

Metal (Aluminum)

Contact window

Thin insulation

Diffusion

Polysilicon

(c)

Fig. 4.6.5 Layout example. (*a*) MOS circuit. (*b*) Layout for (*a*). (*c*) Legend for Figs. 4.6.3 through 4.6.5.

References: Carr and Mize, 1972; Watanabe, 1979; Weinberger, 1967; Losleben and Thompson, 1979.

2. When polysilicon runs underneath a metal connection, make its length underneath the metal as short as possible in order to reduce parasitic capacitances. (The insulation layer between them is relatively thin.)

3. For a network which consists of a cascade of modules, design an appropriate configuration for a module by finding placement such that the layout for the module can be arranged side by side without wasting space.

As an example, the layout for the full adder of Fig. 4.4.2 (as a module of a ripple adder of arbitrary bit length) is shown in Fig. 4.6.6. (Note that some MOSFET gates, such as 8, have nonrectangular shapes in order to save areas.) The layout for the next higher bit position can be obtained by flipping over Fig. 4.6.6. (The two inverters in Fig. 4.4.2 are not shown here but must be laid out at appropriate places where designers want to have them.) Since the ground line can be shared with the adjacent module, the layout is compact. If a modular structure can be used for a network (e.g., the high speed is not necessary and the area shape is appropriate), the layout for the entire network can be compact as with this example, and the layout time is reduced since the same layout for a module can be repeated.

In order to fit networks snugly into allocated areas, we sometimes need to consider the layout shapes of the networks to be placed, such as different logic network configurations or even completely different logic networks with inputs and outputs changed. For example, consider the layouts of an adder. The layout in Fig. 4.6.7a stretches vertically, while the layout in Fig. 4.6.7b stretches horizontally because some MOSFETs need greater channel width. In this figure the ratios show the MOSFET channel width and length (e.g., 30/5 means $W = 30$ and $L = 5$). Actual layouts are not shown in Fig. 4.6.7, but the MOSFETs are placed in the schematics to represent their

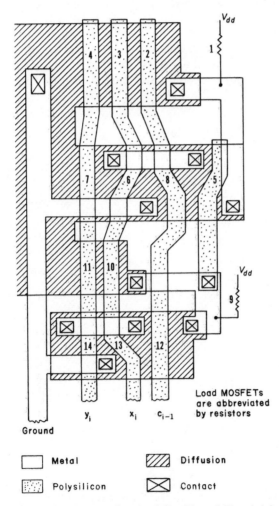

Fig. 4.6.6   Layout for the full adder of Fig. 4.4.2.

relative positions on the chip. The layout of Fig. 4.6.7a is slightly smaller than that of Fig. 4.6.7b. Depending on the area shape available on the chip, either one might be preferable.

4. It is desirable to make the shape of the entire chip as square as possible for maximum packing density.

5. Reduction of the parasitic capacitance in a layout is vitally important for the improvement of speed. Thus we must find a logic network suitable for the reduction of parasitic capacitance.

For example, consider a MOS network to generate a carry in addition. The two networks in Fig. 4.6.8 are identical in terms of the number of MOSFETs and connections, except for the position of the MOSFET with the input $\overline{ab}$ (i.e., in the ground side or the load MOSFET side), representing the same output function $f = ab \lor c(a \lor b)$. But the layout in Fig. 4.6.8a works faster for the worst case, that is, when $\overline{c}$ changes from 1 (high voltage) to 0 (low voltage) while $\overline{ab}$ is 1 and $a \lor b$ is 0, the layout of Fig. 4.6.8a has a smaller area (shaded) to charge up than that of Fig. 4.6.8b. So the layout for Fig. 4.6.8a can generate the carry faster than that of Fig. 4.6.8b. For

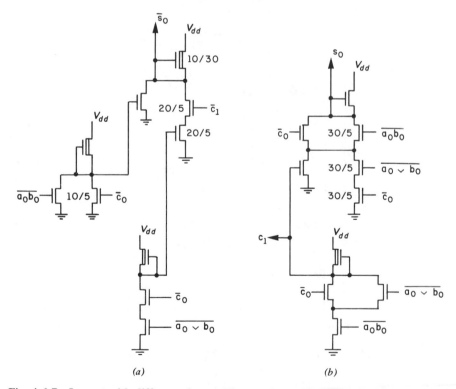

(a)
(b)

Fig. 4.6.7   Layout with different shapes. Figures show *W/L* ratios. (*a*) Layout stretching vertically. (*b*) Layout stretching horizontally.

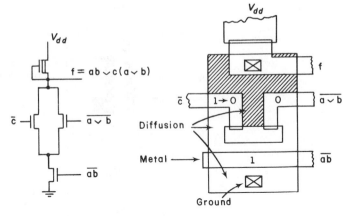

(a)

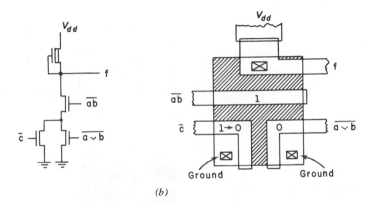

(b)

Fig. 4.6.8 Layouts with different speeds.

other combinations of $\overline{a \vee b}$, $\overline{ab}$, and $\overline{c}$ (i.e., when no carry is to be generated), the layout of Fig. 4.6.8*b* works faster, but logic designers must con-
sider the worst case for high speed, choosing the layout of Fig. 4.6.8*a*.

Logic designers must know the worst case combination of input variable
values for the logic networks that they use.

Notice that the response time of a network is generally different for
different combinations of input values. A MOS cell changes its output value
much faster from 1 to 0 than from 0 to 1, as seen by comparing Figs. 4.3.6*c*
and *d*. Also the response time can be adjusted by changing the dimensions
of the load MOSFETs with the aid of electronic circuit modeling programs,
such as SPICE2.

As we see in some of the above rules, designs of logic networks, electronic
circuits, and layouts are highly interactive for the minimization of chip size
and power dissipation and also for the maximization of speed.

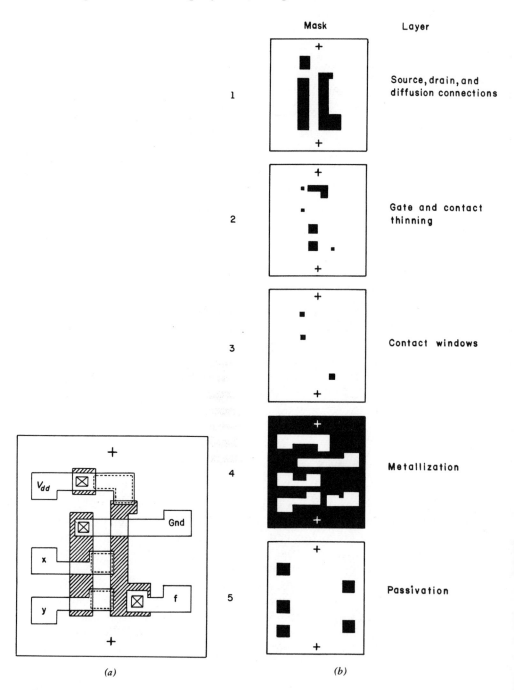

Fig. 4.6.9   Layout and masks. (*a*) Layout for Fig. 4.6.4*a*. Protective circuits are not shown. The layout is not scaled. (*b*) Masks.

After the design of the layout, the masks are prepared. For the inverter of Fig. 4.6.4a, for example, the layout is designed in Fig. 4.6.9a, and then the masks shown in Fig. 4.6.9b are prepared for fabrication. The cross symbols are for the registration of the masks.

4.7   Entire Sequence of Chip Design

Now that we have discussed approaches to logic design, electronic design, and layout design, let us look at the entire sequence of these design stages in conjunction with other design or nondesign stages, considering interplays among them. The entire sequence consists of the following stages:

Architecture design

Logic network design

Electronic circuit design

Layout design

Wafer prototype

Characterization

Unless we have optimized each stage, we cannot have the best chips as the last outcome. Since the integration size of MOS chips is far greater than that of other logic families, each stage and their interplays are far more complex than with other logic families, as discussed in the following. (Other aspects of chip design are discussed in Chapter 10.)

(1)   Architecture Design

Architecture design is the design of a digital system, up to the level of transfer signals among registers (or memories), though the borderline with logic design is usually not clear. In other words, designers must determine a set of instructions, word length, types of adders and other key networks, processor speeds, the number of registers, types of memories, and connection configuration. The architecture must be designed under constraints which are unique to LSI/VLSI, such as the following:

1.   The types of key networks and their interconnections must be chosen such that these networks can be placed on a chip covering the smallest areas, without excessively many crossunders of connections, and that the execution of instructions is fast. For example, if programmable logic arrays (PLAs) or microprograms in read-only memories (ROM) are used for control logic, design change is easy, but a bigger area must be used with lower speed than when control logic is realized in random-logic networks.

2.   Each IC package can have only a limited number of pins. (The pin number is usually less than 100, but it can be as large as 300.) This poses problems. If a complex network can be placed entirely on one chip, the pin number can be within a reasonable limit, but if it is inappropriately divided into many chips, an unusually large number of pins may be needed for intraconnections of the network. Thus if partitioning is not appropriate, only a small integration size can be used, making IC realization of the entire digital system uneconomical and lowering its performance.

When a network must be divided into several chips, duplication of part of the network in different chips may sometimes reduce the number of pins (Hitchcock, 1970; Russo, 1970).

Also, signal transmission among chips can be very slow because of the limited number of pins. If only four pins are available for signal transmission when we need to transmit 8 bits in parallel, we need to use multiplexing (i.e., transmit 4 bits twice) or multivalued signals instead of binary signals. A large parasitic capacitance of connections on a pc board slows down the signal transmission, but the pin number limit aggravates it far more.

Only designers who know the details of the later stages well can design the most appropriate architecture. **Good architecture is essential for the smallest chip size with the best performance. Unless the architecture design is optimized, good work in later stages cannot derive optimal chips, though poor work in later stages will lead good architecture design into poor chips.** In other words, **good chips can be achieved only by good work with all the design stages in good balance, ranging from architecture to layout.**

(2) Logic Design
It is difficult to design logic networks best for later layout, considering all the constraints posed by electronic circuit and layout designs, because these complex constraints are difficult to describe in simple terms. Therefore although some of them can be taken care of partly during logic design, we design logic networks without taking into consideration later stages completely, and then we modify the networks according to the constraints in later stages. Sometimes we need to redesign logic networks totally in order to derive a better layout. In this sense, logic networks, electronic circuit, and layout designs are highly repetitive. In the worst case we need to go back to the architecture design in order to find a better architecture, but experienced designers would not often go through this time-consuming process.

Our ultimate goal is the most compact layout with the highest performance under a prespecified chip cost [determined by Eq. (2.3.1)]. But the use of random-logic networks may be too time-consuming for some chip design projects. So we design the most compact layout under a prespecified design time, or the most compact layout under the constraint of highest performance, spending a great amount of design time. Although this chapter (as well as Chapter 3) has been developed under the latter design approach, there are a spectrum of intermediate design objectives between the above two, and correspondingly there are a large number of different design approaches, as discussed in Chapters 7 and 9. Logic design under the latter design approach is usually done by first ignoring complex constraints and later modifying logic networks with constraints (as discussed in Section 4.2) and also with transitor-level logic design (as discussed in Section 4.5), since apparently there is no better approach. In other words, a top-down design approach is used with modifications at later stages, and, if necessary, we iterate the approach.

Before we proceed to the next design stage (3), the architecture design is usually tested by prototype, using off-the-shelf packages of TTL. (TTL is often chosen because of the availability of a large number of off-the-shelf packages.)

(3) Electronic Circuit and Layout Designs

When designers have finished the logic design, they try to find a best **chip plan**, as mentioned in Section 4.6, by repeating trial plans. Thus each network is allocated a different area on the chip, as exemplified in Fig. 4.7.1. (For the sake of simplicity, border lines between adjacent areas are shown with straight lines. But actually adjacent areas are often interleaved. For example, the 16-bit ALU and register file in the upper middle of Fig. 4.7.1 are not clearly separated.) The electronic circuit design and layout are done almost simultaneously, because unless actual layout is done, the parasitic capacitance, the size of load MOSFETs, and other factors cannot be determined. Also this is the most time-consuming and strenuous stage throughout the entire sequence. (Typically, only 5 to 10 MOSFETs per day can be laid out by one layout designer.) Usually the layout is done by layout designers (i.e., draftpersons specialized in layout drawing), while engineers initially design the electronic circuits modifying the logic networks, design the overall layout configuration, and supervise the layout work of the layout designers, all almost simultaneously. Each layout is verified by computer programs to check whether the layout rules are completely complied with.

Using adders as an example, let us explain the interactive nature of different stages. Look-ahead adders [or carry look-ahead adders, e.g., Muroga (1979, p. 520)] have been considered as the fastest among different types of adders, but this is true only when all gates have the same gate delay. In contrast, ripple adders (or carry ripple adders) have been considered simpler but much slower. However, look-ahead adders realized with MOS cells are actually much slower than ripple adders which have no MOS cells with a large number of fan-out connections due to their modular structure, because some MOS cells in a look-ahead adder have a large number of fan-out connections, and consequently have much slower gate delays. (In contrast, gate delays are much more uniform in the case of TTL or ECL gates since the speed of these gates is much less sensitive to the number of fan-out

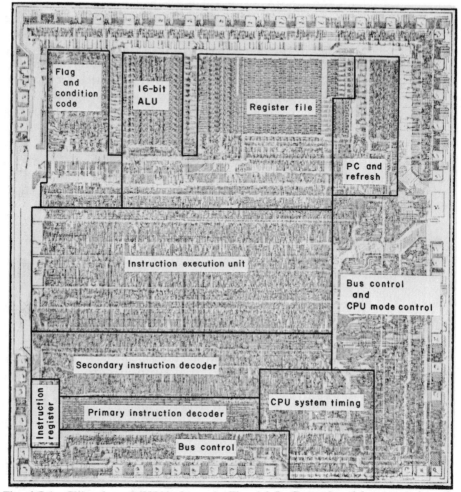

Fig. 4.7.1   Chip plan of Z8000 shown in Fig. 4.2.5. (Reproduced by permission of Zilog, Inc.)

connections than that of MOS cells.) However, if we modify the logic network by adding extra MOS cells to reduce the fan-out connections of some cells, and readjust electronic circuits by increasing the size of the load MOSFETs of MOS cells that have a large number of fan-out connections, we can make the look-ahead adders faster than the ripple adders. In the case of the first commercially successful microprocessor chip, the 8080 of Intel, a ripple adder was chosen as a faster adder than the look-ahead adder, because such a modification of the logic network and readjustment of the electronic circuit was not possible due to the limited chip size. But if we can use a large chip, a look-ahead adder can be used instead as a faster adder, at the expense of increased area. [For chip area comparison, see, for example, Bechade and Hoffman (1980).]

In other words, look-ahead adders can be made slower or faster than ripple adders which have modular structure (i.e., no longer connections), depending on the design of logic networks, electronic circuits, and layouts. (See Remark 4.7.1 for adders with a minimum number of gates and connections.) Generalizing this, **good architecture design is vitally important for chips with small size and high performance, but such chips may or may not be derived, depending on the actual logic, electronic circuit, and layout designs.**

Minimization of the areas for networks, in other words, packing as many networks as possible on a single chip, is also desirable for the following reason. If a network cannot be placed into a single IC chip, part of the network must be placed in other chips, and electric signals must go back and forth between these chips through connections on a pc board. Because driving signals through the pc board lowers the speed and requires more electric power than on a chip, it is preferable to place the entire network on the same chip and let only the outputs of the network (no other part of the network) go through the pc board (usually the outputs of a network can be slowly changing and power-consuming because flip-flops are often connected, the outputs must go long distance, or the outputs must be branched.) Therefore the minimization of areas for networks is meaningful even in terms of speed. Even when the entire network can be placed in a single chip, a reduction of the chip size by appropriate layout results in speed improvement in addition to cost reduction (e.g., 10% reduction in chip size could result in 20% speed improvement).

> *Remark 4.7.1:*   Ripple adders used to be designed by cascading full adders. But it has been proven that if a parallel adder, n bits long, with NOR gates only, is to have a minimum number of gates, the adder must have a modular structure (i.e., no long connections), and each module has many outputs which collectively represent an output carry and many inputs from the module in the next lower bit positions. Furthermore, such adders with a minimum number of NOR gates have actually been designed and have many different configurations (Lai and Muroga, 1979). Also, it has been proven that these adders have a minimum number of connections (Sakurai and Muroga, to be published). (Such ripple adders can be designed with other types of gates also.) Thus roughly speaking, this type of ripple adders (although limited to a specific type of logic gate) is most compact, though the compactness depends on the layout rules. (Notice here that there are a great variety of layout rules. As mentioned in Remarks 4.2.3 and 4.6.1, new processing techniques to reduce the connection areas have been attempted. Also in the case of the gate arrays discussed in Chapter 9, the minimization of the number of gates leads to the most compact chip area if enough connection routing channels are provided. By Procedure 3.4.1, minimal ECL networks can be derived from minimal NOR networks, leading to the most compact ECL gate arrays.)   □

(4) Wafer Prototype

After the layout is finished, masks are prepared and a wafer is actually fabricated for testing. At each of the above stages careful testing is absolutely necessary, but testing by wafer prototype is indispensable. Good experienced designers can finish wafer prototyping within a few iterations. Wafer prototyping is expensive and time-consuming, so too many iterations can be financially disastrous.

(5) Characterization

After the mistakes made in all the previous stages have been corrected to the greatest extent possible, the performance is characterized so that reliable operation is assured over worst case limits of processing. In the case of microprocessor chip design, for example, the execution times of instructions or their combinations are also measured so that user's manuals can be prepared.

Only after the above sequence of stages, where mistakes have been corrected very carefully at every stage, can the production of the designed chip be started. In order to give some idea on the manpower required for each design stage, let us take the design of the general-purpose microprocessor chip Z8000 as an example, which is probably regarded to be the last LSI chip designed in random-logic networks (no regular-structured networks such as ROMs and PLAs) for that integration size, and consequently is the most compactly laid out. Design parameters and man-months of the Z8000 are compared in Table 4.7.1 with those of the previously designed general-purpose microprocessor chips 8080 and Z80. (All are designed by M. Shima and others). The Z8000, which has about 17,500 MOSFETs, as shown in Table 4.7.1a has a computing capability greater than that of the minicomputer PDP 11/45 and slightly less than that of the PDP 11/70, despite its slower logic gates. The functional definition, that is, the architecture design inside the chip, shown in the first row in Table 4.7.1b, is completed jointly by all people concerned, such as overall-system architects (i.e., those who are concerned with the overall system architecture of this chip in relation to many other chips), software people, application people, marketing people, and chip design engineers, because no single person can have expertise knowledge of all these areas. The total time spent by the chip design engineers who go through all the design stages in succession is roughly 40% of the total time spent by all people concerned, as shown in the last column in Table 4.7.1b. Other people, such as the overall system architects, will not participate in the succeeding stages. In contrast, 100% of the logic design and the electronic circuit design stages is done by the chip design engineers. The layout design is most time-consuming, as shown in the third and fourth rows in Table 4.7.1b, although the chip design engineers' time is only 26%. So the design of LSI/VLSI chips of greater integration size, after the Z8000, is increasingly mixed with regular-structured networks with ever-increasing integration size, as discussed in Chapters 7 through 10. For quick design, verifications, and corrections, CAD is becoming vitally important. A number of variations of the above design sequence for shorter development times at the sacrifice of performance or chip size are discussed in Chapters 7 through 10.

Table 4.7.1   Design of General-Purpose Microprocessor chips (Shima, 1979[a])

(a) DESIGN PARAMETERS FOR THE 8080, Z80, AND Z8000

	8080	Z80	Z8000
Date of initial production	1974	1976	1979
Technology	*n*-channel silicon gate (high voltage)	*n*-channel silicon gate (low-voltage depletion load)	Scaled-down *n*-channel silicon gate
Transistor area with scale factor defined as 1.0 for 8080	1	0.70	0.44
Instructions	65	128	414
Chip size, mm^2	22.3	27.1	39.3
Total transistors	4,800	8,200	17,500
Transistors used to implement instruction decoders and microcode generators with random-logic wiring	1,800	3,700	8,600
Percentage of total area	31	36	40
Percentage of usable area	37	44	47

(b) MAN-MONTHS OF DESIGN SPENT FOR THE Z80 AND Z8000

	Z80[b]	Z8000[b]	PERCENTAGE OF TOTAL TIME SPENT BY Z8000 CHIP DESIGNERS
Functional definition	5	15	40
Logic/circuit design	7 S	11 S	100
Layout design	8 S, 10 A	8 S, 30 A	26
Test program debut chip characterization	6 S	10 S, 3 A	65

[a] From Shima, 1979 (Copyright © 1979 IEEE).
[b] S—Senior personnel; A—associate personnel.

Where the design is accomplished with discrete components, or off-the-shelf packages, designers need not consider the manufacturing processes except in special cases, because discrete components are flexible enough in their use (for manufacturing engineers to be able to assemble the components at their disposal). In contrast, the design of LSI or VLSI chips requires good understanding of manufacturing (or fabrication) processes because the ultimate design goals are the preparation of masks, which are essential tools in the manufacturing processes. Thus designers must understand completely how each mask is handled in the fabrication process, including the misalignment problem and processing steps. The layout rules discussed in Section 4.4 form a communication link between designers and process engineers, but these rules can only be prepared under very close cooperation, which in turn requires mutual understanding. For a comparable number of devices, the design of LSI/VLSI chips takes much longer than that with discrete components, since all stages are highly interactive, and best results can be obtained only by a proper balance among them. It has been considered that the best LSI or VLSI design and fabrication can be done only under very tight management of the most talented designers, since anything can go wrong with devastating effects during the very long and complex sequence of design stages and fabrication steps.

Also, since the different stages for logic design, electronic circuit design, layout, and architecture are highly interactive, each designer must go through all these stages, unlike conventional design, where the stages are developed separately by different designers. Or, designers for different design stages must cooperate very closely.

References: Shima, 1979; Moore, 1979; Allan, Apr. 1979; Bloch, 1979; Oakes, 1979; Walker, Feb. 1, 1973; Weiss, 1981.

Problems

4.7.1 In the case of masks for bipolar transistors, transistors that are to be matched in performance must be placed in an appropriate orientation, as explained with Fig. 3.3.5. Discuss the effects on performance, such as speed and current magnitude, of such considerations of orientation when preparing masks in the case of MOS.

4.7.2 As described in step 1 of General Principles in Layout Design in Section 4.6, whenever possible, diffusion or polysilicon connections are run vertically and metal connections are run horizontally. Power supply and ground connections are usually realized in metal and run in parallel with other metal connections. Explain why this is done.

4.7.3 As stated in step 4 of the General Principles in Layout Design in Section 4.6, it is desirable to make the shape of the entire chip as square as possible for maximum packing density, assuming that all pads are placed in the periphery. Discuss why this is so.

4.7.4 Using the layout rules given in Section 4.6, draw a layout for the network of Fig. 4.4.2, minimizing the area. (Do not use paper larger than $8\frac{1}{2} \times 11$ in., since otherwise this would be too time-consuming.) Calculate the chip area size.

4.7.5 Using the layout rules given in Section 4.6, draw a layout for the adder of Fig. 4.4.1*b*, minimizing the area. (Do not use paper larger than $8\frac{1}{2} \times 11$ in.)

Remark: Notice that we need to use load MOSFETs of square shape, unlike the very rectangular layout for the load MOSFET of Fig. 4.4.3 for lower speed.   □

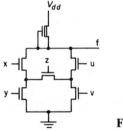

Fig. P4.7.8

4.7.6 Draw a layout for the MOS network of Fig. 4.2.4, using the layout rules given in Section 4.6. Minimize the chip area. (Do not use paper larger than $8\frac{1}{2} \times 11$ in.)

4.7.7

(a) Using the layout rules given in Section 4.6, draw a layout for the latch of Fig. 4.5.1, minimizing the area.

(b) Draw masks for this layout. For the sake of simplicity, the load MOSFETs of depletion mode may be treated as enhancement mode. (Do not use paper larger than $8\frac{1}{2} \times 11$ in.)

4.7.8

(a) Draw a layout for the MOS cell shown in Fig. P4.7.8, using the layout rules given in Section 4.6.

(b) Show a switching expression for the output function f.

(c) Design a MOS network for this function f, using only MOS cells that perform the NOR and NAND operation.

(d) Draw a layout for the network designed in (c).

(e) Compare the size of the layouts obtained in (a) and (d). (Show the size ratio.)

[For the layouts in (a) and (d) do not use paper larger than $8\frac{1}{2} \times 11$ in.]

4.7.9

(a) Design an I^2L network for the following function f.

(b) Design a MOS cell for the following function f.

(c) Draw a layout for each case. For the I^2L network use the dimensions shown in Fig. 3.5.9, and for the MOS cell use the layout rules given in Section 4.6.

$$f = \overline{xy \vee z}$$

4.8 Dynamic MOS Networks

In the MOS networks discussed so far, logic values 1 and 0 are represented by whether or not a current flows from the power supply to the ground through driver MOSFETs. Parasitic capacitances have no significance in the representation of logic values, though the speed is determined by them. These MOS networks are called **static MOS networks**. In this section we discuss **dynamic MOS networks** which work under a different principle.

Logic values 1 and 0 are represented by whether or not a parasitic capacitance is charged in synchronization with clock pulses. Dynamic MOS networks have the advantages of low power consumption and small areas, though the speed is lower than that of the static networks.

Let us discuss the difference between static and dynamic MOS networks, using an inverter as an example. The inverter in static MOS, as discussed in the previous sections, is shown in Fig. 4.8.1a. In order to keep a voltage swing sufficiently large, the ratio of the resistances of load MOSFET T_2 and driver T_1 (when conductive) must be high, as discussed in Section 4.3, so the static inverter is also called a **ratio inverter**. When v_{in} changes from high voltage (logic value 1) to low voltage (logic value 0), parasitic capacitance C_1 must be charged to power supply voltage V_{dd} through load MOSFET T_2, whose resistance is larger than that of T_1, so C_1 is charged slowly. But when v_{in} changes from low voltage to high voltage, C_1 must be discharged through driver MOSFET T_1 whose resistance is much smaller than that of T_2. Thus output voltage v_{out} changes faster from 1 to 0 than from 0 to 1, as shown in Fig. 4.8.1b. Current i_D flows from the power supply to the ground, as long as v_{in} is high voltage.

Two-Phase Dynamic MOS Networks

An inverter in dynamic MOS, which is called a **two-phase ratioless inverter,** is shown in Fig. 4.8.2a. A clock of two phases, ϕ_1 and ϕ_2, is supplied, where each of ϕ_1 and ϕ_2 is connected to two terminals. Notice that ϕ_1 and ϕ_2 are supplied to MOS cells in alternate levels. Also notice that no point of the circuit Fig. 4.8.2a is connected to the ground, but actually one end of the clock generator (i.e., clock power supply) is grounded and the circuit has parasitic capacitances against the ground, as shown by the dotted lines.

Suppose that at time t_0 in Fig. 4.8.2b input v_{in} changes from high voltage (logic value 1) to low voltage (logic value 0), while parasitic capacitance C_1

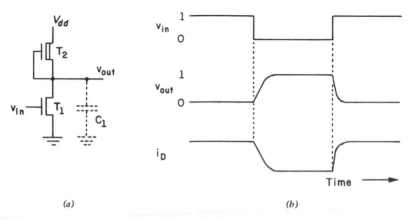

(a)   *(b)*

Fig. 4.8.1 Static inverter.

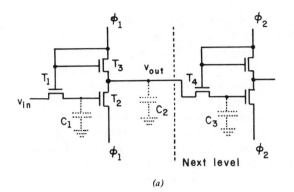

(a)

(b)

Fig. 4.8.2 Two-phase ratioless inverter.

is charged and C_2 and C_3 are discharged. At t_1, clock ϕ_1 makes MOSFETs T_1 and T_3 conductive. (Notice that ϕ_1 is connected to both T_1 and T_3.) C_1 is discharged through T_1 and v_{in} to the ground (of input v_{in}), making T_2 nonconductive. C_2 is charged to high voltage through T_3, and even after the clock pulse of ϕ_1 disappears, C_2 maintains the same voltage. At t_2 the second clock ϕ_2 makes T_4 conductive, transferring the electrical charge from C_2 to C_3. (Notice that ϕ_2 and T_4 are part of a MOS cell in the next level.) Suppose that v_{in} changes to logic value 1 at t_2. At t_3, ϕ_1 makes T_1 and T_3 conductive. C_1 is charged to v_{in}, making T_2 conductive. Although both T_2 and T_3 are conductive, C_2 does not discharge because its voltage is equal to the clock pulse voltage of ϕ_1. At t_4 the clock pulse of ϕ_1 disappears, making T_1 and T_3 nonconductive. But T_2 remains conductive because C_1 remains charged to high voltage. Thus C_2 is discharged through T_2. At t_5 the clock pulse of the second clock ϕ_2 makes T_4 conductive, and consequently C_3 is discharged through T_2. Thus, the inversion of input v_{in} appears at output v_{out}, being also delivered to C_3. Notice that C_2 is always charged by ϕ_1 during its clock pulse and this is called **precharging**.

Notice that only the impulsive current i_ϕ to charge or discharge the parasitic capacitance C_2 flows as shown in Fig. 4.8.2b, compared with current i_D which lasts for a long period of time in Fig. 4.8.1b. Thus the power consumption of the dynamic inverter of Fig. 4.8.2a is far smaller than that of the static inverter of Fig. 4.8.1a.

A dynamic MOS cell for the NAND operation is shown in Fig. 4.8.3. Dynamic MOS networks for other switching functions can be similarly designed. A dynamic shift register is shown in Fig. 4.8.4, where parasitic capacitances C_1 and C_2 perform the roles of master and slave in the case of static flip-flops, respectively.

Since the dynamic MOS network in Fig. 4.8.2 requires two separate trains of clock pulses ϕ_1 and ϕ_2, it is said to have a **two-phase clock** (often abbreviated as 2ϕ clock).

Transistor T_4 (and also T_1) in Fig. 4.8.2a is called a **transfer gate**, since its function is to transfer the electric charge from one parasitic capacitance to another. In this case notice that unless C_3 is much smaller than C_2, the voltage at C_3 may become too low since the electric charge stored at C_2 must be divided by C_2 and C_3 at time t_2 (this is called **charge sharing**). But it is sometimes tricky to control the magnitude of parasitic capacitances at the time of layout. Also the voltage, to which C_3 is charged, tends to be lower than that of C_2 (because T_4 becomes nonconductive when the voltage

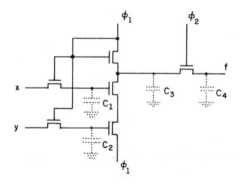

Fig. 4.8.3   Two-phase ratioless NAND cell.

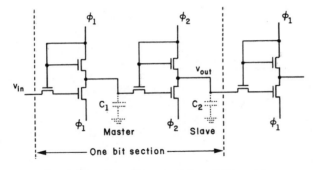

Fig. 4.8.4   Two-phase ratioless shift register.

difference between the gate of T_4 and C_3 is lower than the threshold voltage, as seen in Fig. 4.1.9c).

Next let us discuss two-phase dynamic MOS networks, which are very different from those above. The inverter in Fig. 4.8.5, which is called a **two-phase ratioed inverter**, has a clock of two phases ϕ_1 and ϕ_2 and the direct-current power supply V_{dd}, which the networks in Figs. 4.8.2 through 4.8.4 did not have. Also notice that MOSFET T_2 is grounded. Suppose that at time t_0 in Fig. 4.8.5b the input v_{in} changes from a high voltage (logic value 1) to a low voltage (logic value 0). At t_1, T_2 and T_3 become conductive due to the clock pulse of ϕ_1, and parasitic capacitance C_1 is discharged through transfer gate T_1 and C_2 charged up to a high voltage, which is close to V_{dd} of the direct-current power supply. At t_2 the transfer gate T_4 becomes con-

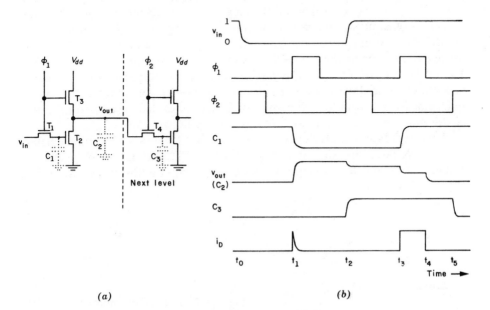

(a)   (b)

Fig. 4.8.5   Two-phase ratioed inverter.

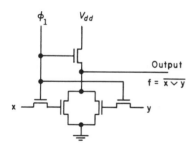

Fig. 4.8.6 Ratioed dynamic NOR.

ductive by the clock pulse of ϕ_2, so the electric charge on C_2 is transferred to C_3. By making C_3 much smaller than C_2 by layout design, the output voltage v_{out} does not become much smaller by this transfer than the voltage at C_2 before the transfer. Suppose that v_{in} changes to a high voltage. At t_3, T_1 and T_3 become conductive by the clock pulse of ϕ_1. In this case C_1 is charged to a high voltage and C_2 is charged to a voltage that is determined by the resistance ratio of T_3 and T_2. This voltage at C_2 cannot be transmitted to C_3 because transfer gate T_4 is nonconductive. (So it does not matter that this voltage at C_2 cannot be as high as its highest voltage at t_1). Notice that during the period from t_3 to t_4 direct current continues to flow from V_{dd} to the ground through T_2 and T_3. Also notice that we do not have the resistance ratio problem of static MOS networks in this case. At t_4, T_3 becomes non-conductive because the clock pulse of ϕ_1 disappears, and then C_2 is completely discharged through T_2. At t_5, T_4 becomes conductive by the clock pulse of ϕ_2, so C_3 is discharged through T_4 and T_2 to the ground. Thus the inversion of input v_{in} appears at output v_{out}, being also delivered to C_3.

Although no current flows from the clock power supply ϕ_1 or ϕ_2 during the above logic operation, current flows from the direct current power supply V_{dd} to the ground for a long period during one clock pulse (from t_3 to t_4), in addition to a short period at t_1. Thus the power consumption is greater than that for the ratioless dynamic MOS networks discussed before, but it is still much smaller than that for static MOS networks.

A ratioed dynamic MOS network for the NOR operation, a shift register, and a S–R latch are shown in Figs. 4.8.6, 4.8.7, and 4.8.8., respectively. While a shift register in static MOS network requires flip-flops, the one in Fig. 4.8.7 does not.

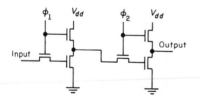

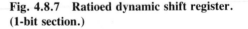

Fig. 4.8.7 Ratioed dynamic shift register. (1-bit section.)

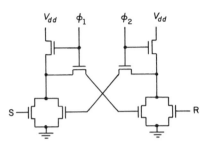

Fig. 4.8.8 **Ratioed dynamic S–R latch.**

Compared with the ratioless networks in Figs. 4.8.2 through 4.8.4, the ratioed networks in Figs. 4.8.5 through 4.8.8 have the following advantages, though they require greater power consumption. The clock power supply is simple because the clocks are connected to MOSFET gates only, requiring almost no power consumption, while in the case of ratioless dynamic MOS networks the clock power supply, which is usually placed on the same chip, requires very large transistors to switch on and off because of its supply of large currents to all MOS cells. Another advantage is the realizability of a complex logic function in only one clock level of ϕ_1, as illustrated in Fig. 4.8.9. In the case of Fig. 4.8.5a (and also Fig. 4.8.6), only one cell consisting of T_1, T_2 and T_3 controlled by ϕ_1 is connected to transfer gate T_4 controlled by ϕ_2. But in Fig. 4.8.9 a small network of MOS cells, all of which are controlled by ϕ_1, can be formed for complex logic operations and is connected to the transfer gate controlled by ϕ_2. Notice that there is no transfer gate required among these MOS cells. This is not possible with ratioless networks. In this case the small network of MOS cells works as a static MOS network during the clock pulse of ϕ_1. (Notice that unlike ratioless networks, the parasitic capacitances are not necessarily precharged during clock pulses.) Consequently, the high ratio between load and driver in each MOS cell is required. Because of this, the networks in Figs. 4.8.5 through 4.8.9 are called **ratioed networks**, even though except for the one in Fig.

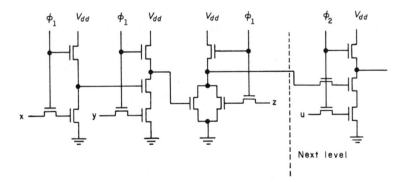
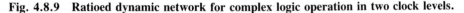

Fig. 4.8.9 **Ratioed dynamic network for complex logic operation in two clock levels.**

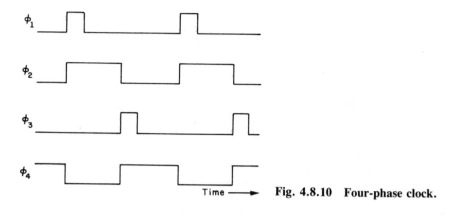

Time ⟶   **Fig. 4.8.10   Four-phase clock.**

4.8.9, they do not have the high-ratio problem as explained with respect to Fig. 4.8.5. Thus the terminology is confusing.

References: Fette, 1971; AMI staff, 1972.

Four-Phase Dynamic MOS Networks

We now discuss dynamic MOS networks with a four-phase clock. In other words, we use four different trains of clock pulses ϕ_1, ϕ_2, ϕ_3, and ϕ_4, as shown in Fig. 4.8.10. (Notice that the waveform for ϕ_4 is simply the complement of ϕ_2.) Using two of them in different combinations (two combinations are not used), we form four different types of MOS cells, as illustrated in Fig. 4.8.11. Each circle denotes a subcircuit of MOSFETs. Figure 4.8.12 shows an example of a type 1 MOS cell in which a subcircuit is shown by the dashed line. The output function of each cell can be calculated as

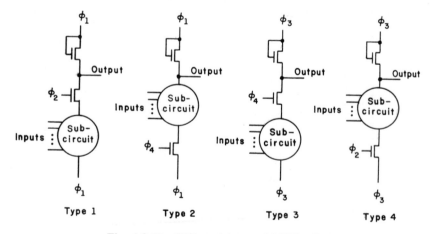

Fig. 4.8.11   Different types of MOS cells.

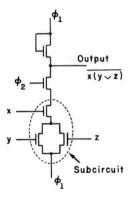

Fig. 4.8.12   Example of type 1 MOS cell.

before. For example, the output function of the cell in Fig. 4.8.12 is

$$\overline{x(y \vee z)}$$

With a simple example let us explain how this works. Figure 4.8.13 is an inverter, based on a type 1 MOS cell. When the clock pulse appears at the first clock ϕ_1, as shown in Fig. 4.8.10, MOSFET T_1 becomes conductive, and parasitic capacitance C is charged up to a high positive voltage of the clock power supply. The clock pulse is present at the second clock ϕ_2 during this charging period and stays there even after the charging time. MOSFET T_2 remains conductive as long as the clock pulse is present at ϕ_2. Then if input x is a high voltage, MOSFET T_3 becomes conductive, and the electric charge stored at C is discharged, making the output voltage low. If x is a low voltage, T_3 is nonconductive, and the electric charge is maintained at C, making the output voltage high. In other words, output f is a high or low voltage, according to whether input x is a low or a high voltage. Thus we have $f = \overline{x}$, and the cell works as an inverter. Other types of MOS cells perform logic operations in a similar manner.

An entire network is constructed by connecting these different types of MOS cells. But in this case the output of a type 1 cell can be connected to the type 2 cell as its input, the output of a type 2 cell can be connected to

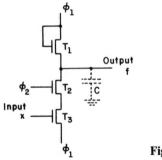

Fig. 4.8.13   Inverter.

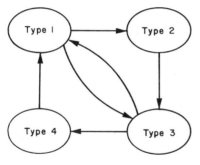

Fig. 4.8.14 **Connectability of the different types of MOS cells in Fig. 4.8.11.**

type 3, and the output of a type 3 cell can be connected to type 4. Then the output of the type 4 cell can be connected to type 1. The output of the type 1 cell can be connected to the type 3 cell, and vice versa. But type 2 and type 4 cells cannot be connected to each other; and also each type cell cannot be connected to itself. The connectability is summarized in Fig. 4.8.14.

As an example of a four-phase dynamic MOS network a shift register is shown in Fig. 4.8.15, where the 1-bit section shown can be repeated to a desired bit length since type 1 and 3 cells can be connected to each other as explained in Fig. 4.8.14.

The four-phase dynamic MOS networks do not have the problem associated with the transfer gates in the two-phase dynamic MOS networks, since no transfer gate is used. But more clock connections are required because of a greater number of clock phases, making layouts cumbersome and larger (a few times larger than the two-phase dynamic networks). Thus the two-phase dynamic networks are usually preferred, except for special cases.

The four-phase dynamic MOS networks have other variations. For example, there is a counterpart to the two-phase ratioed MOS networks, using both direct-current and clock power supplies.

References: Asija, 1977; Fette, Nov. 15, 1971; Yen, March, 1968; Sept. 1968.

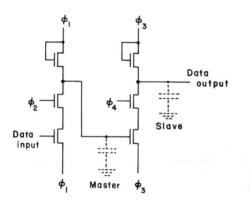

Fig. 4.8.15 **Four-phase ratioless shift register.**

Other Multiphase Dynamic MOS Networks

There are other variations of dynamic MOS networks. Some have more clock phases (e.g., single phase like Fig. P4.8.11 and also three phase and six phase, though three phase and six phase are not popular), and clocks are of no overlapping periods among different phases like ϕ_1 and ϕ_2 in Fig. 4.8.2, or of overlapping periods like ϕ_1 and ϕ_2 in Fig. 4.8.10. Some have different connection configurations (Carr and Mize, 1972; Gosling et al., 1971).

Features of Dynamic MOS Networks

As seen in the above examples, in dynamic MOS networks the logic values 1 and 0 are represented by whether a parasitic capacitance is charged up to a high voltage or discharged, respectively,* in synchronization with clock pulses, during which no current flows from the power supply to the ground through MOSFETs except for a short period. Clock pulses are supplied to all MOS cells. In contrast, in static MOS networks logic values 1 and 0 are represented by whether or not a current flows from the power supply to the ground through driver MOSFETs, and the existence of parasitic capacitances has no significance in the presentation of logic values, though it is important for the network speed. Clock pulses are supplied to some cells in a static MOS network in order to synchronize the propagation of signals over different paths of cells, but they are not supplied to all MOS cells.

By the use of clock pulses in dynamic MOS networks we can eliminate load MOSFETs which are essential in static MOS networks and occupy large chip areas for loads themselves and even more for drivers (though we cannot when we use Fig. 4.8.9), while clock connections add small areas. So dynamic MOS networks are much more compact. The speed of dynamic MOS is determined by the clock frequency, which is in turn determined by the charge and discharge times of the parasitic capacitance. If the parasitic capacitance is small, the charge and discharge times are short, and we can increase the clock frequency, speeding up the operation. But by using a clock, the speed of dynamic MOS cannot be as fast as that of static MOS because the slowest MOS cell is permitted to complete its operations during one clock period. Power consumption is much smaller than that for static MOS because current flows for only a short period during clock pulses. These advantages are gained at the expense of the provision of clocks.

Other Design Problems of Dynamic Networks

The logic design of dynamic MOS networks is essentially no different from that of static networks. In other words, logic networks designed by the procedures outlined in Sections 4.2 and 4.4 can easily be converted into dynamic MOS networks. But in this case we do not have a restriction on

---

* This is in positive logic. The case of negative logic can be similarly treated.

the number of MOSFETs connectable in series (or parallel) as much as in the case of static MOS networks, because the high ratio of the resistances of load and driver MOSFETs required for static MOS networks is not required in this case (except for the type of networks shown in Fig. 4.8.9). Thus we can connect more MOSFETs in a freer configuration, realizing a more complex switching function per MOS cell. But if we try to realize excessively complex functions, the layout becomes cumbersome in view of the layout of clock connections, and many parasitic capacitances and MOS-FET resistances will eventually reduce the speed. In practice at most 4 MOSFETs for high speed and at most 10 MOSFETs for low speed can be connected in series in any path, and at most 10 MOSFETs can be connected in parallel. But the total number of MOSFETs in each cell is probably limited to about 10. (In this sense, Procedure 4.2.1, which tends to yield complex configurations in some cells, may be used more effectively here than in the design of static networks.)

Dynamic networks are feasible with MOS because the inputs to a MOS cell are almost purely capacitive. (Notice that the input resistances are unusually high, about 10^{14} Ω.) There is no counterpart with bipolar logic families where the input resistances of bipolar transistors are low. MOSFETs are used only to charge or discharge parasitic capacitances, and (except for the case of Fig. 4.8.9) they do not have the problem of high resistance ratios for voltage swing in static MOS networks. Thus MOSFETs can be made small due to the freedom of geometries. The charge and discharge times, however, are unpredictable. (They are different with parasitic capacitances at different places on the chip.) Thus we need to use a clock, which requires many clock connections.

Long connection lines are usually not desirable because of excessively large parasitic capacitances. So modular structures are usually preferred. The percentage of areas occupied by connections, due to clock connections, is greater than that given in Table 4.2.1 for static MOS networks. By reducing the number of MOS cells (by procedures such as Procedure 4.2.1) we can reduce not only power consumption but also area size by reducing many clock phases and ground connections (in particular when the number of phases increases).

The selection of clock frequency is closely related to the charge and discharge times of the parasitic capacitances. Since the electric charges must be maintained at parasitic capacitances until they are probed by MOS cells in the next levels, the clock frequency cannot be too low. If it is too low, the electric charges stored at the parasitic capacitances may leak to the ground, because MOSFETs do not have infinitely large resistances even when nonconductive. If the clock frequency is too high, parasitic capacitances cannot be charged to sufficiently high voltages, and once charged, they cannot be discharged completely during a clock cycle. In other words, the MOS networks do not work properly. As we increase the clock frequency, power consumption increases. More precisely speaking, the power consumption is linearly proportional to

$$CV^2 f$$

where C is the parasitic capacitance, V is the clock power supply voltage V_{dd}, and f is the clock frequency. When different phases of a clock are distributed to MOS cells throughout the networks, their timing relationship may change due to different delay times over clock connection lines. So timing is critical in particular for high clock frequencies.

In summary, although the layout is cumbersome because the performance is highly sensitive to parasitic capacitances and many clock connection lines must be laid out, dynamic MOS networks are used, often along with static networks on a chip, when low power consumption is required.

References: Yen, 1969, Oct. 1969; Spencer, 1969; Gosling et al., 1971; Boysel and Murphy, 1970; Moon, 1969; Warner, 1967; Watkins, 1967; Kroeger and Threewitt, March 15, 1974; May 10, 1974; Dirilten, 1972; Karp and de Atley, 1967; Cook et al., 1973; Wallmark and Carlstedt, 1974; Wolfendale, 1973.

Problems

4.8.1 In Fig. 4.8.2 C_2 maintains a high voltage from t_1 to t_4. Find how long the output voltage of the next-level MOS cell maintains a low voltage.

4.8.2 Draw a layout for the shift register of Fig. 4.8.4. For simplicity assume that each MOSFET has a channel width of 10 μm and a length of 5 μm, and use Fig. 4.6.1 for other dimensions. (Do not use paper larger than $8\frac{1}{2} \times 11$ in.)

4.8.3 Analyze the performance of the S–R latch of Fig. 4.8.8 by drawing waveforms at key locations.

4.8.4 Draw a layout for the S–R latch of Fig. 4.8.8 using the design rules described in Problem 4.8.2. (Do not use paper larger than $8\frac{1}{2} \times 11$ in.)

4.8.5 Design a two-phase ratioless dynamic MOS network for a serial adder in double-rail input logic.

4.8.6 Design a two-phase ratioed dynamic MOS network for a serial adder in double-rail input logic.

4.8.7 Explain, with an example, why we cannot realize a complex function with a two-phase ratioless MOS network by connecting a small network to the transfer gate, like the case of ratioed networks shown in Fig. 4.8.9.

4.8.8 Analyze the performance of the four-phase shift register of Fig. 4.8.15 by drawing waveforms at key locations.

4.8.9 Design a four-phase dynamic MOS network for a full adder in double-rail input logic.

4.8.10 Explain, with an example, why connections of different types of MOS cells other than those shown in Fig. 4.8.14 do not work properly.

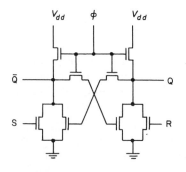

Fig. P4.8.11 Dynamic *S–R* latch.

4.8.11 The *S–R* latch shown in Fig. P4.8.11 is a variation of the one in Fig. 4.8.8, where the two phases ϕ_1 and ϕ_2 are replaced by a single phase ϕ. Analyze the performance of this by drawing waveforms at key locations.

4.8.12. If we reduce parasitic capacitances in dynamic MOS networks, power consumption will be smaller and higher clock frequency can be used. Discuss what the limiting factors are in reducing parasitic capacitances.

4.9 CMOS (Complementary MOS)

A CMOS cell consists of a pair of circuits, one comprising *n*-MOS and the other *p*-MOS, where all MOSFETs are enhancement mode. CMOS, which stands for complementary MOS, means that the *n*-MOS and *p*-MOS circuits are complementary. As a simple example, let us explain CMOS with the inverter shown in Fig. 4.9.1. A *p*-channel MOSFET is connected between the power supply of positive voltage V_{dd} and the output, and an *n*-channel MOSFET is connected between the output and the negative side V_{ss} of the above power supply. The negative side of the power supply is usually grounded, and if so, V_{ss} is grounded. The substrate of *p*-MOS is usually connected to V_{dd}, and in the figure this is denoted by connecting the tip of the arrow to V_{dd}. The substrate of *n*-MOS is usually connected to V_{ss}, and this is denoted by connecting the tail of the arrow to V_{ss}. The connections of the substrates, however, will be omitted in figures from now on for the sake of simplicity. Input *x* is connected to the gates of both *p*-MOS and *n*-MOS.

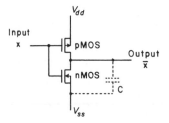

Fig. 4.9.1 CMOS inverter.

When x is a high voltage, p-MOS becomes nonconductive and n-MOS becomes conductive. When x is a low voltage, p-MOS becomes conductive and n-MOS becomes nonconductive. This is the property of p-MOS and n-MOS when the voltages of the input and the power supply are properly chosen. (The voltage of the power supply is usually chosen between 4.5 and 15 volts. The high voltage of the input is somewhat lower than this, and the low voltage of the input is roughly 0 volt.) In other words, when either p-MOS or n-MOS is conductive, the other is nonconductive. When x is a low voltage (logic value 0), p-MOS is conductive, with nonconductive n-MOS, and the output voltage is a high voltage (logic value 1), which is close to V_{dd}. When x is a high voltage, n-MOS is conductive, with nonconductive p-MOS, and the output voltage is a low voltage. Thus the CMOS cell in Fig. 4.9.1 works as an inverter. The p-MOS circuit essentially works as a variable load.

Since either p-MOS or n-MOS is always nonconductive, no current flows from V_{dd} to V_{ss} through these MOSFETs. Consequently when no input changes, the power consumption is simply the product of the power supply voltage V (if V_{ss} is grounded, V is equal to V_{dd}) and a very small current of a nonconductive MOSFET. (Ideally there should be no corrent flowing through a nonconductive MOSFET, but actually a very small current flows. Such undesired very small current is called a **leakage current**.) This is called the **quiescent power consumption**. Since the leakage current is typically a few nanoamperes for $V = 5$ volts, the quiescent power consumption of CMOS is less than 10 nW which is unusually small compared with other logic families.

Whenever the input of this CMOS cell changes, the parasitic capacitance C at the output terminal (including parasitic capacitances at the inputs of the succeeding CMOS cells) must be charged up to a high voltage through the conductive p-MOS. (A peak current can be as large as 0.3 milliamperes.) Then at the next input change the electric charge stored in the parasitic capacitance must be discharged through the conductive n-MOS. Therefore much larger power consumption than the quiescent power consumption occurs whenever the input changes. This dynamic power consumption is given by

$$CV^2 f$$

where C is the parasitic capacitance, V is the power supply voltage, and f is the switching frequency of the input. The power consumption of CMOS is a function of frequency. CMOS consumes very little power at low frequency, but it consumes more than TTL as the frequency increases. But this does not mean that CMOS cannot operate at high power levels. A CMOS cell, if properly designed, can provide as much as 0.5 amperes (*Electronic Design,* Nov. 8, 1978, pp. 74–82).

Let us consider a CMOS cell in which many MOSFETs of the enhancement mode are connected in each of the p-MOS and n-MOS circuits (e.g., the CMOS cell in Fig. 4.9.2). If the p-MOS circuit is regarded as a load in the case of a MOS cell in Sections 4.1 through 4.8, the output function f can be calculated in the same manner as before. Calculate the transmission between the output and V_{ss}, considering MOSFETs as make-contacts of relays. relays. Then its complement is the output function of this CMOS cell. Thus the CMOS cell in Fig. 4.9.2 has the output function $f = \overline{x \vee y}$.

We can prove that if the p-MOS circuit of any CMOS cell has the transmission which is the dual of the transmission of the n-MOS circuit, one of the p-MOS and n-MOS circuits is always nonconductive, with the other conductive, for any combination of input values. (This is left to readers as Problem 4.10.3.) Thus any CMOS cell has the same features of unusually low quiescent power consumption and dynamic power consumption CV^2f, as that of Fig. 4.9.1.

In the CMOS cell of Fig. 4.9.2 the p-MOS circuit has transmission $g^d = xy$, which is dual to the transmission $g = x \vee y$ of the n-MOS circuit, where the superscript d on g means "dual."

The input resistance of a CMOS cell is extremely high and at least 10^{14} Ω. This permits large fan-outs from a CMOS cell without a significant increase of the quiescent power consumption. Thus if inputs do not change, CMOS has essentially no maximum fan-out restriction. Although fan-out increases the parasitic capacitance and consequently reduces the speed, the practical maximum fan-out is 50 or more, which is very large compared with other logic families.

In addition to extremely low power consumption, CMOS has the unique feature of immunity from power supply voltage fluctuation, noise, and temperature change as follows.

A typical plot of output voltage against input voltage of a CMOS cell, that is, a transfer curve, is shown in Fig. 4.9.3 for different conditions of temperature and power supply voltage V. The transfer curves show very little change over a wide range of temperatures. When the power supply voltage V changes, the transfer curve shifts without much changing its shape. Thus when V changes, v_{in} and v_{out} change proportionally, and the CMOS

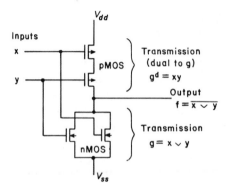

Fig. 4.9.2   CMOS NOR cell.

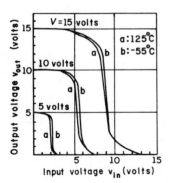

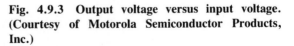

Fig. 4.9.3   Output voltage versus input voltage. (Courtesy of Motorola Semiconductor Products, Inc.)

cell performs logic operations correctly. This results in the capability of the CMOS cell to operate over a wide range of the power supply voltages. This range is wider than that for other logic families, and CMOS is widely used when the power supply voltage is not stable, such as in automobile applications.

The noise immunity of CMOS is high because its transfer curve rises very sharply. If two voltage values close to V volts and 0 volts are chosen to represent logic values 1 and 0, respectively, the CMOS cell can deliver the corresponding output values correctly, even if the input voltages fluctuate within a certain limit due to noise. Thus CMOS qualifies itself for applications in noisy environment, such as for automobiles. The degree of **noise immunity** is defined as the maximum noise voltage that can appear at the input without letting a CMOS cell malfunction from output 0 to 1 or from 1 to 0.

The performance of CMOS is compared with other logic families in Table 4.9.1.

Unfortunately when a CMOS cell has many inputs, its transfer curve shifts, depending on how many of the inputs change values. For example, in the 2-input NAND cell shown in Fig. 4.9.4a the transfer curve for the simultaneous change of the two inputs (1 and 2) is different from that for the sole change of input 1, with input 2 kept at a high voltage. This is because, unlike with static MOS networks where every driver MOSFET in its conductive state must have a much lower resistance than the load MOS-FET in order to have a sufficiently large voltage swing, none of the MOS-FETs in a CMOS cell in its conductive state has a much higher resistance than others (since one of the p-MOS and n-MOS circuits is always conductive, and consequently a sufficiently large voltage swing can be obtained without having a high resistance ratio). For example, the resistance of the p-MOS circuit for the change of only input 1 in Fig. 4.9.4a is twice as large as that for the change of the two inputs 1 and 2; so C is charged in a shorter time in the latter case. Other examples are shown in Fig. 4.9.4. Because of this problem of transfer curve shift, the number of inputs to a CMOS cell is practically limited to 4 if we want to maintain good noise immunity. If the

Table 4.9.1 CMOS Compared to Other Logic Families[a]

	STANDARD TTL	74L LOW-POWER TTL	DTL	9LS LOW-POWER SCHOTTKY TTL	74LS LOW-POWER SCHOTTKY TTL	34000 CMOS WITH 5-V SUPPLY	34000 CMOS WITH 10-V SUPPLY
Propagation delay	10 nsec	33 nsec	30 nsec	5 nsec	10 nsec	35 nsec	25 nsec
Flip-flop toggle frequency	35 MHz	3 MHz	5 MHz	80 MHz	40 MHz	5 MHz	10 MHz
Quiescent power	10 mW	1 mW	8.5 mW	2 mW	2 mW	10 nW	10 nW
Noise immunity	1 V	1 V	1 V	0.8 V	0.8 V	2 V	4 V
Maximum fan-out	10	10	8	20	20	50[b]	50[b]

[a] *Source:* Fairchild, *3400 Isoplanar CMOS Data Book*, 1975, pp. 2–3. (Courtesy of Fairchild Camera and Instrument Corporation.)
[b] Or as determined by allowable propagation delay.

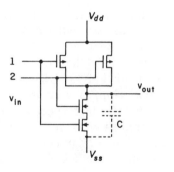

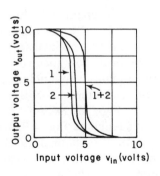

(a)

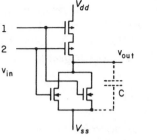

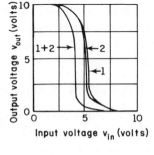

(b)

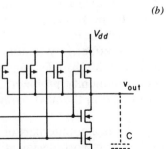

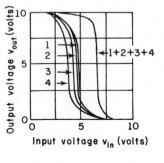

(c)

Fig. 4.9.4   Transfer curves for different numbers of inputs. (Motorola, 1974. Courtesy of Motorola Semiconductor Products, Inc.) (a) 2-input NAND. (b) 2-input NOR. (c) 4-input NAND.

power supply voltage is very stable and we need not worry about noise immunity, the number of inputs to a CMOS cell can be greater.

When we want to raise the speed, the switching speed of the p-MOS circuit should be comparable to that of the n-MOS circuit. Since the mobility of electrons is roughly 2.5 times as high as that of holes, as discussed in

Section 4.1, the channel width W of p-channel MOSFET must be much wider (often 1.5 to 2 times because the width can be smaller than 2.5 times due to different doping levels) than that of n-channel MOSFET in order to compensate for the low speed of p-channel MOSFET if the minimum channel length (the minimum channel length for a p-channel MOSFET is longer than that for an n-channel MOSFET on a CMOS chip) is used in each. [For example, see Figs. 4.9.8c and 4.9.9b and c.] Thus for high-speed applications NAND cells (such as Fig. 4.9.4a) are preferred to NOR cells, because the channel width of p-channel MOSFETs in series in a NOR cell (such as Fig. 4.9.4b) must be further increased such that the parasitic capacitance C can be charged up in a shorter time with low series resistance of these p-channel MOSFETs.

When designers are satisfied with low speed, the same channel width is chosen for every p-channel MOSFET as for n-channel MOSFETs, since CMOS requires already more area than p-MOS or n-MOS and a further increase by increasing the channel width is not desirable unless really necessary.

Logic Design of CMOS Networks

The logic design of CMOS networks can be done in the same manner as that of other MOS logic families, since the n-MOS circuit, with the p-MOS circuit regarded as variable load, essentially performs the logic operation, as seen from Fig. 4.9.2. Design procedures such as Procedure 4.2.1 discussed for static MOS networks can be used more effectively than in the case of static MOS networks, because more than 4 MOSFETs can be in series, unless we are concerned about the noise immunity problem. Also an appropriate use of transmission gates, discussed in the next paragraph, often simplifies networks.

The **transmission gate** shown in Fig. 4.9.5 is a counterpart of the transfer gate of n-MOS, and is used often in CMOS network design. It consists of a pair of p-channel and n-channel MOSFETs whose substrates are connected to V_{dd} and V_{ss}, respectively. The control voltage d is applied to the gate of the n-channel MOSFET, and its complement $\bar{d}$ is applied to the gate of the p-channel MOSFET. If d is a high voltage, both MOSFETs become conductive, and the input is connected to the output. If d is a low voltage, both

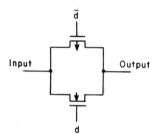

Fig. 4.9.5   Transmission gate.

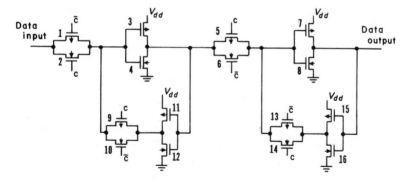

Fig. 4.9.6 *D*-type flip-flop.

MOSFETs become nonconductive, and the output is disconnected from the input, keeping the output voltage at its parasitic capacitance, as it was before the disconnection. Since the input and output are interchangeable, the transmission gate is bidirectional. A *D*-type flip-flop is shown in Fig. 4.9.6 as an example of CMOS circuits designed with transmission gates.

References: Motorola, 1974; Stephenson, 1975; RCA, 1972; Hittinger, 1973; Burns, 1964; *EDN*, Nov. 1, 1972, pp. 20–25; Dec. 1, 1972, pp. 30–35; Jan. 5, 1973, pp. 76–81; Feb. 5, 1973, pp. 38–44; Mar. 5, 1973, pp. 42–48; Apr. 20, 1973, pp. 48–51; Schnable et al., 1978.

Fabrication of CMOS

The fabrication of CMOS is illustrated in Fig. 4.9.7. Because both *p*-MOS and *n*-MOS must be fabricated, more masks are needed than for the previous logic families. Also the chip area is almost twice as large as for *n*-MOS or *p*-MOS because each CMOS cell requires both *p*-MOS and *n*-MOS circuits, and furthermore the **channel stops** to prevent any interaction of MOSFETs due to the formation of parasitic channels add more areas. The channel stops are sometimes called **guard rings** or **guard bands**.

Layout of CMOS

The layout is illustrated in Fig. 4.9.8, which clearly shows how the channel stops add extra areas. Since two different types (i.e., p^+- and n^+-types) of channel stops are required, unlike for the diffusion isolation well for bipolar transistors, we cannot combine adjacent channel stops, as we did in Fig. 3.3.3 for bipolar transistors, as seen in Fig. 4.9.8*c*. Channel stops in some places can be omitted for a certain range of operating voltages (*Electronic Design,* Nov. 8, 1978, p. 75) or for large spacing.

Fig. 4.9.9*b* and *c* shows two different layouts for the 3-input NOR cell

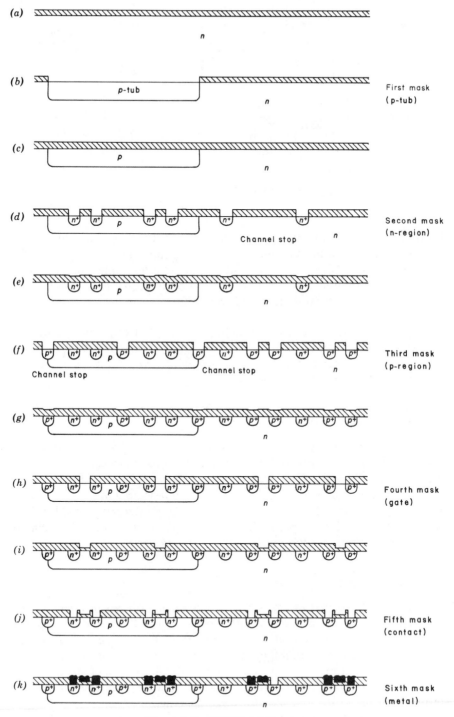

Fig. 4.9.7   CMOS fabrication.

of Fig. 4.9.9*a*. Fig. 4.9.10 shows a layout for the *D*-type flip-flop of Fig. 4.9.6.

References: RCA, 1973; Athanas, 1974.

Improvements of CMOS

In the beginning of CMOS development, CMOS was regarded as an absurd idea by many people because the number of MOSFETs is doubled, requiring a much larger area than *p*-MOS. But with the 4000 series CMOS with metal gates which RCA pursued in the late 1960s, the unique features of the CMOS were recognized. The performance, stability, and cost were substantially improved around 1970 by ion implantation (*Computer Design,* Nov. 1972, p. 48). Then in 1974 the CMOS with self-aligned silicon gates became available, further improving its performance, especially for LSI.

Now the unique features of CMOS, that is, extremely low power consumption (at low switching frequency) and immunity from power supply voltage fluctuation, noise, and temperature change, have been widely recognized and applied in many areas, such as crystal wrist watches and automobile electronics. Meanwhile its shortcomings, namely, the larger chip size than *p*-MOS or *n*-MOS, and consequently the higher prices, as well as the transfer curve shift, have been improved as follows.

Isoplanar CMOS

By applying the Isoplanar technology developed for bipolar transistors, combined with the use of silicon gates, the CMOS size was reduced by roughly one third, and the speed and stability were improved by Fairchild in 1974. Guard rings, which increase the chip size and lower the speed in conventional CMOS with metal gates, as shown in Fig. 4.9.11*a*, are eliminated by oxide isolation (Isoplanar technology) as illustrated in Fig. 4.9.11*b*. In addition, the speed is improved by the self-alignment of the silicon gate and the reduction in the sidewall parasitic capacitances. Other manufacturers also developed similar isolation techniques. All these are sometimes collectively called **selectively oxidized CMOS (SOCMOS)** (*EDN,* June 24, 1981, p. 97).

References: Fairchild, 1977; *Electronics,* Oct. 2, 1975, p. 63; Dec. 7, 1978, pp. 41, 42; Nov. 22, 1979, p. 113; July 17, 1980, pp. 39, 40; July 31, 1980, p. 83; *Spectrum,* Jan. 1980, p. 48; Wollesen, 1979; Winter and Wagner-Korne, 1980; *Electronic Design,* July 19, 1980, pp. 31, 32.

The power consumption of Isoplanar CMOS is compared with that of TTL and ECL in Fig. 4.9.12. Notice that at high frequencies the power consumption is greater than that of ECL.

Because of its reduced size, the **buffered CMOS,** which has improved performance, can be realized by adding a pair of inverters without changing

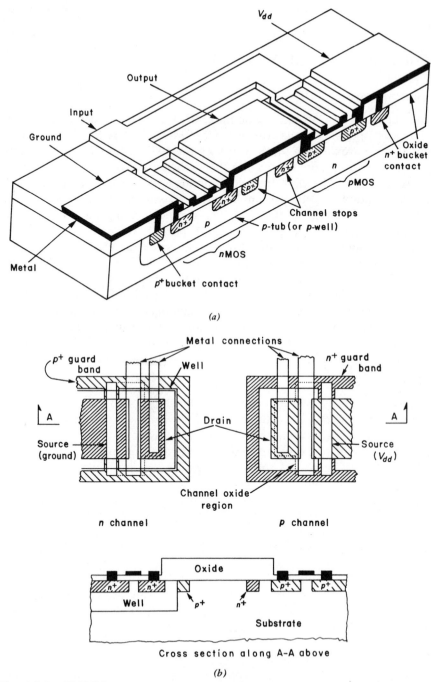

(a)

(b)

Fig. 4.9.8   CMOS layout. (Courtesy of Motorola Semiconductor Products, Inc.) (*a*) **Perspective.** (*b*) **Top and cross-section geometry.** (*c*) **Layout for the inverter of Fig. 4.9.1.**

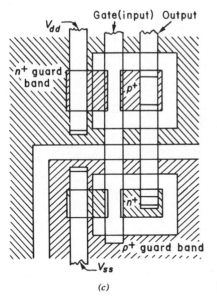

Gate(input) Output

V_{dd}

n^+ guard band

p^+

n^+

p^+ guard band

V_{ss}

(c)   **Fig. 4.9.8**   (*Continued*)

the size much. The NAND cell in Fig. 4.9.4*a*, for example, is buffered, as shown in Fig. 4.9.13*a*. The rise time and fall time of a waveform at output v_{out} in Fig. 4.9.4*a* are different for different combinations of input values, as shown by the thin lines in Fig. 4.9.13*b*. Using the buffered CMOS of Fig. 4.9.13*a*, this is improved, as illustrated by the bold line in Fig. 4.9.13*b*, due to the signal amplification of the pair of inverters in Fig. 4.9.13*a*. The transfer curves are also improved, as shown in Fig. 4.9.13*c*, improving the noise immunity (*Electronics,* Feb. 21, 1974, pp. 89, 90).

The layout of CMOS has other variations, such as closed COS/MOS logic (C^2L), where the drain, channel, and source of each MOSFET are laid out in concentric form (Dingwall and Stricker, 1977; Barnes, Nov. 8, 1978; *Electronics,* July 31, 1980, p. 83). Although *p*-wells are used in the CMOS discussed so far, CMOS with *n*-wells, obtained by exchanging *p*-MOS and *n*-MOS, is also available (*Electronics,* Dec. 4, 1980, pp. 39, 40; Ohzone et al., 1980; Maddox, 1981).

Further Improvements of CMOS

The disadvantages of conventional metal CMOS, compared with *n*-MOS, have been low speed, large chip area, and high cost. In 1973 the CMOS layout was at least 2 times larger and also slower than that of *n*-MOS. Because its structure was more complex, and diffusing the *p*-well takes time, fabrication took longer, requiring extra processes with furnaces, so a fabricated wafer cost twice as much as the *n*-MOS. Because of the larger chip size and its complexity, the yield was low, and a chip cost about four times more. Thus in total CMOS was about eight times more expensive than *n*-

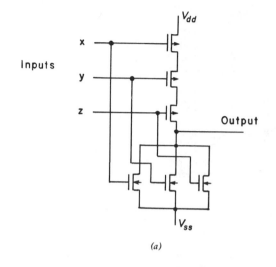

(a)

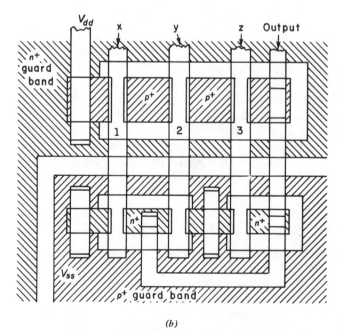

(b)

Fig. 4.9.9   Two different layouts for 3-input NOR cell. (RCA, 1973. Courtesy of RCA Corporation.) (*a*) **3-input NOR cell.** (*b*) **Layout for** (*a*). (*c*) **Alternative layout for** (*a*).

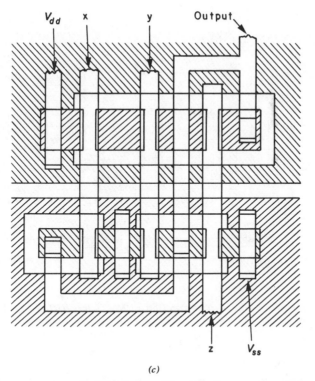

(c)

Fig. 4.9.9 (*Continued*)

MOS. Now because of the above improvements and others, the speed of the CMOS is closer to that of the *n*-MOS. Also the chip area of the CMOS is coming down to being only 20 to 25% larger than that of *n*-MOS. So the cost is also decreasing to about twice that of *n*-MOS. The speed is also improved, and gate delays of 5 to 10 nsec are common (*EDN,* June 20, 1979, p. 110).

As the integration size increases, dissipation of the heat generated in the chip is becoming a serious problem. In this sense CMOS is becoming a vitally important logic family for VLSI.

References: Electronics, Oct. 13, 1977, pp. 65, 66; Oct. 25, 1979, p. 114; Nov. 22, 1979, p. 113; Dec. 20, 1979, pp. 80, 81; Wollesen, 1979; Fisher and Young, 1979; Schnable et al., 1978; Twaddell, June 24, 1981.

CMOS/SOS (CMOS by Silicon on Sapphire)

CMOS built on a sapphire substrate has many advantages over the CMOS built on a bulk silicon substrate as shown in Fig. 4.9.8. It is faster and smaller and requires fewer fabrication steps, but sapphire is more expensive than silicon.

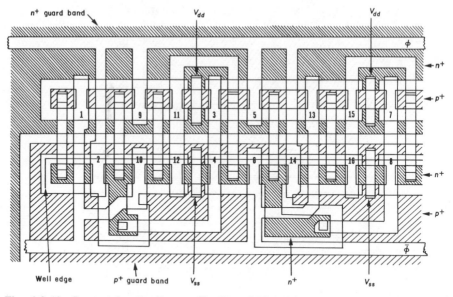

Fig. 4.9.10 Layout for the *D*-type flip-flop of Fig. 4.9.6. (RCA, 1973. Courtesy of RCA Corporation.)

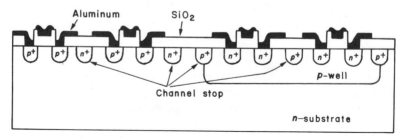

(a)

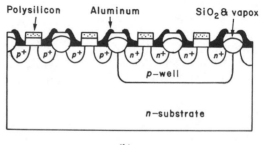

(b)

Fig. 4.9.11 Isoplanar CMOS. (*a*) Conventional metal-gate CMOS structure. (*b*) Isoplanar CMOS structure reduces area by 35%.

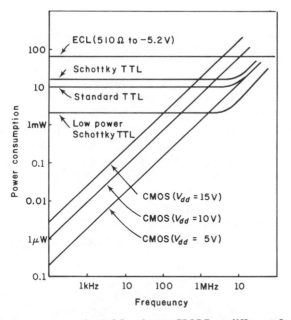

Fig. 4.9.12 Power consumption of Isoplanar CMOS at different frequencies. [Fairchild, *CMOS Data book* **1977. Courtesy of Fairchild Camera and Instrument Corporation.)**

CMOS fabrication on a sapphire substrate is illustrated in Fig. 4.9.14, though there are many variations (Capell et al., 1977). Both *p*-MOS and *n*-MOS are built as separate islands. (In some CMOS/SOS technology MOS-FETs of a mode not discussed in this book are used.) Connections run on the sapphire substrate. Because drain and source diffusion regions are built directly on the sapphire substrate which is a good insulator, parasitic capacitances of conventional CMOS (Fig. 4.9.8) are greatly reduced. (In particular, the parasitic capacitances between the drain and source regions and the ground, which account for almost half the parasitic capacitances of conventional CMOS, become almost negligible.) Also guard rings, which occupy large areas in the case of conventional CMOS, are completely eliminated, allowing *p*-MOS and *n*-MOS to be placed very closely. Thus the speed is two to three times faster. Also the chip area is reduced to one half to one third that of conventional CMOS. The power consumption at high frequency is greatly reduced, yielding a very low delay–power product, unmatched by any other logic family.

Sapphire wafers are several times more expensive than silicon wafers. Another problem is their relatively low yield. Because of these disadvantages, the application of CMOS/SOS has been limited primarily to high-performance areas (*Electronics,* July 31, 1980, pp. 82–84; Aug. 14, 1980, p. 8; Sept. 11, 1980, pp. 106, 108; *EDN,* June 24, 1981, pp. 98, 99).

Because sapphire has poor heat conduction, CMOS/SOS can withstand only one third the power consumption that *n*-MOS or *p*-MOS on silicon wafers can withstand. Thus other logic families have not been built on sapphire, although depletion-load *n*-MOS was tried (*Electronic Design,* September 27, 1979, p. 25).

Channel lengths of 3 to 5 μm are commonly used with delay times of 1 to 2 nsec and 2 to 3 nsec respectively (*EDN,* June 20, 1979, p. 110). Improvements have been continued, and CMOS/SOS with gate delay times of 200 psec for 0.5-μm channel length was reported (Ipri, 1978). However, the maximum speed of any logic family is usually measured by a ring counter, and the speed is greatly lowered in other types of networks where each MOS cell has many fan-out connections, unlike with TTL or ECL.

References: Ipri, 1978; Eaton, 1975; King, 1975; Capell et al., 1977; *Electronics,* Oct. 11, 1973, pp. 82–84; Feb. 21, 1974, p. 89; Jan. 3, 1975, pp. 66, 67; Aug. 18, 1977, p. 93; *Electronic Design,* July 19, 1978, pp. 36, 38; Nov.

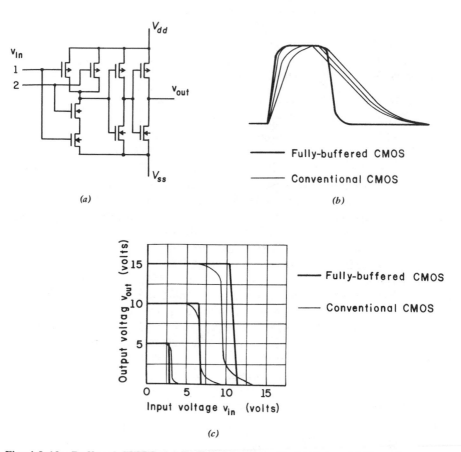

Fig. 4.9.13   Buffered CMOS. (*a*) Buffered NAND cell. (*b*) Delay. (*c*) Transfer curves.

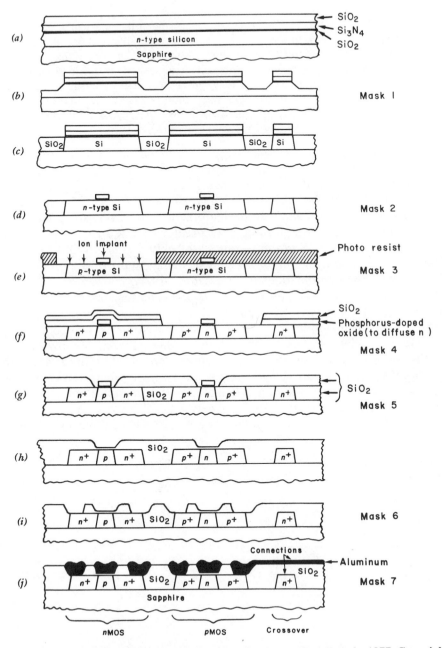

Fig. 4.9.14 Fabrication of CMOS/SOS. (Reprinted from Capell et al., 1977. Copyright © McGraw-Hill, Inc., 1977. All rights reserved.)

22, 1978, pp. 27, 28; Sept. 27, 1979, pp. 25, 26; Smith and Garen, 1978; Ronen and Micheletti, 1975; Stebnisky and Feller, 1980; Tsantes, Sept. 20, 1980; Roy, 1981; *EDN*, Nov. 20, 1980, pp. 70–76.

> ***Remark 4.9.1:***   A MOSFET formed on thick silicon dioxide which is an insulator can have a speed and packing density comparable to those of CMOS/SOS. It had been difficult to make such a MOSFET, but Texas Instruments reported producing it by annealing polysilicon deposited on silicon dioxide by laser beam (*Electronics*, Nov. 22, 1979, pp. 39, 40). Thus a cheap and easily grown silicon chip may replace SOS in the future (*Electronics*, Mar. 1, 1979, pp. 88, 90; Nov. 22, 1979, pp. 39, 40; Nov. 6, 1980, pp. 44, 46; *Electronic Design*, Jan. 18, 1979, pp. 23–25; Lam et al., 1979).   □

Variations of CMOS

In the case of the CMOS discussed so far, each CMOS cell consists of a pair of *p*-MOS and *n*-MOS circuits which realize dual transmission functions, as explained with Fig. 4.9.2. In design practice, however, some variations are often used. For example, Fig. 4.9.15 realizes NOR and NAND. One of the *p*-MOS and *n*-MOS circuits consists of only a single MOSFET. Thus the chip area is reduced, but the speed is slower and more power is dissipated for certain input combinations (Cooper et al., 1977).

By understanding the nature of the tasks to be realized and integrating the design of logic networks and electronic circuits, some tasks can be realized by much simpler circuits than those based on the cell concept. To illustrate this, let us consider the divide-by-three networks shown in Fig. 4.9.16, which are to produce a pulse train of a frequency a third that of the given pulse train. Figure 4.9.16*a* is a static network where three flip-flops are connected by transmission-gate-like CMOS circuits. Even if the input pulse train has uneven periods, this network still works. When an input pulse train has equal the time intervals between adjacent pulses which are

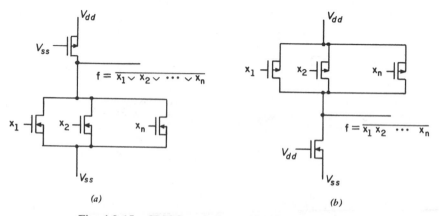

(*a*)   (*b*)

Fig. 4.9.15   CMOS variations. (*a*) NOR. (*b*) NAND.

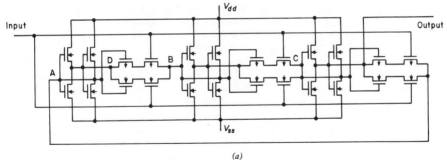

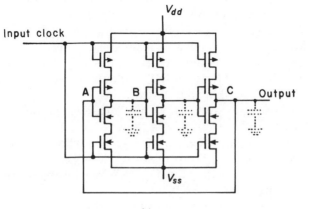

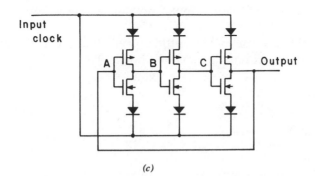

Fig. 4.9.16 **Divide-by-three networks [Reprinted with permission from** *Electronic Design* **(Bingham, 1978). Copyright © by Hayden Publishing Company, Inc., 1978.]** (*a*) **Static divide-by-three network.** (*b*) **Dynamic divide-by-three network.** (*c*) **Dynamic divide-by-three network. (This is used on CMOS/SOS only, and no power supply is required.)**

Table 4.9.2   Truth Table for Fig. 4.9.16

INPUT	A	B	C
1	1	0	0
0	1	0	1
1	0	0	1
0	0	1	1
1	0	1	0
0	1	1	0
1	1	0	0

equal to pulse widths, in other words, when we have a clock of 50% duty cycle at the input, we can use the dynamic CMOS network shown in Fig. 4.9.16*b*. The parasitic capacitances shown by the dotted lines are charged or discharged whenever a clock pulse appears. The electric charges stored at the parasitic capacitances must be maintained until the next clock pulses. (This network works at frequencies as high as 100 MHz.) The network in Fig. 4.9.16*b* requires only 12 MOSFETs, whereas the network in Fig. 4.9.16*a* has 24 MOSFETs. The dynamic network in Fig. 4.9.16*c* is even simpler, using diodes whose realization on a chip is easier than that of MOSFETs, but this network can be used only in the case of CMOS/SOS. Notice that no power supply is needed, because for such a simple network the output voltage drop is minor. The truth table, Table 4.9.2, applies to all of Fig. 4.9.16*a*, *b*, and *c* except that for Fig. 4.9.16*c* the input values must be inverted (Bingham, 1978).

CMOS cells with double outputs, such as those of Fig. 4.9.17, are used by Hewlett-Packard, where different functions f_1 and f_2 are realized by the addition of a diode. By having two different output functions at each CMOS cell, we have greater logic capability; in other words, we can usually design a CMOS network for given output functions with fewer cells (Capell et al., 1977; Mei, 1976).

Since CMOS consumes very little power, dynamic circuitry is not used for the sake of power reduction alone, though this may change as integration

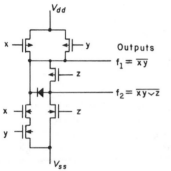

Outputs

$f_1 = \overline{xy}$

$f_2 = \overline{xy \vee z}$

Fig. 4.9.17   CMOS cell with duo outputs (Reprinted from Capell et al., 1977. Copyright © McGraw-Hill, Inc., 1977. All rights reserved.)

size increases. Dynamic circuitry is often used for other unique reasons, as the following examples show.

Domino CMOS, illustrated in Fig. 4.9.18, consists of pairs of a logic CMOS cell and an inverter CMOS cell. The logic CMOS cell in each pair (such as cells 1, 3, and 5) has the p-MOS circuit, consisting of a single p-MOSFET with clock, and the n-MOS circuit, consisting of many n-MOSFETs with logic inputs and a single n-MOSFET with a clock. Each logic CMOS cell is followed by an inverter CMOS cell (such as cells 2, 4, and 6). When a clock pulse is absent at all terminals labeled c, all parasitic capacitances (shown by dotted lines) are charged to value 1 (i.e., a high voltage) because all p-MOSFETs are conductive. Thus the outputs of all inverters become value 0. Suppose that $x = v = 1$ (i.e., a high voltage) and $y = z = u = 0$ (i.e., a low voltage). When a clock pulse appears, that is, $c = 1$, all p-MOSFETs become nonconductive. Then the outputs of cells 1, 2, 3, 4, 5, and 6 become 0, 1, 1, 0, 0, and 1, respectively. Notice that the output of cell 3 remains precharged because its n-MOSFET for u remains nonconductive. Domino CMOS has the following advantages. It has a small area because the p-MOS circuit in each logic cell consists of a single p-MOSFET; it is faster (about twice) than the static CMOS discussed so far because parasitic capacitances are reduced by using a single p-MOS in each cell and each

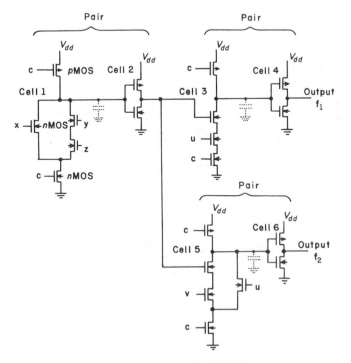

Fig. 4.9.18   Domino CMOS.

logic cell is buffered by an inverter; and it is free of glitches (i.e., transition is smooth) because at the output of each cell a high voltage remains or decreases, but no voltage increases from low to high (Murphy and Edwards, 1981).

Dynamic circuitry is also used to reduce the chip area because of its simple layout geometry for some tasks. Fig. 4.9.19*a* shows a conventional

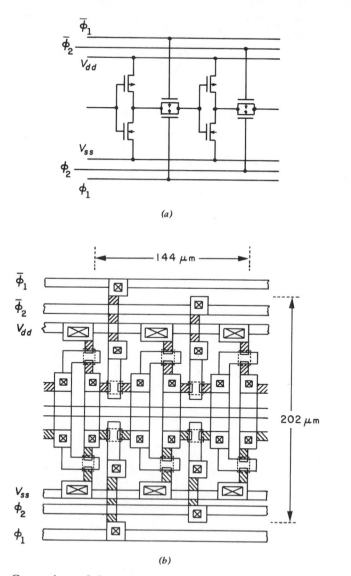

(a)

(b)

Fig. 4.9.19   Comparison of dynamic and conventional CMOS. (© 1973 IEEE.) (*a*) Conventional CMOS shift register. (*b*) Layout for conventional CMOS in (*a*). (*c*) Dynamic CMOS shift register (clocked CMOS). (*d*) Layout of clocked CMOS in (*c*).

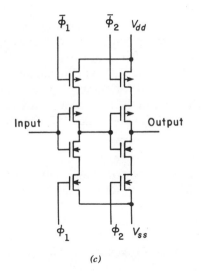

(c)

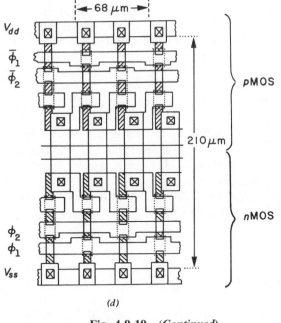

(d)

Fig. 4.9.19 *(Continued)*

static CMOS shift register. (Only 1-bit section is shown. For a shift register of many bits, this section must be repeated.) A two-phase clock ϕ_1 and ϕ_2, shown in Fig. 4.8.5b (waveforms for ϕ_1 and ϕ_2 do not overlap) is supplied. To each cell, ϕ_1 (or ϕ_2) and its complement $\overline{\phi}_1$ (or $\overline{\phi}_2$) are connected. The layout (Fig. 4.9.19d) for the clocked CMOS register (C^{2}MOS) by Toshiba shown in Fig. 4.9.19c occupies a smaller area than that for the conventional CMOS shift register of Fig. 4.9.19b because of its regular structure. Notice that the transmission gages in Fig. 4.9.19a are deleted in Fig. 4.9.19c, although the number of MOSFETs is the same. [This means that the number of MOSFETs is not directly important for compact layout, but the connection configuration is important, though the reduction of the number of MOSFETs is still indirectly important.] (Manabe et al., 1976; Suzuki et al., 1973).

When we use the above variations properly, and possibly others, the chip areas can be reduced with improved performance. Often n-MOS networks are mixed with CMOS networks on the same chip, for the sake of higher speed or smaller area.

> *Remark 4.9.2   Analog Circuits by CMOS:*   CMOS had been considered strictly for digital networks until 1973, when RCA placed CMOS and a bipolar transistor on the same chip (for the CA 3100 video operational-amplifier IC). Now CMOS for digital logic as well as n-p-n bipolar transistors, resistors, capacitors, and diodes for analog circuits can be easily placed on the same chip. Such ICs are sometimes called **BiMOS**. A BiMOS chip can provide as much as 30 volts and 0.5 amperes (*Electronic Design,* Nov. 8, 1978, pp. 74–82). Also by using CMOS alone, we can realize digital logic networks and analog circuits on the same chip, without bipolar transistors. Up to 20 MHz and 30 volts, CMOS can match or beat the performance of all other technologies: n-MOS, p-MOS, bipolar transistor analog circuits, bipolar digital circuits, and I²L.
>
> *References:* Bingham, 1978; Fisher and Young, 1979; Wilenken, 1979.   □

References on CMOS: Athanas, 1974; Ahrons and Gardner, 1970; Lancaster, 1977; Walker, 1975; Stiglianese, 1974; George and Schmidt, 1972; Foltz and Musa, 1972; Karstad, 1973; Garrett, 1973; Gaskill, 1973; Torrero, 1974; Altman, 1975; Nguyen-Huu, 1975; Yen, 1975; Schamis, 1973; Redfern and Jorgensen, 1973; Furlow, 1973; *EDN,* May 20, 1974, pp. 35–40; Aug. 5, 1974, pp. 69–73; Marlowe and Hasili, 1973; *Electronics,* Jan. 13, 1972, pp. 6, 7.

▲ 4.10   Other MOS Technologies

In this section some other MOS technologies are discussed. These technologies were at one time considered strong candidates for future VLSI technology. Today they do not seem to meet such expectations, but appear to find unique applications in other areas. Here we want to discuss them as examples of flexibility and imagination in designing solid-state circuits and also as examples of the difficulty of technological forecast due to the complex nature of IC technology.

DMOS (Double-Diffused MOS)

DMOS was developed by Y. Tarui et al. (1970). A very precise narrow channel can be made by ion implantation, as illustrated in Fig. 4.10.1. (The threshold voltage of the MOSFET can be adjusted by biasing the substrate.) The drift region made of n^--, p^--, or π-type silicon is always depleted (i.e., depleted of electrons), where n^- is lightly doped n-type, p^- is lightly doped p-type, and π is more lightly doped p-type than p^-. Thus when the narrow channel is turned on by the gate voltage, the device switches at very high speed. The extremely narrow channel and self-alignment feature are independent of masks and etching.

Speed can be as high as 1 nsec. But since it is relatively difficult to attain high packing density, D-MOS probably will not be used extensively in LSI but due to its unique feature it may be used in special applications such as fast switching devices with high breakdown voltage and high-power devices. [For example, DMOS is used in Texas Instruments' plasma display driver IC, BIDFET (*EDN*, Sept. 20, 1979, pp. 54, 55).]

References: Allison and Russell, 1971; Cange et al., 1971; Tarui et al., 1970; Ohta et al., 1975; Sigg and Lai, 1977; Torimaru et al., 1978; Kamuro et al., 1980. For power-handling applications, see *Electronic Design,* Mar. 29, 1980, pp. 27, 28; Rodriguez (1980); for advantages and difficulties of DMOS, see *Electronics,* June 9, 1977, pp. 104, 111; Aug. 18, 1977, pp. 92, 94.

VMOS (V-Groove MOS)

VMOS developed by Rodgers and Meindl (1974) is an n-channel MOSFET formed along the wall of a groove. Figure 4.10.2a shows a cutaway view of VMOS, where layers for drain, π-region (i.e., very lightly doped p-region), and channel are formed on a n^+-type silicon substrate which works as a source and is grounded. The wall of this V-shaped groove is covered by a silicon-dioxide insulation layer and then by an aluminum layer which works as a gate of the VMOS, as shown at the far left in Fig. 4.10.3b. A top view of VMOS is shown in Fig. 4.10.2b. Since the p-region stretches as a channel above the substrate, it is easy to form a channel length as short as 1 μm,

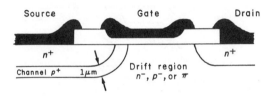

Fig. 4.10.1   DMOS.

which is essential for high-speed MOSFETs, using conventional photolith-ography, because the channel length (i.e., the layer thickness) is independent of mask resolution. Since the MOSFETs are formed on the slopes of the grooves, the packing density of MOSFETs on a chip is high.

Since the source of each VMOS is grounded, each VMOS cell can realize only the NOR operation, as illustrated in Fig. 4.10.3*a* along with the layout geometry in Fig. 4.10.3*c*. Because of this low logic capability, logic networks tend to require more cells than *p*-MOS or *n*-MOS to realize given functions, and this tends to offset the advantage of the small chip area.

Since it is difficult to improve yield (Remark 4.10.1) and VMOS can be replaced by *n*-MOS which has comparable speed and power dissipation with greater logic capability, VMOS probably will not be used in LSI logic net-works except in special cases. But VMOS has the very unique feature of a very heavy current-handling capability with high switching speed, which leads VMOS into many important applications (Remark 4.10.2).

References: Rodgers and Meindl, 1974; Jenné, 1977; Rodgers et al., 1976, 1977; *Electronics,* Sept. 18, 1975, p. 65; Jan. 22, 1976, p. 14; June 23, 1977, pp. 29, 30; Aug. 18, 1977, pp. 92, 100–106; Oct. 27, 1977, p. 93.

Remark 4.10.1:   When VMOS was introduced in 1974, it appeared to be a strongest contender for future LSI technology, and many manufacturers were interested in it. After several years VMOS was abandoned because the unique

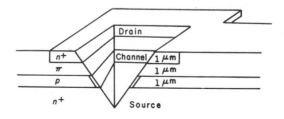

(a)

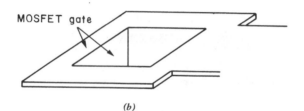

(b)

Fig. 4.10.2   Structure of a VMOS transister. (*a*) Cutaway view. (*b*) Top view.

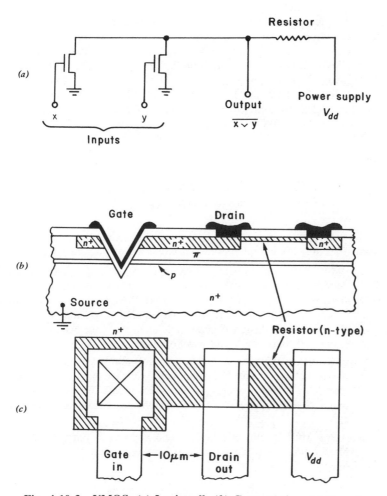

Fig. 4.10.3 VMOS. (*a*) Logic cell. (*b*) Cross section. (*c*) Top view.

structure of VMOS was sensitive to a physical phenomenon called the Kooi effect, and it was difficult to develop a processing technique to alleviate this within a short time. Because of lack of sustained efforts, it is difficult to judge whether it is still a viable technology. The whole story endorses the complexity of IC technology and the difficulty in assessing new technology (*Electronics,* Oct. 11, 1979, p. 42; Dec. 20, 1979, p. 40; *Spectrum,* Apr. 1980, p. 20). ☐

Remark 4.10.2: In VMOS the substrate serves as a source terminal, letting heavy current flow vertically throughout the substrate. No direct current flows through its gate. Thus, VMOS was high input impedance, which a bipolar transistor does not have. At low frequencies VMOS can handle 1,000 volts at 40 amperes, replacing bipolar transistors in many power-handling applications, and can operate at as high as 200 MHz or possibly 4 GHz. It can handle a few hundred watts. VMOS applications include hi-fi audio power amplifiers, broadband high-frequency amplifiers, and also switching power supplies (which convert alternating-current power sources into direct currents at arbitrary voltages, with lower cost, lighter weight, and smaller size than conventional power supplies) (*Electronic Design,* June 7, 1980, pp. 131–137; July 5, 1980, pp. 65–71; Aug. 2, 1980, pp. 83–87; June 11, 1981, pp. 174–180, 182, 184; *Electronics,* June 22, 1978, pp. 105–112; June 22, 1978, pp. 105–112; May 22, 1980, pp. 143–152; *EDN,* June 20, 1977, pp. 71–75; Sept. 20, 1980, pp. 137–142; Feb. 4, 1981, pp. 94–108).□

Problems

4.10.1   Design a static CMOS cell for $f = \overline{x_1 \vee x_2 x_3 \vee x_2 x_4}$ with a minimum number of MOSFETs. Assume that only noncomplemented variables are available as inputs.

4.10.2   Design a full adder in static CMOS, using a minimum number of CMOS cells. Assume that only noncomplemented variables are available as network inputs.

4.10.3   A static CMOS cell consists of two circuits N_1 and N_2, as illustrated in Fig. P4.10.3a, where a and b are connected to the power source of a positive voltage and to the ground, respectively. N_1 consists of p-MOSFETs, and N_2 consists of n-MOSFETs. The same voltages are simultaneously applied to N_1 and N_2 as the inputs $x_1, \ldots, x_n$. N_1 and N_2 are designed such that when the path between b and f is nonconductive, the path between a and f is conductive, and vice versa.

(a)   What relationship exists between the transmission between a and f and the transmission between b and f? Explain why. (Notice that these transmissions are "open" or "closed" states of paths when MOSFETs are regarded as make-contacts of relays. These transmissions are switching functions of x_1, . . . , x_n.)

(b)   When the p-MOS circuit of a CMOS cell is given in Fig. P4.10.3b, design the n-MOS circuit. Also derive a switching expression for f.

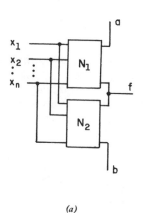

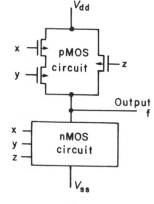

(a)   (b)

Fig. P4.10.3

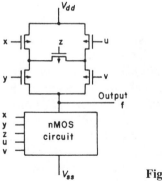

Fig. P4.10.7

4.10.4 Draw a layout for the CMOS NOR cell of Fig. 4.9.2, using the geometries in Figs. 4.9.8 through 4.9.10. (Dimensions other than channel width and length need not be precise.) From the assumption that the CMOS inverter in Fig. 4.9.1 has a p-channel MOSFET with a channel width of 15 μm and a channel length of 5 μm and an n-channel MOSFET with a channel width of 10 μm and a channel length of 5 μm, determine the dimensions for the NOR cell such that the paths in the p- and n-channel MOSFET circuits, when conductive, have no greater resistances than those in the inverter, respectively. Make the layout as compact as possible.

4.10.5 Solve Problem 4.10.4, by replacing 15 μm by 10 μm (i.e., the p- and n-channel MOSFETs have the same channel width of 10 μm). Make the layout as compact as possible.

4.10.6

(a) Design a static CMOS network for $x \oplus y$, using a minimum number of CMOS cells, assuming that only noncomplemented variables x and y are available as network inputs.

(b) Draw a layout for this network, using the geometries in Figs. 4.9.8 through 4.9.10. (Dimensions need not be precise.)

4.10.7

(a) When the p-MOS circuit of a static CMOS cell is given in Fig. P4.10.7, design the n-MOS circuit.

(b) Derive a switching expression for output function f.

(c) Draw a layout for the entire CMOS cell, using the geometries in Figs. 4.9.8 through 4.9.10. (Dimensions need not be precise.)

4.10.8 Explain how the subnetwork between points A and B in Fig. 4.9.16a works, showing waveforms at points A, B, C, and D.

4.10.9

(a) Suppose that $y = u = 1$ and $x = z = v = 0$ in the domino CMOS in Fig. 4.9.18. Find the output values of all cells before and after a clock pulse appears at c.

(b) Derive switching functions at outputs f_1 and f_2.

4.11   Chapter Summary

Historically bipolar logic families were developed first. The progress of MOS logic families, however, has been tremendous. The improvements in speed, size, power consumption, and yield appear to overwhelm bipolar logic families, though bipolar logic families also continue to make improvements. The current trend is toward the use of more complex solid-state circuit structures for MOS for the improvement of performance, and consequently more complex fabrication along with more masks, though this tends to lower yields and raise costs (*Electronics,* Sept. 13, 1979, p. 122; *Electronic Design,* Mar. 29, 1979, pp. 76, 77, 79). In a sense, MOS logic families are approaching bipolar logic families in speed, yield, and cost, though the gap may not disappear completely, and probably each maintains its own unique application areas.

Even with the current status of MOS technology, LSI or VLSI automatically means MOS, since we can pack a much greater number of MOS cells into a single chip due to smaller size and lower power consumption than bipolar logic families. Because of an enormous number of MOSFETs that can be packed in a VLSI chip, the time for design and corrections has become a very serious problem which did not exist for smaller integration sizes, as discussed in later chapters.

Although *p*-MOS is the oldest MOS technology, *n*-MOS is more extensively used. When high speed is important, static *n*-MOS is used, and when low power consumption is important, dynamic *n*-MOS is used, usually realized on the same chips with static *n*-MOS. CMOS is used when the lowest power consumption is important at the sacrifice of cost, size, and speed. CMOS is becoming important for VLSI because of low power consumption.

CHAPTER 5

Comparison of Logic Families and Technological Trends

In this chapter various logic families are compared in terms of electric performance. Since different logic families are increasingly mixed in systems, problems in mixing them are discussed. The current technological trends in realizing faster LSI/VLSI chips are also discussed.

5.1 Comparison of Logic Families

For specific applications, such as toys and watches, we need to choose the most appropriate logic families. But in most digital systems which are more complex, such as computers, we need a mixture of logic families, namely, the most appropriate logic family for each different part of the system. For example, we choose ECL for networks requiring high speed, CMOS for low-speed networks requiring low power consumption, bipolar transistors for analog networks, and so on.

Comparison of different logic families is usually not straightforward; only under given specific conditions may logic families be compared. Otherwise it is practically impossible to choose a best logic family.

Speed

The delay time of each gate or each MOS cell gives some idea of the speed of its logic family, but unless we consider the logic capability of the logic family and integration size, we cannot estimate the speed of networks designed with the logic family.

For example, ECL has a much greater logic capability than NAND of I^2L. Thus ECL networks usually have fewer levels than NAND networks and consequently higher speed than a simple comparison of the delay times of individual ECL and I^2L gates implies. Also, the integration size of a chip must be considered if we want to get an accurate speed comparison. If the integration size of ECL packages is too small, signals must go out to the connections on a pc board. Thus the ECL speed advantage deteriorates. MOS can have a much greater integration size. So the entire network implemented with MOS in a single chip could have a higher speed than the low speed of each MOSFET indicates. Thus a comparison of speed is not easy.

Logic Capability

Comparison of logic capabilities is also not an easy task, because the number of levels, or gates, required for networks varies with different functions, and that number varies with different electronic circuit implementations. Although an intitutive comparison based on the average network size can be made only after the design of many networks, let us give some ideas using simple cases.

1. AND-OR-INVERT has obviously greater logic capability than NOR.

2. The mixture of NOR and OR gates has almost the same logic capability as have NOR gates alone in terms of gate count, unless a maximum fan-in and fan-out is imposed. But if maximum fan-in and fan-out restrictions are imposed, it has greater capability in reducing the level count.

3. The mixture of NOR and AND gates has greater logic capability than have NOR gates alone. Any function of 3 or less variables can be implemented with 6 or less gates in the case of a mixture of NOR and AND gates, and with 7 or less gates in the case of NOR gates only. On the average the use of NOR gates alone requires one more gate than the mixture of NOR and AND gates in the case of these functions (Muroga, 1970).

Delay-Power Product

For comparison of different logic families, a parameter called **delay** (or **speed**)**-power product** is usually used, as mentioned in Section 3.2. The smaller the delay-power product, the more preferable the logic family is, since the same delay time is obtained with smaller power consumption. We are essentially trying to find new logic families with smaller delay-power products. As an example, the delay-power products for different logic families which were in production in 1977 (not including those in exploratory stages) are shown in Fig. 5.1.1 (Shinozaki and Nishi, 1978).

However, since the delay-power product does not consider sufficiently the output load (or fan-out connections) of a gate (or cell), a fair comparison of different logic families is not possible, unless the output load is specified for a delay-power product. The delay-power product of a logic gate is usually measured based on a ring counter of gates. (Usually 10 or more gates are used. If there are less than 10 gates, we obtain a somewhat different delay-power product figure.) In this case each gate has only one fan-out connection. Often the delay-power product deteriorates badly when these gates are used in a network where many fan-out connections are required for some gates. For example, it is reported that an *n*-MOS cell has a gate

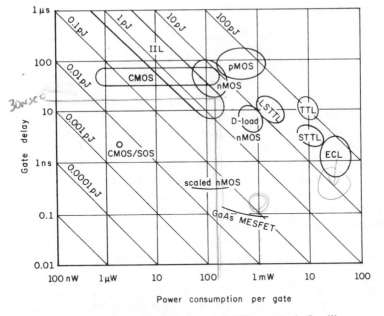

Fig. 5.1.1 Delay–power products of different logic families.

delay of 400 psec in the case of a ring counter, but the gate delay becomes 3 nsec when a chip of random-logic networks with 20,000 MOSFETs is realized. Generally the smaller the delay–power product of a logic family, the more sensitive it is to the output load (usually the parasitic capacitance load). On the other hand, logic families with very high power consumption, such as ECL and TTL, are not sensitive to the output load and can send signals over connections* on the pc board, without greatly reducing speed.

In summary, we cannot determine the electric performance and logic capability of a logic family from the delay–power product alone. We need to realize logic networks for a fair comparison. If speed is excessively lowered by many fan-out connections, we need to design networks using additional gates such that the maximum number of fan-out connections from each gate is reasonably small.

References: Capece, Sept. 13, 1979; Cook et al., 1973.

---

* Connections on the pc board work as transmission lines because parasitic inductances are significantly large and both parasitic capacitances and inductances are distributed over these connections. [See, e.g., Barna and Porat (1973, Sec. 12.2) or Dworsky (1979).] Long connections between logic gates on a chip, with gate delays of 10 psec or shorter, must be regarded as transmission lines, though shorter connections, or any connections between slower logic gates, on a chip are considered to have only parasitic capacitances, as they have been in all previous chapters.

Packaging Problems

Even if architecture, logic networks, electronic circuits, and layouts are superbly designed, the developments of new computers with high performance often collapse because of the packaging problems found with the prototypes. Heat generated in chips cannot be properly dissipated by cooling, or expected speed cannot be attained due to the following reasons. When we need a large number of high-speed networks, the networks must be spread over many chips due to the integration size limit. Consequently connections on pc boards greatly slow down the speed even though the networks inside the chips work very fast. This is becoming a serious problem in the case of subnanosecond speeds. For example, in the Amdahl computer 470 V/6, where about 2,000 LSI chips are used and each chip contains 100 ECL gates with gate delays of 500 to 600 psec, about one third of the propagation delay is in the chips, one third is in the pc boards (each pc board consists of 10 connection layers), and one third is in the interconnecting wires. Here the interconnecting wires among the pc boards are coaxial wires, and signals travel faster on them than on the connections on pc boards (Clements, 1979). Thus even if the speed of the logic gates is improved, the speed of entire computers cannot be improved unless connections outside chips become shorter, and consequently the size of the entire computer becomes smaller. We need to increase the integration size of chips and to shorten connections outside chips. But we have the following diverse packaging problems.

Since each chip can handle at most 4 to 5 W with current technology, the chip is destroyed with excessive heat if the designer tries to pack a large number of transistors to gain speed through greater integration size. The integration size of chips with ECL or TTL gates is limited by this heat dissipation problem. The integration size of chips with p-MOS or n-MOS becomes limited by heat dissipation when a chip contains more than 100,000 MOSFETs. Careful heat dissipation design is required. Thus CMOS will be used more often for LSI/VLSI.

References: EDN, Sept. 5, 1979, p. 38; June 20, 1980, pp. 47–60; Watson, 1980; LaBrie and Miller, 1980; McDermott, May 20, 1980; Steinberg, 1980.

The number of pads (or the number of pins when the chip is encased in a package) that can be provided on a chip is a vitally important factor for the speed of digital systems. A way to partition the entire network into chips is important for reducing the number of pads required. If the entire network is partitioned in an inappropriate way, we need excessively many pads. (For example, if a binary counter network is partitioned into two chips in an inappropriate way, we need many connections among the two chips. But if the counter is partitioned into less and more significant bit sections, the chip for the former section needs only one output connection to the other chip for the latter section.) **Rent's rule**,

$$P = aG^b$$

yields the average number of pads, P, required for the chip of G gates, where a is roughly 4 and b is roughly 0.6. This rule holds well when networks are partitioned into a number of chips (Landman and Russo, 1971; Donath, 1981). When the integration size increases, entire networks or subsystems are more often contained in fewer chips, without being partitioned into a number of chips, and consequently the required number of pads tends to be smaller than the above P. The number of pads which can actually be provided on a chip is limited due to two-dimensional geometry and is often much smaller than the number required for the number of gates on the chip. There can be as many as 150 pads on a chip by placing them on the chip periphery only (Lyman, Mar. 17, 1977). By placing pads throughout the chip, we can even more. IBM introduced a chip with a matrix of 17×17 (i.e., 289) pads with the 4300 series computers, as illustrated in Fig. 2.4.3. (IBM calls them solder bumps instead of pads.) All incoming and outgoing signals must be handled through these pads. When the number of pads is too small for the number of signals to be sent simulataneously, the usual solution is multiplexing. In other words, when we need to send a 16-bit number into or from the chip, for example, we sent a 4-bit number, that is, 1 nibble, four times, where 4 bits is called a **nibble**. [A multivalued signal could be another solution (*Electronics*, Aug. 17, 1978, p. 33).] This obviously reduces the signal transfer speed greatly. When a network requires a large number of pads, the chip area may not be efficiently utilized, effectively reducing the integration size. (In the case of gate arrays discussed in Chapter 9, it is not rare that only half the gates provided on a chip are used because of the pad number limitation.)

In order to improve the speed of a digital system, the entire system must be compact with all connections short so that signal propagation on these connections becomes fast. The connections on pc boards have been shortened by using many layers and smaller dimensions. For example, IBM reduced the system size by using ceramic carriers of 23 connection layers with the 4300 series computers, and the packing density was improved about 40 times over the IBM 370 (*Electronic Design*, Jan. 4, 1980, pp. 39, 40; Mar. 15, 1980, pp. 11, 12). In the mainframe computer 3081 announced in 1980, IBM further improved the packing density by placing 118 chips (about 45,000 gates in total) in one module with a 33-layer ceramic carrier which is cooled by water and helium (*Electronics*, Nov. 20, 1980, pp. 41, 42; *Datamation*, Nov. 1980, p. 43; *IBM JRD*, Jan. 1982).

References: Grossman, Mar. 15, 1979; Mar. 15, 1980; June 21, 1980; *Electronics*, Jan. 17, 1980, p. 195; June 5, 1980, pp. 40, 41; July 3, 1980, pp. 45, 46; Fetterolf, 1980; Masessa and Mohr, 1980; Beall, 1974; Wu, 1978; Chow, 1979; Vacca, 1979; Sullivan, 1978; Lassen, 1979; Amey and Balde,

1980; Southard, 1981; Heller, 1981. For the IBM 4300, see *Electronic Design*, Jan. 4, 1980, pp. 39, 40; Jan. 5, 1980, pp. 11, 12; Vilkelis and Henle, 1979; Clark and Hill, 1980; Werbizky et al., 1979, 1980; Santoni, 1980; Blodgett, 1980.

5.2   Compatibility

Often we need to mix different logic families in a digital system. We need to use logic families with high speed in key networks in the system, but as power consumption and costs are penalized with fast logic families, we use inexpensive slow logic families in other parts of the system. We may need to do this by using different IC packages for different logic families. Also increasingly often, we need to realize different logic families on the same chips, as mentioned with I²L and CMOS, though this is harder to realize (maybe not possible for some combinations of logic families) than the use of different IC packages for different logic families.

Different logic families usually require different power supply voltages and signal voltage levels. (This situation can occur even in the same logic family.) Different power supply voltages can be relatively easily provided, though this is not desirable in many cases. But different signal levels are cumbersome to handle, because whenever signals propagate from one logic family to another, we need interface circuits, such as those shown in Fig. 5.2.1. (There are many other interfaces, depending on the situation.) (Luckey, Dec. 1981; Carr and Mize, 1972, Chapter 6).

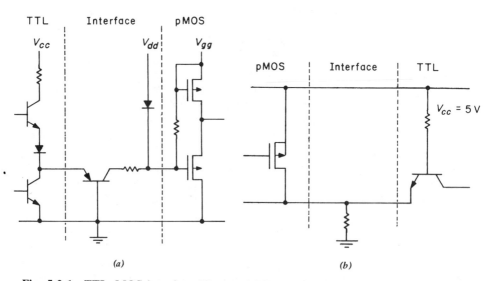

Fig. 5.2.1   TTL–MOS interface (Carr and Mize, 1972, Chap. 6. Courtesy of Texas Instruments, Inc.) (*a*) TTL to MOS. (*b*) MOS to TTL.

Table 5.3.1   Examples of Standard Networks (Availability of Each Item Depends on Logic Families)

Gates
  AND
  NAND
  OR
  NOR
  Exclusive-OR
  Exclusive-NOR
  OR/NOR
  AND-OR-INVERT
  Inverters
  Expanders
  AND/OR select
	Encoders
	  Binary to BCD
	  BCD to binary
	  Priority

	Drivers and buffers
	  Line drivers
	  Lamp drivers
	  Clock drivers
	  Relay drivers
	  High-fan-out buffer

Flip-flops
  R–S
  Latch
  T
  D
  J–K
	Arithmetic functions
	  Ripple adder
	  Look-ahead carry block
	  Subtractor
	  Magnitude comparator
	  Arithmetic unit
	  Ripple-through multiplier

Counters
  Divide-by-10
  Divide-by-12
  Divide-by-16
  Up/down
  7-stage
  14-stage
	Shift registers
	  2-bit
	  4-bit
	  8-bit
	  Multibit (serial)

Decoders
  BCD to decimal
  Excess-3 to decimal
  BCD to 7 segment
  4 to 16 lines
  3 to 8 lines
  2 to 4 lines
	Memories
	  1k RAM
	  4k RAM
	  16k RAM
	  1k ROM
	  4k ROM
	  16k ROM
	  64k ROM
	  Large serial-access read/write
	  CAM or CRAM
	Miscellaneous

5.3.   Family Size

Semiconductor manufacturers usually take many years to develop all kinds of **standard networks** (i.e., frequently used networks) in SSI or MSI off-the-shelf packages. Examples of standard networks are shown in Table 5.3.1. Even if users use custom LSI or microcomputers, they often need these packages. They also need second sources. Therefore the family size of each logic family is an important consideration.

When the performance of a logic family is improved by new processing, it takes many years to renovate the entire family.

Old logic families such as 54/7400 TTL, 4000A CMOS, and 10K ECL have the largest family sizes. So even when custom MOS LSI/VLSI chips are designed, the system design of the chips is usually tested with TTL, which has the largest family size, before designing logic networks with MOS.

When manufacturers want to introduce a new product, they are usually not sure whether the new product will be well accepted or whether they will have to discontinue it if it is not popular. In this case designers can develop the new product within a few months, using off-the-shelf packages, instead of spending one or two years with custom LSI chips, because off-the-shelf packages can be assembled quickly without considering the logic networks inside. Once the product is well accepted with high production volume, designers can replace these off-the-shelf packages later by custom LSI chips to reduce the cost and improve performance.

References: Cushman, 1973; Franson, Feb. 20, 1977.

5.4   Technological Trends

IC technology has made progress, as the chronology in Table 5.4.1 shows. The 1979 status of different logic families is summarized in Table 5.4.2 and Fig. 5.4.1. Logic families have different evaluations at different times. CMOS is such an example and will be more weighed as the integration size increases

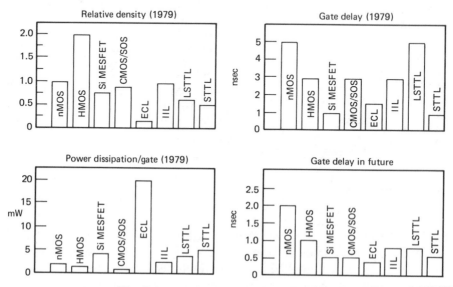

Fig. 5.4.1   Comparison of logic families.

Table 5.4.1   Chronology of IC-Related Events

1823	Silicon was discovered by J. J. Belzelius. C. Babbage designed the analytical engine, incorporating the features of a modern computer.
1853	Silicon crystals were grown by S. C. Deville.
1886	Germanium was discovered by C. Winkler.
1904	Diode vacuum tube was invented by J. A. Fleming.
1906	Triode vacuum tube was invented by L. de Forest.
1939	Pure silicon was explored by DuPont. J. Atanasoff and C. Berry completed the prototype of an electronic digital computer.
1943– 1946	J. P. Eckert and J. M. Mauchly designed an all electronic digital computer, the ENIAC.
1947– 1948	Bipolar transistor was invented by J. Bardeen, W. B. Brattain, and W. B. Shockley.
1951	Field effect transistor was invented. Zone refining was developed by W. G. Pfann.
1956	Oxide masking was developed. IBM introduced the 650, the first commercially successful computer.
1957	Photolithography was developed.
1958	Integrated circuit was invented by J. S. Kilby and R. N. Noyce.
1959	Silicon planar transistor was invented. RTL was invented.
1960	Epitaxial transistor was invented. DTL was invented.
1961	ECL was invented by J. A. Narud. TCTL, a forerunner of TTL, was invented by J. L. Buie.
1962	ECL was introduced by Motorola.
1963	TTL was introduced by Sylvania, and MOSFET was studied by S. Hofstein of RCA.
1964	*p*-MOS IC.
1966	MOS dynamic shift register. Schottky TTL was invented by Y. Tarui and others.
1967	TTL–MSI and ECL with 4 nsec were introduced. CMOS IC was introduced by RCA.
1968	*n*-MOS IC and *p*-MOS LSI were introduced. Intel was formed.
1969	D-load MOS IC and ECL with 1 nsec were introduced. Bubble memory was invented by A. H. Bobeck at Bell Labs.
1970	D-load MOS LSI was introduced. CCD was invented at Bell Labs.
1971	Microprocessor 4004 with 2,250 MOSFETs was designed by M. E. Hoff and M. Shima at Intel. A single-chip calculator device was introduced by MOSTEK. Dynamic RAM of 1k bits was introduced by Intel.
1972	I²L was announced by IBM Germany and Philips.
1973	Low-power Schottky TTL was introduced by Texas Instruments. The first commercially successful 8-bit microprocessor 8080 was announced by Intel.
1974	Electron-beam lithography system was introduced. Dynamic RAM of 4k bits was introduced. The first 5-V only 8-bit microprocessor 6800 was announced by Motorola. The largest selling 4-bit microcontroller TMS 1000 was announced by Texas Instruments. The first CMOS microprocessor 1802 was announced by RCA. The first 16-bit single-chip microprocessor PACE was announced by National Semiconductor.

Table 5.4.1   *(Continued)*

1976	8-bit microprocessor Z80 was announced by Zilog.
1977	Dynamic RAM of 16k bits was introduced.
1978	16-bit microprocessor 8086 was announced by Intel.
1979	16-bit microprocessor Z8000 was announced by Zilog. Dynamic RAM of 64k bits was introduced.
1980	16-bit microprocessor 68000 was announced by Motorola. 32-bit microprocessor iAPX 432 was announced by Intel.

(Wollesen, 1980). We are still exerting strenuous efforts in finding or improving logic families with higher speed, lower power consumption, and lower cost. Improvements have been made in numerous ways, but major efforts include the scale-down of devices (bipolar transistors and MOS-FETs), gallium-arsenide devices, and Josephson junctions. For VLSI the processing technology is becoming very complex and sophisticated. Technological progress often depends on research funds and man-power expended. However, technological accomplishments, as described in the following, usually cannot be transferred immediately to commercial production because of the many manufacturing problems, such as improvement of yield and development of efficient manufacturing facilities, which take often a few to several years after laboratory exploration. [See "Fifty years of achievement: A history" (*Electronics*, Apr. 17, 1980, special commemorative issue) which was also published by the editors of *Electronics* as a book, *An Age of Innovation: The World Electronics 1930–2000* (McGraw-Hill, 1980, 274 pp.). Comprehensive bibliographies were prepared by Agajanian (1976, 1978, 1980).]

Processing Technology

Progress in the processing technology is the motive force of all other VLSI chip design technologies, though ECL and TTL are not considered for VLSI because of limited integration size. For greater integration size we need to reduce the line width (and consequently the channel length). We have to develop new submicron technology. Because of unprecedented minuteness with precision, wafers must be more dislocation-free, wet processing (i.e., processing with liquid chemicals) must be replaced by dry processing using silicide and lift-off, photolithography must be replaced by electron-beam or X-ray lithography, fabrication must be automated with minimum human intervention, and the cleanness of fabrication rooms must be further improved, as mentioned in Chapter 2. Equipment for these is becoming more expensive and fabrication processes are becoming more complex, with an increasing number of masks. Thus the manufacturing costs of chips tend to increase, though speed performance will be improved as well (Noyce, 1977; Block and Galage, 1978; Lattin, 1978; *Electronic Design*, Mar. 29, 1979, pp.

Table 5.4.2 Comparison of Logic Families

PROCESS	RELATIVE DENSITY	POWER DISSIPATION PER GATE (mW)	GATE DELAY (nsec)	
			1979 STATE-OF-THE-ART	FUTURE FORECAST
Si-gate *n* MOS	1.0	2.0	5.0	2.0
HMOS	2.0	1.0	3.0	1.0
Si MESFET	0.6	3.0	0.8	0.5
CMOS/SOS	0.75	0.08	3.0	0.5
ECL	0.1	20.0	0.6	0.3
I²L	0.8	2.5	1.5	0.8
Low-power Schottky	0.7	3.0	6–7	0.8
Schottky TTL	0.6	5.0	2–3	0.5

76–79, 82–86; Grossman, Feb. 5, 1979; Lyman, June 19, 1980; Shah, 1980; Bursky, June 7, 1980; June 11, 1981; Grossman, June 7, 1980; Su, 1981; *Electronics*, Dec. 6, 1979, pp. 46, 48; Barbe, 1980; and Mohammadi, 1981).

Scale-Down of Devices

By scaling down devices we can improve speed and reduce power consumption. In particular, reduction of the channel length of a MOSFET and of the base thickness of a bipolar transistor is important.

As mentioned in Section 4.1, scaled-down *n*-MOS is known as HMOS. But 0.2 μm is estimated as the limit of channel length, and reduction beyond this limit would make a MOSFET inoperable (*Spectrum*, Jan., 1980, p. 45) because of many physical problems (*Computer*, Sept. 1978, p. 11). Since many techniques are used for performance improvements, it is expected that more masks will be used for MOS fabrication, as the recent trend in Table 5.4.3 indicates. (Sumney, 1980; *Electronics*, Jan. 3, 1980, pp. 81–85).

References: Jecman et al., 1979; El-Mansy, 1980; Hayes, 1980; Dennard et al., 1972, 1974; *IEEE JSSC*, Apr. 1979; Hoeneisen and Mead, 1972,

Table 5.4.3 Complexity of MOS Processing[a]

YEAR	LINE WIDTH (μm)	NUMBER OF MASKS		
		SILICON-GATE *p*-MOS	SILICON-GATE *n*-MOS	SILICON-GATE CMOS
1973	7.5	5–7	5–7	8–9
1975	6	6–8	6–8	8–9
1977	5	—	7–9	8–10
1979	3–4	—	8–10	8–10

[a] Reprinted from *Electronics* (Sept. 13, 1979, p. 122). Copyright © McGraw-Hill, Inc., 1979. All rights reserved.

Electronics, Nov. 23, 1978, p. 125; Oct. 25, 1979, p. 105; Jan. 3, 1980, p. 14; Oct. 23, 1980, pp. 44, 46, 114, 116.

The scale-down of bipolar transistors by Isoplanar technology is discussed in Chapter 3. Further improvements, in particular, reduction of the base thickness has been explored to about 0.1 μm, but the performance improvement expected is more complicated than in the case of MOSFETs (Solomon and Tang, 1979; Hoeneisen and Mead, 1972; Gaur, 1979; Hart et al., 1979). As the size of the bipolar transistor gets smaller, it is becoming difficult to place electrodes directly to the transistor, and self-aligned contact windows of polysilicon have been used to avoid this difficulty (Sakai et al., 1979; Sato et al., 1980; Nakashiba et al., 1980).

GaAs MESFET

Since the mobility of electrons is 6 to 10 times faster in gallium arsenide than in silicon (depending on conditions), the gallium-arsenide metal semiconductor field-effect transistor (GaAs MESFET), illustrated in Fig. 5.4.2, has been explored for high-speed logic. The GaAs MESFET also has depletion and enhancement modes. Examples of the GaAs logic gate are shown in Fig. 5.4.3, where each GaAs MESFET is denoted by a symbol similar to that for a MOSFET. (Notice that some MESFETs in Fig. 5.4.3*b* have two gates each.) A GaAs MESFET (depletion mode) with a gate delay of about 100 psec at a power consumption of 1 mW was reported. A gate delay of 50 psec or less can be achieved. The delay–power product is an order of magnitude lower than that for CMOS and two orders of magnitude lower than that for *n*-MOS. But its speed tends to be sharply reduced by the increased output load, as in the case of scaled-down *n*-MOS. GaAs has currently other problems: high cost [GaAs wafers cost $1,000/in.[2] as of 1979, but will become cheaper (*Electronics*, June 21, 1979, p. 66)] and difficult fabrication. When the channel length of *n*-MOS is shorter than 1 μm, parasitic capacitance and resistance are more dominant factors in determining the speed than mobility (*Electronics*, Jan. 3, 1980, p. 14), but GaAs has a much lower capacitance than *n*-MOS because GaAs can be made almost an insulator.

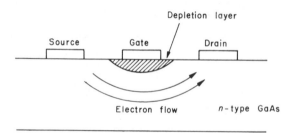

Fig. 5.4.2   GaAs MESFET.

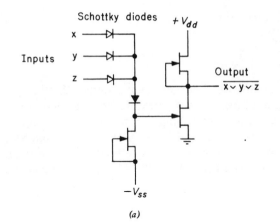

(a)

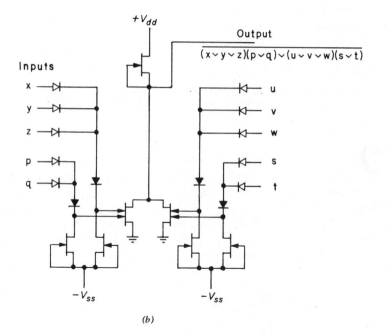

(b)

Fig. 5.4.3 Examples of GaAs logic gate.

References: Dobson, 1981; Welch, 1980; Eden et al., 1978, 1979; Barna and Liechti, 1979; Nuzillart et al., 1976; Van Tuyl et al., 1977; Van Tuyl and Liechti, 1977; Solomon, 1978; Houston et al., 1979; Lee et al., 1980; *IEEE TED*, June 1980; *Electronics*, July 31, 1980, pp. 40, 41; Oct. 9, 1980, pp. 76, 78; Dec. 18, 1980, pp. 40, 41; Jan. 27, 1981, pp. 41, 42; May 19, 1981, pp. 38, 39; *EDN*, Sept. 20, 1978, pp. 44–52; Dec. 15, 1979, pp. 37–50; Oct. 20, 1980, p. 242; Jan. 7, 1981, p. 24; *Electronic Design*, Apr. 12, 1980, pp.

27–29; Dec. 6, 1980, pp. 27, 28; Jan. 8, 1981, pp. 90, 92; DiLorenzo and Khandelwal, 1981.

Physical Limit of Devices

R. W. Keyes of IBM has been extensively investigating the physical limit of devices for digital logic. As the device size is scaled down for high-speed logic, heat dissipation is increasingly difficult, even if water or freon is used as coolant. Thus for high-speed logic, logic devices that work at superconducting temperatures and generate little heat, such as the Josephson junction, look promising (Keyes, 1969, 1975, 1978, 1979; Gaensslen et al., 1977).

Josephson Junction

Josephson junctions, which work at superconducting temperatures, have been explored by IBM and others. Although a special container of liquid helium is required for superconducting temperatures, Josephson junctions have currently the highest speed:

OR operation	40 psec
AND operation	40 to 100 psec (each fan-out adds 14 psec)
16k RAM	15 nsec

(*EDN*, May 5, 1978, pp. 19, 20; *Electronic Design*, Nov. 8, 1979, pp. 19, 20). There are a few different types of Josephson junctions, and some of them (by T. Gheevala) have gate delays of 13 psec for two-intput OR gates at 2.6 μW. The estimated delay–power product of the Josephson junction is shown in Fig. 5.4.4. One disadvantage of Josephson junctions is their much greater sensitivity to temperature (becoming inoperable at high temperature) than the MOS or GaAs logic families.

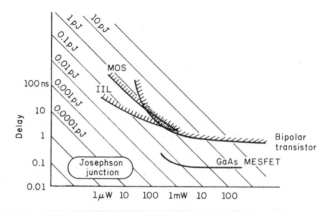

Fig. 5.4.4   Delay–power product of Josephson junction.

IBM is developing a prototype computer of 4,000 Josephson junction gates (*Computerworld*, Feb. 25, 1980, p. 62).

References: Matisoo, 1980; Langenberg et al., 1966; Davidson, 1978; Henkels and Zappe, 1978; Gheevala, 1979; *IBM JRD*, Mar. 1980; Anacker, 1979; *IEEE TED*, Oct. 1980; *Electronics*, May 5, 1981, pp. 48, 57.

> *Remark 5.4.1:* MOSFETs work 2 or 3 times faster at liquid nitrogen temperature than at room temperature. The liquid nitrogen is obtained at much higher temperatures than the superconducting temperature for Josephson junctions, so at such temperatures MOSFETs might be found to be useful in the future (*Electronic Design*, Nov. 8, 1978, p. 34; Gaensslen, 1980).   □

Other Possibilities

Other possible devices have been explored, though no significant feasibilities are reported. Examples are conductive polymer (i.e., plastic that can carry electricity) (*Business Week*, Apr. 14, 1980, pp. 40E–F; *Spectrum*, Dec. 1981, p. 17) and amorphous devices (see Remark 6.6.2) for low cost; and also optical devices (Smith and Tomlinson, 1981) and superconducting devices at much higher temperatures than known to date for high speed.

Problems

5.1 Design an I^2L circuit and a static MOS circuit for the following function, assuming that in both cases only noncomplemented variables are available as network inputs.

$$f = x \lor yz \lor \overline{yu}$$

5.2 When the delay–power product of a logic family is reduced L times, we may be able to pack L times more devices into a chip, keeping the same speed. Discuss whether this is true. (This is an open-ended question, so discuss concisely some possibilities.)

5.3 For each of the following logic families, find the number of masks required and the function of each mask: ECL with high performance, metal-gate n-MOS with enhancement-mode load, metal-gate n-MOS with D-load, silicon-gate n-MOS with D-load, and CMOS.

CHAPTER 6

Memories

One of the greatest impacts of IC technology on logic design is in the use of memories. Semiconductor memories are becoming cheaper, faster, and more compact every year, replacing core memories which were used for years almost exclusively as main memories of computers. The replacement of core memories by semiconductor memories sounds like the simple replacement of a means of realizing an electronic circuit, but it has been causing basic changes in logic design philosophy, as we see in this and the following chapters.

6.1 Features of Semiconductor Memories

Probably the most important feature of semiconductor memories is the compactness, which results from the incorporation of memory cells and peripheral circuits (such as memory address decoders, writing–reading circuits, and memory registers) into a tiny package. In the case of core memories, the peripheral circuits are complex and bulky, generating a considerable amount of heat; so all memories must be centralized and provided with cooling equipment. Also, these peripheral circuits and the cooling equipment considerably increase the total cost of a core memory system. The space required for a core memory is reduced to roughly 1/500 to 1/1000 by a semiconductor memory. The compactness of the memories makes the distribution of memories throughout a digital system feasible. This has caused drastic changes in a variety of ways, which can be summarized as follows:

1. Memories can be placed side by side with logic networks of gates because of their compactness and inexpensiveness. In other words, we can have a combination of gate networks and memories locally everywhere throughout a digital system. This combination presents new possibilities and flexibilities in design, improving performance and reducing cost. As one outcome from this, we can have processors of different sizes and functions throughout the system.

2. Traditionally memories have been used to store programs or intermediate computational results. Semiconductor memories, however, are opening up new usages in addition to the traditional ones. Semiconductor

read-only memories are substituting for logic gates by holding truth tables, minimum switching expressions, or look-up tables. These uses have many variations, and new forms may be invented in the future, depending on the designers' ingenuity.

Semiconductor memories are very compact and easy to lay out because of their regular structure; consequently they are very inexpensive. Thus in many cases, the realization of switching functions with a mixture of logic gates and memories is much more economical than with logic gates alone (i.e., conventional means to realize switching functions). Good logic design now largely depends on how wisely memories are used in logic networks. The potential capability of semiconductor memories in logic design may not have been fully explored yet.

The semiconductor memories to be discussed in this book can be classified as follows:

RAM (Random Access Memory).   Information can be written in and read out of a memory, and any of its address locations can be accessed in any desired sequence with similar access time to each location. Depending on how memory cells are realized by an electronic circuit, RAMs are further classified into two types, static RAM and dynamic RAM.

ROM (Read-Only Memory).   Information can be read out like with RAMs, but cannot be written in as freely as in the case of RAMs. ROMs are further classified into mask-programmable ROM, PROM, EPROM, and EAROM (or EEPROM). (The classification appears confused in the literature.)

Mask-Programmable ROM.   Information is permanently set by masks during fabrication (usually by semiconductor manufacturers), and once set, it cannot be changed. **Mask-programmable ROMs are often simply called ROMs.** (We have to differentiate them from the ROMs in the above paragraph.)

The following types of ROMs are called **field-programmable ROMs,** in contrast to the above mask-programmable ROMs.

PROM (Programmable ROM).   This is a ROM into which users (not just semiconductor manufacturers) can write information only once.

EPROM (Erasable Programmable ROM).   This is a ROM where information in all cells is simultaneously erased by the flood exposure of ultraviolet light, and then new information is written electrically into each cell.

EAROM (Electrically Alterable ROM) or EEPROM (Electrically Erasable Programmable ROM).   This is a ROM into which information can be erased and written electrically. Although EAROM is usually regarded as a synonym for EEPROM, some authors define them differently: EAROM is a ROM where information is electrically erased word by word or bit by bit (i.e., cell by cell), and EEPROM is a ROM where information is electrically erased simultaneously in all cells.

References: JEDEC, 1979; Intel, 1977, 1978; Hnatek, 1979, 1980; Hodges, 1972.

6.2   RAM (Random Access Memory)

A RAM consists of memory cells, an address decoder, a memory register, and a read–write control, as shown in Fig. 6.2.1, though there are many variations in the details of the structure. Memory cells, each of which stores information 0 or 1, are arranged in a matrix form. When an address code $(A_1 A_2 \cdots A_m)$ is given to a decoder, a row of cells, which corresponds to the address code, is selected by an output of the decoder. Then the information stored in these cells is read out into the memory register by the read–write control, or the information stored in the memory register is written into the cells selected.

Depending on the type of electronic circuit used to store information in memory cells, RAMs are classified into static and dynamic memories.

Static RAM

Each cell of a **static RAM** (or simply **SRAM**) consists of a flip-flop. Thus as long as a power supply is connected, stored information is maintained. Figures 6.2.2 and 6.2.3 show examples of a memory cell, though there are many variations. (V_{ss} in Fig. 6.2.3 is usually grounded.) In a matrix of memory cells, each pair of bit-0 and bit-1 lines is shared by all memory cells

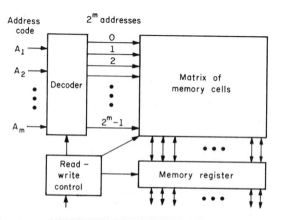

Fig. 6.2.1   Memory structure.

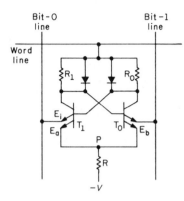

Fig. 6.2.2 Bipolar RAM cell.

in each column, and each word line is shared by all memory cells in each row.

Suppose that we want to write information 1 in the memory cell, i.e., the flip-flop, in Fig. 6.2.2. (Two diodes are simply for performance improvement.) The bit-1 line and the word line are provided with high voltages (about 0.8 volts for the bit-1 line and a much higher voltage for the word line), while the bit-0 line is kept at a low voltage. Then a current flows from the word line to the bit-0 line through resistor R_1 and transistor T_1 because the base voltage of T_1 is higher than that of emitter E_i. (By raising the voltage of the word line, the voltage at P also becomes high, and a greater current flows through E_i than through R, no matter whether a current previously flowed through E_a or E_b. Transistor T_0 becomes nonconductive because of low voltage at its base.) When the voltages of the bit-1 line and the word line are changed back to low voltages, the current flows through R_1 and E_a (switched from E_i to E_a). The flip-flop maintains henceforth this status (i.e., a current through R_1 and T_1, but no current through R_0 and T_0), and this means that the memory cell stores information 1.

When we want to write information 0, the bit-0 line and the word line are provided with the high voltages, keeping bit-1 line at the low voltage. Then

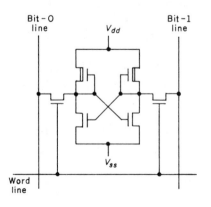

Fig. 6.2.3 Static MOS RAM cell.

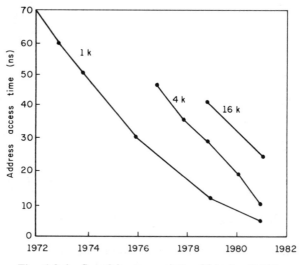

Fig. 6.2.4 Speed improvements of bipolar RAMs.

the flip-flop is set to have a current through R_0 and T_0 only in a manner similar to the above. This means that the memory cell stores information 0.

When we want to read out information stored in the memory cell of Fig. 6.2.2, the voltage of the word line is raised to a high voltage (raising the voltage at P by an increased current through R). Then a current flows from the power supply to the bit-0 line through T_1 or to the bit-1 line through T_0, depending on whether the cell stored 1 or 0, respectively.

We can write or read information in the cell in Fig. 6.2.3 in a similar manner.

Bipolar memories are improving every year in size and speed. For example, the 4k RAM,* Fairchild's 93471, had a chip size of 23,300 mil^2 in 1977 when introduced, and had shrunk by 26% to 17,200 mil^2 in 1978, and again by the Isoplanar S process (mentioned in Chapter 3) to 12,400 mil^2 in 1979, which is about half its original size (*Electronics*, Aug. 30, 1979, pp. 40, 41). As shown in Fig. 6.2.4, the speed of bipolar RAM has been improved by the isoplanar process and scale-down (*Electronics*, Dec. 6, 1979, pp. 139, 140).

Because of their regular structure, static RAMs have a much higher device packing density than random-logic networks, and 8 bits of a static RAM occupy about the same area as one logic gate.

References: Rice, 1979; Glock, 1980; Stinehelfer et al., 1981. For Isoplanar S, see *Electronics*, Aug. 30, 1979, pp. 39, 40; Sept. 13, 1979, pp. 113, 114.

* k in memory-size description denotes 1,024. Thus 2k means 2,048, 4k means 4,096, and so on.

The static MOS memory cell with depletion-mode MOSFET loads in Fig. 6.2.3 (also in Fig. 6.2.5*a*) can be laid out as shown in Fig. 6.2.5*b*. These depletion loads have been replaced by polysilicon resistor loads in Fig. 6.2.5*c*, which are formed on the SiO₂ insulation layer (the resistivities of the polysilicon resistors are adjusted by implanting impurities), as illustrated in Fig. 6.2.6 (*Electronics*, Sept. 13, 1979, p. 112). The use of polysilicon resistor loads has the advantage of reducing the layout size by 40%, as shown in Fig. 6.2.5*d*. (*Electronics*, Apr. 26, 1979, p. 134; June 19, 1980, p. 123). Since polysilicon resistors have much higher resistances than the dif-

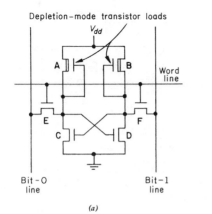

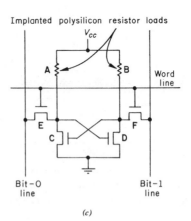

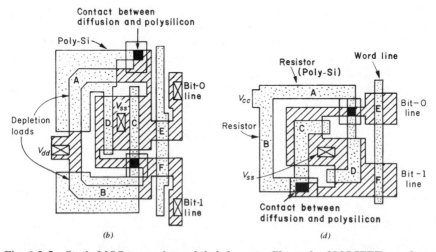

Fig. 6.2.5   Static MOS memories and their layouts. Channels of MOSFETs are formed between diffusion regions underneath polysilicon in (*b*) and (*d*). For example, the L-shaped area *A* in (*b*) forms a channel between two diffusion regions (shaded) in the lower left and upper right locations. (*a*) Cell with depletion loads. (*b*) Layout of (*a*). (*c*) Cell with polysilicon resistor loads. (*d*) Layout of (*c*).

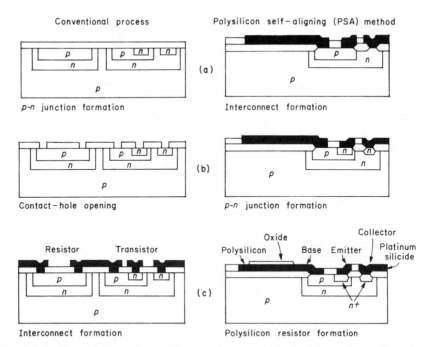

Fig. 6.2.6 **Formulation of polysilicon resistor compared with the formation of conventional diffused resistor. (Reprinted from *Electronics*, Sept. 13, 1979, p. 112. Copyright © McGraw-Hill, Inc., 1979. All rights reserved.)**

fused resistors used in the previous chapters, they cannot be used as loads in MOS logic gates. (Since the current magnitude is limited by the high resistance of the polysilicon resistor load, the speed becomes excessively slow.) However, the high resistance of the polysilicon resistor load is not a problem in the case of static MOS memories. (In writing, only a current that is sufficiently large to change the state of a memory cell needs to be provided from a bit line, and in reading, the memory cell does not change its state.)

References: Capece, Apr. 26, 1979; Huffman, 1979; Okada et al., 1980; *Electronics*, Dec. 6, 1979, p. 126.

The speed of static MOS RAMs has been greatly improved, getting closer to that of bipolar RAMs (Jecman et al., 1979; *Electronics*, Apr. 26, 1979, pp. 131, 132; Dec. 6, 1979, p. 139; Mar. 13, 1980, p. 217). Standby power consumption of static *n*-MOS memories was greatly reduced by Hitachi, Ltd. who used CMOS for peripheral circuits (Minato et al., 1979, 1980).

Dynamic RAM

Each memory cell of a **dynamic RAM** (or simply **DRAM**) keeps information by storing electric charge on a very small capacitance, which is usually called **storage capacitance**. Information 0 and 1 is represented by electric charge and no charge, respectively (or alternatively, no charge and electric charge) on the storage capacitance in a memory cell.

Figure 6.2.7 shows an example of a dynamic RAM, where only one MOSFET is connected to V_{cc} through a storage capacitance C. (V_{cc} is usually grounded.) When we want to write information, a high voltage or no voltage is supplied to the bit line, raising the word line to a high voltage from no voltage. Then the MOSFET becomes conductive, and an electric charge or no charge is stored on capacitance C. When the word line returns to no voltage, the information is kept on C. When we want to read stored information, an electric charge or no charge on capacitance C is read out to the bit line by raising the word line to a high voltage. Thus information 1 or 0 (or alternatively 0 or 1) is read out, respectively.

Since an electric charge stored on capacitance C gradually leaks, information must be written before the charge completely leaks out. In other words, the memory cell must be **refreshed**. The refreshing frequency depends on MOSFET leakage, which increases sharply with temperature. [At 70 °C refreshing is typically required every 2 msec, and at lower temperatures this interval may be increased to hundreds of milliseconds (*Computer Design*, Jan. 1974, p. 69).] For a 16k RAM, for example, the refreshing frequency is about 500 Hz, with about 2.4% of the time spent on refreshing. Compared with static MOS memories, refreshing is an undesirable feature of dynamic MOS memories (Ford et al., 1981).

Because a single MOSFET is used instead of a flip-flop, a dynamic RAM occupies a much smaller area. Thus in the same chip area a dynamic RAM can pack a multiple of the memory cells of a static RAM, that is, about 4 to 10 times more, depending on the technology.

References: Foss and Harland, 1977; Chatterjee et al., 1979; Rideout, 1979; Tasch et al., 1978; Smith and Yu, 1980; Twaddell, Sept. 20, 1980; Sud and Hardee, 1980; Ohzone, 1980; *Electronic Design*, Jan. 18, 1979, pp. 34–44; Leonard, 1980; Chan et al., 1980; Galloway et al., 1981; *EDN*, May 13, 1981, pp. 59–65.

Remark 6.2.1: Usually bipolar transistors are not used for dynamic RAM. But I³L is an exception. Fairchild developed a 4k dynamic RAM with I³L in

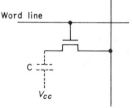

Word line

C

V_{cc}

Bit line   **Fig. 6.2.7 Dynamic MOS memory cell.**

1976 (Quinn et al., 1978; Sander et al., 1976; Bhattacharyya et al., 1980; Antipov, 1980; Penoyer, 1980). □

Physical Limitations of Dynamic MOS RAM Capacity

By packing more memory cells into a chip, memory chips with greater memory capacities have been introduced. If we want to pack more cells into a chip of the same size, the size of each cell must be reduced. An example of such size reduction is given in Fig. 6.2.8. Figure 6.2.8*b* shows a layout for the 4k RAM cell of Fig. 6.2.8*a*, and Fig. 6.2.8*d* shows a layout for the

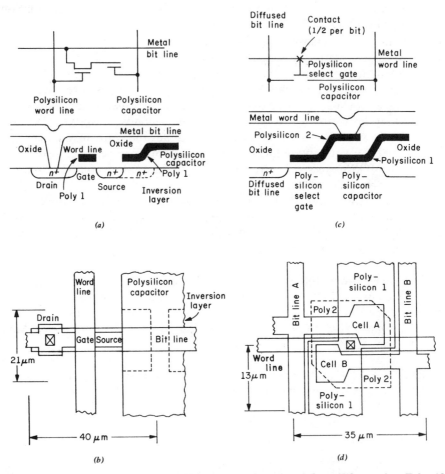

Fig. 6.2.8 Layouts of dynamic MOS RAMs. (Reprinted from *Electronics*, Feb. 19, 1976, pp. 117, 118. Copyright © McGraw-Hill, Inc., 1976. All rights reserved.) (*a*) 4k bit RAM cell. (*b*) Top view of 4k RAM. (*c*) 16k bit RAM cell. (*d*) Top view of 16k RAM.

16k RAM cell of Fig. 6.2.8c. In the latter case the layout size of each cell was reduced by using a different solid-state circuit structure in Fig. 6.2.8c (*Electronics*, Feb. 19, 1976, pp. 117, 118). Consequently when a memory with a large capacity is to be realized, each cell as well as its storage capacitance must be very small. Thus the signal to be read from the cell of a small storage capacitance becomes too small to be detected, and it will become increasingly difficult to realize dynamic RAMs of large capacities (probably 512k bits or more). Undoubtedly entirely new design principles will have to be used, or the use of dynamic RAMs must be given up by switching to static RAMs.

Essentially the same lithographic techniques and design principles have been applied to 256 bit through 16k bit dynamic RAMs. However, for memory packages of greater capacities we have to use new techniques, after having solved the soft error problem discussed in the following. Although electron-beam lithography can be applied to the mask preparation of 64k bit dynamic RAMs, electron-beam lithography, dry etching, and other new techniques must be used in the manufacture of commercially successful dynamic RAMs of 256k bit or greater capacities, because the access time must be fast. Thus we need to use 1-μm technology. Also since the fabrication environment must be almost completely free from dust, an unprecedented level of automation is required, reducing human operations as much as possible. (Since memories are much easier to design than random-logic networks, new technology is usually tested with memories first, keeping RAMs at the leading edge of semiconductor technology.)

References: Leonard, 1980; Coe and Oldham, 1976; Young, Oct. 25, 1978; Lo et al., 1980; Masuda et al., 1980; Bernhard, 1981; Foss et al., 1976.

Soft Errors

Another problem which may limit the increase of RAM capacities is soft errors as reported by M. Woods of Intel in April 1978. Soft errors are caused by alpha particles emitted by the radioactive decay of uranium, thorium, and other radioactive traces present in chip or package materials, and also by alpha particles that penetrate the surface of the chip from the outside. These alpha particles generate a relatively large quantity of electron–hole pairs, discharging at random the electric charge stored in memory cells. This random loss of stored information is called a **soft error**. Soft errors occur with dynamic and static RAMs (and also while charge-coupled device memories to be discussed in Section 6.8). They are most serious with dynamic RAMs and become more critical as memory capacities increase, because as chips become denser, the amount of stored charge decreases (due to smaller storage capacitances), increasing the susceptibility to ionizing radiation. The soft error problem with *n*-MOS static RAMs (*Electronics*, Mar.

15, 1979, pp. 85, 86; Apr. 26, 1979, p. 126) may swing our attention to CMOS static RAMs, because CMOS may offer low power consumption without any increased susceptibility to alpha particles, whereas *n*-MOS becomes more susceptible as its power is reduced (*Electronics*, Apr. 26, 1979, p. 126). One remedy for the soft error problem is the covering of a chip with polyimide, which reduces errors to 1/1000 or less.

References: Katto et al., 1980; Brodsky, 1980; Noorlag et al., 1979; Bossen and Hsiao, 1980; *Electronics*, June 8, 1978, pp. 42, 43; Oct. 25, 1979, p. 106; Nov. 22, 1979, p. 40; Sept. 11, 1980, pp. 40, 41; Feb. 10, 1981, pp. 93, 94; Nov. 22, 1980, pp. 36, 37; Mar. 24, 1981, p. 33; *Electronic Design*, May 24, 1978, p. 37; Feb. 15, 1979, p. 58; Apr. 12, 1979, pp. 27, 28; Sept. 1, 1980, p. 32; Nov. 22, 1980, pp. 36, 37; *EDN*, Nov. 20, 1980, pp. 53, 54.

6.3 RAMs Substituted for Logic Networks

RAMs are not often used in substituting for logic networks or gates, because fixed logic operations are usually required in logic networks, and RAMs lose stored information when the power supply is cut off. Thus ROMs are usually used in substituting for gates, as discussed in the next chapter. Also RAMs are more expensive than ROMs.

The following is an example of RAMs substituting for logic networks.

Shift Registers Realized with RAMs and Counters

RAMs are at least several times denser than logic gates in terms of chip area [about 8 bits of a RAM occupy the same area as one logic gate (*Electronic Design*, Oct. 11, 1980, p. 33), and probably twice more than this, depending on technology], although RAMs are slower than logic gates. Thus registers, in particular shift registers, can be more compactly and economically realized with RAMs than with logic gates when we need many long registers. Also, shift registers realized with RAMs are sometimes more convenient than those with gates, as explained in the following.

Let us illustrate this with the shift register of 4 bits shown in Fig. 6.3.1. When a clock pulse arrives, all the contents in the shift register in Fig. 6.3.1*a*

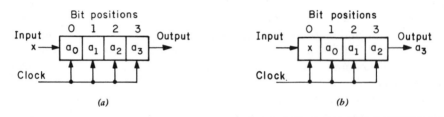

(a)   (b)

Fig. 6.3.1 Shift register. (*a*) Before a clock pulse. (*b*) After the clock pulse.

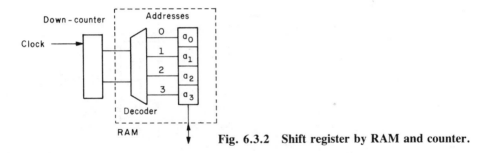

Fig. 6.3.2   Shift register by RAM and counter.

are shifted right by one bit position, as shown in Fig. 6.3.1b. In bit position 0, x enters and a_3 is sent out from bit position 3.

The same performance can be obtained by the combination of a RAM of four 1-bit words and a 2-bit down-counter, as shown in Fig. 6.3.2. Suppose that the current count in the counter is 0, and correspondingly the RAM is in address 0. When the next clock pulse arrives, the count decreases by 1 and the RAM is in address 3. Then a_3 is read out. Write x in this address. For the next pulse both Figs. 6.3.1 and 6.3.2 read out a_2, and so on. In the case of Fig. 6.3.1b, x in bit position 0 can be read out at the fourth clock pulse after (b). In the case of Fig. 6.3.2 also, x can be read out at the fourth clock pulse.

If we use an up-counter instead of the above down-counter, the network in Fig. 6.3.2 realizes a left shift register instead of the above right shift register.

Compared with a shift register with gates, this shift register realized with a RAM and a counter is inexpensive and compact, though speed may be reduced and it is not possible to simultaneously write all the bits of numbers into the RAM or read them out. (In other words, this cannot work as a loadable shift register.)

In the case of a shift register with gates, if we want to read an arbitrary bit position in the shift register, we must shift many times. But in the case of the shift register in Fig. 6.3.2, any bit position can be read out instantly by simply replacing the count by a desired address. Also, if we have a RAM with multiple-bit words, a number of shift registers can be simultaneously realized, as shown in Fig. 6.3.3, which contains $2m$ shift registers. Two shift registers for $(a_{01}, a_{11}, a_{21}, a_{31})$ and $(b_{01}, b_{11}, b_{21}, b_{31})$ do not necessarily have shift cycle 8, since any bit position in any shift register can be instantly read out.

Instead of having a counter, we can store the next count in each address, and when each address is read out, the next count is supplied to the decoder. By this approach the counter can be eliminated, and any counting sequence (i.e., counting in any order) can be realized by simply changing the next counts in the memory.

References: Wyland, Jan. 5, 1974.

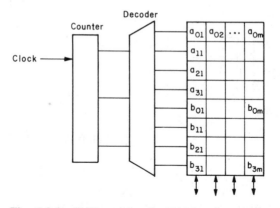

Fig. 6.3.3   **Shift registers by RAM and counter.**

Registers which are not shift registers are also increasingly often realized by RAMs rather than with logic gates, since RAMs are several times denser on a chip. In this case each memory word is used as a register. Because of this, and also because of the ease of distributing many memories throughout the chip, architecture based on the memory-to-memory transfer of data is increasingly used, rather than conventional architecture based on the register-to-register transfer for computers with centralized memories (Scrupski, 1980; Hughes and Chappell, 1980).

6.4   Mask-Programmable ROM

The packing density of a mask-programmable ROM, or simply ROM, can be at least several times (roughly 10 times) as high as that of a static RAM (i.e., in the area for a 16k static RAM, for example, we can pack roughly 160k ROM without difficulty). So the packing density of a ROM is higher (roughly two or three times) than that of a dynamic RAM, and a ROM is much easier to manufacture than a dynamic RAM because of its simpler storage mechanism, as we see later. Thus a ROM, in the case of high-volume production, is much cheaper than a static or dynamic RAM [the cost of a ROM per bit is lower by 4 to 8 times than that of a static RAM (*Computer*, Oct. 1980, p. 29)], and when computer programs are completely debugged and need not be rewritten during computation, there is no point in using RAMs. Programs (or microprograms) can be packed into a much smaller space of a ROM with lower cost (i.e., as firmware). If we want to pack programs into as small a space as possible (as with terminals or calculators), we had better use ROMs to store the permanent part of the programs. Another advantage of using ROMs for software is the convenience that a ROM cartridge with software stored (often called **silicon software** or **solid-state software**) can simply be inserted into a calculator or computer, with metal contacts or pins, but without any mechanically moving equipment, whereas if software is to be stored in a RAM, the RAM must be loaded with software, usually using a mechanically moving equipment such as paper or a magnetic tape reader, which are unreliable and expensive, requiring maintenance.

As we see in later chapters, ROMs are used in many different ways other than the implementation of simple firmware (or silicon software).

Structures of ROMs

A ROM implemented with diodes is shown in Fig. 6.4.1. This is called bipolar type, since bipolar transistor circuits (TTL or ECL) are used. In Fig. 6.4.1a an address code is divided into two decoders. One of the horizontal lines is chosen by the first decoder, that is, only one of the horizontal lines is supplied with a positive voltage, but all the other horizontal lines have 0 volts. Then only vertical lines that are connected to the chosen horizontal line through diodes (other diodes have incomplete connections) get the positive voltage. Then the second decoder sends the output of one of the sensing amplifiers to the memory output. Thus only one bit information, 0 or 1, is read out of this memory.

In contrast to the ROM of Fig. 6.4.1a, which is called **bit-organized** (or **bit-oriented**), the ROM of Fig. 6.4.1b, which has only one decoder, is called **word-organized** (or **word-oriented**), since the ROM has all the bits of a word as its outputs. The decoder chooses only one horizontal line. The vertical lines, which are connected to that line by diodes, produce a high voltage, and all the other vertical lines produce a low voltage. Thus the stored information is read as outputs $f_1, f_2, \ldots, f_t$ in word form.

The terminals, **chip enable** (or **chip select**), in Fig. 6.4.1 suppress the outputs of the ROM chips when the terminals are given signal 0 and enable them when the terminals are given signal 1.

[The structures of RAMs are also classified into bit-organized and word-organized RAMs (or bit-oriented and word-oriented RAMs), and can have chip enable terminals.]

Figure 6.4.2a shows a ROM implemented with MOS. This works in the same way as Fig. 6.4.1b. For a chosen horizontal line, only vertical lines which are not connected to that horizontal line by MOSFETs have high voltages, assuming that n-MOS are used with a positive-voltage power supply.

The connection of the gates of MOSFETs can be realized by Fig. 6.4.2b or c. In Fig. 6.4.2b, if the silicon dioxide insulation layer underneath the metal gates is thin enough, as shown by the dashed lines, MOSFETs are formed. If a portion of the layer is very thick, a horizontal metal connection cannot function as a MOSFET gate, and a MOSFET is not formed. In Fig. 6.4.2c the dashed-line rectangles show the areas of thin silicon oxide insulation layers. If the horizontal lines are extended over these areas, MOSFETs are formed in the dashed-line rectangles.

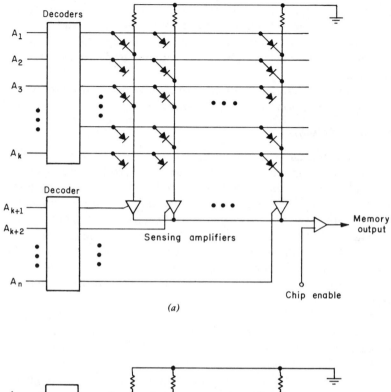

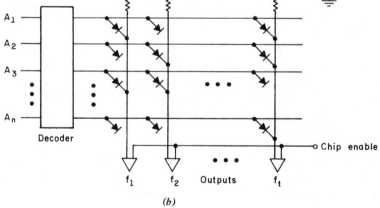

Fig. 6.4.1   ROM with diodes. (*a*) **Bit-organized structure.** (*b*) **Word-organized structure.**

Figures 6.4.1 and 6.4.2*c* can be prepared by changing only the connection mask, leaving other masks intact. Also Fig. 6.4.2*b* can be prepared by changing only the gate mask. (Recall that in the case of random-logic networks, all masks must be custom-made.) The custom preparation of only one mask makes ROMs very economical. A mask-programmable ROM is

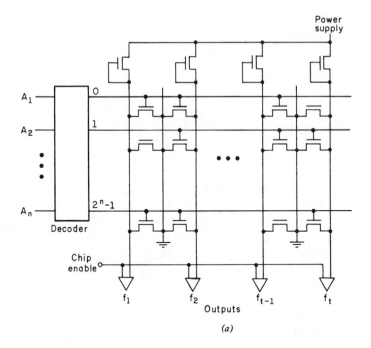

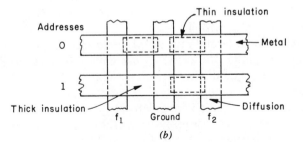

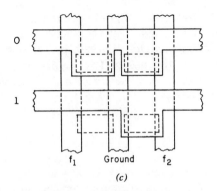

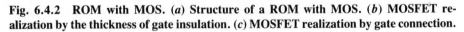

Fig. 6.4.2   ROM with MOS. (*a*) **Structure of a ROM with MOS.** (*b*) **MOSFET realization by the thickness of gate insulation.** (*c*) **MOSFET realization by gate connection.**

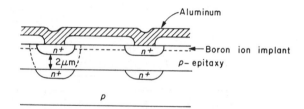

Fig. 6.4.3   ROM in three dimensions (Lohstroh and Slob, 1978. Copyright © 1978, IEEE.)

also called **mask-programmed ROM** or **custom-programmed ROM**. Thus when the production volume is high, ROMs are the most economical among all memories (static RAMs, dynamic RAMs, PROMs, EAROMs, and others). Although mask-programmable ROMs available in the market may not necessarily be the fastest (due to lesser competition among semiconductor manufacturers than for memories in mass production such as RAMs), they can be manufactured with higher speed. Bipolar ROMs available in the market are faster with read times of about 20 nsec to 50 nsec or up, while MOS ROMs have read times of 200 nsec to 500 nsec. (Bipolar ROMs are mostly implemented with TTL circuits. If ECL circuits are used, the read time can be further reduced.) With a greater memory capacity the read time gets slower. Faster ROMs are coming (e.g., a 64k MOS ROM with 100 nsec and a 1k ECL ROM with 10 nsec).

When electronic circuits for ROM decorders, sense amplifiers, and other circuits are dynamic, these ROMs are called **dynamic ROMs** and have the advantage of low standby power consumption. Otherwise ROMs are called **static ROMs**.

Improvements of the packing density of ROMs have been explored. One approach is the use of 3 different sizes for a MOSFET by changing the size of the dashed-line rectangles in Fig. 6.4.2*b* (i.e., forming 4 different sizes in total, including no MOSFET). By identifying the current magnitudes from these MOSFETs by sense amplifiers, each memory cell can store 2 bits, instead of 1 bit in Fig. 6.4.2, at the sacrifice of yield in chip fabrication. (Stark, 1981; *Electronics* Oct. 9, 1980, p. 39; Feb. 24, 1981, pp. 100–103). Another approach is the use of two layers of lines where information is stored by the presence or absence of a conductive path between the two layers at each cross point (Remark 6.4.1).

PLAs, a special type of ROMs, are discussed in the next chapter.

Remark 6.4.1:   A ROM in a three-dimensional structure has been explored as shown in Fig. 6.4.3. Connection lines in the n^+-regions in the lower layer run perpendicular to the figure. Each n^+-region in the upper layer is connected or not connected to the n^+-region underneath in the lower layer by a conductive path, which is provided, as desired. The chip area is roughly halved (Lohstroh and Slob, 1978; *Electronics*, Oct. 23, 1980, p. 141).   □

References on ROMs: Dussine, 1971; Barnes, July 5, 1978; Gunn et al., 1977; Gray, 1977; Stewart, 1977; Ludwig, 1980; *Electronics*, Mar. 30, 1978, pp. 94–107; May 27, 1978, pp. 39, 40; July 20, 1978, p. 6; Wilson and Schroeder, 1978; *Electronic Design*, July 5, 1978, pp. 40–44.

6.5 PROM (Programmable Read-Only Memory)

For small-quantity production of a ROM, custom preparation of even a single connection mask may be too expensive, or delivery time from a semiconductor manufacturer may be too long. Hence **field-programmable ROMs (FPROM)** are available from some semiconductor manufacturers. FPROMs are ROMs where users can set information by themselves. Although there are many different types of FPROMs, those into which information can be written only once (permanently) are called **programmable ROMs (PROM)**.

A commonly used structure of PROMs is illustrated in Figs. 6.5.1*a* and 6.5.2*a*, with their cross sections shown in Figs. 6.5.1*b* and 6.5.2*b*, respectively. At every intersection of word and bit lines, a bipolar transitor and a fuse in Fig. 6.5.1*a*, or a diode and a fuse in Fig. 6.5.2*a* are connected in series. To store information, users blow the fuses at desired intersections by feeding in heavy currents. On a chip, special electronic circuits are provided to supply these heavy currents, adding 40 to 50% more area on the chip.

The fuses in early commercially available PROMs were made of nichrome. But these fuses were found to have the **growback problem**, where disconnected fuses reconnect after some time. (Fusion of nichrome takes place without oxygen under a passivation layer and blown-up nichrome powders crystallize under an electric field, eventually reconnecting the fuse.) (Intel, 1978).

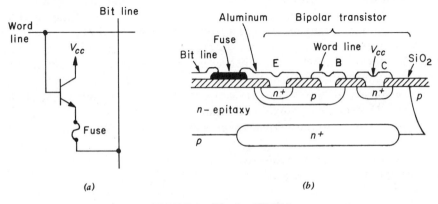

(a)   (b)

Fig. 6.5.1 Bipolar PROM.

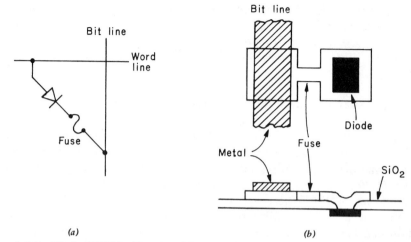

(a)   *(b)*

Fig. 6.5.2   Diode PROM. (Reprinted by permission of Intel Corporation. Copyright © 1977.)

In order to solve the growback problem, fuses are made of polysilicon by Intel, as shown in Fig. 6.5.2. (The Intel 16k bit PROM 3636-1 has a maximum access time of 65 nsec.) Other manufacturers use fuses made of titanium-tungsten, platinum silicide, or others. PROMs have other structures with different materials (*Electronics*, Oct. 23, 1980, p. 141).

References: Wallace and Learn, 1980; Nelson, May 5, 1980; Fukushima, 1979; Tanimoto et al., 1980; *Electronic Design*, June 7, 1978, pp. 155, 156; July 5, 1978, pp. 40–44; July 5, 1979, p. 31; Electronics, Feb. 24, 1982, p. 184.

6.6   Erasable ROM

In erasable ROMs information can be rewritten, unlike with ROMs and PROMs, where information cannot be changed once written. Since writing new information usually requires high voltage and a long writing time, writing is not frequently done. In this sense, erasable ROMs are mostly used for reading information, while information can be occasionally rewritten. The time to retain written information varies but has been improved to a few years or longer. In this sense, erasable ROMs are often called **nonvolatile memories**. Depending on slightly different working principles and structures, there are many different types of erasable ROMs as follows; each has variations.

The **erasable programmable ROM (EPROM)** is a ROM in which information stored in all memory cells is simultaneously erased by flood exposure of ultraviolet light, and then new information is electrically written into each cell. The **floating-gate avalanche-injection MOS (FAMOS)** is shown as an example of EPROM in Fig. 6.6.1, where a floating gate which is completely insulated by SiO_2 from the outside, is placed underneath an ordinary gate. The FAMOS is used as each memory cell. Information can be written as follows. First all memory cells are exposed to ultraviolet light for minutes, and the electrons, which are stored in the floating gates of some cells by previous writing, are excited by the ultraviolet light and discharged to the substrate. As a result, no cell stores electrons at its floating gate, that is, every cell stores information 0. Then a positive voltage (about 25 to 40 volts), which is higher than the normal operating voltage (about 5 volts), is applied for a long time (on the order of milliseconds) to the gate of the memory cells where we want to store information 1. In these cells electrons penetrate into the floating gates through a thin insulation layer from the substrate. Writing information in this manner, the presence or absence of electrons in the floating gate represents information 1 or 0, respectively. Since the floating gate is surrounded by the insulation layer, the electrons stay inside the floating gate, once trapped. Actually the electrons trapped in the floating gate gradually leak out. In other words, information is gradually lost. Some manufacturers claim that information is retained for at least 10 years. Reading can be done in a much shorter time without destroying stored information. (The reading time is somewhat longer than that of the memories discussed in the previous sections.) When electrons are present in the floating gate, the path between drain and source is nonconductive because a positive voltage applied to the gate is masked by the electrons. When no electrons are present in the floating gate, the path becomes conductive by a positive voltage applied to the gate.

In the case of **electrically alterable ROM (EAROM)** and **electrically erasable programmable ROM (EEPROM)**, previous information need not be erased by the exposure of ultraviolet light, but electric charges are injected to or repelled from memory cells by applying high voltages. The **metal-nitride-oxide semiconductor (MNOS)**, shown in Fig. 6.6.2, is an example of EE-PROM, where information in all cells is simultaneously erased electrically. Unlike with FAMOS, which has a floating gate, an electric charge is trapped in the boundary between the silicon nitride layer and the thin silicon dioxide layer. The **floating-gate tunnel oxide (FLOTOX)** is an example of EAROM where information in each bit is erased electrically, though MNOS can be used in this manner also.

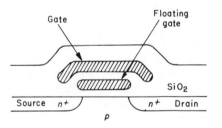

Fig. 6.6.1 FAMOS.

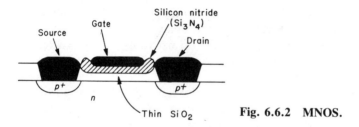

Fig. 6.6.2   MNOS.

Erasable ROMs can be used in a variety of ways. One important application is the test of a program with erasable ROMs in order to completely debug the program before it is permanently stored in ROM for high-volume production.

References: Horiuchi and Katto, 1978; Cayton, 1977; Tsantes, Jan. 5, 1980; Des Roches, 1980; Greenwood, 1980; Woods, 1980; Klein et al., 1979; Gerson, 1980; Ogdin, Nov. 20, 1978; Woods, 1978; Barnes, July 19, 1978; Franson, 1978; PRO-LOG, 1977; Hagiwara et al., 1979; Johnson et al., 1980, Feb. 28, 1980; Reddy, 1978; Bursky, No. 22, 1980; Uchiumi and Makimoto, 1981; Twaddell, Jan. 21, 1981.

Nonvolatile Logic Networks

Nonvolatile logic networks are logic networks combined with nonvolatile memories (i.e., erasable ROMs) such that the status of the networks is retained even after the power supply is cut off. For example, the last count of a counter is stored in nonvolatile memories when the power supply is cut off, and when the power supply is connected, the counter resumes counting from that count. Such counters can be used as odometers, and are much cheaper and more reliable than (electro-)mechanical odometers. Also timers which measure the cumulative connection time of electric appliances to power supplies can be realized by nonvolatile logic networks (*EDN*, Dec. 15, 1978, pp. 35, 37; *Electronics*, May 24, 1979, pp. 90, 92; July 3, 1980, p. 39; Plessey, 1981).

Remark 6.6.1:   The **nonvolatile RAM (NVRAM)**, such as Toshiba's, is a static RAM combined with nonvolatile memory, such as MNOS. This eliminates the drawback of RAMs (which have much faster read times than ROMs), that is, the loss of data upon the removal of power (Leonard, 1980).   □

Remark 6.6.2:   Memories with amorphous material have been explored. Amorphous material has no crystals, like glass but unlike silicon substrates discussed so far, which are made from single crystals. Thus amorphous memories can be made possibly much cheaper than semiconductor memories. Energy Conversion Devices, Inc., of S. R. Ovsinsky has spearheaded the efforts, calling their devices Ovonic memories. Solar cells, erasable microfilms, and switching devices made of amorphous material have also been explored (Henish, 1969; Adler, 1977; Horninger, 1975; Shanks, 1974; Seidman, 1981; Shanks and Davis, 1978; *Business Week*, Apr. 5, 1976, pp. 44–46; *Forbes*, Aug. 4, 1980, pp. 35, 36).   □

6.7   Comparison of Semiconductor Memories

The performance and cost of semiconductor memories are compared in Table 6.7.1, although a precise comparison is not possible due to technological changes and product variations. For example, the read or write time is significantly slowed down when the memory capacity of a chip increases. Furthermore, MOS static RAMs are getting closer to bipolar static RAMs in manufacturing complexity as speed improves (*Electronics*, Apr. 26, 1979, p. 129). Also, production of new memories with improved performances may be discontinued due to low yields. This table is simply to give rough ideas (*Electronics*, Apr. 26, 1979, pp. 129, 132; Jan. 18, 1979, pp. 34–44; May 22, 1980, p. 120; *Electronic Design*, July 19, 1980, pp. 83–92, 101–118, 127–138).

Selling prices may not reflect manufacturing costs. When competition is fierce in a market, selling prices are low. In this sense, the selling prices of mask-programmable ROMs are difficult to predict, because mask-programmable ROMs are custom-made and selling price depends on the customer–vender relationship and the production volume in each individual case.

Comparison of Memories and Gate Networks

The use of RAMs as substitutes for logic networks is discussed in Section 6.3. Registers and shift registers are probably the only case of such a substitution, and RAMs are used mostly for storing software and data, because RAMs are most flexible, that is, it is easy to change information. ROMs are less flexible, but because of lower cost and permanence of information storage, ROMs can be used in many different ways, as discussed in the next chapter. As seen in Figs. 6.2.2, 6.2.3, 6.4.1, 6.4.2, 6.5.1, and 6.5.2, RAMs, ROMs, and gate networks are all composed with logic gates or devices, that is, the ROM in Fig. 6.4.2*a* consists of NOR gates, the memory cells in Figs. 6.2.2 and 6.2.3 are flip-flops, and the ROM in Fig. 6.4.1 and the PROMs in Figs. 6.5.1 and 6.5.2 are not different from the decoder networks used in logic design [i.e., Exercise 2.10 in Muroga (1979)]. The only differences are whether gates or devices are arranged in regular matrix form or in irregular configuration, and also whether gates or devices represent special switching functions (i.e., NOR and flip-flops) or more complex ones. These differences bring in a significant difference in size, cost, speed, and flexibility. Because of the regular configuration of memories, we can pack more devices, and layout is easy, but because of long connections for bit and word lines, memories are slow due to parasitic capacitances and high resistances of

Table 6.7.1 Comparison of Semiconductor Memories

	ROMs				RAMs		
	MASK-PROGRAMMABLE ROMs		PROMs	EPROMs	MOS		BIPOLAR
	MOS TYPES	BIPOLAR TYPES	BIPOLAR TYPES	MOS WITH ULTRA-VIOLET ERASE	DYNAMIC	STATIC	STATIC
Memory capacity per chip	Highest	Medium	Low	High	High	Medium	Lowest
Read time	200–500 nsec	10–60 nsec	30–60 nsec	200–500 nsec	100–400 nsec	30–400 nsec	5–100 nsec
Write time	—	—	—	—	100–400 nsec	30–400 nsec	5–100 nsec
Programming time	—	Highest	Long	Very long	—	—	—
Cost for low volume	High	Highest	Low	Low	Lowest	Medium	Highest
Cost for high volume	Lowest	High	High	Medium			

268

Table 6.7.2 Comparison of Memories and Random-Logic Gate Networks

MEMORY TYPES	LAYOUT	SIZE (bit or gate)	COST (bit or gate)		SPEED	FLEXIBILITY	USES FOR
			FOR LOW PRODUCTION VOLUME	FOR HIGH PRODUCTION VOLUME			
RAM	Regular	Medium (bit)	Medium (per bit)	High (per bit)	Low	High	Software or shift registers
ROM	Regular	Small (bit)	Low (per bit)	Medium (per bit)	Medium	Medium	Firmware, look-up tables, or logic operations
Random-logic network	Irregular (time-consuming)	Large (gate)	High (per gate)	Low (per gate)	High	Low	Logic operations

these connections. Despite many different ways of storing information in ROMs, as discussed in the next chapter, information is generally not packed as much as in the case of random-logic networks, that is, ROMs have lower **information packing density** on devices than random-logic networks, though the former has higher **device packing density** on a chip than the latter. But if we want to change information, that is, logic operations, it is far more time-consuming to do so in the case of gate networks. Because of greater logic capability (i.e., each logic gate, in particular a MOS cell, can express a much more complex function than gates or devices in each memory cell), random-logic networks are most compact for entire networks (though gates alone can be much larger than memory cells), and consequently are fastest and cheapest for high production volume. But for low production volume, random-logic networks are most expensive because logic design and, in particular, layout in irregular and compact configurations are time-consuming and expensive. We see these differences in detail in the succeeding chapters. Table 6.7.2 roughly summarizes the differences.

Since semiconductor memories have been added as a new dimension in logic design, the appropriate use of all kinds of memories and gates in calculators up through mainframes, such as the IBM 370, the DEC 11/60, and the VAX-11/780, is a vitally important consideration for logic designers in order to improve the cost/performance ratio. For example, RAMs are used as cache memories, registers, or shift registers, and ROMs are used as look-up tables for floating-point processors, instruction decoders, or microcode storage. More sophisticated uses are coming (*Electronics*, Oct. 27, 1977, p. 110; Conklin and Rodgers, 1978; Langer and Dugan, 1978).

6.8   Other Memory Types

Other memory types, which are not discussed in detail or at all in this book, are listed in the following. Some of them are classified by their functions, such as associative memories, and others by their hardware realizations, such as optical disks. Memories such as magnetic tapes, magnetic disks, and magnetic core memories, which are extensively used, are not discussed in this book because they are less related to LSI/VLSI chip design.

CCD (Charge-Coupled Device)

CCD was invented in 1970 at Bell Labs. Electrodes are placed on the insulation layer atop a silicon substrate, and a multiphase clock is applied to these electrodes alternately. Electrons in the substrate are attracted to the electrodes to which a clock pulse of positive voltage is applied. As time elapses, electrons move sequentially underneath these electrodes. Thus a serial semiconductor memory can be realized based on CCD. Access time is slower by two orders of magnitude than for semiconductor RAMs. CCD can also be used as an image sensor.

References: Amelio, 1974; Howes and Morgan, 1979; Kosonocky and Sauer, 1975; Sequin and Tompsett, 1975; *Electronics*, Jan. 4, 1979, pp. 85, 86; *EDN*, Sept. 20, 1980, p. 125; Beynon and Lamb, 1980; Barbe, 1980.

CCD can be used to implement logic operations. For example, OR and AND operations can be realized by arranging electrodes of different sizes (Zimmerman et al., 1977; Allen, 1977; *Electronic Design*, July 19, 1978, p. 104; Zimmerman and Barbe, 1977; Dao, 1981; Kerkhoff et al., 1981; Agrawal and Agrawal, 1981).

Magnetic Bubbles

Magnetic bubbles were invented by A. H. Bobeck of Bell Labs. in 1967. A thin epitaxial magnetic film is formed on a nonmagnetic substrate. When a direct-current magnetic field of appropriate strength is applied perpendicular to the magnetic film, small cylindrical magnetic domains of at most a few micrometers in diameter are formed within the magnetic film. They are called **magnetic bubbles** and are polarized in the opposite direction in the sea of the rest of the magnetic film polarized in the direction of the externally applied magnetic field. On the surface of the magnetic film, small pieces which are magnetically soft (usually permalloy) are deposited. When a rotating magnetic field (i.e., alternating-current magnetic field) is applied in parallel to the magnetic film, in superposition to the direct-current magnetic field, magnetic bubbles move along permalloy pieces. When a loop is formed by equally spaced permalloy pieces, magnetic bubbles continue to move along the loop indefinitely. If we have a bubble or no bubble corresponding to signal 1 or 0, respectively, along this loop, then this loop works as a serial memory by reading or writing information at one place on the loop. Thus the access time is of the order of a few milliseconds, which is much slower than for semiconductor RAMs. When the power supply is cut off, the magnetic bubbles stay in the positions occupied at the time of interruption, and no information is lost as long as the direct-current magnetic field is supplied by an external permanent magnet.

References: Bobeck and Scovil, 1971; Chester, 1980; Nicolino, 1980; Eschenfelder, 1981; *Digital Design*, Dec. 1980, p. 39; *Electronics*, Mar. 13, 1980, p. 33; June 16, 1981, pp. 93, 94; Bobeck et al., 1975.

Logic operations can be realized by magnetic bubbles on the same chip as memories by using appropriate configurations of permalloy pieces, though the speed is low. This is convenient since data can be processed at least partially during storage (Minnick et al., 1972; Chang et al., 1973; Sandfort and Burke, 1971; Lee et al., 1974; Bongiovanni and Luccio, 1980; Chung et al., 1980).

Josephson-Junction Memories

This is a RAM realized with Josephson junctions mentioned in Chapter 5. A fast memory with large capacity can be realized (Henkels and Zappe, 1978; Brown et al., 1979; Faris, 1979; Henkels and Greiner, 1979).

Optical Disks

This is a rotating disk used with a laser beam. Information can be permanently written in by making holes on the surface of the disk by a laser beam with high intensity. For reading, a laser beam of low intensity is used. The beam is reflected from the disk when no hole is present, but is diffused when a hole is present. There are variations of writing and reading mechanisms. We can have a ROM with 100 billion bits per disk or far more at an extremely low price. (The price of a disk is expected to be on the order of tens of dollars.) This makes the per-bit cost of an optical disk far lower than that of a magnetic tape or disk (Kenney et al, 1979; Laub, 1981; *Electronic Design*, Dec. 20, 1978, pp. 23, 24; Aug. 2, 1980, pp. 28, 30; June 25, 1981, p. 36; Aug. 6, 1981, p. 16; *Electronics*, May 5, 1981, pp. 97–102; June 16, 1981, pp. 89, 90).

Associative Memories

An associative memory is a special type of RAM where if some bits of a word are given to this memory, all words that have these bits (0 or 1) in the specified positions are read out. This is also called **content addressable memory (CAM)** (Hodges, 1972; Leonard, 1971).

6.9   Comparison of All Memories

Let us compare semiconductor memories with other types of memories in order to find where semiconductor memories stand in the entire spectrum of memories.

Figure 6.9.1 compares the access times and system costs of all memories. System costs are the total system costs after assembly of memory cells with peripheral circuits, pc boards, cabinets, and cooling equipment. So system costs are higher than nonsystem memory costs, namely, memory cell costs. In the case of magnetic core memories, a magnetic toroidal core costs only about 0.1 cent/bit, but its system cost is several times higher because the peripheral circuits are expensive and bulky. In the case of semiconductor memories the difference is very small unless the memory capacities are very

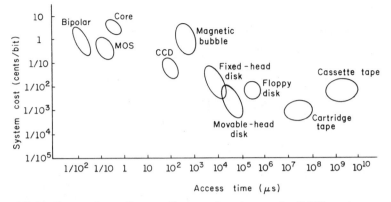

Fig. 6.9.1 **Comparison of access times and system costs of different memories.**

small. For comparison, costs of dynamic RAMs are shown in Fig. 6.9.2. Since the cost per bit of semiconductor memories has been decreasing considerably, magnetic core memories are less used now, although the memory retention of magnetic core memories, when the power supply is cut off, is more reliable and longer. [For this reason, magnetic core memories are still used sometimes (*Electronic Design*, Mar. 31, 1981, pp. 48, 50).]

The disks and tapes shown in the right half of Fig. 6.9.1 are mechanical memory equipment and are used only for very large memory capacities when semiconductor memories are too expensive.

Remark 6.9.1: Semiconductor memories and magnetic bubble memories are often manufactured with redundant memory cells. These memories have been used after disconnecting faulty cells. A less than 10% increase in chip area by

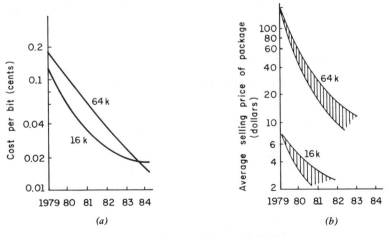

Fig. 6.9.2 **Costs of dynamic RAMs.**

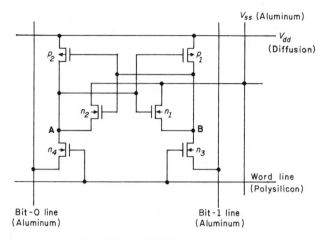

Fig. P6.4 CMOS memory cell.

redundancy improves the yield by at least 30% (*Spectrum*, Jan. 1980, p. 48; Jan. 1982, p. 51; *Electronic Design*, Jan. 8, 1981, p. 94; Donaldson, 1980; *Electronics*, Sept. 13, 1979, pp. 90, 91; Jan. 20, 1980, p. 33; Dec. 4, 1980, pp. 108, 110; Jan. 3, 1981, pp. 196, 198; June 2, 1981, p. 44; July 28, 1981, pp. 117–130). □

References on Memories: Hnatek, 1975, 1977, 1979, 1980; Matick, 1977; Capece, Oct. 25, 1979; Theis, 1978; White, 1980; Riley, 1971; Marshall, 1976; Gossen, 1979; Luecke and Mize, 1973; *Electronics*, Oct. 25, 1971, pp. 82–86; Foss et al., 1976; Eimbinder, 1971.

Problems

6.1 Describe how to write and read information into or from the memory cell of Fig. 6.2.3.

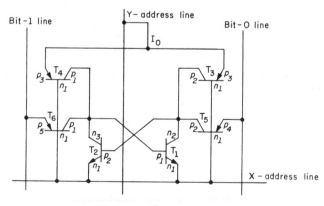

Fig. P6.5 I²L memory cell.

6.2 Design a layout for the memory cell of Fig. 6.2.3, using the design rules given in Section 4.6. (Dimensions need not be precise.)

6.3 Design a layout for the memory cell of Fig. 6.2.2, using the design rules given in Section 3.3. (Dimensions need not be precise.)

6.4 A CMOS memory cell is shown in Fig. P6.4. Discuss how this memory cell works.

6.5. Fig. P6.5 shows a memory cell based on I^2L. Design a layout of this I^2L memory cell, using p_3 as the injector. Then show a cross section.

6.6 When the bit length of an address code is too long for the decoding scheme of Fig. 6.4.1b with a single decoder, we can use the two-decoder scheme of Fig. 6.4.1a. If the bit length is still too long for this two-decoder scheme, devise a decoding scheme to handle this situation.

6.7 Using n-**MOSFET**s and **MNOS**, design an S-R flip-flop circuit in which previously stored information can be restored when the power supply comes back after interruption.

CHAPTER **7**

ROMs in Logic Design and Task-Realization Algorithms

In the previous chapters we have discussed the very basic aspects of LSI/VLSI technology, that is, electronic circuits, logic design procedures, and layouts, which are pertinent to different logic families. In this case the size and performance of a designed chip are more influenced by a higher level design than by a lower level one, though each design level is important, that is, they are influenced more by electronic circuit design than by layout design, and more by logic design than by electronic circuit design. **As discussed in this and the succeeding chapters, the size and performance of LSI/VLSI chips are more influenced by higher level designs, such as task-realization algorithm design and architecture design, than by logic design, though the LSI/VLSI technology directly influences the lowest level, that is, the layout.**

As hardware becomes inexpensive and smaller because of LSI/VLSI technology progress, tasks which used to be processed by software are increasingly being realized by hardware. Because of this, it is vitally important to devise task-realization algorithms (which should be differentiated from algorithms or procedures for logic or electronic circuit design) and architectures (which are for a specific task and accordingly more diversified than the architecture for a general-purpose processor), which are most appropriate for the given tasks to be realized by LSI/VLSI chips. If such a selection is inappropriate, we shall not be able to minimize the chip size or maximize performance. This is a new problem, since complex tasks, which formerly were too expensive for hardware realization, can now be realized by LSI/VLSI chips. Algorithms and architecture, which were appropriate for the design of conventional computers with centralized core memories and discrete components, may therefore not be appropriate for LSI/VLSI realization.

In this chapter task-realization algorithms are discussed in Sections 7.5 and 7.6, particularly in connection with the use of ROMs for logic design, while task-realization algorithms must be considered even in the case of logic design with gates (as discussed in later chapters).

The use of ROMs in logic design is probably more unique than logic design with gates, and thus is discussed in detail in this chapter. So ROMs are opening up new unique dimensions in logic design, as they become more

276

compact, less costly, and faster. In this chapter let us discuss how to use ROMs in logic design.

7.1 Motivations for the Use of ROMs in Logic Design

In the early days of the computer development when RAMs were expensive, ROMs were used to store important constants such as π and e for numerical computation. More recently electronic telephone exchanges used ROMs to store semipermanent data. ROMs had been used in microprogramming of the IBM 360 and other computers. But until 1969 engineers saw little need for ROMs in general applications other than the above, since RAMs stored with desired patterns could always do the job of ROMs. This view was justifiable at that time, because ROMs at that time were built from cores, transformers, capacitors, or resistors (e.g., capacitance versus no capacitance, a wired magnetic core versus an unwired core, or other means were used to distinguish between 1 or 0), and consequently no real advantage of ROMs over RAMs in cost and performance existed (Dussine, 1971).

Now because of progress in semiconductor technology, ROMs are compact and inexpensive, and ROMs have been extensively used in logic design and many other applications for the following reasons:

1. ROMs have a device packing density at least a few times higher than RAMs (roughly 10 times higher than static RAMs and 2 to 3 times higher than dynamic RAMs), and their electronic realization is less complicated. So ROMs cost a few to several times [about 4 to 8 times (*Computer*, Oct. 1980, p. 29)] less than RAMs. Thus we often store permanent data or permanent portions of software in ROMs rather than RAMs.

It is important, however, to notice that the high device packing density of ROMs (or RAMs) does not mean smaller areas for ROMs (or RAMs) to realize given tasks (i.e., switching functions or more complex job functions such as floating-point operations) than the areas for random-logic gate networks. This is because tasks usually require many more memory bits than logic gates. In other words, **ROMs (or RAMs), despite their higher device packing density, usually have lower information packing density (a large number of memory bits is redundant) than logic gates, where each gate carries a lot of information, performing much more complex logic operations than each memory cell does.** (Registers and shift registers discussed in Section 6.3 are exceptions.)

2. Layouts of ROMs are regular and repetitive. Thus layouts of ROMs are much less time-consuming than random-logic gate networks, for which a compact layout must be made, tailored to each network. As explained in Table 4.7.1, the layout of random-logic networks is extremely time-consuming, and even more so as the integration size increases. By using ROMs, the layout time is greatly reduced, roughly by an order of magnitude, thus reducing also the chip design expenses. This is also a great advantage when technology changes. In other words, ROMs based on the most advanced technology can be developed in a short time. The device packing density is much higher than that of random-logic gate networks, and about 80 bits of a ROM occupy the same area as one logic gate.

3.   Since the layouts of ROMs are completely independent of the information to be stored, design changes can be made easily and, once information is prepared and debugged, the same information can be used even if solid-state technology for ROMs changes. When information to be stored is changed, it can easily be changed on ROMs.

4.   Since the memory cells in a ROM are repetitive and regular, it is easy to test stored information.

The main disadvantage of ROMs compared with random-logic gate networks, is their low speed, because long connections are arranged in matrix form in the case of ROMs, no matter whether or not portions of some connections are necessary, and their parasitic capacitances and high resistances make ROMs slower than random-logic networks. In the case of low production volume, ROMs cost much less than random-logic networks because the design and layouts of random-logic networks are time-consuming and consequently expensive, and because only a connection mask must be custom-made in the case of ROMs. In high-volume production ROMs cost more because of the more compact layouts of random-logic networks, due to the higher information packing density explained in (1) above.

ROMs have been used extensively in computers, from calculators up through mainframe computers. As discussed throughout this chapter, ROMs and their variations, such as PROMs, PLAs, and FPLAs, are used in logic design in many different forms. Since it appears that there is no systematic way to use ROMs in logic design and usage depends on the designers' imagination and skills, many examples of different applications are discussed in this chapter. This may be the best way to explain how to use ROMs in logic design. Each ROM application, that is, a ROM realization of a task (or logic function), needs to be compared with logic gate realizations of the same task in terms of speed, cost, and size (or chip area) before a final decision as to its adoption in a digital system. The logic gate realizations of the tasks whose ROM realizations are discussed in this section are not discussed here since they are extensively discussed in many other books. ROMs are also extensively used in storing programs such as interpreters and compilers for BASIC, PASCAL, FORTRAN, PL/M, and others. These firmware applications are also not discussed in this book because they are software problems rather than logic design problems. They are straightforward as LSI/VLSI realizations, and there are many other publications on the subject (*Electronics,* Oct. 27, 1977, p. 110; Langer and Dugan, 1978; Carroll, 1977; Conklin and Rodgers, 1978).

Choice of Mask- or Field-Programmable ROMs

When ROMs are used in logic design, we can usually use mask- or field-programmable ROMs, but we have to choose the appropriate one, considering cost, speed, and delivery time, as follows.

When customers order mask-programmable ROMs, semiconductor manufacturers usually charge several hundred dollars for each mask to be custom-made, with delivery times of a few to several weeks. The price that a customer must pay is given by the following formula:

(total price of each mask-programmable ROM)

$$= \text{(price of each ROM)} + \frac{\text{(mask charge)}}{\text{(production volume)}}$$

In a field-programmable ROM a fuse occupies an extra area in each memory cell, and a special electronic circuit to supply heavy currents for the blowing of fuses requires a large area. Thus the price of each field-programmable ROM is slightly higher than that of a mask-programmable ROM with the same memory capacity, when the memory capacity is small; it is much higher than that of a mask-programmable ROM when the memory capacity is large. Therefore unless the production volume is at least 1,000 (because of the mask charge of several hundred dollars), mask-programmable ROMs are not economical for customers. When the production volume is extremely high, however, the selling price of each mask-programmable ROM may be negotiable, because the profit margin of a mask-programmable ROM is generally set higher by manufacturers than that of a field-programmable ROM, which is squeezed by competition. (Semiconductor manufacturers prefer to make field-programmable ROMs because of greater income due to extremely high production volume.)

Of course, customers must use field-programmable ROMs (or RAMs) until information is completely debugged.

References: Allison and Frankenberg, 1976; Percival, June 1972, June 8, 1972; Henle and Ho, 1969; Tang, 1971; Thurber, 1971; Nichols, 1967; McMullen, 1974; Byrd and Jenkins, 1974; McDowell, July 5, 1974; July 20, 1974; Wyland, Sept. 1974; *Electronics,* Oct. 26, 1978, p. 157; Monolithic Memories, Application Note 110.

7.2   Generation of Waveforms and Characters

Let us start with simple but somewhat novel applications of ROMs.

Generation of Binary Waveforms

Some logic networks (e.g., dynamic MOS networks in Chapter 4) and memories in computers need binary waveforms for timing, as shown in Fig. 7.2.1*a*. Such sequences of binary signals used to be generated by complex

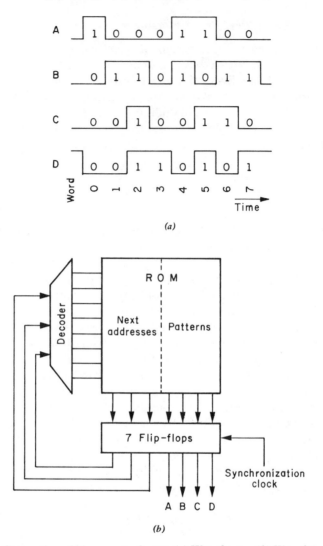

Fig. 7.2.1   Generation of binary waveforms. (*a*) Waveforms. (*b*) Waveform generation by a ROM.

electronic circuits which consist of counters and *J–K* flip-flops. ROMs, however, offer a novel means to generate binary waveforms, as shown in Fig. 7.2.1*b*.

As an example, suppose that a minicomputer has 32 memory reference instructions, each of which has different timing requirements on signal lines *A* through *D*, as shown in Fig. 7.2.1*a*. We want to design a timing generator based on a clock. All instructions require 8 clock cycles. This can be realized by storing words 1000, 0100, 0111, and so on, in addresses 0, 1, 2, . . . , 7 of a ROM, respectively. In these addresses, the next addresses are also stored in the left half of the ROM in Fig. 7.2.1*b*. The outputs of the ROM are transferred into flip-flops in synchronization with the clock, eliminating waveform discontinuities (so-called glitches) during the transitions. It is easy to find and correct design mistakes.

Reference: Monolithic Memories, Application Note 110.

Generation of Analog Waveforms

Analog signals, such as the waveform shown in Fig. 7.2.2*a*, can be generated using ROMs. First a waveform is sampled at a constant time interval, as shown in Fig. 7.2.2*a*. The amplitude at each sampling time is stored in a ROM as a binary number, by pulse-code modulation (PCM). Whenever necessary, these binary numbers stored are converted into the original waveforms, as shown in Fig. 7.2.2*b*. There are many applications, such as storing human voice and musical sounds in this manner.

Information condensed from the original waveforms, however, is often stored in ROMs instead of the original waveforms. Many compression methods, which are more complex than PCM, are used. For example, in the case of speech synthesis, phonemes and related information can be stored.

ROM chips storing speech have been extensively used in the consumer electronics industry. Speak-and-Spell and language translators developed

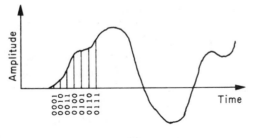

(*a*)

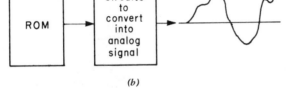

(*b*)

Fig. 7.2.2 Generation of analog waveform. (*a*) Sampling of waveform. (*b*) Generation of waveform.

by Texas Instruments store speech in ROMs. Speaking dolls and toys have been developed, and telephone operator standard responses to customers are stored in ROMs.

References: Wiggins and Brantigham, 1978; Burstein et al., 1979; Teja, 1979; Smith and Crook, 1981; Ahrens et al., 1981; *EDN,* Oct. 20, 1980, pp. 57, 59; *Business Week,* May 19, 1980, pp. 44D–F; July 28, 1980, pp. 68A, E; *Electronic Design,* May 28, 1981, pp. 35, 36; *Computerworld,* June 23, 1980, p. 65; *Electronics,* Mar. 27, 1980, pp. 39, 40; April 24, 1980, pp. 42, 44; May 22, 1980, pp. 95–105; Feb. 10, 1981, pp. 78, 80, 118–121.

> *Remark 7.2.1:* **Voice mail** is an important outcome of the progress of computer and memory technology. When a telephone subscriber calls another subscriber and finds him absent, the human voice message of the former is stored in a memory (magnetic tapes, disks, or RAMs) at a telephone exchange and then will be automatically delivered to the latter when the latter is available. When this new technology of voice mail is refined for memory requirement reduction and voice quality improvement, ROMs also may be used as part of the entire system (*Business Week,* June 9, 1980, pp. 80–84; Aug. 24, 1981, pp. 90F, H; *Electronics,* Apr. 21, 1981, pp. 99, 100). □

Character Generation

For digital systems, characters such as numerals, alphabet letters, arithmetic operation symbols, and Chinese characters need to be displayed on CRT (cathode ray tube) displays, printers, or LED (light-emitting diode) displays. These characters can be stored in ROMs.

Alphanumeric characters are stored in matrix form, using 5 columns and 7 rows (or more columns and rows), as illustrated in Fig. 7.2.3. Chinese characters need much greater numbers of columns and rows. In order to save memory space, instead of literally storing characters, many compression schemes have been devised and sometimes used. In this case electronic circuits and logic networks to restore the original characters are needed.

References: Mrazek, 1973; Bratt, 1973; Otsuka, 1978, 1979; *Electronics,* Sept. 2, 1976, p. 151; Millman and Halkias, 1972; Titus et al., 1979; Leventhal, 1977.

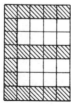

Fig. 7.2.3   **Letter E stored in a ROM.**

Table 7.3.1   Truth Table

x_1	x_2	$\cdots$	x_n	f_1			$\cdots$	f_m
0	0	$\cdots$	0	0	0	1	$\cdots$	0
0	0		0	1	0	1		1
0	0		1	0	1	0		1
0	0		1	1				
1	1		1	0				
1	1	$\cdots$	1	1				

7.3   Combinational Logic Networks

Switching functions $f_1, f_2, \ldots, f_m$ of n variables $x_1, x_2, \ldots, x_n$ are usually realized by logic gate networks. But as the cost of ROMs decreases, they can be conveniently realized by a single ROM, by storing the truth table of these functions, given in Table 7.3.1, in the ROM, as illustrated in Fig. 7.3.1. In this case each combination of values of the variables $x_1, x_2, \ldots, x_n$ is used as an address to access the ROM.

As a simple example, let us consider the hardware realization of the conversion of a Gray code to a binary code, as shown in Fig. 7.3.2a, where the pair of Gray code words that correspond to any pair of binary code words with consecutive values differ in only one bit position. If this conversion is done in a logic gate network, it requires 3 Exclusive-OR gates, as shown in Fig. 7.3.2b or a total of 15 AND, OR, and NOT gates, since each Exclusive-OR gate consists of 5 AND, OR, and NOR gates, as illustrated in Fig. 7.3.2c. But if the conversion is done in a single ROM, it requires 16 4-bit words, or a total of 64 bits. In other words, a single ROM of only 64 bits, where each Gray code word is used as a memory address, can replace 15 gates; so in this case each gate is equivalent to approximately

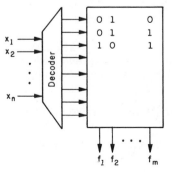

f_1 f_2 f_m **Fig. 7.3.1   Truth table in a ROM.**

Gray				Binary			
0	0	0	0	0	0	0	0
0	0	0	1	0	0	0	1
0	0	1	1	0	0	1	0
0	0	1	0	0	0	1	1
0	1	1	0	0	1	0	0
0	1	1	1	0	1	0	1
0	1	0	1	0	1	1	0
0	1	0	0	0	1	1	1
1	1	0	0	1	0	0	0
1	1	0	1	1	0	0	1
1	1	1	1	1	0	1	0
1	1	1	0	1	0	1	1
1	0	1	0	1	1	0	0
1	0	1	1	1	1	0	1
1	0	0	1	1	1	1	0
1	0	0	0	1	1	1	1

(a)

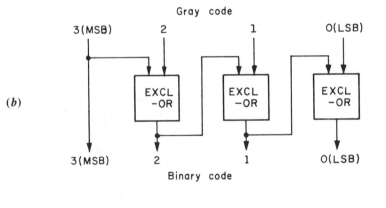

(b)

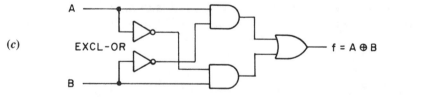

(c)

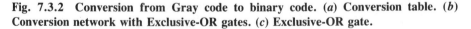

Fig. 7.3.2 Conversion from Gray code to binary code. (*a*) **Conversion table.** (*b*) **Conversion network with Exclusive-OR gates.** (*c*) **Exclusive-OR gate.**

4 bits. (The equivalence ratio of the number of ROM bits per gate varies depending on the task to be realized.)

In many cases information stored in a ROM, such as character tables (in Section 7.2) and look-up tables (in later sections), can be regarded as truth tables. So the truth table is an important concept. When *n*, the number of variables, is small, the use of a ROM as a truth table is a convenient means to realize logic functions. But when n increases, the number of rows increases exponentially (i.e., 2^n), and the size of the ROM will grow out of hand (i.e., such a large ROM cannot be manufactured). Also even if the size is manufacturable, a ROM with a large memory capacity has very slow access time, with high cost. If we use more than one ROM, mixed with gates (or wired-logic), however, we can often reduce the memory space requirement drastically with higher speed, as discussed in the following sections. Although some advantages of ROMs discussed in Section 7.1 may diminish, such as the ease in changing information, this is sometimes a useful technique.

> **Remark 7.3.1:** It is an interesting question how many bits of ROM are equivalent to a gate, but there is no simple answer. According to [Monolithic Memories (Application Note 110)], which converted into ROMs a dozen SSI and MSI packages of TTL gates, such as a parity generator, BCD-to-decimal converter, and 2-bit adder, roughly 37 bits of ROM are required per logic gate, if two somewhat unusual cases are excluded. This figure probably serves only to give a rough idea, because the conversion highly depends on the designer's skills, the functions to be converted, and also whether a single ROM or a ROM network is used. □

Networks of ROMs

By realizing networks of ROMs, possibly mixed with gates (or wired-logic), we can reduce the total memory space requirement. To do this, we need to investigate given switching functions in order to find whether any properties of them can be utilized for appropriate network synthesis.

Let us consider the AND function of the variables x_1, x_2, x_3, x_4 as a simple example. This function $x_1 x_2 x_3 x_4$ can be realized by the truth table stored in a single ROM of 16 bits. But this function can be realized by the network of 3 ROMs shown in Fig. 7.3.3, where two ROMs in the first level realize the functions $x_1 x_2$ and $x_3 x_4$, which serve as inputs to the ROM in the second level. In this case the total memory space requirement is 12 bits. Actually it is impractical to use ROMs for the AND function of only 4 variables, but this simple example illustrates the basic technique. For an AND function of a larger number of variables, the difference in the memory space requirements increases, as shown in Table 7.3.2 for a small number of variables.

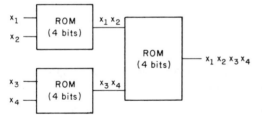

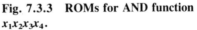

Fig. 7.3.3   ROMs for AND function $x_1 x_2 x_3 x_4$.

Table 7.3.2   Difference in Memory Space Requirements

NUMBER OF VARIABLES FOR AND FUNCTION	MEMORY SPACE REQUIREMENT (BITS)	
	SINGLE ROM	CASCADED ROMs IN 2 LEVELS
4	16	12
5	32	16
6	64	20
7	128	24

Another simple example is shown in Fig. 7.3.4. When a function of 12 variables can be realized by the cascade of two ROMs, as shown in Fig. 7.3.4 (it is to be noticed that this is not always possible, depending on the functions), the total memory requirement is only 2,048 bits, whereas the function requires $2^{12} \times 4 = 2^{14} = 16,384$ bits for a single ROM.

The above examples illustrate that when the memory size of a single ROM becomes unrealistically large, networks of ROMs may make the memory size of each ROM realistically small and the total memory space requirement much smaller than the single ROM realization, depending on functions.

A more complicated example is shown in Fig. 7.3.5. In this case the two outputs of the ROMs in the first level are tied down as a Wired-OR and connected as a single input to the ROM in the second level. The Wired-OR reduces the number of the inputs to the second-level ROM from eight to seven, so the memory size of the second-level ROM is halved. Generally by a appropriate use of gates (including Wired-ORs or Wired-ANDs), we can reduce the memory size.

The above technique can be applied when the designer finds unique properties of switching functions in given truth tables.

Reference: Kvamme, 1970.

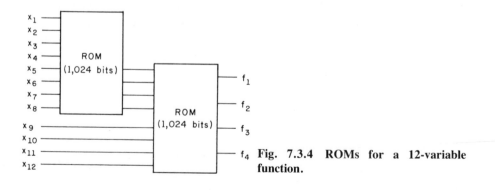

Fig. 7.3.4 ROMs for a 12-variable function.

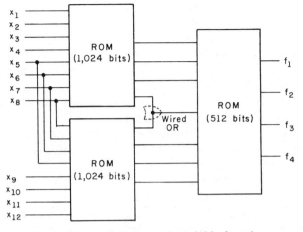

Fig. 7.3.5 ROMs for a 12-variable function.

Design of 1-out-of-*n* Detector

As a more complex example, let us discuss the design of a 1-out-of-*n* detector with ROMs, in order to compare it with other design approaches, such as logic gate network design.

A 1-out-of-*n* detector is a network whose output is 1 when exactly one of the *n* inputs to this network assumes value 1, and otherwise 0. This is used to detect whether or not *n* transmission lines carry erroneous signals (i.e., if the number of 1's in these lines is more or less than 1, they carry erroneous signals).

Let us design this with three different approaches and then compare the results, as follows.

First, let us design a 1-out-of-*n* detector using a loopless gate configuration, as shown in Fig. 7.3.6. This network works fast, but it requires *n* inverters, at least *n* AND gates, and at least one OR gate. When *n* increases, the number of gates increases because of maximum fan-in restriction.

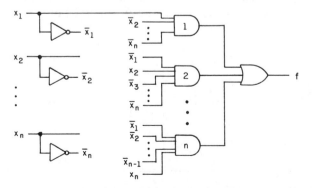

Fig. 7.3.6 1-out-of-*n* detector using a loopless gate network.

The second approach is based on the combination of a loadable shift register and a 2-bit counter, as shown in Fig. 7.3.7. Inputs $x_1, \ldots, x_n$ are loaded into the shift register. Then in synchronization to a clock, the bits in the shift register are serially fed into the counter which counts the number of 1's in $x_1, \ldots, x_n$. When the counter shows 1, the detector output shows 1, and otherwise 0. [The 2-bit counter with a feedback line from the inverter to gate 1 can do this, no matter how large n is. In other words, if the counter shows output (0 1), the detector output f is 1. If the counter shows (0 0), the detector output f is 0. Also if the counter shows (1 0), the detector output is 0, and gate 1 is inhibited to transmit the shift register output to the counter in order to prevent the counter overflow. In other words, when the number of 1's is more than one, the detector output is always 0.] This approach requires considerably fewer gates than the first approach, as n increases. [This is an interesting example of combinational network designed with loops. Even if a function can be realized with a combinational network without loops, as shown in Fig. 7.3.6, its network realization with loops, as shown in Fig. 7.3.7 (i.e., the counter contains loops) requires fewer gates.] Because of the serial counting operation, the response time gets slower as n increases if the clock rate is fixed (or the clock rate must increase as n increases, requiring expensive high-speed gates).

The third approach is based on the use of ROMs. A simple-minded approach is to store the output values of the detector in a single ROM, using input values $(x_1, \ldots, x_n)$ as addresses, as shown in Fig. 7.3.8. The ROM in this case needs to have the memory capacity of $2^n \times 1$ bits. The detector's output f is read out of the ROM.

When n is large, this approach requires a ROM with a large capacity. The following approach with many ROMs with a smaller capacity is preferred, because a ROM with a large capacity is expensive or unrealizable. The detector is constructed with ROMs with t inputs, each as illustrated in Fig. 7.3.9 with $t = 8$ as an example. At the output of each ROM with a memory capacity of 2^t bits, where $t < n$, three states in t inputs are distinguished: all 0's, a single 1, and two or more 1's. Since the all-0's state can be identified from knowledge of the other two states, each ROM has only two outputs for the other two states, namely, one output which is 1 only when the number

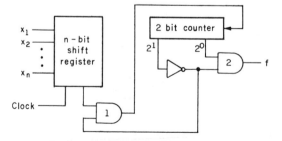

Fig. 7.3.7   1-out-of-n detector using a shift register and counter.

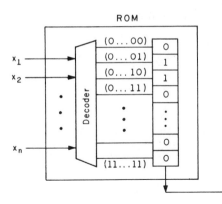

Fig. 7.3.8 **1-out-of-***n* **detector using a single ROM.**

of 1's in the *t* inputs is two or more, and the other output which is 1 only when the number of 1's is exactly 1. (In the case of *t* = 8, each ROM has 512 × 2 bits memory capacity, where the 2 bits are for two outputs.)

The first level in Fig. 7.3.9 consists of ⌈*n/t*⌉ ROMs. The first outputs of ROMs, which represent the state of the "the number of 1's is two or more," are collected by OR gates in groups of *t* in the second level. The second outputs of ROMs are connected to ROMs in the second level, in groups of *t*, each of which determines whether the number of 1's in its *t* inputs is exactly one, or two or more.

In the succeeding levels this is repeated. In the second, third, . . . , levels

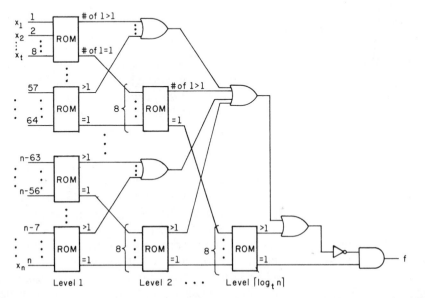

Fig. 7.3.9 **1-out-of-***n* **detector using ROMs. (Reprinted from Kobylar et al., 1973. Copyright © McGraw-Hill, Inc., 1973. All rights reserved.)**

Table 7.3.3 Comparison of Four Approaches to the Implementation of the 1-out-of-n Detector

APPROACH	n				
	10	50	80	800	5000
Gate network (Fig. 7.3.6)					
Number of levels	3	5	5	7	9
Response time (nsec)[a]	30	50	50	70	90
Number of IC packages[b]	3	311	737	71,369	2,781,056
Shift register and counter (Fig. 7.3.7)					
Number of levels	13	53	83	803	5,003
Response time (nsec)[a]	130	530	830	8,803	50,003
Number of IC packages[b]	5	12	12	84	504
ROMs (Fig. 7.3.9 with ROMs)					
Number of levels	5	5	6	6	7
Response time (nsec)[a]	200	200	240	240	280
Number of IC packages[b]	7	12	19	133	810
PLAs (Fig. 7.3.9 with PLAs)					
Number of levels	3	5	5	6	7
Response time (nsec)[c]	120	200	200	240	280
Number of IC packages[b]	3	9	10	62	385

[a] A switching time of 10 nsec per level of gate or shift is assumed.
[b] An AND gate with 10 inputs, an OR gate with 10 inputs, 10 NOTs, a shift register of 10 bits, a single counter, a ROM with 8 inputs and 2 outputs [memory capacity of 512 bits and matrix size of $(8 \times 2 + 2) \times 256 = 4,608$], or a PLA with 16 inputs and 2 outputs is assumed for each IC package.
[c] An access time of 40 nsec is assumed for a ROM and a PLA.

we need $\lceil n/t^2 \rceil$ ROMs, $\lceil n/t^3 \rceil$ ROMs, ..., respectively. The AND gate in the last level receives an inhibiting signal through the chain of OR gates when the ROMs in the first level receive two or more 1's in the inputs.

The total number of t-bit ROMs required for a 1-out-of-n detector is

$$\sum_{i=1}^{\lceil \log_t n \rceil} \lceil n/t^i \rceil$$

A comparison of the number of IC packages and the response times for the above three approaches is shown in Table 7.3.3 for different values of n. (The case with PLAs to be discussed in Section 7.7 is also shown.) The first approach with a loopless network of gates is the fastest. But it requires the largest number of IC packages, and accordingly it is the most expensive. The second approach is the most economical, but the slowest. The third approach with ROMs is between these two approaches. It is slower than the first approach, but not much. It is much faster than the second approach. In terms of the number of IC packages, it is more expensive than the second approach, but not much. (The cost of a ROM of 1,024 bits would not be much different from that of an IC package of a gate with 10 inputs.) But it is much less expensive than the first approach. The advantages of the third approach with ROMs are obvious. When the number of IC packages is reduced, labor, time, and material for assembling (fewer pc boards may be used) are reduced, so saving in cost is further enhanced, and manufacturing time is shortened. Another important advantage of the reduction in the number of IC packages is reliability. Since IC packages replace soldering of many connections, the reliability is greatly improved. Testing of the network is also easier because of the reduction of the number of connections and the systematic configuration.

When we lay out each of the above approaches on a single chip, instead of using off-the-shelf packages, the differences among them would be further pronounced. In the case of the random-logic gate network of Fig. 7.3.6, connections occupy a significant part of the chip area. As the number of gates increases, the number of connections and the average connection length increase rapidly, and connections occupy a greater area. Also layout is tedious and time-consuming because of the irregular configuration of random-logic gate network. The ROM network approach in Fig. 7.3.9 does not have these problems. Its layout can be done in a much shorter time, and the chip size can be smaller (though the chip size is not necessarily smaller when the total memory space requirement is larger). If the ROMs in Fig. 7.3.9 are replaced by PLAs to be discussed later, the chip size will be further reduced.

Reference: Kobylar et al., 1973.

7.4 Sequential Logic Networks

When a transition-output table of a sequential network is given, the sequential network can be realized easily by storing the transition-output table in a ROM.

Suppose that a transition-output table, such as Fig. 7.4.1a, is given. Then a sequential network which realizes this can be implemented as shown in Fig. 7.4.1b, using x_1, x_2, y_1, y_2 as an address into the ROM. The ROM stores next values of the internal variables Y_1, Y_2 as next addresses of the ROM, along with the network outputs Z_1, Z_2.

It is easy to use this approach and to change its memory contents, though it is not necessarily a best approach.

References: Perrin et al., 1972; Muroga, 1979.

Counters with ROMs

When the output functions of sequential networks are specifically given, the above general design approach can be simplified or modified.

As a simple example, a binary counter is realized in Fig. 7.4.2. When the current count (0 1) at address (0 0) is displayed at the outputs, suppose that one clock pulse (or input pulse) comes in. Then the current count (0 1) is used as the next address, and count (1 0) stored in the address (0 1) is displayed at the outputs. And so on.

Similarly, other types of counters such as up–down counters and Gray-code counters can be designed.

References: Kvamme, 1970; Farr, 1976; Nichols, 1967.

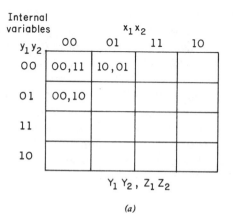

(a)

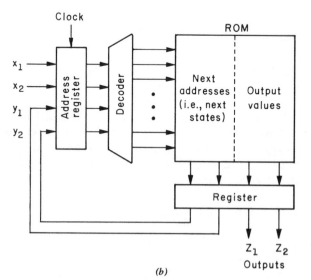

(b)

Fig. 7.4.1   ROM realization of a sequential network. (*a*) **Transition-output table.** (*b*) **Sequential network with ROM.**

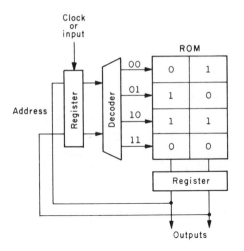

Clock
or
input

ROM

00	0	1
01	1	0
10	1	1
11	0	0

Address

Register

Decoder

Register

Outputs

Fig. 7.4.2   Counter with a ROM.

7.5   Code Conversion

In the case of simple tasks, such as addition, a task can be expressed easily with switching functions which can be realized with a gate network of reasonable complexity and size. For example, by using the carry and sum as output functions, a simple network can be easily derived for addition. In the case of more complex tasks, such as code conversion, trigonometric function calculation, and data sorting, switching functions appropriate for achieving a given task are often not obvious. Thus we have to devise an appropriate task-realization algorithm such that the task can be achieved by processing data by an appropriate configuration of subnetworks in an appropriate processing order specified by the devised algorithm, and then the output functions for each subnetwork are realized by a gate network. In other words, design must take the following sequence:

1.   Task-realization algorithm.   If a task to be realized becomes complex, we often need to devise a sophisticated task-realization algorithm, as we see later in the case of square-root or other calculations.

2.   Network (or system) configuration design.   An appropriate configuration of many subnetworks with appropriate output functions must be devised, such that the algorithm devised in step 1 can be carried out with maximum performance or minimum cost.

3.   Design of each subnetwork.   Each subnetwork defined in step 2 is designed with gates and memories.

Here step 2 could be called architecture design tailored to the algorithm in step 1, and step 3 is a consequent logic design.

In general, a number of different task-realization algorithms are conceivable for a given task, so we need to try and compare the final networks obtained from these algorithms. Often this is not a simple task. In order to illustrate this, code conversion is discussed in this section, and advantages and disadvantages of the use of ROMs are analyzed.

Code conversion among different number representation codes, such as from binary numbers to binary-coded decimal (BCD) numbers, or from Gray-code numbers to BCD numbers, is important in digital systems, because binary numbers are convenient for processing inside digital systems, but other numbers such as BCD are convenient outside for human interpretation or analog-to-digital conversion of signals. Code conversion was once a formidable task but became much easier with the availability of off-the-shelf packages.

In the following, let us discuss the conversion from binary numbers to BCD numbers as a typical example.

Conversion by a Random-Logic Gate Network

The conversion from binary numbers to BCD numbers is shown in Table 7.5.1. Using this as a truth table, we can derive a switching expression for each BCD digit. For example,

$$B_0^0 = \bar{x}_6\bar{x}_5\bar{x}_4\bar{x}_3\bar{x}_2\bar{x}_1 x_0 \lor \cdots \lor \bar{x}_6\bar{x}_5 x_4\bar{x}_3\bar{x}_2\bar{x}_1 x_0 \lor \cdots$$

After simplifying these expressions, we can design a random-logic gate network with these expressions as its outputs. Although this may yield the fastest network, the network will be very complex. Interconnections of gates will be extremely complex, and even if the entire network is realized on a

Table 7.5.1   Conversion from Binary Code to BCD Code

BINARY NUMBER							BCD								DECI-MAL
							TENS				UNITS				
x_6	x_5	x_4	x_3	x_2	x_1	x_0	B_3^1	B_2^1	B_1^1	B_0^1	B_3^0	B_2^0	B_1^0	B_0^0	
0	0	0	0	0	0	0	0	0	0	0	0	0	0	0	0
0	0	0	0	0	0	1	0	0	0	0	0	0	0	1	1
		·							·				·		·
		·							·				·		·
		·							·				·		·
0	0	1	0	0	0	1	0	0	0	1	0	1	1	1	17
		·							·				·		·
		·							·				·		·
		·							·				·		·
1	0	0	0	0	1	0	0	1	1	0	0	1	1	0	66
		·							·				·		·
		·							·				·		·
		·							·				·		·

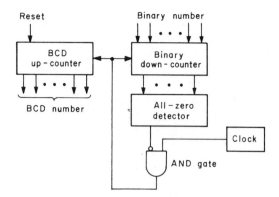

Fig. 7.5.1 Count-comparison method.

chip, layout will be excessively time-consuming. When the number of BCD digits increases, the difficulty increases quickly, because the number of rows in Table 7.5.1 increases exponentially, making this approach impractical.

Conversion by the Count-Comparison Method

The count-comparison method is based on a binary down-counter and a BCD up-counter. As shown in Fig. 7.5.1, a binary number to be converted is loaded into the binary down-counter, while the BCD up-counter is reset to the all-zero state. A clock then toggles both counters through the AND gate until the binary counter reaches the all-zero state. Since the BCD counter will have been increased the same number of times as the binary counter, the BCD counter will contain the desired BCD number.

This approach can be realized with the simplest network, but because of its extremely slow speed, this approach is also impractical.

Conversion by a ROM Network

Let us realize the conversion with ROMs. A straightforward approach is to realize the truth table in Table 7.5.1 by a single ROM, but the memory capacity requirement becomes excessive, along with decreasing speed, as the number of BCD digits increases. So let us realize the conversion with many ROMs, as shown in Fig. 7.5.2. One of four ROMs, each of which has 32 8-bit words, is enabled by the 1-out-of-4 decoder controlled by the 2 bits in the positions of weights 2^5 and 2^6. Then all other bits of a given binary number are used as an address to access the enabled chip, and then the BCD value stored in the accessed memory location is read out. In this case four outputs in this value are used for the value of the "units" position of the BCD number, and the remaining four are used for the value of the "tens" position. In order to convert a binary number to a 2-digit BCD number, 0 through 99, the first ROM contains the BCD values 0 to 31; the second

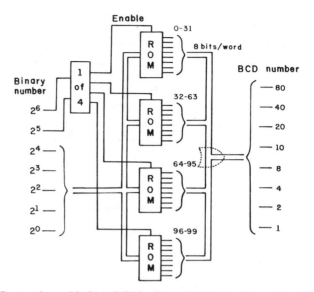

Fig. 7.5.2 Conversion with four ROMs. Each ROM has 32 words with 8 bits each.

ROM, 32 to 63; the third ROM, 64 to 95; and the fourth ROM, 96 to 99. The outputs of the ROMs are Wired-ORed in eight output positions, though only one Wired-OR is shown in Fig. 7.5.2 for the sake of simplicity. For example, when the binary number 0010101 (i.e., decimal number twenty-one) is given, the first ROM is enabled, and the BCD value 0010,0001 is read out of the memory location at address 10101.

By carefully examining the property of this conversion, the ROM network in Fig. 7.5.2 can be simplified, as shown in Fig. 7.5.3. Every odd or even binary number is also an odd or even BCD number, respectively. Thus the least significant bit of a binary number can be directly used as the least

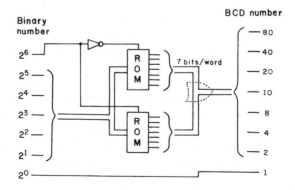

Fig. 7.5.3 Improvement of Fig. 7.5.2 with two ROMs. Each ROM has 32 words with 7 bits each.

significant bit of the corresponding BCD number. By this direct connection of an input to the corresponding output, the number of ROMs is reduced to half, as shown in Fig. 7.5.3, along with reduced memory space. In other words, we need to store 50 BCD values with 7 bits per value. The first three outputs of the two ROMs are for the "units" of the BCD number, the least significant bit of a binary number is the fourth output directly, and the next four outputs are for the "tens" of the BCD number. Also the 1-out-of-4 decoder can be replaced by the inverter of Fig. 7.5.3. Thus the memory space requirement of $4 \times 32 \times 8 = 1,024$ bits for Fig. 7.5.2 is reduced to $64 \times 7 = 448$ bits in Fig. 7.5.3. This means that even a simple property of the given task can simplify a hardware realization drastically.

The conversion approach for a 2-digit BCD number discussed above can be extended to a 3-digit BCD number, a 4-digit BCD number, and so on. But the memory space requirement increases rapidly. For example, to convert to 3-digit BCD numbers 000 through 999, 500 BCD values must be stored with 11 bits each, and to convert to 4-digit BCD numbers, 5,000 BCD values must be stored with 15 bits each. Thus we need to use a different approach, mixing with gates, as follows.

As an illustrative example, let us consider the conversion of a 13-bit binary number to a 4-digit BCD number. This can be realized with ROMs and adders, as shown in Fig. 7.5.4. The idea behind this realization is far more complex than that behind the previous cases, so let us outline it.

The seven lower order bits with weights 2^0 through 2^6 are converted to a BCD number, using the method described above. (Two ROMs with 32 8-bit words are shown combined into a single ROM with 64 8-bit words here.) The next six higher order bits with weights 2^7 through 2^{12} are used as an address to the left-hand ROM. This ROM needs to have 64 words with 15

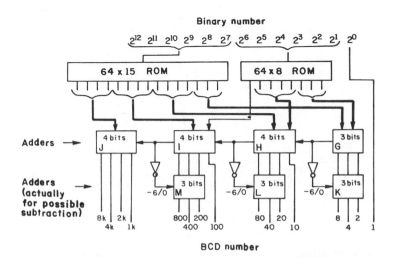

Fig. 7.5.4 Conversion of 13-bit binary number to a 4-digit BCD number.

bits per word, because 6-bit address inputs require 64 words, and a 4-digit
BCD number (each digit except the "units" digit requires 4 bits) requires
a total of 15 bits. (In contrast, the right-hand ROM requires 64 words with
8 bits per word, as can be seen easily.) Since the address inputs to this ROM
have weights 2^7 through 2^{12}, the first address of this ROM contains zero,
the second address contains BCD number 128, the third address contains
BCD number 256, and so on, with the increment of 128. The last address
contains BCD number 8,064.

BCD numbers read out of two ROMs cannot be directly added by a binary
adder, because any digit in a BCD number has a carry to a next higher digit
position only when that digit is equal to or greater than ten, as illustrated
in Fig. 7.5.5a. When a carry is generated from any BCD digit by adding 6
to that digit, this means that that BCD digit is equal to or greater than 10.
Thus in each address of the left-hand ROM in Fig. 7.5.4, an excess-6 BCD
number (i.e., a BCD number where 6 is added to each digit) is stored instead
of an ordinary BCD number. (In the example of Fig. 7.5.5b the excess-6
BCD number 6 7 8 14 in decimal representation is shown instead of the
original BCD number 128 in Fig. 7.5.5a. After this excess-6 BCD number
has been added to the ordinary BCD number from the right-hand ROM, and
carries, if any, are added to the next higher order digits, 6 is subtracted from
each BCD digit whenever no carry is generated from that digit, as illustrated

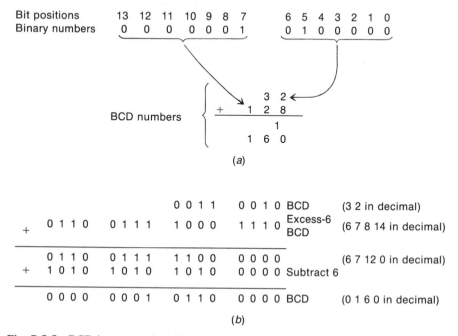

(a)

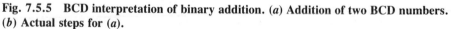

(b)

**Fig. 7.5.5   BCD interpretation of binary addition. (*a*) Addition of two BCD numbers.
(*b*) Actual steps for (*a*).**

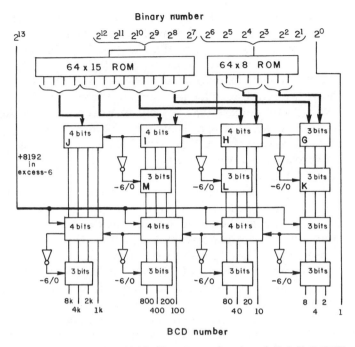

Fig. 7.5.6 Conversion of a 14-bit binary number to a full 4-digit BCD number.

in Fig. 7.5.5*b*. (In Fig. 7.5.5*b* 6 is not subtracted from the least significant digit, because a carry is generated from this digit.)

Based on the above algorithm, the conversion network is realized in Fig. 7.5.4. Adders K, L, and M are 3-bit adders, because subtraction or no subtraction of 6 does not change the least significant bit of each BCD digit. Since we are dealing with a maximum number 8,191, the BCD digit for "thousands" cannot have a value greater than 8, and a binary adder J alone is sufficient, eliminating the adder that would be named N. Also the BCD numbers stored for the adder J need not be in excess-6. Thus Fig. 7.5.4 requires ROMs of 1,472 bits and 24 full adders. If a single ROM is used, $2^{13} \times 16 = 131,072$ bits (or 65,536 bits if the last bit is separate) are required in total, and if the approach in Fig. 7.5.3 is used, a total of $4,096 \times 15 = 61,440$ bits is required.

If we need to have a full 4-digit BCD number converted from a 14-bit binary number, which actually has a maximum value of 16,384, at least two approaches are conceivable as follows.

The first approach is to add the value 8,192, the weight for the fourteenth bit, to the results of converting a 13-bit binary number to a BCD, as shown in Fig. 7.5.6, which contains the network of Fig. 7.5.4. The value of the fourteenth bit, 2^{13}, expressed in excess-6 representation (i.e., 6 is added to each digit in the BCD number representation of 8,192) is added into a 15-bit adder (in the second level from the bottom in Fig. 7.5.6). Eight adders, that is, 27 full adders, are added, with a total of 51 full adders.

The second approach is to expand the ROM of 64 15-bit words in Fig. 7.5.6 to the ROM of 128 15-bit words, as shown in Fig. 7.5.7. In other words, the ROM of 128 15-bit words consists of ROMs of 32 15-bit words (totally 8 ROMs) and a 1-out-of-4 decoder for the chip-enable, like Fig. 7.5.2. In total, 10 ROMs, 27 full adders, and a 1-out-of-4 decoder are required. In terms of package count, this approach is better than that of Fig. 7.5.6. However, if it is assumed that ROMs cost about 10 times as much as full adders, this approach has a relative cost of $(10 \times 10) + (1 \times 27) = 127$, whereas Fig. 7.5.6 has a relative cost of $(6 \times 10) + (1 \times 51) = 111$. Here the relative cost of 10 is arbitrary; this is mentioned to facilitate the comparison of different approaches.

When we want to expand the conversion to handle a 16-bit binary number, at least two approaches are conceivable as follows. One is to expand the left-hand ROM to a ROM of 512 16-bit words (using a 1-out-of-16 decoder), and some gates to generate the fifth BCD digit, as shown in Fig. 7.5.8. The second approach is to use the last 3 bits to address ROMs of 32 24-bit words and add BCD digits to the results obtained from the first 13 bits, as shown in Fig. 7.5.9. This requires more ROMs and 30 more adders.

The different approaches above are compared in Table 7.5.2. As the number of bits in a binary number increases, the cost increases, assuming that each ROM costs 10 times each adder. Thus it is important to attempt to reduce the cost by carefully examining the nature of conversion.

References: Tarbox, 1971; Hemel, 1969.

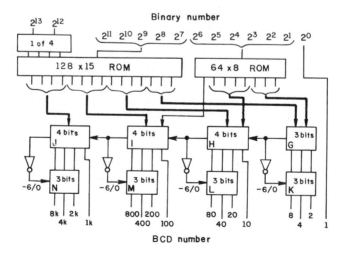

Fig. 7.5.7   Alternative to Fig. 7.5.6.

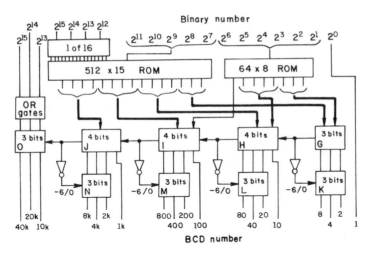

Fig. 7.5.8 **Conversion of a 16-bit binary number to a 5-digit BCD number.**

Conversion by the Add-3 Method

The add-3 method is a clever algorithm to convert a binary number to a BCD number (Conleur, 1958). As illustrated in Table 7.5.3, the method repeats the following two steps:

1. Shift the number left by one bit position.

2. If the binary number representation at any BCD digit is equal to or greater than 5, add 3 to that BCD digit.

At the third iteration in Table 7.5.3, the units digit after shift (i) is 7 (greater than 5), so 3 is added to this digit. At the fourth iteration, 3 is added to the units digit. At the sixth iteration, the tens digit is 6, so 3 is added to that digit. At the seventh iteration, 3 is added to the units digit. When the given binary number is completely shifted out of the original bit positions

Table 7.5.2 Comparison of Different Approaches

FIGURE NUMBER	7.5.3	7.5.4	7.5.6	7.5.7	7.5.8	7.5.9
Number of bits in a binary number	7	13	14	14	16	16
Number of ROMs[a] A	2	6	6	10	34	9
Number of full adders B	0	24	51	27	30	54
Package count $C = A + B$	2	30	57	37	64	63
Relative cost $D = 10A + B$	20	84	111	127	370	144

[a] It is assumed that we use a ROM package of 32 8-bit words.

Table 7.5.3 The Add-3 Method[a]

ITERATION	ACTION	HUNDREDS				TENS				UNITS				b_7	b_6	b_5	b_4	b_3	b_2	b_1	b_0	
														1	1	1	1	1	1	1	1	= (255)
1 (i)	Shift									0	0	0	1	1	1	1	1	1	1	1		
2 (i)	Shift									0	0	1	1	1	1	1	1	1	1			
3 (i)	Shift									0	1	1	1	1	1	1	1	1				
(ii)	Add 3 to units									1	0	1	0	1	1	1	1	1				
4 (i)	Shift					0	0	0	1	0	1	0	1	1	1	1	1					
(ii)	Add 3 to units					0	0	0	1	1	0	0	0	1	1	1	1					
5 (i)	Shift					0	0	1	1	0	0	0	1	1	1	1						
6 (i)	Shift					0	1	1	0	0	0	1	1	1	1							
(ii)	Add 3 to tens					1	0	0	1	0	0	1	1	1	1							
7 (i)	Shift	0	0	0	1	0	0	1	0	0	1	1	1	1								
(ii)	Add 3 to units	0	0	0	1	0	0	1	0	1	0	1	0	1								
8 (i)	Shift	0	0	1	0	0	1	0	1	0	1	0	1									
		(2)				(5)				(5)												

BCD number

---

[a] Reprinted by permission from Linford, 1970. Copyright © Computer Design Publishing Corporation.

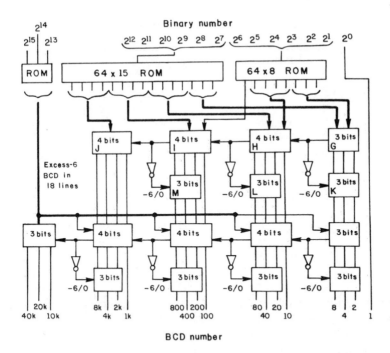

Fig. 7.5.9   Alternative to Fig. 7.5.8.

at the eighth iteration, the BCD number converted from the given binary number has been obtained at the bottom of Table 5.7.3.

The reason for using the above steps can be explained as follows. When a binary number 1000 (i.e., decimal number 8), for example, is shifted left by one bit position, or equivalently multiplied by 2, the binary number 10000 (i.e., decimal 16) is obtained. But decimal 16 is expressed as 0001 0110 in BCD. Thus 6 must be added to the binary 10000 (i.e., decimal 16) in order to obtain the corresponding BCD number 0001 0110. The same result can be obtained by adding 3 to the binary 1000 before shifting, instead of the above operation. 3 must be added prior to shifting whenever the binary number representation at any BCD digit is equal to 5 or more, corresponding to 10 or more after the shifting (i.e., multiplication by 2).

The above algorithm, the add-3 method, can be realized in many different manners. The add-3 operation in step 2 can be done with the truth table in Table 7.5.4 stored in a ROM or with the network in Fig. 7.5.10, instead of the combination of a network for detecting 5 or greater numbers and an adder. The shifting operation in step 1 can be done by a shift register or connections of the outputs of gates to the inputs of gates in the next higher bit positions (a so-called **barrel shifter**, as shown in Fig. 7.5.11). Figure 7.5.11 shows a network for converting a 12-bit binary number to a 4-digit BCD number, where each rectangle can be realized by a ROM storing Table

Table 7.5.4   Add 3 if a Binary Number Is Equal to or Greater than 5

INPUTS				OUTPUTS			
x_3	x_2	x_1	x_0	z_3	z_2	z_1	z_0
0	0	0	0	0	0	0	0
0	0	0	1	0	0	0	1
0	0	1	0	0	0	1	0
0	0	1	1	0	0	1	1
0	1	0	0	0	1	0	0
0	1	0	1	1	0	0	0
0	1	1	0	1	0	0	1
0	1	1	1	1	0	1	0
1	0	0	0	1	0	1	1
1	0	0	1	1	1	0	0
1	0	1	0	1	1	0	1

7.5.4 or the gate network in Fig. 7.5.10, and the interconnections between the levels in the tree are shifted one bit to the left.

The network of Fig. 7.5.11 with the gate network of Fig. 7.5.10 is slower but less complex than the random-logic gate network approach, and much faster with reasonable complexity than the other approaches. The count-

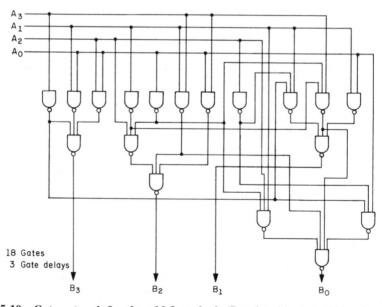

Fig. 7.5.10   Gate network for the add-3 method. (Reprinted by permission from Linford, 1970. Copyright © Computer Design Publishing Corporation.)

comparison method is the slowest and can be easily on the order of milli-seconds for conversion of a 16-bit binary number. The network based on the add-3 method (like the one in Fig. 7.5.11) can be roughly a few thousand to some ten thousand times faster, though the use of a shift register tends to be slower by a few times than the gate network in Fig. 7.5.10. The networks of ROMs and adders, like those in Figs. 7.5.4 through 7.5.9, are slightly slower (i.e., almost comparable, but possibly a few times slower than the fastest realizations based on the add-3 method). Thus the network based on the add-3 method is a reasonable compromise between speed and complexity. Therefore it is very important to devise the most appropriate task-realization algorithms for given tasks. Also the conversion by software on a general-purpose computer is a few hundred to a few thousand times slower, due to repeated access to a memory, than are the realizations based on the add-3 method. This endorses the importance of software–hardware tradeoff considerations.

There are other algorithms. (For example, 2^i for each i is converted to a BCD number, and then these BCD numbers are added, depending on whether these 2^i's are contained in a given binary number to be converted.)

For the conversion from a BCD number to a binary number, these al-gorithms, including those discussed above, can be used with modifications tailored to this conversion.

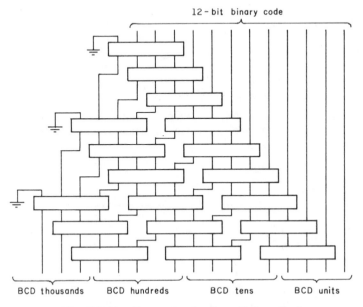

Fig. 7.5.11 Conversion by the add-3 method.

References: Linford, 1970; Raphael, 1973; Schmid, 1973; Anderson, 1969; Phadnis and Joshi, 1978; Stetson, 1977; *EDN*, Jan. 5, 1977, p. 46; Bredason, 1974; Brockman, 1977, Herrick, 1980; Reece, 1980; Varsos et al., 1978.

Problems

7.5.1. Discuss a best way to connect ROMs to minimize the total memory requirement, when the AND function $x_1 x_2 \cdots x_n$ is to be realized. (The case of $n = 4$ is illustrated in Fig. 7.3.3.) The number of levels need not be 2. If many networks have the same minimum total memory space requirements, choose the fastest network.

7.5.2 For the function f given in Table P7.5.2, design a network of ROMs only with a minimum total memory space requirement.

7.5.3 Although the ROM network of Fig. 7.3.9 contains OR gates, is it possible to design a ROM network for the 1-out-of-n detection without any gates? Each ROM is not restricted to have exactly two outputs. If you find it possible, show your answer, concisely explaining why your network works.

7.5.4 Design a network to count the number of 1's in 100 input variables x_1, x_2, $\ldots$, x_{100}, using ROMs, possibly along with any types of gates. Minimize the total memory space requirement, using the fewest gates. The count is to be displayed by outputs $z_7, z_6, \ldots, z_1$, where z_1 is the least significant bit. Then design this network with gates only of any type, but without ROMs.

7.5.5 Design an up–down counter with a single ROM. The ROM has an input to specify up- or down-counting.

7.5.6 Using the conversion approach illustrated with Figs. 7.5.2 and 7.5.3, design a network to convert a binary number to a 4-digit BCD number with ROMs, possibly along with as few gates as possible. The 4-digit BCD number must be shown up to 9999. (The binary number in the inputs assumes up to the corresponding values.) But do not use adders (though Figs. 7.5.4 and 7.5.6 through 7.5.9 contain adders). In this case minimize the total memory space requirement.

7.5.7 Design a ROM network to convert a 10-bit binary number to a BCD number, using gates, in a manner similar to Fig. 7.5.4. Minimize the total memory space required.

Table P7.5.2

x_1	x_2	x_3	x_4	f
0	0	0	0	0
0	0	0	1	1
0	0	1	0	1
0	0	1	1	0
0	1	1	1	1
1	0	1	1	1
1	1	1	0	0

7.5.8 Design a ROM network to convert a binary number expressing days (its maximum is 365 days) to a special BCD number that shows month and day. Try to reduce the memory space requirement as much as possible.

7.5.9 Devise a procedure to convert a BCD number to a binary number, modifying the add-3 method explained in this section.

7.5.10 Devise a procedure to convert a decimal number to a binary number.

7.6 ROMs for Calculations

ROMs are often useful for the calculation of some mathematical functions, such as trigonometric functions and square roots. When such calculations are to be done often, the use of ROMs as look-up tables yields results faster than calculation by software, though more sophisticated algorithms using ROMs are sometimes preferable, as discussed later. Even for simple calculations, such as addition and multiplication, ROMs as look-up tables are sometimes convenient because we can avoid time-consuming layout, though addition and multiplication are usually realized by random-logic gate networks because of their importance in terms of speed in digital systems.

In general, when we want to derive a best realization for a given task, we need to devise and compare different task-realization algorithms and the corresponding realizations in addition to those discussed in this section. Arithmetic operations or other calculations realized with random-logic gate networks, which are usually faster but more expensive and more time-consuming to design, should be compared. But they are not discussed in this book, because there are many books available for them along with related algorithms. [See, for example, Swartzlander (1980); Hwang (1979); Blaauw (1976); Oberman (1979); Schmid (1974).]

7.6.1 Addition

Because of their utmost importance in practically every digital system, adders have to be realized as the fastest and most compact network, and they are usually realized with random-logic gate networks. Such realizations are not discussed here because many other books on the subject are available. Instead, realizations of adders with ROMs are discussed in order to illustrate how ROMs are used for various tasks. Adders based on ROMs may sometimes be useful when we need to design adders with new IC processing technology, since ROMs can be prepared in a much shorter time than the layout of random-logic gate networks.

As an example, let us design an 8-bit adder with ROMs. In other words, we want to add two binary numbers $(A_8 A_7 \cdots A_2 A_1)$ and $(B_8 B_7 \cdots B_2 B_1)$. If we want to use this adder sometimes as a subtractor, we need to consider carry input C_0. Thus a ROM must have 17 inputs, that is, $2^{17} = 131{,}072$ words, and each word must be a sum consisting of 9 bits. (In contrast, a ROM for the multiplication of two 8-bit binary numbers must have 16 inputs, i.e., $2^{16} = 65{,}536$ words.) The ROM requires $131{,}072 \times 9 = 1{,}179{,}648$ bits in total memory space. But if the 8-bit addition is considered as a cascade of 4-bit addition, as illustrated in Fig. 7.6.1, then the total memory space requirement is reduced. Each ROM now has only 9 inputs, so it has $2^9 = 512$ words with 5 bits per word. Thus the ROM network in Fig. 7.6.1 requires a total of 2×512 (words) $\times 5$ (bits per word) $= 5{,}120$ bits. This is much less than $1{,}179{,}648$ bits for a single ROM.

It is to be noticed that the delay time is not necessarily doubled since the access time is usually smaller in a ROM of small capacity than in one of large capacity.

> **Remark 7.6.1.1:**   Adders with a minimum number of NOR gates and a minimum number of connections derived in Lai and Muroga (1979) and Sakurai and Muroga (to be published) have the most compact layout at least in the case of gate arrays (discussed in Chapter 9). See Remark 4.7.1.   □

7.6.2 Multiplication

Multipliers are important not only for digital computers, but also for other applications, such as fast Fourier transforms and digital filters. In general multipliers are also realized with random-logic gate networks like adders, but realizations with ROMs are more meaningful than the ROM realization of adders, because multipliers in logic gate networks are more complex than adders, and accordingly their layout is much more time-consuming.

Multiplication can be done with a shift register and an adder by repeating the following procedure. A multiplicand is multiplied by each bit of a multiplier, starting with the least significant bit, and is shifted left by 1 bit and added to the previous sum. This approach yields the simplest but slowest realization. Another approach is so-called **array multipliers** or **parallel multipliers**. Products of a multiplicand and all bits of a multiplier are formed simultaneously, each product is shifted in bit positions by a connection (i.e.,

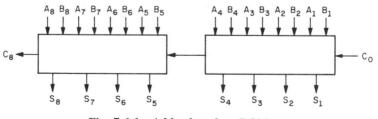

Fig. 7.6.1   Adder based on ROMs.

a connection to a next higher order bit gate, instead of shifting by a shift register), and then all these products are added simultaneously. This approach yields the fastest but the most complex realization. These two approaches have many variations in their realizations, as discussed in the literature.

Multipliers based on ROMs, discussed in the following, represent a reasonable compromise between the above two approaches. In other words, they are faster than those based on the first approach but slower than those based on the second, and they are more complex than those based on the first approach but less complex than those based on the second.

If we try to use a single ROM, the memory requirement is in general excessively large. For example, if two binary numbers of 8 bits each are to be multiplied, a ROM must have 16 inputs and accordingly $2^{16} = 65,536$ words. Each product consists of 16 bits. Thus the total memory space requirement is $65,536 \times 16 = 1,084,576$ bits. This is excessively large for a single ROM.

In order to avoid the use of a single ROM, each of the multiplier and multiplicand is divided into two parts, each 4 bits long, as follows:

$$A_8 = A_4 + (\Delta A)_4$$
$$\text{with } A_4 = \text{XXXX } 0\,0\,0\,0$$
$$(\Delta A)_4 = \qquad \text{XXXX}$$
$$B_8 = B_4 + (\Delta B)_4$$
$$\text{with } B_4 = \text{XXXX } 0\,0\,0\,0$$
$$(\Delta B)_4 = \qquad \text{XXXX}$$

where each X represents 0 or 1. Then the original multiplication can be expressed as follows:

$$A_8 B_8 = [A_4 + (\Delta A)_4][B_4 + (\Delta B)_4]$$
$$= [A_4 B_4] + [A_4(\Delta B)_4] + [(\Delta B)_4 B_4] + [(\Delta A)_4(\Delta B)_4]$$

These four products and their sum can be realized in four 2048-bit ROMs and five 4-bit adders, as illustrated in Fig. 7.6.2, where zeros are shown in order to indicate relative bit positions, but they are not actually connected.

References: Hemel, 1970; Stenzel et al., 1977.

For multipliers in logic gate networks, see the references in Muroga (1979, Sec. 9.2) and also see Swartzlander (1980); Twaddel (July 20, 1980); Bucklen et al. (1981); Rollenhagen and Wild (1978); Gajski (1980). For random-logic layout of array-type multipliers the approach of Baugh and Wooley (1973) is useful. Array-type multpliers by Lai and Muroga (to be published) and

Yu and Muroga (to be published) have fewer gates than conventional ones and have a more compact layout in gate arrays (to be discussed in Chapter 9), depending on the width of the routing channel.

Division can be realized in a similar manner. In other words, by splitting a binary number into shorter numbers, we can avoid excessive memory space requirement on ROMs.

7.6.3 Square Root

Let us discuss a ROM realization of the calculation of square root, as an example that is more complex than addition or multiplication and consequently requires a sophisticated task-realization algorithm. In this example an appropriate algorithm is essential for deriving a good ROM realization.

In general-purpose mainframe computers the square root of N is not used

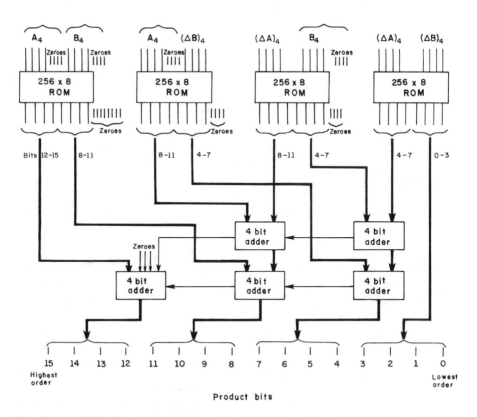

Fig. 7.6.2 Multiplier with ROMs. (Reprinted from Hemel, 1970. Copyright © McGraw-Hill, Inc., 1970. All rights reserved.)

by general users as often as addition or multiplication, so it is calculated by software. Iterative methods, such as the Newton–Raphson method, are used, so computation is slow, even with the fastest mainframes. The Newton–Raphson method is the iterative use of the formula

$$S_{k+1} = \frac{1}{2}\left(S_k + \frac{N}{S_k}\right)$$

where $\lim_{k\to\infty} S_k = \sqrt{N}$. By using a network of discrete gates, which realizes a square-root calculation algorithm, computational speed is improved by an order of magnitude, mainly because the decoding time of instructions in software and the slow access to a main memory are eliminated. LSI/VLSI realizations of a square-root calculation algorithm can further speed up the computation because of shorter connections inside a chip. Probably a square root can be calculated in many different ways by devising algorithms tailored to LSI/VLSI realizations, as algorithms for LSI/VLSI realizations can be very different from those for software processing, which are essentially based on serial operations in general-purpose computers. Square-root computation can be done with random-logic gate networks, but here we discuss computation with ROMs in order to show an appropriate use of ROMs.

We can store $\sqrt{N}$ for each value of N in a single ROM in order to use the ROM as a look-up table. For example, if we want to find the 5-bit square root for a given 10-bit value, we need a ROM of 2^{10} 5-bit words, that is, a total of $2^{10} \times 5 = 5{,}120$ bits. But as the magnitude of the values increases, the total memory space requirement increases faster than exponentially. For the 10-bit square root of a 20-bit value we need a ROM of 20^{20} 10-bit words, that is, $2^{20} \times 10 = 10{,}485{,}760$ bits. Thus the look-up table approach with a single ROM is obviously impractical.

If we look up the ROM in reverse order, we can avoid the above difficulty. In other words, the value N is stored in a ROM with $\sqrt{N}$ as its input, whereas above $\sqrt{N}$ is stored in the ROM with N as its input. Then we repeatedly look up the ROM on a trial-and-error basis until we hit the given N as a memory word, finding the memory address $\sqrt{N}$ as the answer.

Suppose that we want to find the 8-bit square root R_8 of a 16-bit value N_{16}. By this new approach the N_{16} are stored in a ROM with the R_8 as inputs; in other words, we have a ROM with $2^8 = 256$ words and 16 bits per word, or a total of 4,096 bits, whereas in the case of storing the R_8 with the N_{16} as inputs, we need a ROM with 2^{16} words and 8 bits per word, or a total of $2^{16} \times 8 = 524{,}288$ bits. The iterative search for the 8-bit square root R_8 of a given 16-bit number N_{16} begins with using a trial root of $R_8 = 0$ as the address to the ROM. Then this trial square root is repeatedly increased by 1 until we find a number equal to or greater than the given number N_{16} as a word in the memory. Then the trial root is the true square root R_8.

This can be further sped up by the following search method, which is essentially binary search.

Procedure 7.6.3.1   Calculation of Square Root

1. As the first trial square root, that is, the first address to the ROM, use the binary number 1 0 0 . . . 0, where the most significant bit is 1 and all other bits are 0.

   If the memory word for this address exceeds the given number N, the most significant bit is changed to 0. Otherwise this is kept at 1.

2. Change the next lower bit from 0 to 1 and use this binary number as the next trial square root, that is, as the address to the ROM. If the memory word exceeds the given number N, this bit is changed back to 0. Otherwise this is kept at 1.

3. Repeat step 2. When all the bits of the binary number are processed, the binary number which results is the desired square root $\sqrt{N}$.   □

For example, the number N whose square root is to be found is 280 (in decimal), that is, N_{16} = 0000 0001 0001 1000 (in binary). Processing of the above procedure is illustrated in Table 7.6.1 along with successive trial square roots. In the fourth row the square of 16 is only 256, and consequently the 1 is retained in the fourth bit position. The square root for 280 has been found to be 17. This is correct in this case, since the actual square root of 280 is 16.733, correctly rounded off to 17.

When the number N whose square root is to be found is 270, the same successive trial square roots given in Table 7.6.1 will be used. The result, however, will be incorrectly rounded because the actual square root of 270

Table 7.6.1   Example of Procedure 7.6.3.1

ADDRESS		MEMORY WORD	COMPARE
TRIAL SQUARE ROOT R_8	DECIMAL	N_{16}	N_{16} to 280
1 0 0 0 0 0 0 0	128	16,384	>
0 1 0 0 0 0 0 0	64	4,096	>
0 0 1 0 0 0 0 0	32	1,024	>
0 0 0 1 0 0 0 0	16	256	<
0 0 0 1 1 0 0 0	24	576	>
0 0 0 1 0 1 0 0	20	400	>
0 0 0 1 0 0 1 0	18	324	>
0 0 0 1 0·0 0 1	17	289	>

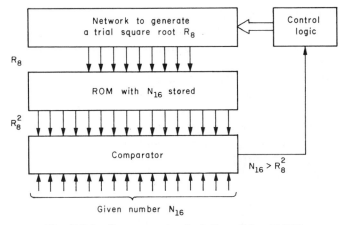

Fig. 7.6.3   Square-root calculation with a ROM.

is 16.432. This should be rounded off to 16, instead of 17 found in Table 7.6.1.

To avoid this rounding error, we can store $(R_8 - \frac{1}{2})^2$ in the ROM instead of $R_8{}^2$, that is, N_{16}. Then a memory word is always closer to the true square of R_8 than to the square of $(R_8 - 1)$, because $(R_8 - \frac{1}{2})^2$ and $R_8{}^2$ differ by $(R_8 - \frac{1}{4})$, while $(R_8 - \frac{1}{2})^2$ and $(R_8 - 1)^2$ differ by $(R_8 - \frac{3}{4})$. If we use this ROM with $(R_8 - \frac{1}{2})^2$ stored, the above procedure yields 17 and 16 as square roots of 280 and 270, respectively.

The above approach can be realized with a ROM, a comparator, and a control logic network, as shown in Fig. 7.6.3.

When the number N whose square root is to be found consists of many bits, the memory space requirement for the above ROM may become excessively large. In this case we can use many ROMs and reduce the total memory space requirement by partitioning the entire number N into shorter numbers as $N = N_1 + N_2$.

References: Hemel, 1970, 1972; Egbert, 1977; Kostopoulos, 1972; Ramamoorthy et al., 1972; Metze, 1965.

7.6.4   Trigonometric Functions

In the case of general-purpose mainframe computers, trigonometric functions, such as sin x, cos x, and tan x, are not vitally important for general users, so they are calculated by software, using series expansions such as

$$\sin x = x - \frac{x^3}{3!} + \frac{x^5}{5!} - \frac{x^7}{7!} + \cdots$$

or more efficient rational approximations. These computations are slow because of instruction decoding and repeated memory access. When scientific calculators became widely available, hardware realizations to calculate trigonometric functions were adopted since computation by software was found too slow.

The use of ROMs may be appropriate for storing a numerical value table such as the values of trigonometric functions, since a ROM can be used as a look-up table. In this case we can reduce the memory space requirement greatly by utilizing unique properties of the table contents, or equivalently by detecting any regularity in the distribution of 0's and 1's in the tables. In other words, if the same sequence of 0's and 1's appears repeatedly in some successive rows in the tables, we can eliminate it by reorganizing the tables. In a sense we eliminate redundant information. We can thus reduce the memory requirement in a manner different from the case of the square root where the memory space requirement was reduced by modifying the square-root algorithm but not by detecting the redundant information in the memory contents. Thus if we want to derive good realizations for given tasks in general, we have to utilize appropriately the intrinsic properties of the individual tasks.

Suppose that we want to calculate the value of $\sin x$ for $0 \leq x \leq \pi/2$, where x is given 15 bits and $\sin x$ is to be in 16 bits (accurate to 14 bits).

If the values of $\sin x$ for all possible values of x are to be stored in a single ROM, the ROM must have 2^{15} 15-bit words, or a total memory space of $2^{15} \times 15 = 491,520$ bits. Let us try to reduce the memory space requirement by using many ROMs.

Let us divide the argument x into the more significant part M and the less significant part L. Then each of them is divided again into more and less significant parts, such that

$$\sin x = \sin (M + L) = \sin M \cos L + \cos M \sin L$$

$$= \sin (MM + ML) \cos (LM + LL) + \cos (MM + ML) \sin (LM + LL)$$

We take MM in 6 bits, and ML, LM, and LL in 3 bits each.

For any small value of y, the value of $\cos y$ is 1. [In other words, the entries in the numerical value table for $\cos (LM + LL)$ are the same 1 for the entire range of the value of $(LM + LL)$. Thus the table contains redundant information.] We can then use the approximation

$$\cos(LM + LL) = 1$$

We can also use the approximations

$$\sin(LM + LL) = \sin LM + \sin LL$$

$$\cos(MM + ML) = \cos(MM + \delta MM) = \cos[(1 + \delta)MM]$$

where $\delta = 2^{-7}$. Thus we obtain

$$\sin x = \sin(MM + ML) + \cos[(1 + \delta)MM](\sin LM + \sin LL)$$

This is realized by the network shown in Fig. 7.6.4 with 4 ROMs and 6 4-bit adders. ROM 1 generates the 8 most significant bits of sin($MM + ML$), and ROM 2, the 8 least significant bits. ROM 3 generates cos[(1 + δ)MM] sin LM in 8 bits and ROM 4 generates cos[(1 + δ)MM] sin LL in 6 bits. It is to be noticed that only ROM 4 has 6 outputs, while all other ROMs have 8 outputs. So the total memory space requirement is (3 × 2^9 × 8) + (2^9 × 6) = 15,360 bits, which is a significant reduction compared with 491,520 bits for a single ROM.

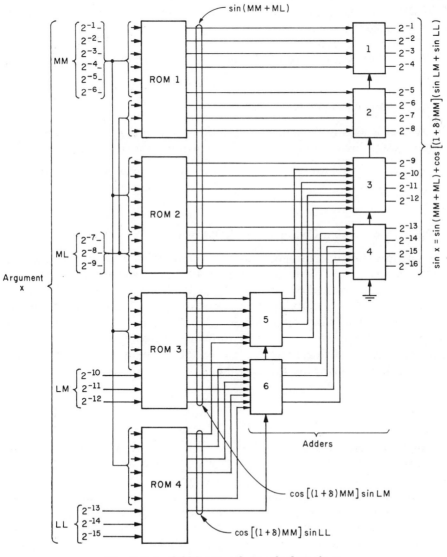

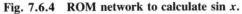

Fig. 7.6.4   ROM network to calculate sin *x*.

Adders 5 and 6 add the outputs of ROM 3 and ROM 4 to produce $\cos[(1 + \delta)MM]$ $(\sin LM + \sin LL)$, which is added to the 8 least significant bits of $\sin(MM + ML)$ in adders 3 and 4. Adders 1 and 2 are simply to add the carry output of adder 3 to the 8 most significant bits of $\sin(MM + ML)$.

When many trigonometric functions or transcendental functions need to be calculated, special processors dedicated to these functions are often realized, as discussed in Chapter 10. Unlike the ROM realization discussed above, these processors have all the inside facilities of general-purpose computers, such as control logic, memories, and gate networks, tailored to the given tasks, though in small scale. Also a unique algorithm, called CORDIC, which is appropriate for the calculation of these functions, is often used in designing such processors.

References: Schmid, 1974; Hemel, 1969; Shapiro, 1979; National Semiconductor, Mar. 1972.

7.6.5 Other Functions

The use of ROMs in the calculation of other functions, such as exponential functions and logarithmic functions, has also been investigated.

References: Voyer, 1974; Cantor, 1962; Wang, 1968; Marino, 1972; Shi, 1976; Pogge, 1977; Code, 1976; Schmid, 1974; Swartzlander, 1980; Oberman, 1979.

7.6.6 Concluding Comments on ROM Networks

Although the layout time required for random-logic network realizations of given tasks is greatly reduced by ROM realizations, realizations by networks of many ROMs instead of single ROMs, usually reduce the ROM memory space requirement greatly. But this requires careful and probably imaginative investigations of task-realization algorithms that make ROM networks compact, still maintaining good performance. This, however, reduces the advantage of using single ROMs, that is, the flexibility, in design changes or corrections, though the speed of the networks is often higher than single ROMs.

7.7 PLAs and Variations

A **programmable logic array (PLA)** is a special type of ROM, though its usage is completely different from that of ROMs discussed so far. [PLA was invented by R. Proebsting at Texas Instruments (*Electronics,* Oct. 28, 1976, p. 82).] MOSFETS are arranged in a matrix on a chip, as illustrated in Fig. 7.7.1*a*. The connections between the MOSFET gates and vertical or horizontal lines are set up by semiconductor manufacturers during fabrication according to customers' specifications. Since for the connections only one mask need to be custom-made, PLAs are inexpensive when the production volume is high enough to make the custom preparation cost of the connection mask negligibly small. Because of low cost and design flexibility, PLAs are extensively used and will be used more in LSI/VLSI chips, such as microprocessor chips, watch chips, and toy chips.

A ROM with n address inputs has a decoder with 2^n outputs, as shown in Fig. 6.4.2a, whereas a PLA does not. A PLA consists of an **AND array** and an **OR array**, as illustrated in Fig. 7.7.1a, such that information is stored in the form of switching expressions, whereas in the case of a ROM information is stored in the form of a truth table. Thus usually we can pack more information into a PLA due to redundancy reduction than into a ROM of comparable size, as explained later.

When MOSFET gates are connected, as denoted by the large dots in Fig. 7.7.1a, we get $\overline{x\bar{y}\bar{z}}, \overline{\bar{x}z}, \overline{xyz}$ at the outputs P_1, P_2, P_3 of the AND array, respectively, since P_1, P_2, and P_3 represent the outputs of NAND gates, if negative logic is used with n-MOS. (Negative logic is used because of the convenience in deriving disjunctive forms for the output functions f_1, f_2, and f_3. If positive logic is used, as we did in Chapter 4, then P_1, P_2, and P_3 represent the outputs of NOR gates, and f_1, f_2, and f_3 are expressed in conjunctive forms.) Then the outputs f_1, f_2, f_3 of the OR-array also represent the outputs of NAND gates with P_1, P_2, P_3 as their inputs. Thus,

$$f_1 = \overline{P_1 P_3} = \overline{P_1} \vee \overline{P_3} = x\bar{y}\bar{z} \vee xyz$$

$$f_2 = \overline{P_2} = \bar{x}z$$

$$f_3 = \overline{P_1 P_2} = \overline{P_1} \vee \overline{P_2} = x\bar{y}\bar{z} \vee \bar{x}z$$

(It is to be noted that the AND-array contains inverters right after inputs x, y, z, whereas the OR array does not.) Therefore the two arrays in Fig. 7.7.1a represent a network of NAND gates in two levels. Being well known in logic design [see Muroga (1979, Sec. 6.1)], this is interpreted as a network of AND gates in the first level and OR gates in the second (output) levels, as illustrated in Fig. 7.7.1b. This is the reason why the upper and lower matrices are called AND and OR arrays, respectively. The vertical lines which run through the two matrices in Fig. 7.7.1a are called the **product lines**, since they correspond to the product terms in disjunctive forms for the output functions f_1, f_2, and f_3. Thus any combinational network (or networks) of AND and OR gates in two levels can be realized by a PLA.

Sequential networks can also be easily realized on a PLA, as shown in Fig. 7.7.2. Some outputs of the OR array are connected to the inputs of master–slave flip-flops, whose outputs are in turn connected to the AND array as its inputs. More than one sequential network can be realized on a single PLA, along with many combinational networks. Minimization of the number of flip-flops is important, because flip-flops occupy a chip area of significant size, and consequently not many flip-flops (usually J–K master–slave flip-flops) are usually provided in a commercially available PLA chip.

In many PLAs the option of an output f_i or its complement $\overline{f}_i$ is provided in order to give flexibility, as illustrated in the lower right-hand corner of Fig. 7.7.2. By disconnecting one of the X's at each output, we can have either f_i or $\overline{f}_i$ as output. When f_i has too many products in its disjunctive form and cannot be realized on a PLA, its complement $\overline{f}_i$ may have a sufficiently small number of terms to be realizable on the PLA, or vice versa.

In commercially available PLA packages the number of inputs is usually somewhere between 10 and 30, the number of outputs is also 10 to 30, and the number of product lines is between 50 and 150. Larger PLAs are also often used.

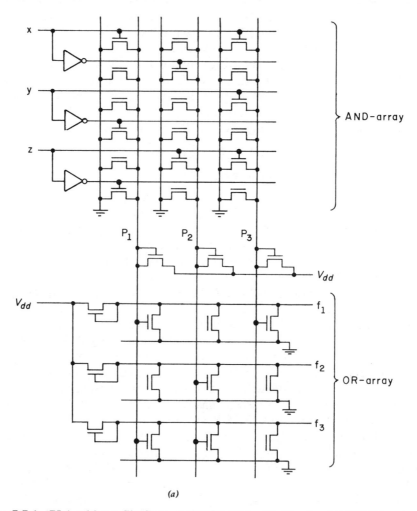

(a)

Fig. 7.7.1   PLA without flip-flops. (a) MOS PLA. (b) Two-level AND-OR network which is equivalent to (a).

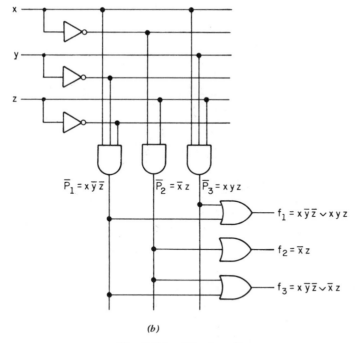

(b)

Fig. 7.7.1 *(Continued)*

The number of product lines in a PLA is roughly limited to 150, because beyond that number, each horizontal line gets too long, with an increase in parasitic capacitance. Then if a majority of the MOSFET gates provided is connected to this horizontal line, the input (or its inverter) has too many fan-out connections on this horizontal line. Similarly, the total number of horizontal lines cannot be too large. In other words, the array size of a PLA is limited because of speed considerations. In contrast, the size of a ROM can be much larger since we can use more than one decoder, as illustrated in Fig. 6.4.1*a*, or use a complex decoding scheme [e.g., Matick (1977, Chap. 4)].

A **field-programmable PLA (FPLA)** is also available. In the FPLA shown in Fig. 7.7.3 a user can set up a dot pattern by burning short titanium-tungsten fuses connected in series with diodes or bipolar transistors by temporarily feeding excessive currents. In this realization a special electronic circuit to blow fuses must be provided in addition to the PLA arrays, and this adds roughly 40% to the entire area. Fuses also occupy some extra area. In large-volume production FPLAs are more expensive due to this extra size than PLAs, but when users need a small number of PLAs, FPLAs are much cheaper and convenient, since users can program FPLAs by themselves inexpensively and quickly. In contrast to FPLAs, the PLAs discussed above are called **mask-programmable**.

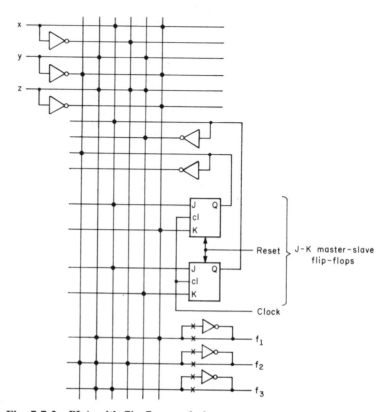

Fig. 7.7.2   PLA with flip-flops and also output-complementation choice.

In contrast to the above FPLAs based on fuses, FPLAs whose undesired connections are disconnected by laser beam are available. In this case the chip size is smaller than that of the above FPLAs since the electronic circuits to blow fuses are not necessary, but special laser equipment is necessary.

The response time of PLAs and FPLAs is roughly 35 to 70 nsec.

References: Texas Instruments, Application Bulletin CA-158; Rockwell, 1973; Cline, 1978; Texas Instruments, Feb. 1979; National Semiconductor, 1973; Cavlan, 1975, 1976, 1979.

PLAs combined with nonvolatile memories are proposed (Wood et al., 1981).

Logic Design with PLAs

Minimization techniques for multiple-output switching functions known in switching theory [e.g., Muroga (1979, Chap. 5)] can be used to minimize the size of a PLA. If the number of AND gates in a two-level AND-OR network (i.e., the number of distinct multiple-output prime implicants in disjunctive forms) for the given output functions is minimized, we can minimize the number t of product lines. Thus the array size $(2n + m)t$ of a PLA is minimized when the PLA has n inputs, m outputs, and t product lines. (n and m are given.) Also, if the total number of connections in a two-level AND-OR network is minimized as the secondary objective, as we do in the minimization technique of a multiple-output switching function, then the number of dots (i.e., connected intersections of the product lines and the horizontal lines) in the PLA is minimized. Then the chances of faults due to bad connections of MOSFET gates at these dots have been minimized. Therefore the derivation of a minimal two-level network with AND and OR gates by the minimization techniques known in switching theory is very important for the minimal and reliable design of PLAs.

However, the minimization of the number of connections in a minimal two-level AND-OR network may not be as important as the minimization of the number of AND gates, though it tends to reduce the power consumption, because the chances of faulty PLAs can be greatly reduced by careful fabrication of chips. But the PLA size is determined by the number of AND gates and cannot be changed by any other factors. Also instead of

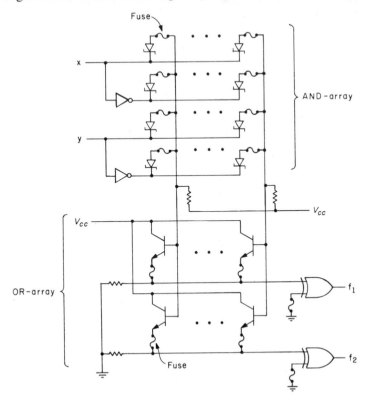

Fig. 7.7.3   FPLA.

making connections (i.e., dots) as they become necessary on a PLA, a PLA is sometimes prepared by disconnecting unnecessary connections by laser beam after it has been manufactured with all MOSFET gates connected to the lines. In this case the chances of faults may be reduced by increasing the number of connections (i.e., dots) in the two-level AND-OR network.

Let us show an example. Suppose that functions are given as in Table 7.7.1. Then we can obtain minimal expressions:

$$f_1 = x_1\bar{x}_3\bar{x}_4 \lor x_2\bar{x}_4 \lor \bar{x}_1\bar{x}_3x_4 \lor \bar{x}_1\bar{x}_2x_3$$

$$f_2 = \overline{\bar{x}_1x_2\bar{x}_4 \lor x_1\bar{x}_3x_4 \lor \bar{x}_2x_3}$$

$$f_3 = \bar{x}_1\bar{x}_3x_4 \lor \bar{x}_2x_3 \lor x_1\bar{x}_3x_4 \lor x_1x_2\bar{x}_3$$

The corresponding PLA is shown in Fig. 7.7.4 with 9 product lines. Thus the array size is $11 \times 9 = 99$.

For comparison with a PLA, the MOS realization of a ROM is shown in Fig. 7.7.5. The upper matrix is a decoder which has 2^n vertical lines if there are n input variables. The lower matrix stores information by connecting or not connecting MOSFET gates. Figure 7.7.5 actually realizes the same output functions (in negative logic) as those in Fig. 7.7.1. The AND array in Fig. 7.7.1a is essentially a counterpart of the decoder in Fig. 7.7.5, or the decoder may be regarded as a fixed AND array with the maximum number 2^n of product lines. The AND array in Fig. 7.7.1a has only 3 vertical lines, whereas the decoder in Fig. 7.7.5 has 8 fixed vertical lines. This indicates the compact information packing capability of PLAs. But the decoder lacks

Table 7.7.1   Example

x_1	x_2	x_3	x_4	f_1	f_2	f_3
0	0	0	0	0	1	0
0	0	0	1	1	1	1
0	0	1	0	1	0	1
0	0	1	1	1	0	1
0	1	0	0	1	0	0
0	1	0	1	1	1	1
0	1	1	0	1	0	0
0	1	1	1	0	1	0
1	0	0	0	1	1	0
1	0	0	1	0	0	1
1	0	1	0	0	0	1
1	0	1	1	0	0	1
1	1	0	0	1	1	1
1	1	0	1	0	0	1
1	1	1	0	1	1	0
1	1	1	1	0	1	0

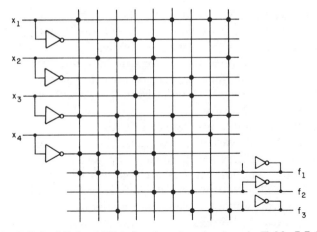

Fig. 7.7.4 Minimal PLA for the example given in Table 7.7.1.

MOSFETs at half of the intersections of the vertical and horizontal lines (no MOSFETs at fixed positions), so the area occupied by the decoder may be reduced by closer packing. Also more than one decoder or a complex decoding scheme may be used. So precise comparison is difficult, though

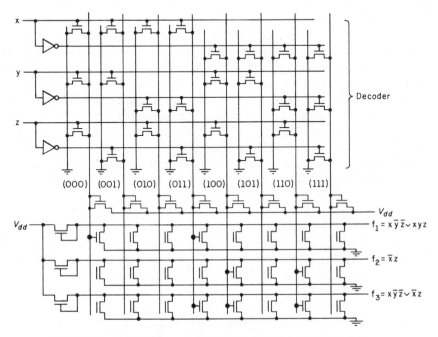

Fig. 7.7.5 ROM that corresponds to the PLA in Fig. 7.7.1.

PLAs are smaller than ROMs, and the packing advantage of PLAs varies depending on functions. If we construct a ROM that realizes the functions of the PLA of Fig. 7.7.4, in a manner similar to Fig. 7.7.5, the decoder consists of 8 horizontal lines and 16 vertical lines, and the lower matrix for information storage consists of 16 vertical lines and 3 horizontal lines. Thus the ROM to realize Table 7.7.1 requires the array size of $16 \times (8 + 3) = 176$, compared with 99 in Fig. 7.7.4.

Generally the size difference between PLAs and ROMs sharply increases as the number of input variables increases. For some functions of 12 variables, for example, PLAs can be smaller by an order of magnitude than ROMs. For many functions of a greater number of variables, the size difference can be easily two orders of magnitude or far greater.

A PLA, however, cannot store some functions, such as $x_1 \oplus x_2 \oplus \cdots \oplus x_n$ for large n, since 2^{n-1} product lines are required and the number of these lines is excessively large for a PLA. (The horizontal lines become too long with excessive fan-out and parasitic capacitance.) However, we can store these functions in a ROM with an appropriate decoding scheme.

Of course, storing a truth table without worrying about conversion into switching expressions is convenient, though it makes the ROM size bigger than the PLA size.

Minimal two-level networks of AND and OR gates for the minimization of the PLA size can be derived by minimization techniques, if a function to be minimized has either at most several variables, or many more variables but with a simple relationship among its prime implicants [as discussed in Muroga (1979, Sec. 5.3)]. But otherwise we have to be content with near-minimal networks instead of minimal networks. In many cases, efforts to reduce the PLA size, even without reaching an absolute minimum, result in significant size reduction. Also CAD programs have been developed with heuristic minimization techniques, such as the one by Hong et al. (1974), which is the first powerful heuristic procedure drastically different from conventional minimization procedures. The minimization of the PLA size generally is not an easy task. It appears that different procedures are required for different ranges of functions. Recently a substantial improvement in minimization time has been achieved in deriving even absolute minimization for a certain range of functions (Cutler and Muroga, to be published).

References on minimization techniques: Muroga, 1979, and references contained in its Chapters 4 and 5; Dietmeyer, 1971; Hong et al., 1974; Roth, 1978, 1980; Cutler, 1980; Cutler and Muroga (to be published); Young, 1978; Young and Cutler, 1978; Sherwood, 1977; Bricaud and Campbell, 1978; Sasao, 1978; Sasao and Terada, 1979; Kang and vanCleemput, 1981; O'Donovan and Lind, 1976.

When a single PLA is not sufficient, we can use the following techniques.

1. When many functions for the same input variables are to be realized, the inputs of two (or more) PLAs may be shared by the same input variables, as shown in Fig. 7.7.6*a*. Then all outputs of the two PLAs can be used to express these functions.

2. When the number of product terms for functions exceeds the limit for a single PLA, the inputs of two (or more) PLAs may be shared by the same input variables. Then each output of one PLA and the corresponding output of the other PLA may be connected to an OR gage (or Wired-OR, depending on the electronic circuit realization of the PLA outputs) such that the products from these two outputs together constitute a function f_i, as illustrated in Fig. 7.7.6*b*.

3. When the number of product terms for functions exceeds the limit for a single PLA, and conjunction of the two disjunctive forms $(Q_1 \vee Q_2 \vee \cdots \vee Q_s)(R_{s+1} \vee \cdots \vee R_q)$ for a function f makes the total number of products q smaller than the total number p of products $(P_1 \vee \cdots \vee P_t) \vee (P_{t+1} \vee \cdots \vee$

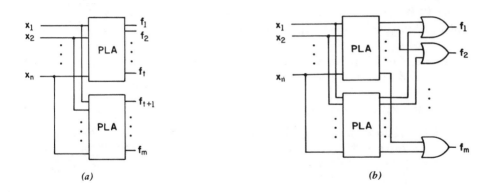

(a)   (b)

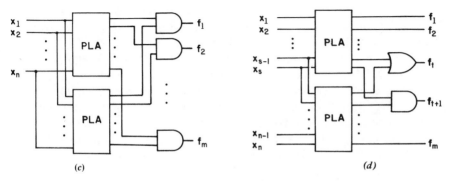

(c)   (d)

Fig. 7.7.6 Different configurations for connecting PLAs.

P_p) for f, we can realize these functions as shown in Fig. 7.7.6c by replacing the OR gates in Fig. 7.7.6b by AND gates. In other words, by using Fig. 7.7.6c we can use smaller PLAs than those in Fig. 7.7.6b.

4.   As illustrated in Fig. 7.7.6d, two PLAs (or more) may have the inputs partially shared by the same input variables (i.e., x_{s-1} and x_s are shared by these PLAs but x_1, x_2, x_{n-1}, x_n are not), and the outputs may be a mixture of the above cases 1, 2, and 3. (f_1, f_2, f_m are not connected to any gates, but some outputs from these PLAs are connected to the OR gate for f_t and the AND gate for f_{t+1}.)

Of course, we may use more complex configurations of PLAs, possibly with feedback loops from some outputs to inputs.

5.   When we need synchronization in order to avoid hazards, one of the inputs of a PLA may be used as clock c so that all products in output functions contain c as their literal.

Like in the case of ROM networks (mentioned in Section 7.6.6), we can in general reduce the chip areas by designing PLA networks appropriately, using PLAs as building blocks of networks, instead of a single large PLA, but flexibility in design changes or corrections is somewhat sacrificed.

References: Mrazek and Morris, 1973, 1977; Hemel, 1976; Vodovoz, 1975; Glasser and Penfield, 1980; Sasao and Terada, 1980.

Decoded PLA

If input variables are partitioned into groups and input variables in each group are decoded before being connected to a PLA, this is called a **decoded PLA**. An example for four variables which are partitioned into two groups is shown in Fig. 7.7.7. Decoded PLAs were introduced by IBM and generally have a greater design capability than the original PLAs discussed so far.

Because each group need not contain the same number of input variables and also different groups may share some input variables, there are a large number of possible partitions of given input variables, and it is not easy to find a best input partition. So input variables are usually partitioned into disjoint groups with two variables each, as exemplified in Fig. 7.7.7.

In order to show the advantage of the decoded PLA over the ordinary PLA of Fig. 7.7.1a, let us consider an output function S_2 whose horizontal line has dots at its intersections with the second and third vertical lines in Fig. 7.7.7. The second vertical line (from the left) has dots at the intersections with the output lines of decoders for $x \lor y$, $x \lor \bar{y}$, $\bar{x} \lor y$, $u \lor \bar{v}$, and $\bar{u} \lor v$. The third vertical line has dots at the intersections with the output lines for $\bar{x} \lor \bar{y}$, $u \lor v$, and $\bar{u} \lor \bar{v}$. Thus function S_2 can be expressed as

$$S_2 = (x \lor y)(x \lor \bar{y})(\bar{x} \lor y)(u \lor \bar{v})(\bar{u} \lor v) \lor (\bar{x} \lor \bar{y})(u \lor v)(\bar{u} \lor \bar{v})$$

by ANDing the dotted decoder outputs in the AND array and the ORing the

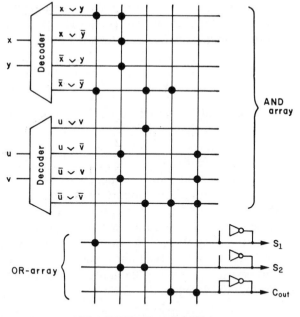

Fig. 7.7.7   Decoded PLA.

products in the OR array. In this expression we have a product of alterms (i.e., $x \vee y$, $x \vee \bar{y}$, . . .) for each vertical line, instead of a product of literals (i.e., $x_1, \bar{x}_1, x_2, \bar{x}_2$, . . .) for each vertical line in the case of Fig. 7.7.1a. In other words, each vertical line can express a more complex function than a single product term. Thus a decoded PLA can realize given functions with no more vertical lines than an ordinary PLA, such as the one in Fig. 7.7.1a.

As an example, let us design a 2-bit adder with a decoded PLA. As illustrated in Fig. 7.7.8, x and y are lower order bits of augend and addend, respectively, and u and v are the next higher order bits. S_1 and S_2 are sum bits, and C_{out} is the carry from the most significant bit position. Then these output functions are expressed as follows:

$$S_1 = x \oplus y$$

$$S_2 = xy \oplus (u \oplus v)$$

$$C_{\text{out}} = uv \oplus xy(u \oplus v)$$

Fig. 7.7.8   A 2-bit adder.

We need to compare these functions and their complements, in their disjunctions of products of alterms, in order to find a best combination of expressions such that the decoded PLA size is minimized. Thus we have

$$S_1 = x\bar{y} \vee \bar{x}y = (x \vee y)(\bar{x} \vee \bar{y})$$

$$S_2 = xy\,\overline{(u \oplus v)} \vee \overline{(xy)}(u \oplus v)$$

$$= (\bar{x} \vee y)(x \vee \bar{y})(x \vee y)(u \vee \bar{v})(\bar{u} \vee v) \vee (\bar{x} \vee \bar{y})(u \vee v)(\bar{u} \vee \bar{v})$$

$$\overline{C}_{\text{out}} = (\bar{u} \vee \bar{v})(\bar{x} \vee \bar{y}) \vee (\bar{u} \vee \bar{v})(\bar{u} \vee v)(u \vee \bar{v})$$

(C_{out} can be used as well, yielding the same number of vertical lines, though we need one more dot.) Based on these expressions, we have obtained the decoded PLA shown in Fig. 7.7.7. It is important to notice that the realization of the adder with a decoded PLA requires only 5 vertical lines because the decoded PLA permits a parity function on a single vertical line, which is not possible with an ordinary PLA. If an ordinary PLA, such as the one of Fig. 7.7.1a, is used, we need 9 vertical lines, based on disjunctive forms of multiple-output prime implicants. For example, S_2 above needs 6 vertical lines based on the following disjunctive form:

$$S_2 = xyuv \vee xy\bar{u}\bar{v} \vee \bar{x}u\bar{v} \vee \bar{x}\bar{u}v \vee \bar{y}u\bar{v} \vee \bar{y}\bar{u}v$$

Although decoders add small areas and slight gate delay, decoded PLAs appear to have smaller areas than ordinary PLAs.

If standard networks such as adders are packed in PLAs with minimum size (they are often called **macro PLAs**), we can use them without redesigning each time. This is convenient for large standard networks whose minimization is time-consuming (Weinberger, 1979; Schmookler, 1980; Cook, Ho and Schuster, 1979; Golden et al., 1980; Logue et al., 1981).

References: Fleisher and Maissel, 1975; Wood, 1975; Eichelberger and Lindbloom, 1980; Sasao and Terada, 1979 1980; *IEEE*, Oct. 1977.

Folded PLA

In general the percentage of used MOSFETs in a PLA (i.e., their MOSFET gates are connected) is very low, though decoded PLAs probably have higher percentages than the conventional PLAs of Fig. 7.7.1a. [One estimate is 10% in AND arrays and 4% in OR arrays in the conventional PLAs of Fig. 7.7.1a and also in the decoded PLAs of Fig. 7.7.7 (*IEEE TC*, Sept. 1979, p. 602).] In particular, when PLAs are available as off-the-shelf packages, the number of inputs, outputs, and product lines is preset, so the percentage of used MOSFETs greatly varies, depending on the functions to be realized. In this sense, folded PLAs developed by IBM represent a significant refinement of conventional PLAs and decoded PLAs, enhancing not only information packing densities but also design flexibility.

A **folded PLA** packs four PLAs into a single form, as illustrated in Fig. 7.7.9, where the dashed lines show borders among four PLAs. These borders (i.e., the dashed lines) are custom-made by cutting metals lines, depending on the functions to be realized. Although Fig. 7.7.9 shows four decoded PLAs of Fig. 7.7.7, the outputs of each decoder can be changed to x, $\bar{x}$, y, and $\bar{y}$ from $x \vee y$, $x \vee \bar{y}$, $\bar{x} \vee y$, and $\bar{x} \vee \bar{y}$, respectively, by disconnecting metal lines inside each decoder. In this case any output can be changed (e.g., from $x \vee y$ to x), independently of the changes of other outputs. In other words, we can have a decoded PLA (Fig. 7.7.7), a conventional PLA (Fig. 7.7.1a), or something between them.

The four PLAs are surrounded by bus loops of metal lines. Each bus loop can be used as many separate connections by cutting it into separated pieces.

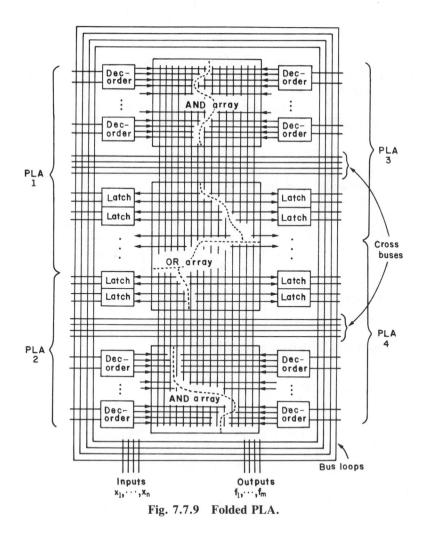

Fig. 7.7.9 Folded PLA.

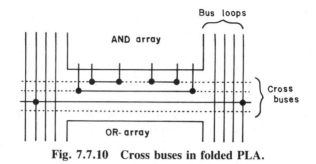

Fig. 7.7.10   Cross buses in folded PLA.

By cutting these bus loops at appropriate places, any variable inputs x_1, ..., x_n or the outputs of latches can be connected to the inputs of any decoder, and also the outputs of any latches can be used as the PLA output functions f_1, ..., f_m.

The cross buses shown in two places in Fig. 7.7.9 can be used for connecting many product lines in an AND array, as illustrated in Fig. 7.7.10, where the dotted line portions of the metal lines are deleted. A cross bus also can be used for connecting many bus loops.

Each latch can be a J–K master–slave flip-flop, a gated latch, or a latch with an AND gate, simply by changing the metal line configurations inside the latch.

With these possibilities for custom design by changing only metal line configurations, the design flexibility is greatly enhanced in folded PLAs to improve information packing density. (We can see the improvement by observing that by permuting the product lines and input lines in Fig. 7.7.1a, we can have unused MOSFETs in the right-hand corners. Then these MOS-FETs can be used in another PLA.) Metal lines are provided in two layers, so two masks must be custom-made, unlike one mask for conventional PLAs.

References: Cox et al., 1976; Wood, 1979; Hachtel et al., 1980; Paillotin, 1981.

Advantages and Disadvantages of PLAs

PLAs, like ROMs which are more general, have the following advantages over random-logic gate networks.

1.   There is no need for the logic design of random-logic gate networks, that is, the time-consuming design of gate networks is generally not necessary and the even more time-consuming layout is eliminated. Thus the design time is greatly shortened, design checking is easy, and design change is also easy.

2.   Since only the connection mask (or two masks for folded PLAs) needs to be custom-made and, as mentioned in (1), there is usually no need for logic design, this is a very inexpensive approach.

3. Since the layout is far simpler than that for random-logic gate networks, and it is less affected by solid-state circuit technology because of the simple matrix configuration, design information is almost immune to technological changes. In other words, we can use previous design information with ease but without change, even if a new solid-state (or electronic) circuit technology is introduced.

Of course, although the logic design and the layout of random-logic gate networks are much more tedious and time-consuming, the random-logic gate networks have higher speed and occupy smaller chip areas than PLAs or ROMs. (In other words, the size of the PLAs can be reduced by deleting unused portions of connections and shortening long connections by relocating MOSFETs inside the PLAs.) The bits-to-gate ratios between PLAs and random-logic gate networks are roughly estimated to be between 10 and 40 bits per gate and, in optimum cases, 8.5 bits per gate. (Donath, 1974). Also when random-logic gate networks are redesigned with PLA chips, usually more PLA chips are required with increased chip sizes. (For each random-logic network, in most cases roughly one PLA, and in some cases, two or more PLAs are required with increased chip sizes.) Also with large production volumes, random-logic gate networks are cheaper than PLAs or ROMs.

PLAs have the following advantage and disadvantage, compared with ROMs:

1. For storing the same functions or tasks, PLAs can be smaller; generally the size difference sharply increases, as the number of input variables increases.

2. PLAs cannot store complex function (i.e., those that contain excessively many product terms.)

Applications of PLAs

Considering the above advantages and disadvantages, PLAs have numerous unique applications. Because of the compactness and ease of design change and check, PLAs are used in the control logic of microprocessor chips and calculators (control logic of computers is complex and requires many changes, even during its design). Many simple microprocessor chips contain single PLAs, and more complex ones contain many of them (e.g., Motorola's microprocessor chip 68000 contains a dozen PLAs). Also PLAs are used for code conversions, microprogram address conversions, decision tables [see Muroga (1979, Sec. 3.7)], bus priority resolvers, and memory overlay.

When new merchandise is to be manufactured in small volume or test marketed, PLAs are a natural choice. For example, new intelligent terminals are designed and manufactured with PLAs, possibly mixed with gate networks and ROMs. When new functions need to be added, we can do so easily with PLAs. When the terminals are well received and do not need further changes, PLAs can be gradually replaced by random-logic gate networks for economy and speed. Also PLAs are sometimes used in wrist watches for adding special features. PLAs will be used more extensively in VLSI.

References: Cavlan and Cline, 1976; Cook et al., 1979; Reyling, 1974; Roberts, 1980; Kaestner, 1975; Wimmer, 1981; Bentley, 1981; *Electronic Design,* June 21, 1973, pp. 28–34; Oct. 25, 1976, pp. 164–169; *EDN,* Nov. 20, 1976, p. 55; *Electronics,* Dec. 9, 1976, p. 107; July 17, 1980, pp. 143, 144, *Spectrum,* Jan. 1977, p. 50.

Other References on PLAs: Priel and Holland, 1973; Hebenstreit and Horninger, 1976; May and Schiereck, 1976; Ayres, 1979; Patil and Welch, 1979; Gutman, 1979; Kambayashi, 1979; Fischer, 1980; *Electronic Design,* Oct. 25, 1973, p. 67; *EDN,* Oct. 15, 1978, pp. 52, 65, 66; *Electronics,* July 6, 1978, p. 46. [Though the PLA in Fig. 7.7.1*a* is based on MOSFETs connected in parallel, PLAs based on MOSFETs connected in series are proposed, with two arrays like ordinary PLAs (Lin, 1981) and with a single array for a single output (Suzuki et al, 1973)].

▲ 7.8   Field-Programmable Array Logic

FPLAs and PROMs are convenient and inexpensive when users set information in memories by themselves. Also FPLAs (or PROMs) are less expensive than mask-programmable PLAs (or ROMs) for small production volumes, though for high production volumes mask-programmable PLAs (or ROMs) are much less expensive. In particular, when designers want to use mask-programmable PLAs (or ROMs) but their design is not completely debugged, they should try their design ideas with FPLAs (or PROMs) and then switch to mask-programmable PLAs (or ROMs) only after debugging is complete, because if a semiconductor manufacturer is already working on mask-programmable PLAs (or ROMs), sudden interruption of the work due to the discovery of design mistakes is unprofitable for both the manufacturer and the customer.

As the above advantages of field programmability have been recognized by designers, field programmability has been extended to other types of IC packages as follows.

A **programmable array logic (PAL)** is an FPLA where the OR array is not programmable. In other words, in a PAL the AND array is field-programmable but the OR array is fixed, whereas in a FPLA, both arrays are field-programmable. The advantage of PALs is the elimination of fuses in the OR array in Fig. 7.7.3 and of special electronic circuits to blow these fuses. (These circuits are not shown in Fig. 7.7.3.) Since these special electronic circuits and the programmable OR array occupy a very large area, the area is significantly reduced in a PAL. Since single-output two-level networks (i.e., many AND gates in the first level and one OR gate as the network output) are needed most often in design practice, many single-output two-level networks which are mutually unconnected are placed in a PAL package. Based on different combinations of the numbers of inputs and networks and also the complementation of outputs, many different IC packages (more than one dozen) are commercially available, such as packages of 8 AND-OR networks with 10 inputs, 6 AND-OR networks with 12 inputs, and 8 AND-NOR networks with 10 inputs. Some packages provide flip-flops, output suppression capability, or an additional small network for arithmetic processing. There are other extentions of field programmability.

Significance of the Use of FPAL Packages

Fuses on these field-programmable array logic (FPAL) packages, including the FPLA packages described in Section 7.7, have replaced the soldering of the pins of IC packages in the holes of pc boards. In digital systems many nonstandard networks are still used because switching functions which designers require are too diverse to be standardized by semiconductor manufacturers. When MSI, LSI, and VLSI off-the-shelf packages for standard networks, or microprocessors and their peripheral networks, are assembled on pc boards, many nonstandard networks are usually required for interfacing other key networks or for minor modifications, and they require many SSI packages and discrete components (e.g., an SSI package with only four gates sits next to a microprocessor package with 5,000 gates), occupying a significant share of the areas on pc boards (in terms of package count, roughly from 20 to 50%). FPAL packages can replace them, reducing the area (or package count) to one fourth or less, though the number of gates per FPAL chip (also per FPLA chip) is limited to a few hundred due to the large area occupied by fuse blowing circuits and fuses (*Electronics,* July 3, 1980, p. 119). (For example, a PAL package can replace 4 to 10 SSI and MSI packages in a typical application, though the exact number depends on individual situations.) If we consider related factors such as reductions of cabinet size, power consumption, and fans, the significance of this reduction is further appreciated. Also when logic design is not finalized and needs to be changed often, these FPAL packages can reduce expenses and time for repeatedly redesigning and remaking pc boards.

PALs are used in minicomputers such as the Naked Mini 4/10S of Computer Automation (*Electronics,* May 24, 1979, p. 117) and the Eclipse MV/8000 of Data General. [The parts count is reduced by a factor of 30:1 or

50:1 (*Electronic Design,* July 5, 1980, pp. 33, 34; *Electronics,* May 22, 1980, p. 133).]

References: Monolithic Memories, 1978; Durniak, 1979; Cavlan, 1979; Cavlan and Durham, July 5, 1979; July 19, 1979; Bursky, Wescon 1979, July 5, 1980; Birkner, 1978, Aug. 16, 1978, 1979; Twadell, 1980; Kroeger and Tozun, 1980; National Semiconductor, 1979; Alsing et al., 1980; *Electronics,* July 31, 1980, p. 128; June 2, 1981, pp. 49, 50; *Electronic Design,* Jan 4, 1977, p. 35; Jan 18, 1977, p. 101; Apr., 30, 1981, pp. 38, 40.

▲ 7.9 Multilevel Gate Array (MGA)

Combinational networks (including those inside sequential networks), which can be realized with PLAs, are restricted to only two levels. Hence for some functions the PLAs have large sizes and consequently are uneconomical and slow. In other words, some functions require many more gates in two-level networks than in multilevel networks [see Muroga (1979, Sec. 5.4)], and also, as the PLA size increases, the number of unconnected intersections (i.e., no dots) increases, wasting chip area. For example, the parity function $x_1 \oplus \ldots \oplus x_n$ requires $2^{n-1} + 1$ gates in a two-level network, though the number of gates required for it in a multilevel network is only linearly proportional to n. So this function can be realized in a multilevel network more compactly than in a two-level network.

If a multilevel network is realized in PLA-like matrix form on a chip, functions in general can be more compactly realized on a chip than with a PLA, and time-consuming layout for the random-logic network realization (i.e., the functions are laid out without placing gates and connections in the matrix form) can be avoided, maintaining some of the advantages of PLAs. Let us call such a multilevel gate (network) realization a **multilevel gate array (MGA)**.

Figure 7.9.1 shows an example of the MGA. The multilevel NAND gate network shown in Fig. 7.9.1a is realized in the MGA in Fig. 7.9.1c, where a MOSFET is placed at each intersection of horizontal and vertical lines. It is to be noticed that MOS cells (i.e., NAND gates with n-MOS in negative logic) realizing f_1, A, and f_2 stretch vertically in Fig. 7.9.1c, while all other MOS cells stretch horizontally. The first row of MOSFETs constitutes one MOS cell. In the second row there are three MOS cells. The third row constitutes one MOS cell. Here it is to be noticed that despite input $\bar{z}$ at gate 3, z instead of $\bar{z}$ must be connected in the second row of MOSFETs in Fig. 7.9.1c because the MOSFETs constituting gate 3 are arranged vertically under f_1, and z cannot be directly connected to gate 3. Thus an inverter is required between input z and gate 3, as illustrated in the equivalent network of Fig. 7.9.1b. Similarly we need inverters between f_1 and gate 4. Figure 7.9.1d shows the MGA of Fig. 7.9.1c, abbreviated with dots and crosses,

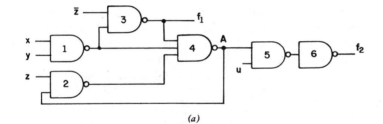

(a)

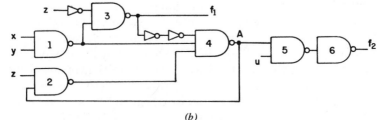

(b)

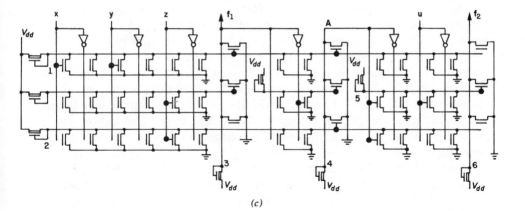

(c)

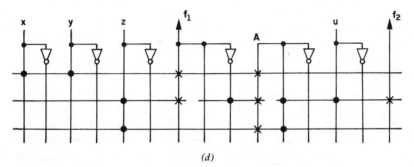

(d)

Fig. 7.9.1 MGA. (a) Multilevel NAND gate network. (b) Equivalent network of (a). (c) MGA for (a). (d) MGA expressed with dots and crosses.

where dots denote connecting the gates of MOSFETs to vertical lines and crosses denote connecting the gates of MOSFETs to horizontal lines. In Fig. 7.9.1*d* the second horizontal line has two breaks, whereas the first and third horizontal lines stretch without any break.

Any multilevel NAND network can be laid out in an MGA in this manner by placing alternately MOS cells stretching horizontally and those stretching vertically and breaking lines at appropriate points. Flip-flops can be formed inside a MGA, whereas flip-flops are placed outside the arrays in PLAs, as illustrated in Fig. 7.7.2. Since an MGA is a multilevel network, it is generally smaller than a two-level network. Even though inverters are shown in Fig. 7.9.1*b*, they are not really added because this is simply connecting to one of a pair of vertical lines. Any given multilevel network can be minimized for the most compact MGA without being converted into a network of AND and OR gates, unlike PLAs where two-level NAND gate networks can always be converted into two-level AND-OR networks for the minimization of their switching functions. [MGAs can be designed by the heuristic design methods of NAND networks discussed in Muroga (1979, Chap. 6). The "transduction method" mentioned there is also a powerful means.] Unlike PLAs where the locations of the AND and OR arrays are fixed, the locations of horizontally or vertically stretching MOS cells and breaks can be determined only after completion of the logic design. This means that MOSFETs and ground lines must be horizontally or vertically placed, considering the contacts with metal or polysilicon (or diffusion) lines. (Very long lines should be avoided if we do not want to lower the speed.) In this sense, MGAs are less flexible than PLAs, but still retain the advantage of simple layout (i.e., all MOSFETs are placed in matrix form, and every connection line is straight, unlike the jig-saw-puzzle-like configuration of the layouts of random-logic networks).

Since all masks are custom-made, unused MOSFETs can be eliminated, reducing mask preparation time. Also whenever necessary, vertical and horizontal lines can be directly connected, using contacts (i.e., skipping vertically arranged MOSFETs).

In the **associative logic matrix** by Greer (1976) the combination of certain fixed numbers of horizontally stretching MOS cells, vertically stretching MOS cells, and then horizontally stretching MOS cells, arranged in cascade, is treated as one block. (We may have more than one type of blocks.) The MGA in Fig. 7.9.2 consists of three such blocks (i.e., three associative logic matrices). This special type of MGA can avoid the disadvantage of the general MGAs such as that of Fig. 7.9.1, in which locations of horizontally or vertically stretching MOS cells cannot be determined until the logic design is finalized, and it also has the advantage of requiring the custom design of only one connection mask, making the associative logic matrix programmable. But there are greater chances of wasting part of it, compared with general MGAs like that of Fig. 7.9.1. If associative logic matrices were in off-the-shelf packages, the preparation of associative logic matrices with

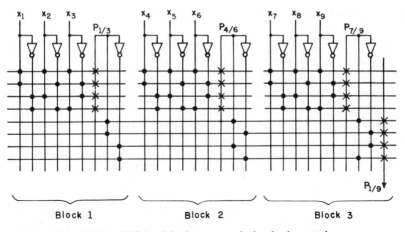

Fig. 7.9.2   MGA with three associative logic matrices.

different numbers of inputs and outputs (or configurations) would alleviate this to some extent.

▲ 7.10   Use of Memories in Computers

Although the use of semiconductor memories started with calculators and microcomputers which require compactness, low cost, and low power consumption, semiconductor memories have been extensively used in minicomputers such as the PDP 11/60, in superminicomputers such as the VAX-11/780 (Conklin and Rodgers, 1978), and even in mainframe computers (*Electronics,* May 24, 1979, pp. 120–123), improving computer speed and reducing cost. For example, look-up tables for a fast floating-point processor, ROM instruction decoding, and microcode are used in the PDP 11/60 (Electronics, Oct. 27, 1977, p. 110). Since the cost of ROMs can be as low as one tenth of the cost of RAMs, extensive use of ROMs can reduce the cost of computers.

ROMs and their variations (e.g., PLAs) have been used throughout digital systems. The advantages of using them are vividly manifested, in particular in the control logic of computers. Control logic is the most complex and time-consuming part in designing computers, no matter whether computers are microprocessors, minicomputers, or mainframes. [As we need to design more complex microprocessors in VLSI chips, the design of control logic will be much more complex and time-consuming than that of any other part in the chips. See, for example, Tredennick (1981).] Throughout the entire design stages, or even after delivery to customers, the control logic needs to be changed or corrected. Thus ROMs or their variations, whose stored information is easy to change, are probably most widely used in control logic (in particular in microprocessors) as an ideal means to cope with this difficulty.

Control logic can be realized with random-logic gate networks. If so, this is called **hard-wired control logic** and has the fastest speed, though design and change are cumbersome and time-consuming.

The use of ROMs for control logic suggested by M. V. Wilkes in 1951, called **microprogrammed control**, is convenient. Although it is slower than hard-wired control, it is less expensive, and the instruction set of a computer can be easily changed by the change of ROMs. Microprogrammed control was first commercially adopted by IBM in the system 360 computers.

An instruction of a computer program is read out from the main memory of the computer. Then the initial address of microinstructions, which constitute this instruction, is provided to the ROM address register shown in Fig. 7.10.1. The information stored in this address is then read out of this ROM. The information consists of two parts, a ROM address for the next microinstruction, and the current microinstruction, that is, a set of signals to control gates in the other part of the computer. By sending the latter part to the remaining part of the computer, some gates are opened or closed such that arithmetic or other operations are performed or information in one register is sent to another. The former part is combined with the information that represents the status of the other part of the computer, and is sent to the ROM address register. This process repeats. Realization of microprogrammed control has many variations.

PLAs can be mixed with ROMs. PLAs can specify the first microinstruction address or can make patch errors in ROMs.

Programs stated in microinstructions are called **microprograms** (or **microcode**). Software stored in ROMs is called **firmware**.

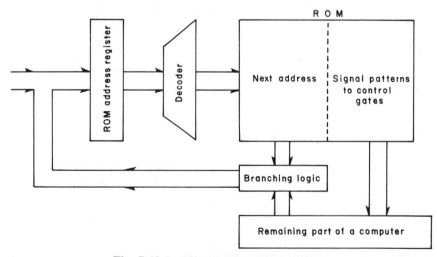

Fig. 7.10.1   Microprogrammed control.

A computer program (or so-called software) is a sequence of instructions which perform all together a specific task. Each instruction can be executed by a microprogram or hard-wired control, as mentioned above. Complex tasks, such as floating-point operations, can be directly microprogrammed as firmware (i.e., stored in a ROM), instead of being programmed with instructions of the computer (i.e., in software). Such firmware realization of software is usually a few times faster than the original software. But a random-logic network realization of the task can be much faster than the original software, possibly 100 times, 1,000 times, or even more, depending on the individual task and the way to realize it, as discussed in later chapters.

References: Husson, 1970; National Semiconductor, 1972, 1974; Mc-Dowell, 1974; Clare, 1973; Reyling, 1974; Miles, 1976; Petrale, 1979; Travis, 1976; Johnson, 1976; Leung et al., 1977; Carroll, 1977; Kaestner, 1975; Cavlan and Cline, 1976; Perrin et al., 1972; Mitchell and Holland, 1978; Phister, 1979; Buchwald, 1977; Coleman, 1980; Durniak, 1979; Franson, 1977; Grasselli and Montanari, 1970.

7.11 Chapter Summary

As discussed in Chapter 6, the main differences between logic gates and semiconductor memories in terms of electronic realizations are whether or not logic gates or, in a broader sense, devices are arranged in matrix form, and also whether or not gates, or devices, perform complex logic operations. Thus memories are special cases of logic gate networks. (In a sense, microprogramming with ROMs may be interpreted as logic design where only connections among transistors whose locations are predetermined are to be determined.) But the above two differences make a big difference in the reduction of layout time and also in the ease of changing information. So as illustrated in this chapter with many examples, if tasks are complex, such as complex software, and consequently require frequent changes due to debugging or updating, they are stored in RAMs. If tasks become simpler, such as completely debugged microprograms, and require changes less frequently, they are stored in ROMs (PLAs are excluded), as illustrated in Table 7.11.1. If tasks become further simpler such as the control logic of microprocessors, and require almost no change, they are realized in PLAs. If tasks are even simpler, such as switching functions, and require no change, they can be realized in random-logic gate networks. Therefore **there is no clear border line between logic gates and memories in realizing given tasks**.

As we move from RAMs to logic gates, the speed increases because of an increase of logic capability and also because of decreasing parasitic capacitances and resistances of interconnection lines due to the area decrease of the gate network. (In other words, tasks realized in firmware can be realized by PLAs with smaller areas. Since about 95% of the devices in a

Table 7.11.1   Tradeoffs

TASK	COMPLEX ──────────────→			SIMPLE
Realization	RAM	ROM PLA MGA		Logic gate
Logic capability	Flip-flops	NOR (or NAND)		Negative (or complex) function
Structure		Regular		Random logic
Ease of change	Easy ────────────→			Difficult
Speed	Slow ────────────→			Fast
Cost				
Low volume	Low ────────────→			High
High volume	High ←────────────			Low

PLA are not used, the PLA area can be greatly reduced by deleting these unused devices and freely bending connections, with devices relocated. Only NAND gates are used in a PLA. So by replacing these NAND gates by negative gates, the PLA size can be further reduced because the number of gates and the number of connections are reduced, and wasted area can be squeezed out even more. The area for random-logic gate networks thus derived occupies less space than the original PLA or firmware.) Also changing information which has been stored or realized becomes more difficult because information is converted into more sophisticated forms. In other words, in RAMs or ROMs the original software is simply stored, but in PLAs, information is converted into minimal switching expressions (without the use of parentheses). In the case of logic gate networks, information for task realization is converted into gate networks in a more complex manner (essentially based on minimal switching expressions, using parentheses). The cost increases with low production volumes because the time needed for logic design and layout is longer. But with high production volumes the cost becomes less because the initial expenses for logic design and layout are less dominant than the manufacturing cost, which decreases due to smaller chip areas [as shown with Eq. (2.3.1)].

Since random-logic networks can easily be faster by an order of magnitude (or two or more orders of magnitude if appropriate task-realization algorithms are used) than RAM realizations of given tasks, the tasks are often realized by a mixture of random-logic networks and memories, where random-logic networks are used in the key parts that require high speed, even though the layout time for random-logic networks is longer by an order of magnitude than for memories.

The realization of tasks with a mixture of memories and logic gates is a new dimension in the designers' thinking. There are so many different types of memories, such as RAMs, ROMs, PLAs, and MGAs, that clever and appropriate use of them is becoming a vitally important part of logic design. Also as ICs are becoming extremely cheap, particularly when high-volume production is justifiable (like for microcomputers), we can improve the cost-to-performance ratio of computers by IC implementation of increasingly more complex tasks, even though these tasks are not encountered as often as addition or multiplication. All these facts mean that, unlike with conventional logic design where given simple tasks are expressed in switching expressions and then realized with logic gates, **appropriate algorithms must be developed before system design and logic design such that not only logic gates but also all kinds of memories are properly, efficiently, and economically used throughout the system.** If we want to have best results, this is a formidable task, requiring imagination, cleverness, and diversified backgrounds. This trend has just started.

Problems

7.1   Using ROMs and adders like in Fig. 7.6.2., design a multiplier network to multiply two numbers of 10 bits each. Each ROM can have 10 inputs.

7.2.   Design a network for the control logic in Fig. 7.6.3, using any types of gates, and also the other networks along with a flow chart to such an extent of detail that the operational sequence of all the networks can be easily understood with the flow chart.

7.3.   Design a PLA (mask-programmable) of 3 input variables and 2 outputs, with I^2L gates. (Show a layout.) The outputs need not be provided with a choice of complementing. Then compare it with a MOS PLA in terms of speed and chip area, and discuss other conceivable problems.

7.4   Design a full adder with the following means:

    (a)   A single PLA as shown in Fig. 7.7.1a. (Express with dots.)

    (b)   A single MGA as shown in Fig 7.9.1d. (Express with dots and crosses.)

7.5   Design a 2-bit up–down counter, using the PLA shown in Fig. 7.7.2, where the flip-flops are J–K master–slave flip-flops. Make the array size as small as possible.

7.6   Design a 1-out-of-n detector network with PLAs and gates only, instead of ROMs and gates as in Fig. 7.3.9. Also describe how each PLA should be programmed.

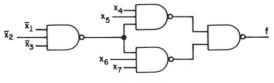

Fig. P7.7

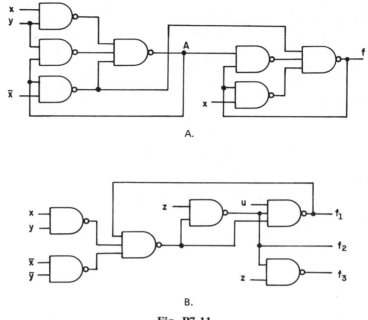

A.

B.

Fig. P7.11

7.7   Realize the function f realized by the logic gate network shown in Fig. P7.7 with a single PLA (express with dots), as small as possible. Also realize this logic gate network with a single MGA (express with dots and crosses). Then show the size of each case (i.e., the product of the number of vertical lines and the number of horizontal lines).

7.8   Realize the following three functions with a single decoded PLA with two-input decoders only. Also, realize them with a single ordinary PLA.

$$f_1 = x \oplus y \oplus u, \qquad f_2 = x \oplus u \oplus v, \qquad f_3 = x\bar{u} \vee y\bar{v}$$

Then compare the sizes of these two realizations (i.e., calculate the product of the number of horizontal lines and the number of vertical lines).

7.9   Rewrite the MGA network of Fig. 7.9.2 into a multilevel NAND network.

7.10   Discuss advantages and disadvantages of MGAs.

7.11   Realize one of the networks shown in Fig. P7.11 with a MGA (express with dots and crosses).

CHAPTER 8

Computer-Aided Design (CAD)

Computers have been extensively used in all stages of the design and development of an LSI/VLSI system, starting from system specifications to test of prototypes, as illustrated in Fig. 8.1. At one time **design automation (DA)** had been attempted, but the effort is now called **computer-aided design (CAD)**, since complete design automation without human intervention is a formidable task which we cannot achieve even with the fastest computers. The use of computers with human intervention for complex critical decisions is a more meaningful approach. Any procedures for design, simulation, verification, or test (and also device modeling in some cases) are called CAD programs when they are converted into computer programs as aids, though those which do not require human intervention are also often called CAD programs.

The objectives of developing CAD programs are to shorten the design and development time of LSI/VLSI chips (as well as systems with these chips), to minimize design mistakes, to facilitate design changes, and to shorten the time for design verifications and tests. As chips and digital systems become increasingly complex, the design and development time is increasing very sharply, requiring more manpower, and expenses for design and development are getting proportionately greater. So reducing the time for design and development by CAD is vitally important to produce competitive products, because it can reduce the product cost, particularly in low production volumes [see Eq. (2.3.1)]. Testing and debugging are extremely important since the debug time occupies an increasing percentage of the design and development time. For example, a chip with 100,000 devices takes roughly 60-man-years to design and another 60 to debug, and chips with more devices will take a greater percentage of time for debugging. So even a 10% reduction would be significant. As a matter of fact, increasingly many LSI/VLSI chips cannot be designed without CAD (Breuer et al. 1981; Yamada, 1981).

CAD programs can be either of great value or of little value, depending on the tasks in the design for which CAD programs are used. CAD programs such as SPICE2 are an indispensable tool for electronic circuit simulation, for example, but for some other complex tasks, CAD programs are of little value. Sometimes appropriate CAD programs cannot even be developed.

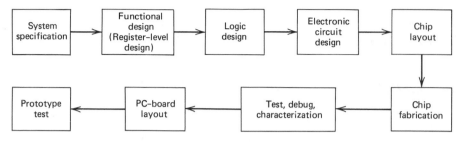

Fig. 8.1   Different stages in design and development of an LSI/VLSI system.

In the following, let us survey the CAD programs appropriate to the different design stages shown in Fig. 8.1.

8.1   System Specification, Functional Design, and Logic Design

When we want to design a digital system, we need to specify the system performance. This is called **system specification**. Then the system must be broken down into subunits or registers. So we have a **functional design**, which specifies the functional relationships among subunits or registers. Computer systems can be best designed based on registers. It is difficult to design them based on a level higher or lower than registers (Bell 1976). Architecture usually means functional design and system specification, often including part of the subsequent logic design. Then we need the **logic design** of networks which constitute subunits or registers.

System specification, functional design, and logic design are still the most difficult stages for the use of CAD, due to their complexity, throughout the entire design sequence. Thus they are largely done by hand. Although logic design by CAD programs was attempted in the past, the outcome was disappointing, unless short design time was much more important than network size. [For example, logic networks designed by the design automation program ALERT required roughly 160% more gates than those designed by humans (Friedman and Yang, 1970).] For very large integration sizes, however, the situation may change because the design time is getting out of hand for human designers as the integration size increases (as shown in Table 4.7.1, logic design is a very time-consuming part of chip design for professional designers, who only supervise more time-consuming layouts), and also because the difference is becoming smaller due to CAD procedure improvements by human interaction (Darringer and Joyner, 1980; Darringer et al., 1980). (The transduction method mentioned in Remark 4.2.2 is very different from conventional logic design procedures and is a candidate for such logic CAD.) Also computer languages to facilitate system specification, functional design, and in particular control logic (in terms of microprograms) have been explored (Datamation, Aug. 1973; Evangelisti et al., 1977; van Cleemput, *Proc. 16th DA Conf.* 1979; Zimmerman, 1980).

8.2 Logic Simulation

When a system architecture or logic networks are designed, performance and errors are checked by CAD programs. This is called **logic simulation**, since CAD programs check by simulation whether the designed systems or networks are realized as desired. In this case the CAD programs usually check also whether the networks contain hazards. Some logic simulation programs are available to the public or can be purchased. Currently logic networks of up to tens of thousands of gates can be simulated, but if larger networks are to be simulated or more complete simulation is to be made, logic simulation can be excessively expensive. Logic simulation is important because any logic design errors discovered after realizing chips are time-consuming and expensive to correct, due to the repetition of electronic circuit design, layout, and prototype wafer fabrication.

References: Breuer and Parker, 1981; Ellenberger and Ng, 1981; Muehldorf and Savkar, 1981; Szygenda and Thompson, 1975; MacDougall, 1970; Breuer, 1972, 1975; Kani, 1979; Chang et al., 1974; Lake, 1970; Lewin, 1977; Biancomano, 1977; Bening, 1979; Wilcox and Rombeek, 1976; Kawato et al., 1979; Sasaki et al., 1980; Muroga, 1979, references in Sec. 9.6.

8.3 Partitioning and Chip Design

After the logic design has been completed, the entire network must be **partitioned** into networks such that each network is **assigned** to an IC chip with as few connections among chips as possible. Since the number of pins that can be provided at each chip is limited, the partitioning must be done appropriately so that a group of gates with their interconnections **clustered** go into a single chip as much as possible (Mennoue and Russo, 1974; Kodres, 1972; Kernighan and Lin, 1970; Kurtzberg, 1962).

Then we need to go to the internal design of each chip. The entire chip area is divided into areas such that a different network is placed in each area. In this case we have to allocate networks to these areas such that the entire chip area is minimized, keeping high-speed performance. Since with this **chip plan** designers do not know precisely how much area each network requires (the area size of each network can be determined only after electronic circuit and layout designs), only experienced designers can make a good estimation. If, after electronic circuit and layout designs, designers find that some networks do not fit the allocated areas, they repeat the chip plan, electronic circuit design, and layout design (as mentioned in Section 4.7).

8.4   Electronic Circuit Simulation and Analysis

Logic networks have to be converted into electronic circuits. When designers specify electronic circuit requirements such as speed, power supply voltage, types of logic operations, and signal level tolerances, it is desirable to have CAD programs which automatically design electronic circuits meeting all requirements, and specify parameters such as dimensions of transistors and magnitudes of currents. This is difficult. The current practice is as follows. Before layout, designers design electronic circuits, temporarily assuming the sizes of some circuit parameters, analyze the circuits by CAD programs, and then modify the circuits based on the analysis. After having obtained more specific values for circuit parameters from actual layout, designers finalize the design of the electronic circuits, analyzing and simulating the laid-out circuits by CAD programs. (The above procedure is often iterated.)

For this electronic circuit simulation and analysis, CAD programs perform complex numerical analysis calculations of nonlinear differential equations which characterize electronic circuits. Since we need to finish calculation within a reasonable time limit, keeping the required accuracy, many advanced numerical analysis techniques are used. The CAD programs usually yield the analysis of transient behavior, direct-current performance, stationary alternating-current performance, temperature, signal distortion, noise interference, sensitivity, and parameter optimization of the electronic circuits.

After CAD programs such as TAP, ECAP (by IBM), and NET 1 had been developed for SSI and MSI during the period from 1964 to 1972, more powerful CAD programs for LSI, such as ASTAP (by IBM) and SPICE (by Nagel and Pederson, 1973; Nagel, 1975), have been developed since 1973. Currently many CAD programs with different features are available, such as CIRCUS 2, SPICE2, ISPICE, ITAP, NET II, SUPER*SCEPTRE, and UCCAP. Most of them typically can handle electronic circuits of up to a few hundred transistors. In particular SPICE2 and its modified versions are used extensively throughout the industry.

An LSI/VLSI chip based on MOS can contain many more transistors, and the CAD programs mentioned above are not fast enough. Thus a CAD program, MOTIS, was developed by Bell Labs in 1975. It has high speed by being tailored to the transient analysis of electronic circuits of MOSFETs at the sacrifice of accuracy. MOTIS can handle up to 1,000 MOS cells, and MOTIS-C developed by Fan et al. (1977) can handle up to 2,500 MOS cells. Though the application of these two programs is limited to the transient analysis of MOS circuits only, computational efficiency is improved by two orders of magnitude, handling larger electronic circuits by an order of magnitude, compared with the CAD programs mentioned in the previous paragraph. MOSTAP, which is similar to MOTIS can handle logic simulation also.

References: Chua and Lin, 1975; Jensen and McNamee, 1976; Director, 1974; Kaplan, 1975; Yasutoshi and Kani, 1978; Biancomano, 1977; Branin et al., June 1971, Aug. 1971; Blattner, 1976; Bowers et al., 1976; Weeks et al. 1973; Chawla et al., 1975; Kozak et al., 1975; Mahoney et al., 1974; Hodges and Pederson, 1974; Greenbaum et al., 1973, 1974; Cranswick et al., 1980; Branin, 1980; Ruehli, 1979, 1981; Getreu, 1978; Musgrave, 1979; Tanabe et al., 1980.

> **Remark 8.1:** CAD programs to analyze the performance of a single or at most a few transistors are vitally important when we need to introduce new fabrication techniques or new devices. These programs for **device modeling** require advanced knowledge of solid-state physics and numerical analysis, and they are not necessary for logic or electronic circuit design once fabrication techniques and devices to be used have been determined. So they are not discussed in detail in this book. The development of new memories with higher performance is essentially the exploration of new fabrication techniques and new devices, so these programs are very important.
>
> Device modeling programs are used to determine the solid-state structure of devices (or transistors) so that ICs based on these devices can be manufactured with reasonable production yields during fabrication. Based on mathematical models of the devices, which characterize the physical behavior of carriers and potential distribution, nonlinear partial differential equations are solved. Because of long computation times due to the complexity of the problem, at most a few devices can be handled, even with advanced numerical analysis techniques. Device modeling programs such as CADDET, MINIMOS, GEMINI, DTRAN, SITCAP, MODMAG, and TRANS 2D were developed. Since the performance of the device is dependent on the processing techniques used during fabrication, programs for **process modeling** are important. Process modeling programs such as SUPREM-II were developed.
>
> *References:* Antoniadis et al., 1978; Toyabe and Asai, 1979; Greenfield and Dutton, 1980; Shichman and Hodges, 1968; Dutton and Antoniadis, 1979; Barbe, 1980; Selberherr et al., 1979; Browne and Miller, 1979. □

8.5 Layout

The layout for random-logic networks is the most time-consuming stage throughout the entire sequence of LSI/VLSI chip design shown in Fig. 8.1, though it is done mostly be draftpersons. Thus as discussed in the next chapter, different design approaches (so-called semi-custom design), such as those based on gate arrays and cell libraries, to shorten layout time, have been conceived in addition to random-logic networks, and designers must choose one based on a tradeoff between layout time and quality (i.e., size and performance) of the finished chips. Different CAD programs for these design approaches have been developed as discussed in the next chapter.

After having finished the layout, designers usually check by CAD programs whether the layout conforms to the layout rules.

Then designers make prototype wafers and test them by electronic measurement, by visual examination, and also by software. The development of efficient test programs is very important, as discussed in the following. If any mistakes are discovered, designers must correct them in the functional design, logic design, electronic design, or layout, and make prototype wafers again. They must repeat this until no more mistakes are discovered. If repeated too many times, it is excessively time-consuming and expensive.

Placement and Routing

When the chip design is completed, different chips must be placed on a pc board, and interconnections among chips must be routed properly so that the entire board area is minimized, maintaining maximum speed performance. (If some important connection becomes too long, the speed of the entire chip may be slowed down. If the switching time of each gate is 4 nsec, for example, the maximum length of each interconnection is usually limited to a few inches.) **Placement** of the chips is very important, since with improper placement, best routing cannot be obtained (i.e., if chips with clustered interconnections among them are placed apart, these interconnections become long, occupying large areas, and slowing propagation time). Placement can be done best by human eye inspection or by highly interactive CAD programs.

References: Hanan and Kurtzberg, 1972; Khokhani and Patel, 1977; Weindling, 1964; Kamikawai et al., 1976; Hanan et al., 1976; Feurer et al., 1980; Fiduk et al., 1980.

Routing must be automated with CAD programs because connections inside and outside chips are now excessively numerous in digital systems, making routing by hand impractical. Many procedures for routing have been developed. The following are typical methods, which can be used for routing interconnections among chips on a pc board.

CAD procedures for the placement and routing inside the chips of semicustom design are essentially the same (or similar) as those discussed here, as mentioned in Chapter 9. (Since area minimization is more critical with chips than with pc boards, human routing is usually mixed with CAD programs, based on the following methods for chip design.)

Procedure 8.5.1   Maze Running Method

Suppose that we want to connect two points A and B on a pc board (or a chip), when other points C through F are already connected, as shown in Fig. 8.5.1. The entire area is uniformly divided into rectangles (or squares), and in each blank rectangle a number, which shows the distance (expressed in the number of rectangles) from point A, is entered, starting from the rectangles next to A and moving toward point B. When there are obstacles such as already existing interconnections among points C through F, we can reach point B from different directions with different numbers by taking detours. The route that ends with the smallest number at B is the shortest path between A and B.   □

We can always find the shortest path by this method, which is also called **Lee's algorithm**. But when the entire area is large, a large memory and long computation time are required. (For example, if we use a square with each side 5 μm long, a chip with each side about 250 mil contains more than one million squares.) So this algorithm is usually combined with other methods.

References: Lee, 1961; Akers, 1967, 1972; Pape, 1974; Hoel, 1976; Kamikawai et al., 1976.

Procedure 8.5.2   Line Search Method

Suppose that we want to connect two points A and B, as shown in Fig. 8.5.2. At point A a vertical line and a horizontal line (broken lines) are drawn with each end bounded by an existing line (solid lines) or by the border of the entire area. Then a horizontal or vertical line that intersects any of these lines but can stretch beyond the ends of these lines is drawn. For example,

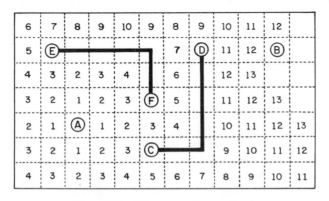

Fig. 8.5.1   Maze running method.

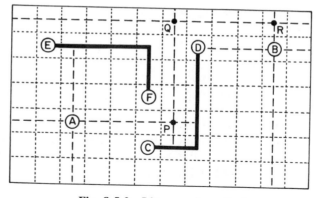

Fig. 8.5.2   Line search method.

by moving toward point P on the horizontal line from A, we can try to stretch vertically, and finally at P we can stretch upward.

If we repeat this from A and B and two lines from both sides meet, we have obtained a connection such as $A-P-Q-R-B$ in Fig. 8.5.2. The connection is not necessarily the shortest.   □

Computation is short, with small memory requirements, when a connection can be found (though it may turn many times). Usually a small percentage of the connections need to be modified manually later. But for a very large number of connections, hand routing of even a small percentage can be very time-consuming (*Electronic Design*, Sept. 1, 1979, p. 45) and a special hardware processor is proposed (Hong et al., 1981).

References: Mikami and Tabuchi, 1968; Hightower, 1969, 1973, 1974; Soukup, 1978, 1979; Hitchcock, 1969; Hashimoto and Stevens, 1971; *Electronics*, May 10, 1971, pp. 76–79; Nov. 24, 1977, p. 175; Jan. 19, 1978, pp. 102–107; June 19, 1980, pp. 82–84; Oct. 9, 1980, pp. 146, 147; Kuo and Magnuson, 1969; Mattison, 1974; Mori et al., 1980; Heller et al., 1977, 1978; Tsukiyama et al., 1981; Geyer, 1971.

8.6   Design Verification and Testing

As the integration size of LSI/VLSI chips becomes larger, design verification and testing at each design stage is vitally important, because any errors which sneak in from the previous design stages are more difficult to find and more expensive, since once found, we need to redo the previous design stages. As the integration size increases, the test time increases very rapidly, so it is crucial to find a good way to test within as short a time as possible, though it appears very difficult to find good solutions. Complete test and design verification with software or hardware (i.e., computers specialized in testing) by manufacturers is usually impossible, and often only long-term use of the chips leads to the discovery of design mistakes or malfunctions [e.g., the pocket calculator that was recalled by its manufacturer after a customer found a design mistake (*Electronics,* Nov. 6, 1972, p. 28)].

References: There are numerous papers on the subjects of design verification and testing, for example, Chang et al., 1970; Friedman and Menon, 1971; Roth, 1980; Breuer and Friedman, 1976; Liguori, 1974; Muroga, 1979, references in Sec. 9.6; Rozeboom and Crowley, 1976; Rozeboom, 1976; Muehldorf, 1980; Harrison et al., 1980; Eichelberger and Williams, 1977; Eichelberger et al., 1978; Berglund, 1979; Krohn, 1977; Grason and Nagle, 1980; Lindsay and Preas, 1976; Hayes and McCluskey, 1980; Akers, 1976, 1980; Batni and Kime, 1976; Williams and Angell, 1973; Barraclough et al., 1976; Yen, 1969, Oct. 1969; *Electronics* Mar. 15, 1971, pp. 68–71; Aug. 30, 1971, pp. 44–47; Oct. 3, 1974, p. 39; Nov. 27, 1975, pp. 108–113; Jan. 22, 1976, pp. 100–105; Jan. 27, 1981, pp. 42, 44; many papers in Proc. 14th DA Conf., 1977; Funatsu et al., 1975; Nair et al., 1978; McCluskey and Wakerly, 1981; Rosenberg and Benbassat, 1974; Yoshida et al., 1977.

8.7   Problems of CAD Programs

In the preceding, only CAD progams for typical aspects of LSI/VLSI system design are discussed. But for numerous other aspects CAD programs have been developed. We have to understand that CAD can be very effective or ineffective, depending on which aspects and how CAD is applied. For repetitive and simple tasks, such as logic simulation, CAD is highly effective, doing tasks far faster and more errorfree than human designers. For complex tasks, such as functional design, CAD is not appropriate (though simulation is effective once the functional design is set). Also, for some types of problems, preparation of the input data to CAD programs is cumbersome and time-consuming, often much more than CAD processing time. CAD is good for local minimization, but it is not appropriate for global minimization, being inferior to human visual observation. (For example, CAD programs can minimize the layout area fairly well locally, but not globally. The limited scope of CRT screens aggravates this.)

When the integration size increases, in particular in the case of VLSI, the number of variables or data a CAD program must handle increases very rapidly, getting out of hand easily. For example, when we need to design a digital system which requires 10 LSI chips with 10,000 gates on each chip, we need to handle about 1,000,000 squares on a chip, and a total of 10,000,000 squares for the entire system. This presents tremendous difficulties in processing time and memory space for partitioning, placement, and routing. Also testing is extremely time-consuming for the total of 100,000 gates, even if we do not attempt complete tests. Thus the use of appropriate procedures or algorithms, along with data structures, is important for improving computational efficiency. When tasks are of a combinatorial–mathematical nature, computational efficiencies can be improved easily by a hundred, a thousand or more times by choosing appropriate procedures and data structures. In general it may make a drastic difference in the computational efficiency of CAD programs whether or not the programs are developed with a good knowledge and appropriate applications of basic mathematical algorithms (see, e.g., Knuth, 1968, 1969, 1973; Liu, 1968, 1977; Aho et al., 1974; Preparata and Yeh, 1973; and, for computational efficiency of switching algebra manipulation, Muroga, 1979), and probably with the development of new algorithms tailored to the given problems. [In particular, for certain types of CAD programs, use of the branch-and-bound method (see, e.g., Reingold et al., 1977; Horowitz and Sahni, 1978), combined with these algorithms, may greatly improve the efficiency of the combined algorithms. The branch-and-bound method is a procedural principle rather than a finely defined algorithm, and the development of appropriate branching and bounding operations is vitally important for its computational efficiency. For example, the efficiency of a PLA minimization program for a small number of variables was improved by 30 to 100 times by developing appropriate branching and bounding operations and a data structure as well, as mentioned in Muroga, 1979, Remark 4.6.2]. When we use interactive CAD programs, we often cannot use them effectively unless we understand well the algorithms on which the CAD programs are based.

Practically all CAD programs must be developed in-house, because CAD developments and data for most tasks are proprietary, and commercial CAD programs are only available in limited areas. Even if we find good commercial CAD programs, we often need to modify and update them, and this requires a significant effort (Schindler, 1977).

The development of useful CAD programs is very time-consuming (Nash and Willman, 1981). Consequently we need to develop technology-independent CAD programs, because otherwise the programs must be developed again from scratch whenever technology changes. In many cases, however, technology-independent CAD programs are less effective than technology-dependent ones. So prudent decisions are necessary as to which ones we develop.

8.8 CAD Data Base Systems and CAM

For a number of years manufacturers have been developing numerous CAD programs. In order to utilize CAD programs, many manufacturers are expanding data base management systems where a large number of terminals (graphic or nongraphic) are distributed in-house so that users can access appropriate CAD programs easily and design their chips and systems on the CRT screens. Also they can quickly send processed data to other users. The engineering design system (EDS) of IBM is an example of such a CAD system (*Electronics,* July 14, 1981, pp. 93, 94). High-level languages, which are appropriate for description and processing, are used. It appears essential for the survival of manufacturers to develop convenient and efficient data base management systems for CAD when numerous new products contain LSI/VLSI chips and the technology is becoming more complex each year. Such large-scale CAD systems, however, often present inconveniences to users by frustrating them with very complex manuals, contradicting the original objective of providing design convenience to users. So it is probably not a simple task to develop convenient and extensive CAD systems.

Computer-aided manufacturing (CAM), that is, computer programs to control manufacturing machines such as electron beams for mask preparation, industrial robots, and numerically controlled milling machines, are also essential for the future of manufacturers. As a means for improving productivity, CAM as well as CAD are important not only for computer or semiconductor manufacturers, but also for many other manufacturers, such as the automobile and aircraft manufacturers. (Remark 8.2) These manufacturers are also developing data base management systems for CAD and CAM, which are often combined.

References: Wiemann, 1979; Spence, 1979; Waxman, 1979; Korenjack and Teger, 1975; O'Neill et al., 1979; van Cleemput et al., 1978; Faster, 1975; Breuer, 1975; Hassler, 1974; Falk, 1975; Carmody, 1978; Ebbinghaus and Kreilkamp, 1978; Rosenberg, 1980; van Cleemput IEEE, 1979; Powell, 1980; *Electronics,* May 24, 1979, p. 129; July 31, 1980, pp. 73–80; Aug. 28, 1980, pp. 109, 110; Claiborne, Jan. 21, 1982; Donlan, 1980; *New York Times,* Jan. 18, 1981, Business Sec. p. 7.

Remark 8.2: CAD is important not only for chip or computer design, but also for other design problems. CAD, practically unknown in aircraft development in 1971, has produced sleeker planes with far less air drag. In 1979 McDonnell Douglas probably had the leading design, the work of a 55-person SST (supersonic transport, such as the Concorde) team, the industry's largest. The McDonnell Douglas computers, which can conjure up an SST design in 1 week, versus the 6 months it took to design an airplane 8 years ago, have produced a spaceship-like SST with a lift-to-drag ratio nearly 40% better than that of the Concorde or the 1971 Boeing design (*Business Week,* Mar. 19, 1979, p. 130E).

At General motors also, CAD has been used extensively. On a CRT screen the body of a car is designed, its mechanical strength is tested, and any weak spots found are fixed before the car is actually built. It is expected that about 90% of all the new machines in General Motors' manufacturing and assembly plants will utilize CAM by 1990.

Drug manufacturers, such as Merck, are designing new drugs on CRT screens by computer.

References: Olson, 1978; Moses, 1978; Appleton, 1977; Decker, 1978; Groover, 1980; Nevins and Whitney, 1978; Besant, 1981. □

8.9   References on CAD

There are numerous publications on CAD. Many of them appear in the proceedings of design automation conferences held annually by the ACM. In addition to the publications referred to in this chapter, the following may be consulted: van Cleemput (1976, 1977); Allan, (1975); Breuer (1966, 1977).

Problems

8.1   Suppose that we are given a chip with each side of 250-mil length, and we want to use the maze running method (Procedure 8.5.1) for routing when the minimum line width and the minimum spacing are L μm each. Make an approximate estimation of the maximum memory requirement for $L = 5$ μm and 3 μm.

8.2   Show an example of the situation where the line search method (Procedure 8.5.2) does not yield a good connection, requiring manual modification, while the maze running method (Procedure 8.5.1) can give the shortest connection.

CHAPTER **9**

Full-Custom and
Semi-Custom
Design Approaches

Because LSI/VLSI chips have been widely used in diversified applications, such as toys, games, computers, and home appliances (as discussed in Chapter 1), we have very diversified motivations in designing chips. When semiconductor manufacturers want to introduce the most powerful microprocessor chip which they expect to be produced in high volume, they can justifiably spend many years for the development, by a number of designers, of the most compact chip with the best performance based on random-logic networks. In many other cases, however, we cannot employ this full-custom design approach, because the chip cost will be too high due to low production volume [as seen from Eq. (2.3.1)], or a good timing of product introduction will be missed due to long design time. In these cases, depending on a combination of different levels of production volume, speed, power consumption, chip size, man-power, etc., a number of semi-custom design approaches are available. The reduction of layout time and logic design time is essential, and consequently use of regular-structured networks (such as ROMs, PLAs, and MGAs), availability of appropriate CAD programs, and logic design methodology (such as logic design based on minimal sums, logic gate networks, and software on a microcomputer) are key issues. In this chapter full-custom and semi-custom design approaches are compared with design approaches with discrete components or off-the-shelf packages, and the advantages and disadvantages in each approach are discussed.

9.1 Different Design Motivations

As the applications of digital systems expand into a large number of areas and products, these digital systems have been designed with a wide spectrum of different motivations. Accordingly we have to design a digital system with different criteria, considering tradeoffs among high speed, small chip size, design time, ease of design change, and possibly others. Naturally we have to choose an appropriate design approach among uses of discrete components, off-the-shelf, semi-custom, or full-custom IC packages, in-house or vendor design, and possibly others. The choice is not easy in many cases.

Let us look at some examples of different design motivations.

A logic designer who is working on an electronic measurement instrument in a laboratory may want to have an interface network to connect the measurement instrument to a microcomputer. Since this network is for one-time use, the designer would not order from an IC manufacturer a custom-designed LSI. This would be astronomically expensive because only one LSI package is to be manufactured. The design would be made in-house, using off-the-shelf packages.

Next let us consider the case of mainframe manufacturers. Suppose a firm wants to design a large computer with high speed within a few years. For high speed, ECL is a natural choice with current technology. Designers cannot use off-the-shelf IC packages, because if they did, the speed of their computers would not be much different from that of their competitors' computers. They would develop their own version of ECL with improved speed. Because of the proprietary nature of design, they will want to design in-house. The chip sizes need not be minimal for the sake of cost, because many other parts of the computer, such as memories, peripherals, and overhead expenses, are more expensive than all the logic networks, and accordingly chip areas need not be minimized. But if the chip sizes are too large, the speed of the networks is sacrificed, and the yield becomes too low, so the chips must be reasonably small. If every chip is designed with a full-custom design approach, design becomes too expensive, because they need a large number of different chips, each of which is in small production volume. (Notice that the production volume of each mainframe model is much lower than that of off-the-shelf microprocessor chips.) A semi-custom design approach is a compromise. The computer requires many logic networks of medium or slow speed, in addition to the above ECL networks. These lower speed networks can be off-the-shelf packages purchased from the outside, while some of them can be designed by outside venders. Often designers are required to use only certain IC logic families due to a company-wide standardization of electronic circuits for logic gates. (They want to cut down on production cost by standardization throughout their company.) Because of the complexity of computer design, the computer manufacturer must use a large number of CAD programs in nearly every aspect of design. Because there are numerous interconnections within and among chips and on and among pc boards, CAD programs for routing are extensively used.

The design of LSI or VLSI chips, which a semiconductor manufacturer wants to sell by the millions as off-the-shelf packages, would be the other extreme to the mainframe case above. Since the initial design and development cost can be spread over a large number of chips to be manufactured, the manufacturing cost of a single package is the dominant factor in the selling price of each package [i.e., the first term in Eq. (2.3.1) is much smaller than the second term]. Hence designers want to minimize the chip size with a full-custom design approach, using as many random-logic networks as possible. (Even a 10% size increase would be very significant financially.) Semiconductor manufacturers also use CAD programs extensively in many design stages, but probably less extensively than do mainframe computer manufacturers (or at least CAD is emphasized in different design aspects), because the major concern of the semiconductor manufacturers, chip size minimization, cannot be taken care of by CAD programs as well as it can be by hand design.

The situation of toy or watch manufacturers is very similar to the above. Even if the cost of an LSI chip is only a small percentage of the cost of the entire product, manufacturers want to save pennies by making the chip size as small as possible, since the entire toy or watch to be manufactured in very high production volume must be inexpensive in a highly competitive market.

Between the extreme cases explained above, there is a wide spectrum of different situations, resulting in different motivations. An example of such a situation is as follows. When a semiconductor manufacturer needs to design calculators with slightly different functions, the most economical approach is the incorporation of all the modifications into a single mask or into very few masks (usually for connection changes of PLAs and ROMs), keeping all other masks unchanged, because masks are time-consuming to prepare and expensive, and each calculator model may not justify the full custom design of all the masks (*Business Week,* Oct. 7, 1971, p. 50; Roberts, *Proc. 17th DA Conf.,* 1980).

Significance of Reduction of Design Time

The reduction of the design time not only results in the reduction of the design expenses, if the same number of people are involved, but it also has vitally important marketing aspects. In general a manufacturer who introduces a product first can grab a lion's share of the market and earn much greater total profits during the entire product life than latecomers, as illustrated in Fig. 9.1.1. The latecomers can only get the remaining customers and also must suffer price erosion. If the market entry of the second product is delayed longer, the situation is further aggravated. Thus quick design may be much more important than deliberate design, depending on the type of products (Davidow, 1974; Kidder, 1981). If products incorporate LSI/VLSI chips, such as intelligent terminals, these products will have small production volumes, at least initially, and the cost of these chips will be high [according to Eq. (2.3.1)]. But these chips need to be designed and developed quickly at the sacrifice of chip size or performance. Even if expected production volume is sufficient to justify full-custom design, many products need to be introduced quickly into the market. But high-volume off-the-shelf IC packages for semiconductor manufacturers, such as microprocessors, are exceptions, because chip size (consequently cost) or performance are important selling points.

LSI/VLSI chip design and, in particular, layout are time-consuming. Each draftperson can lay out 5 to 10 devices per day for a wide variety of layout techniques. [In contrast, an average programmer can produce about 10 lines of code per day (Allison 1980)]. This includes the time to draw, check, and correct a layout. Of course, precise estimation of layout productivity is difficult because productivity can differ by about 5 times among different persons and also because random-logic networks in modular structures (such as counters) are less time-consuming than nonmodular random-logic networks, while memories are less time-consuming than these modular logic networks. (Many general-purpose or dedicated processor chips, including those for pocket calculators, contain 30 to 40% random-logic networks on a chip, with memories for the remainder. Roughly speaking, the layout for random-logic networks is 10 times more time-consuming than the layout of the memory portion.) Figure 9.1.2 shows the trend of man-years required for most advanced LSI/VLSI chip design. The man-power required for LSI/VLSI chip design is rapidly increasing as the integration size increases. Thus the design time is getting longer because we cannot design LSI/VLSI chips of greater size by an increasingly larger number of people. (If we assign too many people, good chips could not be designed because of communication difficulties during design.)

If the design time is too long, manufacturers will not be able to introduce the new product, missing a chance to enter the market forever. Thus the design time must usually be at most a few years. In particular, when the first product is greatly successful in market acceptance, competitors are pressured to introduce second products in a shortest time (Kidder 1981).

The consideration of the design time and cost is a very unique problem of LSI/VLSI system design and, without it, it is not possible to design LSI/VLSI chips because design techniques and methodology are determined by them. In the case of conventional manufacture with discrete components,

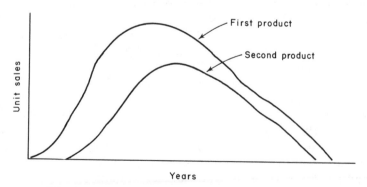

Fig. 9.1.1   Sales advantage of first entry to market.

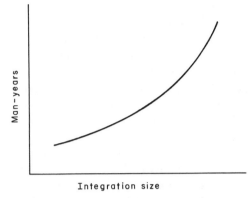

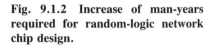

Fig. 9.1.2 Increase of man-years required for random-logic network chip design.

there is essentially no choice in the design approaches, and the design time is reasonably short (because the packing of components is not as stringent as that on chips, the replacement of components for design changes or corrections is quick and inexpensive, the number of components is generally much less, and other aspects are also simple). What chip designers do in packing devices on an LSI/VLSI chip is analogous to what factory workers on the assembly line used to do in assembling discrete components in the case of conventional manufacture. Thus, chip designers inherit the time and cost consideration which production engineers used to do for assembly workers.

In the following sections we discuss semi-custom design approaches to reduce the time and cost of chip design, compared with off-the-shelf packages and full-custom design approaches. Henceforth full- and semi-custom designs will be collectively called **custom design**.

9.2   Design with Discrete Components and Off-the-Shelf Packages

When digital systems are to be manufactured in low production volume, or frequent design changes are expected, they are usually designed with discrete components and off-the-shelf packages. (The design with off-the-shelf microprocessor chips is discussed later because of the different nature in task realization.)

When the design of a digital system with discrete components and off-the-shelf packages is compared with the use of custom-design LSI/VLSI packages, the initial investment per unit for the former is generally much lower than that for the latter. But as the production volume increases, the manufacturing cost per unit for the former [i.e., the second term in Eq. (2.3.2)] is much higher than the cost per package for the latter [i.e., the second term in Eq. (2.3.1)]. If a digital system is designed mostly with SSI packages with 4 gates each, for example, we have to add the cost of many other things to the average cost of a gate, as listed in Table 9.2.1, since a large number of packages must be assembled on large pc boards in cabinets, and large power supplies, which are expensive, as well as cooling equipment, must be provided due to high power consumption (Monolithic Memories, Application Note 110). Thus in the case of design with discrete components and off-the-shelf packages, the system cost per gate is a few times higher than the raw cost of a gate. If the same system is designed with custom-design packages, all these additional items are unnecessary or insignificant, and the system manufacturing cost per gate is a fraction of a penny. (When the system requires several or fewer off-the-shelf SSI or MSI packages, there is no advantage in designing with a custom-design LSI/VLSI package.)

Thus the manufacturing cost per unit for the design with discrete components and off-the-shelf packages is much higher than the cost if custom-design packages are used, but with low production volumes it is still an insignificant factor in the unit cost formula, Eq. (2.3.2) [i.e., compared with the first term in Eq. (2.3.2)].

It appears difficult to find the threshold production volume, under which the design with discrete components and off-the-shelf packages should be used. In the case of large computers, where logic networks are only part of the computers, it is complicated since the cost-to-performance ratio and the costs of other parts, such as memories, must be considered. The number of units manufactured for each model of mainframe computer is usually on the order of hundreds to thousands. (The numbers are much smaller for some models, such as the IBM 360/195, which is on the order of tens.) As the price has been decreasing since the late 1970s, the number has increased, on the order of tens of thousands for some mainframes. But these numbers are still much smaller than those of many other products. Many mainframe computers and minicomputers were designed with discrete components and off-the-shelf packages until the late 1970s (Eberlein 1979).

Table 9.2.1   SSI System Manufacturing Cost per Package[a]

IC (4 gates/package)	$0.25
PC board	0.27
Incoming inspection	0.07
Component insertion	0.05
Board test	0.18
Connections	0.05
Capacitors	0.04
Wiring	0.12
Power supply ($1.00/watt)	0.05
Cabinetry and fans	0.17
	$1.20

[a] Courtesy of Monolithic Memories, Inc.

9.3 Full-Custom Design Approach

Among many design approaches, the full-custom design approach, using random-logic networks, yields the highest speed or the most compact chip, whichever we want to achieve.

The full-custom design approach requires the largest initial investment compared with the other design approaches. The logic design of random-logic networks is more time-consuming than that of other cases, such as PLA minimization by deriving minimum switching expressions. The layout, however, is the most time-consuming part of the full-custom design approach, and far more time-consuming than other approaches since devices and interconnections must be packed as close as possible by hand. Everything is so closely packed that the correction of errors is cumbersome and time-consuming. (If one device is omitted by mistake, designers may have great difficulties in squeezing it into a chip.) Also because of the random nature of the layout configuration of devices and interconnections on a chip, it is time-consuming to check the design. When we want to design an LSI chip of 3,000 gates, for example, layout means the handling of 300,000 rectangles per chip, assuming 100 rectangles per gate. Thus layout is cumbersome. Also layout can be done only at a slow pace, with only 5 to 10 MOSFETs a day by one person, if the time to draw, check, and correct the layout is included. Thus for a large LSI chip, say, 10,000 MOSFETs, at least several people must be assigned to the design and layout of the chip, taking 20 to 40 months easily. In this case the entire chip area is divided into subareas, each of which is designed and laid out by a smaller group of people. Well over one million dollars would easily be required as the initial investment. When the integration size increases, the expenses go up sharply.

CAD is of little help for chip area minimization, because global minimization, by designing networks that match the shapes and sizes of usable subareas of the chip, can be done better by human visual inspection than by CAD. But for other aspects, such as logic simulation, layout verification, and testing, CAD has been used extensively.

Actually, a chip whose entire area consists of only random-logic networks is becoming rare when the integration size is large.* The area percentage of random-logic networks in LSI chips, which is 30 to 40% of the entire area in many chips designed in the late 1970s (the remainder is regular-structured networks, i.e., RAMs and ROMs) will decrease in VLSI chips with greater

---

* The Z8000, which does not contain any regular-structured networks (i.e., no memories or PLAs) would be the last LSI chip with random-logic networks alone for that large an integration size (about 17,500 MOSFETs). But the use of regular-structured networks appears to have existed in industry for many years, as exemplified by the 8-bit single-chip microcomputer SC/MP introduced by National Semiconductor Corporation in 1974, which consisted of predominantly regular-structured networks (including 4 PLAs).

integration sizes to reduce layout time [e.g., *Proc. 17th DAD Conf.,* 1980, p. 345]. (RAMS and ROMs are definitely regular-structured networks, but the border line between **regular-structured networks** and **random-logic networks**, defined in Section 4.2, is actually vague. Although PLAs are a special type of ROMs, they are sometimes regarded as random-logic networks, but they should be called regular-structured networks because of the ease of layout. Modular networks based on random-logic network modules, such as counters, may be called either random-logic or regular-structured, but they are usually called random-logic networks.) Also parts of random-logic networks often have a modular structure, such as counters and ripple adders (cascading full adders). The layout time of memories can be about 10 to 15 times faster than for random-logic networks. Among random-logic networks those in a modular structure can be laid out faster than others, since the layout of one module can be repeated. Another advantage of using regular-structured networks is the reduction of logic design time by extensively using CAD programs, such as minimization programs for PLAs, or by microprogramming with ROMs. It is, however, important to notice that even if regular-structured networks are more extensively used on a chip, random-logic networks are still important for functions that are critical for the performance of the chip. So random-logic networks will be used at least partly in VLSI chips and also will continue to be used possibly predominantly or totally in LSI chips of smaller integration sizes.

The most compact chip derived by the full-custom design approach means that if the production volume is very large, that is, at least 100,000 or definitely 1,000,000, the full-custom design approach yields the lowest cost (*Electronic Design,* Aug. 3, 1972, p. 42). But if the production volume is low, that is the most expensive approach because of the large initial investment. Because of the compact size, the best performance in terms of speed and power consumption can be obtained.

The off-the-shelf microprocessor chips, such as the 8080, Z80, Z8000, 6800, 68000, and TMS 1000, were developed by semiconductor manufacturers with this approach, since they wanted to have the best performance and the smallest chip size, and yet they expected large production volumes, probably on the order of ten millions or a quarter to a half billion for some microprocessor chips during their product lives (*EDN,* Nov. 5, 1980, pp. 98, 99).

References: Lattin 1979; Kroeger and Tozun 1980; *Electronics,* Feb. 10, 1981, p. 100.

Symbolic Layout

As the integration size increases by LSI/VLSI technology progress, the layout time for random-logic networks is becoming a serious bottleneck of LSI/VLSI chip design. In order to reduce time, a CAD approach called

symbolic layout, such as SLIC, STICKS, and others, has been developed by many firms. (The term "symbolic layout" is used with other connotations in semi-custom design approaches discussed later. So the readers should be cautioned not to be confused.) Although this is not complete design automation, if a designer simply specifies the positions of diffusions, metal con-

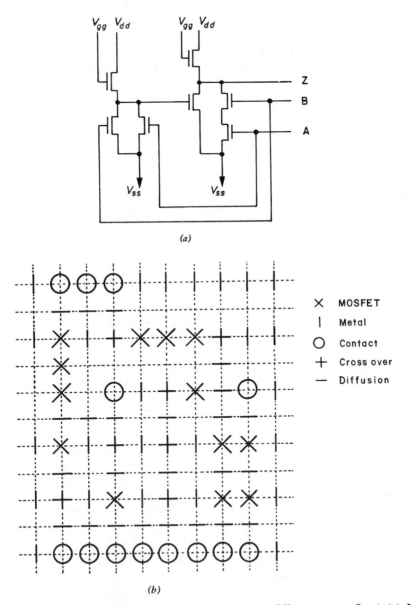

(a)

(b)

Fig. 9.3.1 Symbolic layout. (Courtesy of American Microsystems, Inc.) (*a*) **MOS network.** (*b*) **Symbolic layout for (*a*).**

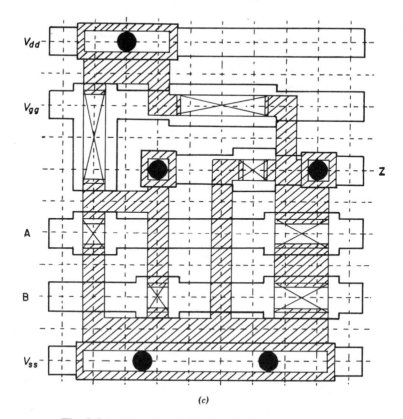

Fig. 9.3.1   (*continued*) (*c*) **Topological layout for (*b*).**

nections, contacts, MOSFETs, and crossovers, without worrying about complex details of layout rules, the symbolic layout program automatically makes a layout in compliance with the layout rules, packing devices as closely as possible. As an example of a symbolic layout program, Fig. 9.3.1 shows **symbolic layout of ICs (SLIC)** by AMI (American Microsystems, Inc.). A symbolic layout for the MOS network in Fig. 9.3.1*a* is done by a designer, as shown in Fig. 9.3.1*b*, where × means a MOSFET, 1, 0, and − mean a metal layer, a contact window, and a diffusion region, respectively, and + means a crossover. Then the SLIC program prepares the topological layout shown in Fig. 9.3.1*c*. (Some symbolic layout programs such as STICKS automatically try compaction, though others such as SLIC do not.) Since a designer has to prepare the symbolic layout of Fig. 9.3.1*b*, the layout is still time-consuming, but the layout time is greatly reduced. Although it is difficult to obtain reliable figures, it is estimated that the layout time can be reduced by 2 to 3 times, and if the time for repeated corrections is considered, by about 10 times. The chip size designed by the symbolic layout cannot be smaller than hand-drawn chips and may be larger by 15 to 30%, or even 50 to 100%. (These figures may not be reliable, because they depend on whose hand design is compared. Even in the case of good designers and draftpersons it is said that there is easily a 5 to 10% difference in the sizes of hand-drawn chips.) But global minimization by the use of symbolic layout on a CRT screen is questionable. In this sense the design based on symbolic layout should be called **pseudo-full-custom design**, though it is often called full-custom design.

The details of symbolic layout are highly dependent on wafer processing technology, so whenever a new processing technology is developed, a symbolic layout program has to be updated, although some firms are trying to develop symbolic layout programs to minimize the influence of technology.

References: Nance, 1978; Larson 1978; Gibson and Nance, 1976; Williams, 1978; Rosenberg, 1980; Cho et al., 1977; Marshall and Waller, 1980; Loosemore, 1980; *Electronics*, July 20, 1978, pp. 125–128; Dec. 21, 1978, p. 64; Apr. 10, 1980, p. 40.

9.4   Semi-Custom Design Approaches

When a full-custom design approach is too time-consuming, or not justifiable financially due to low production volume, and when off-the-shelf packages are too bulky (and consequently too expensive because of the many additional pc boards and large power supplies, or too slow because of long connections), then the semi-custom design approaches are a compromise solution.

Currently the semi-custom design approaches utilize gate arrays, cell library, and others including PLAs. Since PLAs are discussed in Chapter 7, they are not discussed here. Any logic network can be designed with PLAs only, but this is not usually done.

9.4.1   Gate Arrays

A **gate array** is an LSI chip on which gates are placed in matrix form without connections. But strictly speaking, cells, each of which consists of unconnected devices, are arranged in matrix form, and each cell can realize one of certain types of gates by connecting these devices. Then by connecting these gates, networks can be realized. Only 2 or 3 masks for connections and contacts have to be custom-made, instead of all the masks as for full-custom design. Also because only the connection layout, along with the placement of gates, needs to be considered, CAD can be effectively used, greatly reducing the layout time. Thus the design with gate arrays is very inexpensive and quick, compared with the full-custom design. Gate arrays of ECL and TTL have been extensively used in mainframes and minicomputers (*Business Week,* Apr. 21, 1980, pp. 134D–J).

Gate arrays are also called **master slices** or **uncommitted logic arrays**.

Before discussing advantages and disadvantages of gate arrays, we show simple examples of TTL gate arrays.

TTL Gate Arrays

The 112-gate array offered by Motorola is a typical example based on TTL gates (Motorola, 1972). On a single chip 112 cells, each of which consists of three bipolar transistors and five resistors, are arranged in a matrix without connections among them, as shown in Fig. 9.4.1. Of these cells, 98 have 4 inputs each, and the remaining 14 cells have 8 inputs each. The topological layout of a cell with 4 inputs is shown in Fig. 9.4.2. Resistors R_3, R_4, and R_5 are connected in series with common taps to reduce the required area. The topological layout of a cell with 8 inputs is the same, except that the input cluster is elongated to provide the additional emitters, consequently increasing the cell size. The expander shown in Fig. 9.4.3 can be realized, using some of the components of Fig. 9.4.2. Several electronic circuit var-

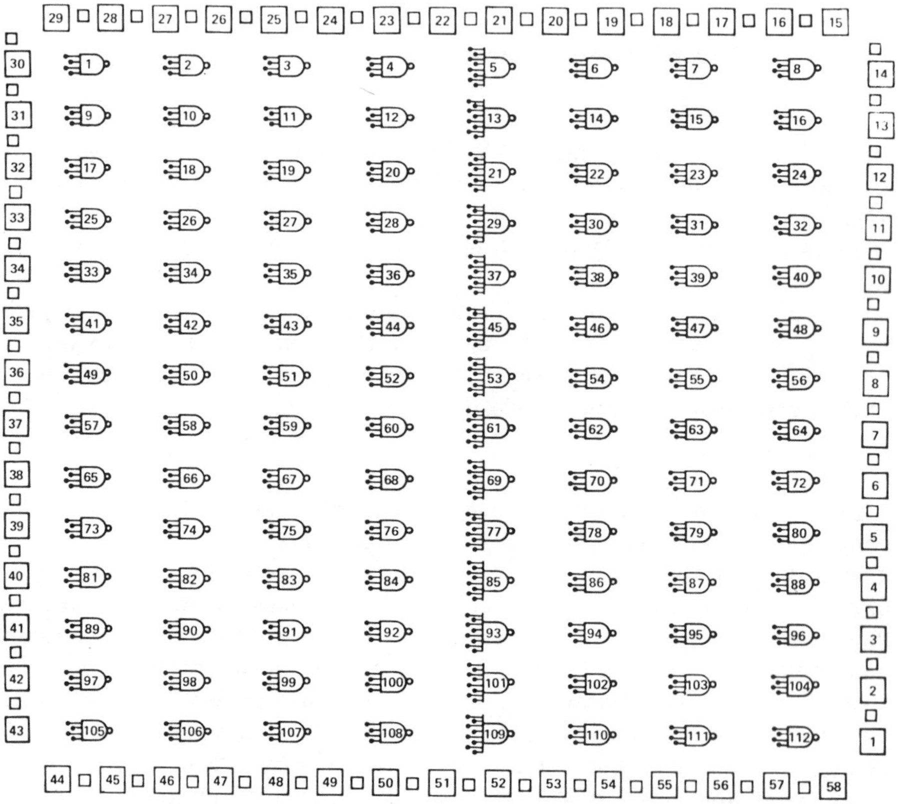

Fig. 9.4.1 **112-gate TTL array. Each gate symbol denotes a cell. (Courtesy of Motorola Semiconductor Products, Inc.)**

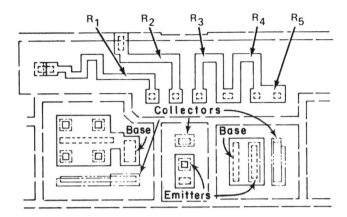

Fig. 9.4.2 **Topological layout of a cell with 4 inputs. (Courtesy of Motorola Semiconductor Products, Inc.)**

iations are possible for each cell by modifying metal connections, as shown in Figs. 9.4.4, 9.4.5, and 9.4.6:

Low-Power Gate (Fig. 9.4.4). The low-power gate is used only to drive other gates in the chip (i.e., as an internal gate).

High-Power Gate (Fig. 9.4.5). The high-power gate may be used as an internal gate with a number of fan-out connections or an external gate. (It should be located near the ground bus on the third metal layer whenever possible.)

High-Power Gate with Totem Pole (Fig. 9.4.6). The high-power gate with totem pole is constructed by using two cells. It is used when we need to send out signals over long connections on a pc board, or when we have a large number of fan-out connections.

As explained above, a user can design TTL networks by making connections among the cells. The AND-OR-Invert can be realized by connecting expanders to the expander nodes of gates, or by tying together the outputs of gates (except high-power gates with totem pole).

In 1978 a gate array of 704 Schottky TTL gates was adopted by IBM in the system/38 series and 4300 series computers (*Electronics,* Mar. 15, 1979,

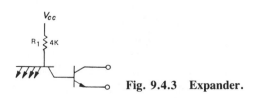

Fig. 9.4.3 **Expander.**

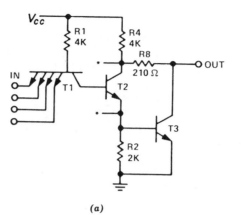

(*a*)

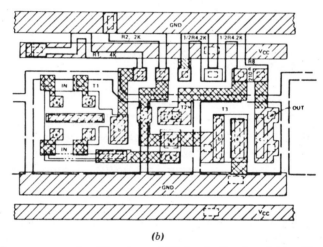

(*b*)

Fig. 9.4.4   Low-power gate. (Courtesy of Motorola Semiconductor Products, Inc.)
(*a*) **Electronic circuit (∗ denotes nodes for expanders).** (*b*) **Layout for (*a*).**

pp. 105–108). Digital Equipment Corporation and other manufacturers are also using TTL gate arrays (Armstrong 1981; *Electronic Design,* Nov. 22, 1980, pp. 39, 40; Cane 1980).

In 1980 an experimental gate array of 7,640 TTL gates on a chip of size 7 × 7 mm was announced by IBM, which is to realize an 8-bit version of the IBM 370/138 central processor unit (cpu) (which has a 32-bit architecture) by using 5,000 of the 7,640 TTL gates on this chip. Let us describe the statistics, which show important characteristics of gate arrays, since such statistics are rarely published. The processing technology used for this gate array is almost identical to the one developed for the IBM 4300 series computers, with the features of recessed oxide isolation and three metal layers for interconnections. For this, however, the line width is scaled down from 3 to 2.5 μm. The array has 96 rows, 4 of which contain receiver gates (to receive signals from outside the chip), and 92 columns, including 12 for driver gates (to send signals off the chip), as shown in Fig. 9.4.7. Only 65% of the 7,640 TTL gates are used to realize the IBM 370 cpu, and the unused 35% provides additional channels for interconnections. Each gate used provides 5 horizontal channels on the first metal layer and 10 vertical channels on the second metal layer. Each unused gate provides additional 4 horizontal channels. The third metal layer is used mainly for power supply and input/output connections. The placement of logic gates on the gate array by a CAD program took about 200 minutes on an IBM 370/model 168. The routing by a CAD program took about 90 minutes for all 11,000 connections, except 68 which were manually routed. But manual routing of these 68 and checking

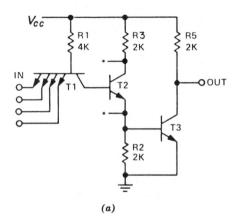

(a)

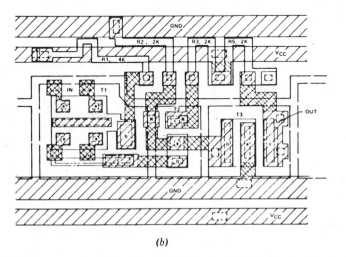

(b)

Fig. 9.4.5 High-power gate. (Courtesy of Motorola Semiconductor Products, Inc.)
(a) **Electronic circuit (* denotes nodes for expanders).** *(b)* **Layout for** *(a)***.**

took about 1 month. The placement and routing (including the use of CAD programs) took a total of two months, and this is much faster than the total manual placement and routing. The total interconnection length is 5.8 m, compared with 2.2 m (typical) for the 704-TTL gate array used for the 4300 and System/38 series. Since there are 4,437 routes on the chip, the average length of a connection is 5,800/4,437 m, that is, 1.3 mm. The statistics on the connection length, parasitic capacitance, and gate delay (when the gate is turned off) are shown in Fig. 9.4.8. Using a parasitic capacitance of 0.63 pF/mm for the first metal layer and of 0.2 pF/mm for the second, the capacitance statistics obtained are shown in Fig. 9.4.8*b*. About 6% of the routes have a parasitic capacitance greater than 3.0 pF, and gates that have such a large capacitance are provided with 4-kΩ output resistors, instead of 8 kΩ for other gates, in order to improve the speed performance by increasing the output current. Even with 4-kΩ resistors, 6% of the gates have a turn-off delay more than twice the average, as shown in Fig. 9.4.8*c*. But if 8-kΩ resistors were used, these gates would have a turn-off delay greater than three times the average (*Electronics,* Oct. 9, 1980, pp. 139–147).

References: EDN, Jan. 7, 1981, pp. 13, 14; Khokhani et al., 1981.

Advantages and Disadvantages of Gate Arrays
Gate arrays have the following advantages. Even when processing technology or electronic circuitry is changed, only one cell of devices (or a few different cells) needs to be carefully designed and laid out, and this layout can be repeated on a chip. After designers have designed logic gate networks (though this is still very time-consuming, the minimization of the number of gates under constraints such as maximum fan-out would be a designer's primary concern, and connections are a much less significant problem),

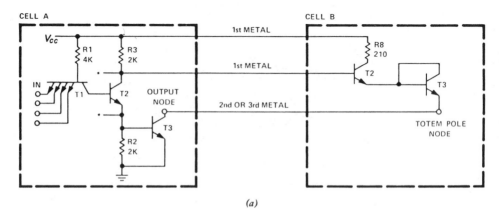

(*a*)

Fig. 9.4.6   High-power gate with totem pole. (Courtesy of Motorola Semiconductor Products, Inc.) (*a*) **Electronic circuit (* denotes nodes for expanders).**

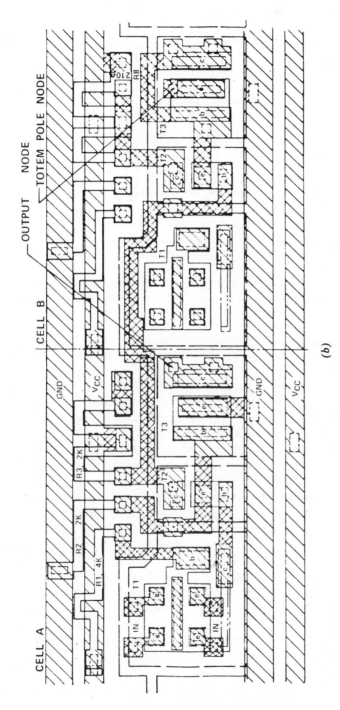

Fig. 9.4.6 (*Continued*) (*b*) Layout for (*a*).

(*b*)

371

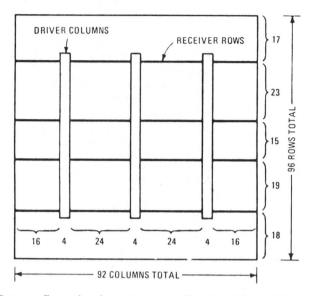

Fig. 9.4.7 Gate configuration in gate array. (Reprinted from *Electronics*, Oct. 9, 1980, pp. 139–147. Copyright © McGraw-Hill, Inc., 1980. All rights reserved.)

CAD programs automatically do the placement of logic gates and the routing of connections on a chip, though in complex cases placement, which is more difficult than routing, must be done manually by designers, possibly using a placement library of standard networks. (Usually a few percent of the connections cannot be processed by CAD programs and must be routed manually. Thus when the number of connections is very large, even a few percent means a large number of connections. So it is important to reduce this percentage. The manually processed 68 connections in the example in the previous paragraph constitute only 0.6% of the total connections.) It is to be noted that because the gate positions are prefixed on the array, CAD for placement and routing becomes much easier than otherwise. For the above reasons the layout time is greatly reduced, shortening the design time, and consequently design expenses. (Delivery time by vendors is usually several to a dozen weeks.) Only a few masks for connections and contacts must be custom-made for each special case, and all other masks are common to all cases, spreading the initial investment for the preparation of all these common masks over all cases. Thus gate arrays are cost effective in as low a production volume as hundreds or thousands. [In the case of a 32-bit computer VAX-11/780, the design time and cost through the use of gate arrays are reduced to 15% of those for full-custom design (Armstrong, 1981), and only 5 among 13 masks need to be custom-made (Cane, 1980).] Speed is improved over the network realizations with discrete components, SSI packages, or MSI packages because interconnections among gates are short-ened on the average (most interconnections are inside gate array chips rather than on pc boards). Some networks, which are expected to be changed frequently before final design, can be realized with connections made outside, using gates inside the chip as discrete gates, though this tends to slow down the speed and to increase the number of unused gates. Also the power consumption is reduced.

The first disadvantage of gate arrays is the large chip size because large spacing must be provided for connections between every pair of adjacent columns of gates, and also between every pair of adjacent rows of gates. A second disadvantage is the possibility of a high percentage of unused gates for the following reasons. Depending on the kinds of networks or which parts of a large network are placed in a gate array chip, all spacings provided for connections can be used up (by taking a detour if the shortest paths are filled up by other connections), or all the pins of the chip can be used up

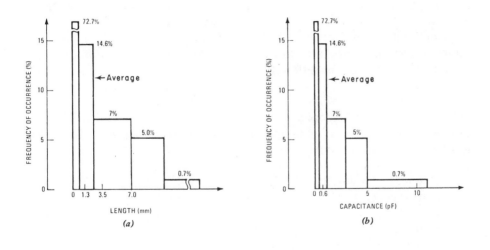

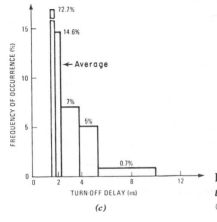

Fig. 9.4.8 Statistics. (Reprinted from *Electronics*, Oct. 9, 1980, pp. 139–147. Copyright © McGraw-Hill, Inc., 1980. All rights reserved.)

by incoming and outgoing connections. In either case many gates may not be used at all, and less than half the gates on a chip are used in some cases. Because of these disadvantages, the average size of a gate array chip is easily four or five times as large as that of a full-custom designed chip, or it can be even greater, for the same networks. The cost difference would be greater (the cost is not necessarily linearly proportional to chip size) for the same production volume. Another disadvantage is the difficulty of keeping gate delays uniform. [If delay times of gates are not uniform, the network tends to generate spurious output signals, as discussed in Section 6.7 of Muroga (1979).] In the case of full-custom design, the increase of gate delay by long or many output connections of a gate can be reduced by redesigning the transistor circuit. But such a precise adjustment is not possible in the case of gate arrays.

Responding to a variety of different needs of users in terms of speed, power consumption, cost, design time, ease of change, and possibly others, a large number of different gate arrays are commercially available from semiconductor manufacturers or are used in-house by computer manufacturers. Different numbers of gates are placed on chips, with different configuration capabilities for different IC logic families. Gate arrays of not only TTL but also ECL, TRL, IIL, *n*-MOS, CMOS, and CMOS/SOS are available with different logic capabilities. Since layouts are important for the capability of gate arrays, components are laid out in many different configurations by different manufacturers, as unique features of their gate arrays. In the following, let us outline some of them.

References: Twaddell, Apr. 5, 1980; Cox and Davis, 1980; Posa, Sept. 25, 1980; Kroeger and Tozun, 1980; Hartmann, 1981.

ECL Gate Arrays

Although the first commercial gate arrays made by Fairchild with DTL appeared in 1967, the significance of gate arrays was not widely recognized until ECL gate arrays were first used extensively by Amdahl Corporation in its 470 series computers in 1975. The 470 series computers were plug-compatible to the IBM 370 series computers which were some of the fastest general-purpose computers then available, except those for scientific computation. (**Plug-compatible** means that all software developed for these IBM computers can be run on the 470 series computers without modifications.) The 470 series computers outperformed the IBM computers in speed (about two times in the case of the 470/V6) by adopting ECL gate arrays, since the IBM computers did not use such large-scale integration. Amdahl's gate array chip consists of 100 ECL gates with 84 pins, and there are about 1,200 transistors and resistors per chip. (The switching time of each gate is 500 to 600 psec.) The 470/V6 computer contained about 2,000 gate array chips with 110 different chip types. Instead of having to generate more than 1,400 different masks, as would have been the case with a full-custom design approach, Amdahl actually needed less than 400 different masks with 3 custom-made masks for each chip type. (Each chip required a total of 13 masks, including the common masks.) Thus gate arrays are much more economical than the full-custom design approach, although they are not as fast as the latter. But their speed is much faster than that with IC chips of smaller integration sizes, judging by the fact that in an Amdahl computer about one third of the propagation delay is inside the chips and the rest outside. 42 chips are assembled on a multiple-chip carrier, and then 60 such carriers are mounted on cabinet frames (Clements 1979; Wu, 1978).

Since each model of a mainframe computer is manufactured in hundreds or thousands (some models were in tens), the full-custom design approach is not economically justifiable, though the speed improvement would be substantial. After Amdahl Corporation had adopted LSI based on the gate array semi-custom design approach in mainframes with commercial success, many computer manufacturers followed suit. In particular, IBM's adoption of gate arrays in 1978 and 1979 had a significant influence on popularizing the use of gate arrays in the design of computers for medium production volumes.

In 1978 and 1979 IBM started to use TTL and ECL gate arrays in mainframe computers. The number of masks per chip to be custom-made is reduced to 2 to 4 masks, from 12 to 16 masks for a full-custom designed chip (*1980 Electro*, 22/3, p. 1). Thus the cost-to-performance ratio of these computers is greatly improved. It is estimated that IBM cut its manufacturing costs by 75% or more by using gate arrays. Not only are inspection and assembly costs lower, but the costs of power supplies, cabling, and other system components are all substantially reduced (*Business Week*, Apr. 21, 1980, p. 134I).

The ECL gate array was adopted by IBM in the 4300 series computers. There are 748 internal cells in 34 rows and 22 columns. Since up to 2 gates can be realized at each cell, the chip contains up to 1,496 logic gates. Around the internal cells (i.e., in the periphery of the chip) there are 88 receiver circuits, 64 high-power driver circuits, and 2 reference voltage generator circuits. An internal cell is shown in Fig. 9.4.9. From this, ECL gates such as the one shown in Fig. 9.4.10 can be realized. By skipping connections,

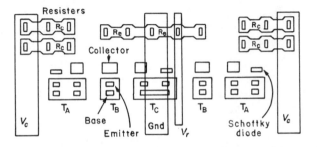

Fig. 9.4.9   Internal cell. (Blumberg and Brenner, 1979. Copyright © 1979, IEEE.)

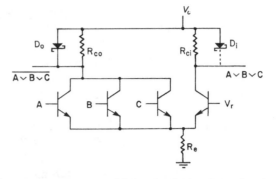

Fig. 9.4.10   ECL logic gate with clamped Schottky diodes. Schottky diodes are clamped only for collector-dotting. (Blumberg and Brenner, 1979. Copyright © 1979, IEEE.)

some transistors may not be used. Also Schottky diodes can be clamped when collector-dotting is used, as illustrated in Fig. 9.4.10. Three metal layers are used. Instead of placing pads in the periphery of the chip, as in conventional chips, solder bumps are placed in 17 rows and 17 columns throughout the surface of the chip, increasing the number of pads by 289 beyond the number of pads that a conventional chip can afford on its periphery.

References: Pomeranz et al., 1979; Blumberg and Brenner, 1979; Jen and Nan, 1980; Khokhani and Patel, 1977; *Electronics,* May 24, 1979, p. 118; Dec. 4, 1980, pp. 41, 42.

Interdesign, Inc., a subsidiary of Ferranti Electronics Ltd., offers many different gate arrays, including linear circuits and Ferranti's CML gate arrays.

Let us discuss a CML gate array made by Ferranti, where current mode logic (CML) is a variation of ECL. [Usually CML has the same feature as ECL in current steering from one transistor to another, but Ferranti's CML switches current on and off (*Spectrum,* Jan. 1980, p. 47).] The gate array chips come in a range of 100 to 1,000 gates of about 8 to 10 components per gate. Figure 9.4.11 shows a typical cell with 5 bipolar transistors, 3 resistors,

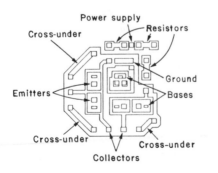

Fig. 9.4.11   CML cell. (Ramsey, 1980. Copyright © 1980 IEEE.)

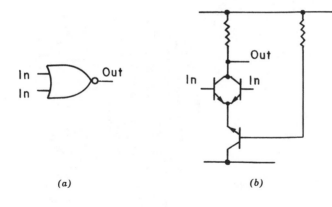

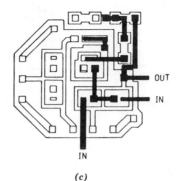

Fig. 9.4.12 2-input NOR gate and layout. (Ramsey, 1980. Copyright © 1980, IEEE.) (*a*) NOR gate. (*b*) Electronic circuit for (*a*). (*c*) Layout for (*a*).

and 3 crossunders. Figure 9.4.12 shows a 2-input NOR gate with a switching time of 8 nsec and a layout. Figure 9.4.13 shows a *D*-type latch and a layout (*Electronic Design,* Mar. 31, 1981, p. 211).

References: Ramsey, 1980, 1981; Slater and Cox, 1979; *Electronics,* Mar. 18, 1976, p. 69; Dec. 7, 1978, p. 65.

The **Macrocell Array** by Motorola is actually an extension of gate arrays discussed so far in the sense that in the case of the Macrocell Array, transistor-level logic design such as series-gating is possible, whereas in the case of gate arrays, logic design is based on logic gates and series-gating is difficult if not impossible. The Macrocell Array chip consists of 106 cells, organized as shown in Fig. 9.4.14. These cells are classified into three types, 48 major cells, 32 interface cells, and 26 output cells. Each major cell consists of two independent half cells, an upper half and a lower half. Each cell consists of unconnected transistors and resistors arranged in a fixed configuration.

Major cells are for logic capability and are placed in the internal area of

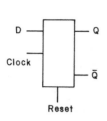

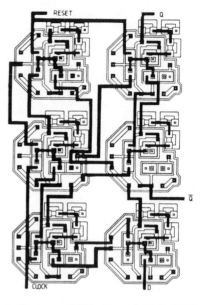

Fig. 9.4.13 *D*-type latch and layout. (Ramsey, 1980. Copyright © 1980, IEEE.)

$$\frac{(12.7)(43.8)}{(96)} = 9$$

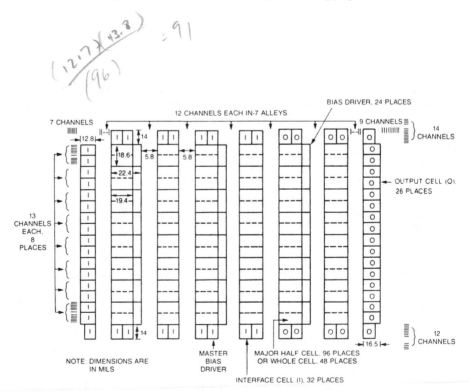

Fig. 9.4.14 Macrocell array. (Courtesy of Motorola Semiconductor Products, Inc.)

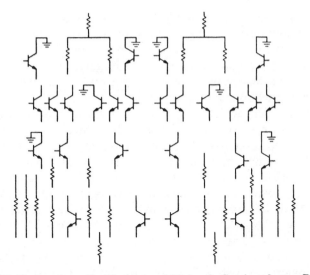

Fig. 9.4.15   Half of major cell. (Courtesy of Motorola Semiconductor Products, Inc.)

the chip. Each major cell consists of 52 bipolar transistors and 48 resistors, half of them shown in Fig. 9.4.15. ECL circuits can be realized by freely connecting these components, for example, by series-gating, collector-dotting, or emitter-dotting, as illustrated in Fig. 9.4.16. The power of transistor-level design, such as series-gating, is demonstrated in the 4-input Exclusive-

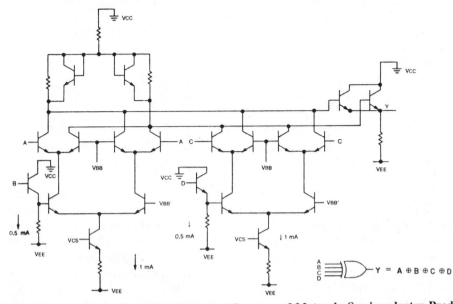

Fig. 9.4.16   4-input Exclusive-OR gate. (Courtesy of Motorola Semiconductor Products, Inc.)

OR gate in Fig. 9.4.16, for example. If this function is realized with ECL gates only without series-gating, eight 4-input AND gates and one 8-input OR gate in double-rail input logic would be required, along with 40 connections. [If only NOR gates are used, 10 NOR gates and 29 interconnections among them are required in single-rail input logic (Lai 1976).] But in Fig. 9.4.16 only 5 connections are required.

Each interface cell, which is for input interfacing, consists of 17 bipolar transistors and 13 resistors. Each output cell, which is to send high-power signal off the chip, consists of 15 bipolar transistors and 16 resistors. In addition to these cells, about 500 IIL gates are placed around the periphery of the chip, though they are not shown in Fig. 9.4.14. These I^2L gates can be used for diagnostic and chip control logic as an option to users.

Configurations of unconnected components to form standard networks such as flip-flops, adders, and decoders are stored in a computer memory and called **macros** (or **macrocells**). More than 100 macros are available. With a graphics terminal and a plotter, a customer can speed up the design by freely using macros, which is facilitated by this CAD program of Motorola. The Macrocell Array can contain an equivalent of 1,192 gates if full adders and latches are used in all the cells.

There are 94 vertical routing channels realized by the first metal layer and 108 horizontal routing channels realized by the second metal layer. Unlike gate arrays which require a large number of channels to interconnect individual gates, the Macrocell Array requires fewer channels, allowing high chip utilization.

Using Isoplanar technology, the switching time of a major cell is 0.9 to 1.3 nsec. The chip can handle up to 4 W.

Compared with the design with SSI or MSI packages, use of the Macrocell Arrays yields a reduction of up to 50:1 in package count, with a power consumption reduction of up to 5:1.

The Miniarray with a power dissipation of 2 W is a scaled-down version of Macrocell Array.

References: Prioste, 1980; Prioste et al., 1979; King, 1980; Blood, 1981, Sept. 8, 1981; Ranada, 1979; MECL 10K Macrocell Array Design manual, Motorola. For the Miniarray, see *Electronics,* Apr. 10, 1980, pp. 46, 48.

ECL gate arrays are available from other manufacturers, usually with CAD programs. For example, the F200 ECL gate array based on Isoplanar II and the F300 gate array (8 times more complex than the F200) based on Isoplanar S, which have transistor-level logic design capability, are available from Fairchild, and CAD programs for them are available through Control Data Corporation's Cybernet time-sharing network. A gate array with a 128-bit RAM is available from Siemens. The F300 is powerful, having PLAs, RAMs, and testing capability (i.e., the LSSD, testing method of sequential networks, proposed by IBM) on some chips of F300. For example, the Cray-

1, which is the fastest computer built in the late 1970s, consists of 450,000 ECL gates in 225,000 SSI packages (each package contains two 5-input ECL gates). It is estimated that the number of packages could be reduced to 150 by using the F3000, which would contain 3,000 gates each.

References: Goodman and Owens, 1980; Goodman, 1980; Owens, 1981; Hively, 1978; Ranada, 1979; Braeckelmann et al., 1979; Offerdahl, 1978; Posa, Sept. 25, 1980; Yano et al., 1980; *Spectrum,* Jan. 1981, p. 42.

Schottky TRL Gate Arrays

Megalogic announced by Motorola in 1975 is an LSI design approach based on gate arrays of the Schottky TRL (discussed in Section 3.5), supported by LSI packages for often used networks such as "error pattern register," "2-of-8 touch tone decoder," and "MOS dynamic memory refresh logic circuit." It is said that 335 SSI/MSI packages for a typical tape-drive controller can be replaced by 15 bipolar LSI packages in Megalogic, reducing the system cost by 30 to 50%. As a nucleus of Megalogic, there are two gate arrays of different sizes as follows:

XX	160	160 gates	48 pads	3 fan-in	5 fan-out	480 mW/chip
XC	400	400	74	3	5	1200 mW/chip

Both consist of Schottky TRL gates with 3 inputs. A typical gate delay is 25 nsec when used inside a chip.

Each cell consists of four resistors and one transistor, unconnected as shown in Fig. 9.4.17. To achieve a high packing density, the gates are configured as a cell pair, as shown in Fig. 9.4.18*a*. The basic cell metal layer for Fig. 9.4.18*a* is shown in Fig. 9.4.18*b*, where the shaded areas indicate optional paths which can be used for connections after the gate components have been connected. The cell pairs are arranged, as shown in Fig. 9.4.19, in 10 rows of 10 cell pairs (or 20 gates) in the case of the 400-gate array, which has a total power dissipation of 1200 mW. A 160-gate array is also available. A library of macros contains good connection configurations for

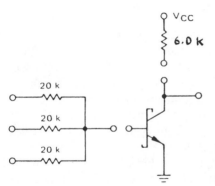

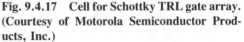

Fig. 9.4.17   Cell for Schottky TRL gate array. (Courtesy of Motorola Semiconductor Products, Inc.)

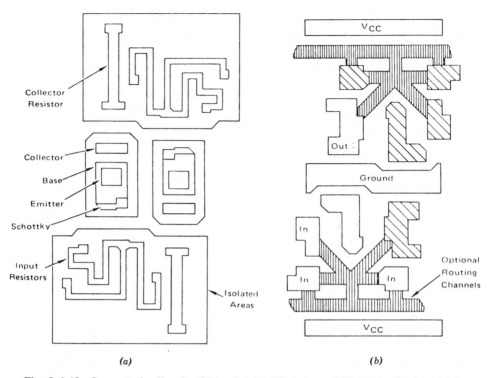

(a) *(b)*

Fig. 9.4.18 **Layout of cell pair of Fig. 9.4.17. (Courtesy of Motorola Semiconductor Products, Inc.)** *(a)* **Cell pair.** *(b)* **First metal layer for** *(a)***.**

standard networks such as flip-flops. Special standard networks, such as error-checking networks for cassette or floppy disk memories and data communications, need not be realized with the above gate arrays in Megalogic, but they are available as off-the-shelf LSI packages in TTL, TRL, or IIL, which can be connected to the above gate arrays.

Reference: Motorola, 1975.

I²L Gate Arrays

Gate arrays of I²L gates are available from many manufacturers. Figures 9.4.20 and 9.4.21 show I²L gate arrays of Exar Integrated Systems, Inc., where the switching time of I²L gates is in the range of 15 to 100 nsec, depending on the gate current value, and 3 masks are custom-made. The D-type flip-flop in I²L shown in Fig. 3.5.6 is laid out in Fig. 9.4.20. Figure 9.4.21 shows the configuration of the entire gate array where 288 I²L gates are shown in the center. An I²L gate array from Signetics contains 2,000 gates (*Electronics,* May 25, 1978, pp. 41, 42).

The F9480 I³L gate array introduced by Fairchild in 1980 consists of 4,000 I³L gates, 80 large-fan-out drivers and tristate Schottky-TTL input/output

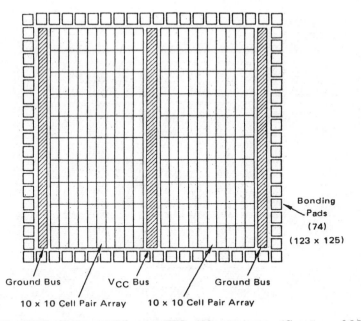

Fig. 9.4.19 **Configuration of Schottky TRL 400-gate array. (Courtesy of Motorola Semiconductor Products, Inc.)**

buffers, which interface between I^3L and TTL voltage levels. With Isoplanar technology the switching time of an I^3L gate is improved to 6 nsec at a gate current of 100 μA, but still slower than ECL. I^3L is good in terms of delay–power product and packing density.

Gate arrays of integrated Schottky logic (ISL) are also available. An ISL gate has a typical switching time of 3.5 nsec, whereas I^2L and low-power Schottky TTL gates have typical switching times of 15 and 10 nsec, respectively. The delay–power product of ISL is typically 1 pJ, whereas those of I^2L and low-power Schottky TTL are 4 and 20 pJ, respectively. It is

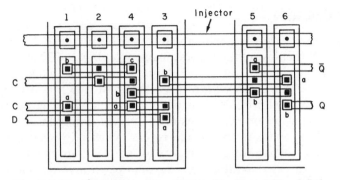

Fig. 9.4.20 **I^2L layout of the D-type flip-flop of Fig. 3.5.6.**

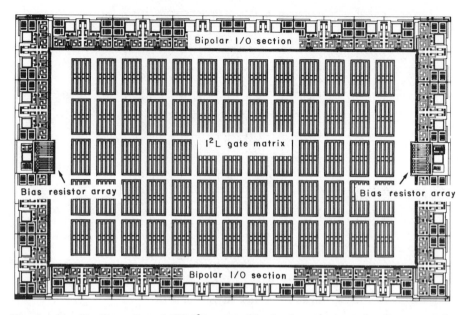

Fig. 9.4.21   Configuration of 288 I²L gates. Bipolar input/output circuits occupy the periphery. (Courtesy of Exar Integrated Systems, Inc.)

claimed that one ISL gate array of 1200 gates can replace more than 50 MSI packages at 20 to 50% less cost.

Gate arrays of Schottky transistor logic (STL) are available from Texas Instruments, with a typical switching time of 2.5 nsec and efficient CAD.

References: O'Neill, 1979; Exar, 1979; Crippen and Hingarh, 1980; Lau, 1979; Davis, 1981; Deutsch and Glick, 1980; Hightower and Alexander, 1980; Brown, 1980; Krautner, 1978; Hightower and Horton, 1980; Katz, 1980; Hingarh, 1980; Chin et al., 1980; Hightower and Roberts, 1981; *Electronics,* Jan. 13, 1981, pp. 41, 42; Horton et al., 1981.

CMOS Gate Arrays

CMOS gate arrays are also commercially available from many manufacturers in slightly different layout forms. As an example, Fig. 9.4.22 shows a CMOS gate array (with metal gate) from International Microcircuits, Inc. The *p*-channel and *n*-channel MOSFETs (in a group of two MOSFETs and also a group of three MOSFETs) laid out in horizontally long rectangles are arranged in two rows on both sides of the power supply lines V_{dd} and V_{ss}. The dotted lines indicate thin oxide layers. Vertically long rectangles arranged in each row are for crossunders for connections among MOSFETs. Figure 9.4.23 shows two alternative layouts of an inverter and one layout of a 2-input NOR gate. Figure 9.4.24 shows a layout of a *D*-type master–slave flip-flop using crossunders. Gate arrays of different sizes ranging from 60 to 3,500 gates are available from this firm.

RCA has developed a CMOS/SOS gate array of 800 gates, with delay time of 2.5 nsec each.

References: Kroeger and Tozun, 1980; Allen, 1979; Posa, Sept. 25, 1980; Noto et al., 1980; Ashida et al., 1979; Young, 1981; Goldsmith, 1981; Chen et al., 1981; *Electronics,* Sept. 11, 1980, p. 33; Sept. 25, 1980, p. 154; Jan. 13, 1981, pp. 163–166; Jan. 27, 1981, pp. 44, 47; Apr. 7, 1981, p. 62; May 19, 1981, pp. 86–88.

n-MOS Gate Array

One cell of an *n*MOS gate array available from Interdesign, Inc., is shown in Fig. 9.4.25*a,* which realizes one depletion-mode load MOSFET and 4 enhancement-mode driver MOSFETs in Fig. 9.4.25*b*. The entire gate array, called Monochip MD-A, contains 224 of these cells arranged in 8 rows in the center of the chip. The 38 cells for input/output buffers are placed along the periphery among 40 pads. Each cell can realize either a static or a

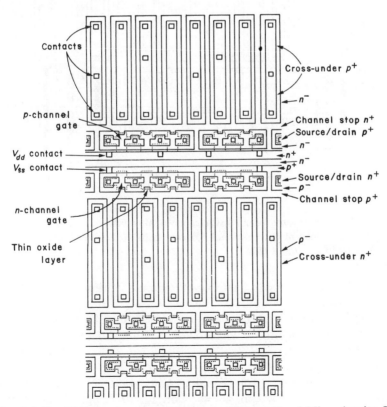

Fig. 9.4.22   CMOS gate array. (Courtesy of International Microcircuits, Inc.)

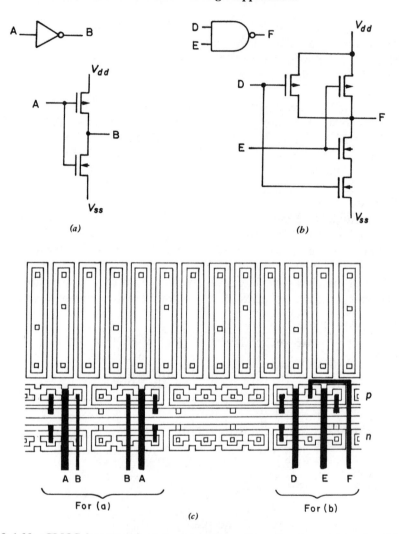

Fig. 9.4.23   CMOS layouts of an inverter (two alternative layouts) and a 2-input NAND gate. (Courtesy of International Microcircuits, Inc.) (a) Inverter. (b) NAND gate. (c) Layouts for (a) and (b).

dynamic MOS circuit. As an example of logic realization, Fig. 9.4.26 shows a layout of a 4-input NOR gate.

The gate array of 920 DMOS gates by Mitsubishi Electric has a different arrangement for connections, where DMOS is a special type of *n*-MOS, as discussed in Section 4.10.

References: Stephan, 1978; Tanaka et al., 1981; Nakaya et al., 1980; *Electronics,* Feb. 16, 1978, pp. 46, 48; Oct. 12, 1978, pp. 67, 68; Tomisawa et al., 1978.

General Comments on Gate Arrays

Many other gate arrays with different features are available from manufacturers, many of which are specialized in custom-design services rather than being semiconductor or computer manufacturers. These firms are developing CAD programs to facilitate the design.

Many factors influence the overall efficiency of a gate array, such as the number and performance of transistors available at each cell, the number of high-power gates to send the signal off the chip, the number of pads, routing channel space, and the number of available crossunders. Tradeoff considerations are not simple. For example, if routing channels between adjacent rows or columns of cells are wide enough, connection routing is

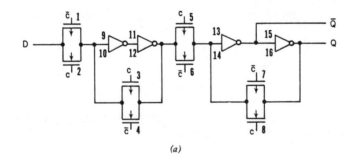

(a)

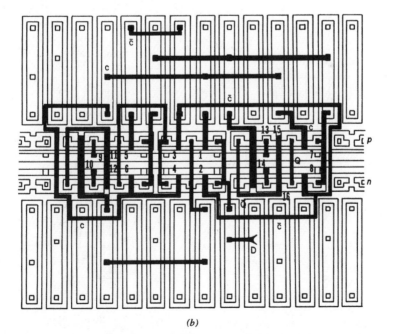

(b)

Fig. 9.4.24 **CMOS layout of *D*-type master–slave flip-flop. (Courtesy of International Microcircuits, Inc.)** (*a*) *D*-type master–slave flip-flop. (*b*) Layout for (*a*).

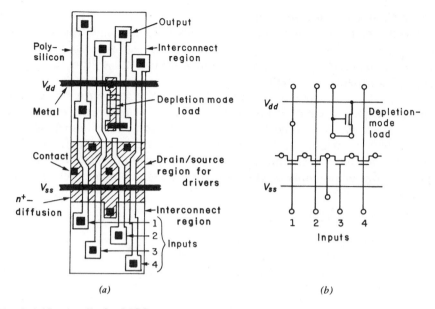

Fig. 9.4.25   A cell of *n*-MOS gate array. (Courtesy of Interdesign, Inc.) (*a*) Cell of *n*-MOS gate array. (*b*) Electronic circuit of (*a*).

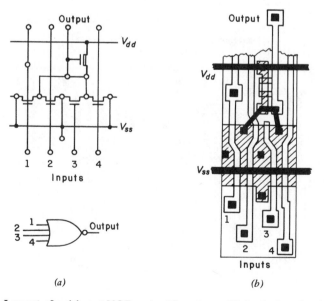

Fig. 9.4.26   Layout of a 4-input NOR gate. (Courtesy of Interdesign, Inc.) (*a*) 4-input NOR gate. (*b*) Layout for (*a*).

facilitated, but packing density is sacrificed. If routing channels are too narrow, many channels are filled up with connections, and then we have many detours, slowing the speed, or many gates may not be used. Also many applications need abundant pads, by which packing density is sacrificed. The number of pads required for the number of gates on a chip can be found by Rent's rule, described in Section 5.1. (This means that the ratio of the number of pads to the number of gates is about 1:10 for a 2,000 gate array.) Since it is difficult to increase the number of pads on a chip drastically, we need to add useful networks and memories that are already compactly laid out on a chip, if we want to have more powerful gate arrays in the future. Another possibility is the use of cells that perform complex logic operations, replacing the gates that perform simple logic operations such as NAND. [One example along this line is the use of 3 MOSFETs connected in series, as illustrated in Problem 9.5, though this may not be appropriate for high speed (Tomisawa et al., 1978). Another example is the use of universal logic gates, which can perform one out of a certain range of logic operations, selected by appropriate input connections of each cell (Hurst, 1980).]

The efficiency and also the quality of placement and routing of CAD programs are very important. When many detouring connections result with CAD application, these connections must be routed by hand later. Even if the percentage of such connections is very small (say a few percent), their number is very large for large gate arrays, and the manual correction is time-consuming.

References: Twaddell, Apr. 5, 1980; Hartmann, 1981; Hartmann et al., 1981; Franson, Feb. 5, 1977; Posa, Sept. 25, 1980; Kroeger and Tozun, 1980; Kamikawai et al., 1976; Preparata, 1972; Hage and Krueger, 1980; Armstrong, 1981; Ueda et al., 1978; Pitts, 1981; Cane, 1980.

9.4.2 Cell-Library Design Approaches

Compact layouts for basic gates (such as NOR gates) and standard networks are designed by good designers with time-consuming efforts and are stored in computer memories (i.e., magnetic disks or tapes). Such a collection of layouts is called a **cell library,** and the layouts are called **cells.** Once a cell library is ready, any designer can call up specific cells on the screen of a CRT terminal linked to a computer. By arranging them and adding connections among them by the use of a light pen or keys, he can make a layout of the entire network. When the layout is complete, photomasks are automatically prepared by computer. Such a design approach is called the **cell-library design approach**.

Polycell Design Approach

When every cell has the same height, though its width may be different, this approach is called the **polycell design approach (or standard-cell** (or **array)**

design approach, or **building-block design approach**). Although it used to be called the cell-library approach, and often still is called so, it is now called the polycell or standard-cell design approach in order to avoid confusion with the words "cell" and "library" from those which are used in Section 9.4.1, Gate Arrays.

Polycell is commercially available from Motorola and is a typical example with a library of about 100 MOS cells. (In this case Polycell is a tradename.) Micromosaic of Fairchild is another commercial example. Bell Labs has a very extensive polycell layout system with CAD programs, such as LTX for in-house use. This is not only for n-MOS, but also for TTL, ECL, and I²L with thousands of cells. A layout of a chip with 500 cells takes about 1 week. Figure 9.4.27 shows an example of a polycell layout.

Routing is the major problem in the polycell design approach and is greatly facilitated by CAD programs such as LTX. Routing is mainly done in each channel between two adjacent rows of cells, as shown in Fig. 9.4.27.

Connections from one routing channel to others can be done through narrow passages, which are provided between cells in each row. In some cases connections are allowed to go through cells horizontally or vertically (without going through the narrow passages provided) in order to reduce wiring areas. In this case detailed routing rules on where in each cell connections are allowed to go through must be incorporated in the CAD programs.

There is a great variety in the quality of cell libraries in terms of compactness, performance, and the number of available cells, depending on the design effort made in creating libraries. If designers want to keep up with technological progress, cell libraries need to be updated frequently. Unless a number of designers use it, it may not be economically justifiable to have an extensive library and to update it frequently.

Although the preparation of a cell library takes time (probably more than a dozen man-months for a small library), layout time of chips is greatly reduced by the polycell design approach (probably by an order of magnitude). This is a great advantage. But the chip size is sacrificed. Compared with the full-custom layout of random-logic networks, the chip size is at least 2 times larger, and sometimes can be as much as 4 times larger or more, though it is much smaller than the size by a gate array, depending on the sophistication of cells. In order to minimize wasted area among cells laid out, all cells are prepared with equal height, keeping their inputs and outputs at the top and bottom sides. Then cells for simple functions become very thin and those for complex functions become squat. Thus the area in each cell is not efficiently utilized. Also interconnections among cells take up a significant share of the chip area.

Unlike gate arrays where only masks for connections and contacts are to be custom-made, here all masks must be custom-made, so the polycell design approach needs expensive initial investment, despite its advantage of smaller chip size. However, since the layout is highly automated with

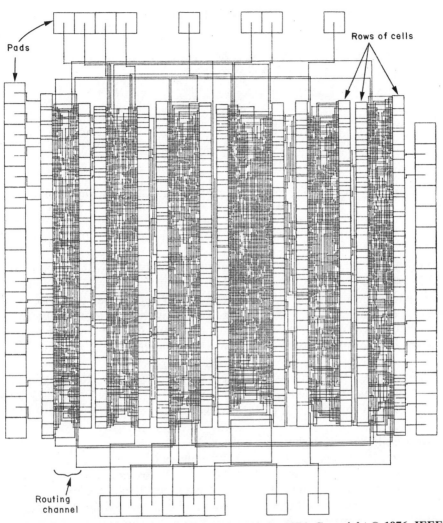

Fig. 9.4.27   A polycell layout by LTX. (Persky et al., 1976. Copyright © 1976, IEEE.)

interactive CAD programs for placement and routing, all the masks are not
as expensive as those by the full-custom design (or the pseudo-full-custom
design assisted with symbolic layout), and consequently the polycell design
approach is favored in many cases. In this sense the polycell design approach
is not as purely semi-custom design as the gate array design approach. But
the approach is not purely full-custom design either, since the chips are not
as compact as those by the full-custom design. Thus like the design approach
with symbolic layout, it might be called pseudo-full-custom design approach,
though some people actually call it a full-custom design approach.

The polycell design approach is cost effective when the production volume

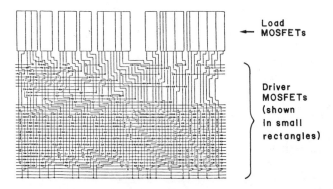

Fig. 9.4.28   Layout of dynamic MOS cells. (Shirakawa et al., 1980. Copyright © 1980, IEEE.)

is more than 100,000, since the initial investment costs $20,000 or more, which is much lower than for the full-custom design (*Electronics,* Jan. 24, 1974, pp. 67, 68). Although the polycell design approaches were not very successful commercially, they have been extensively used in-house by some manufacturers of digital systems, which do not require the chip compactness attained by the full-custom design approach (*Spectrum,* Jan. 1977, p. 40).

References: Motorola, 1971; Persky et al., 1976, 1977; Kozawa et al., 1974; Falk, Jan. 1977; Feller et al., 1980; *Electronics,* Sept. 25, 1972, pp. 78, 79; *EDN,* Oct. 15, 1972, p. 14; Crocker et al., 1977; Fairchild, 1972.

General Cell Library Design Approaches
The cell library design approaches, using cells of different shapes and sizes, can reduce the chip size more than the polycell design approach, because by keeping the same height, the area of each cell is not necessarily minimized, and by keeping all connections among cells in routing channels, the connection area may not be minimized. Moreover, by using a hierarchical approach based on this, in other words, by treating many cells as a building block in a higher level, and many such building blocks as a building block in a next higher level, and so on, we can further reduce the chip area because global area minimization can be treated better, even though this is done on a CRT screen. But this is much more time-consuming than the polycell design approach, and the development of efficient CAD programs is harder. It appears to be difficult to make the difference of chip area from full-custom designed chips within about 20%, though the areas of full-custom designed chips greatly vary with designers, and like the case of symbolic layout, comparison is not simple. As the integration size increases, this approach will be more favored, and the difference would become smaller.

References: Kani et al., 1976; Preas and Gwyn, 1978; Sato et al., 1978; Chiba et al., 1981; Lauther, 1980.

9.4.3 Other Design Approaches

There are numerous ways of doing layouts, and it is difficult to classify all of them. In the following let us describe some methods that do not belong to the previous design approaches.

Dynamic MOS cells can be arranged in a row, as shown in Fig. 9.4.28, since the layout of MOS cells is less constrained than that of static MOS cells (where the channel width of each MOSFET needs to be properly adjusted). All load MOSFETs (whose MOSFET gates are connected to a clock)

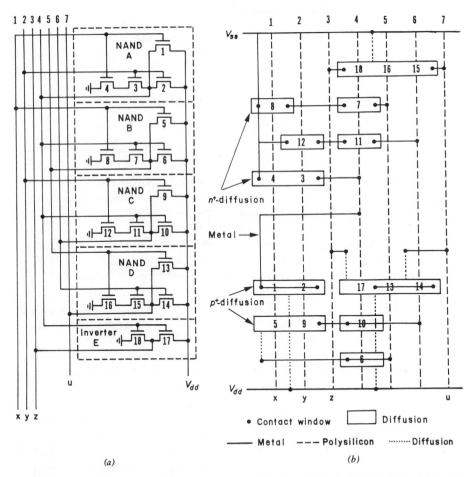

(a) (b)

Fig. 9.4.29 Gate matrix. (Lopez and Law, 1980. Copyright © 1980, IEEE.) (*a*) Logic network. (*b*) Gate matrix for (*a*).

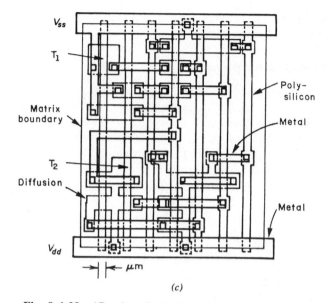

Fig. 9.4.29 *(Continued)* *(c)* **Layout of gate matrix *(b)*.**

are lined up in the top row, and input lines run horizontally across the driver MOSFETs. In this case each cell (i.e., each logic gate) need not be a rectangle with straight-line sides, but it can have a vertically long shape with irregular sides and equal height. A CAD program produces such a layout which is 20 to 30% wider with equal height than those designed by hand. A few months of hand layout for some networks can be reduced to a few minutes by CAD processing, though feeding input data to the CAD program takes several hours (Shirakawa et al., 1980).

The **gate matrix** of CMOS developed by Bell Labs is illustrated in Fig. 9.4.29. The CMOS network of four 2-input NAND gates and one inverter shown in Fig. 9.4.29*a* is schematically laid out in the gate matrix in Fig. 9.4.29*b*. The columns of the gate matrix are realized with polysilicon (shown in dashed lines) and diffusion (shown in dotted lines), and, whenever necessary, with metal. The rows are realized with diffusion (shown with rectangles) and metal (shown in solid lines). Each MOSFET is formed underneath a polysilicon inside a rectangle. The MOSFETs of NAND gate *A* are placed on columns 1 and 2 with the *n*-channel in series (MOSFETs 3 and 4) and the *p*-channel in parallel (MOSFETs 1 and 2) in Fig. 9.4.29*b*. The output of this NAND gate goes to column 4, which also serves as the MOSFET gates of inverter *E* (MOSFETs 17 and 18). The series connection of *n*-channel MOSFETs 15 and 16 of NAND gate *D* is via the diffusion row, because they are in adjacent columns, but in NAND gate *B* the series connection of *n*-channel MOSFETs 7 and 8 is through metal in order to cross over columns 2 and 3. The actual layout done by a CAD program is shown in Fig. 9.4.29c. (Whenever necessary, the channel width can be increased to a multiple of the minimum, as indicated by T_1 and T_2.) As seen in this example, all MOSFETs that have common inputs are placed in the same columns, yielding high packing density. Also because of the matrix structure, checking of layout errors is easy, and new processing techniques (i.e., new layout rules) can be easily and quickly adopted without redoing the layout. A layout of 20,000 MOSFETs was actually carried out (Lopez and Law, 1980).

There are many other approaches or variations, and new approaches are surely coming [e.g., Leblond et al. (1980)].

9.5   Comparison of Different Design Approaches

As seen in this and the previous chapters, there is an almost continuous spectrum of different design approaches from off-the-shelf packages, PLAs, semi-custom design approaches, to the full-custom design approach, depending on the regularity of device arrangement in the layout. Naturally, comparison of these approaches from many different criteria such as performance, design time, and chip size is very complex. Also different design approaches are often mixed on the same chip, further complicating the comparison. Reliable comparison data are rarely available. But here we will try to give some idea of the advantages and disadvantages of the different design approaches.

9.5.1   Off-the-Shelf Package Design Approaches

Among all off-the-shelf packages, let us consider only discrete components, SSI packages, MSI packages, and possibly some LSI packages with relatively small integration sizes, excluding off-the-shelf microprocessor packages, since off-the-shelf microprocessor packages, which are processors for general purposes or for somewhat specialized purposes (still general enough to justify high-volume production), have different features, as discussed later. (It is to be noticed that off-the-shelf packages in forms other than general-purpose microprocessors or processors dedicated to frequently encountered special tasks, such as floating-point operations, are usually not available in VLSI or LSI with relatively large integration sizes, because the production volume is not high enough to justify them economically.)

The major advantage of off-the-shelf packages (discrete components, SSI, or MSI) is the ease of design change or faulty part replacement. But there are many disadvantages, such as low reliability, high cost, and bulkiness. When off-the-shelf packages are assembled on pc boards and further into cabinets, the overall system costs make a substantial difference because of additional costs, as exemplified in Table 9.2.1. When we use LSI or VLSI packages, all additional costs, such as those for pc boards and fans, become zero or insignificant, but we still have to consider test costs, which are even higher because of more stringent test requirements. Thus LSI or VLSI packages are cost effective for high-volume production which justifies high initial investment.

9.5.2   Custom Design Approaches

After reviewing PLAs which were detailed in Chapter 7, let us discuss other custom design approaches.

Use of PLAs

PLAs, which are mask-programmable, are easy to change. As discussed in Section 7.7, PLAs are also cheaper than the custom design approaches discussed in this chapter, which require 2 or more custom-made masks. PLAs require only 1 custom-made mask for connections (the folded PLAs require two custom-made masks for metal connections), and CAD programs can be applied easily. The design time is greatly reduced by the use of PLAs [e.g., it was 5 times faster with the use of a PLA in the design of HP pocketable calculators (*Spectrum*, Feb. 1974, p. 37)] for the following reasons. Designers can use algebraic minimization, which in turn can be processed by CAD, instead of the more time-consuming design of logic gate networks; designers need not consider the placement of gates and the routing of connections (thus the design of PLAs is immune to any change of layout rules due to new processing technology); testing PLAs is easy; and the design can be changed easily in case of mistakes or function changes. Compared with the custom design approaches discussed in this chapter, however, PLAs have lower speed and also lower packing density. [Typically only 10% or less of the MOSFETs in the entire PLA are utilized, though the folded PLAs have higher percentages (Wood, 1979).] Thus not all the networks in a digital system are usually designed with PLAs only, though PLAs are a powerful tool as part of the entire system, in particular in its control logic. The full-custom design approach is very time-consuming, typically taking more than 1 year, but if PLAs are used in the control logic, a number of different custom design chips with high performance can be made quickly by changing only one connection mask for the PLAs, though these chips cannot have drastically different performance and functions. [Texas Instruments development many different models from one microcomputer chip by changing one custom design mask for PLAs and ROMs. By using a total of about one dozen custom design masks, the firm introduced several hundred products in the 4-bit microcomputer family TMS 1000 (Roberts, 1980; Fischer, *Proc. 17th DA Conf.*, 1980).]

Other Design Approaches

An approximate comparison of different design approaches in terms of design time and chip area (shown in logarithmic scale) is given in Fig. 9.5.1.

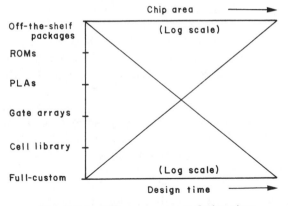

Fig. 9.5.1 Chip area versus design time.

For each design situation, designers must choose the most appropriate approach, considering tradeoffs between design time (which is closely related to design cost) and chip area (which is related to manufacturing cost and performance).

The full-custom design has the following disadvantage. Design changes and corrections of LSI/VLSI packages are very time-consuming and expensive. When mainframes were designed with discrete components and SSI packages, corrections of a few to several thousand connections, or probably more, after a few years' use of the computers by customers were common. (Complete correction often took years of use.) Since in the case of LSI/VLSI packages correcting mistakes after completion is very expensive and time-consuming, test and correction are repeated carefully at every design phase, but still design mistakes are often discovered after LSI/VLSI packages are delivered to customers. Testing by many selected customers, prior to high-volume production, is becoming a common business practice, and is called **sampling**; it usually takes several months or longer. [As a matter of fact, customer found design mistakes in one high-performance pocket calculator during use, and consequently all calculators of this model were recalled, resulting in considerable damage to the manufacturer (*Business Week*, Dec. 23, 1972, p. 26).] When the integration size increases, chances of design and manufacturing mistakes increase more sharply, increasing time and expenses for test and verification.

This problem is greatly alleviated by CAD and also by methodological simplicity of semi-custom design approaches.

Design and layout (including test and verification) for LSI/VLSI packages are very time-consuming, though layout is far more time-consuming for large integration sizes. The full-custom design approach is most time-consuming, and a few years and several good designers are easily required for the development of a state-of-the-art LSI chip with about 10,000 MOSFETs (roughly 2,000 to 3,000 gates). The development cost of a full-custom design chip can be roughly calculated from the required manpower and facilities. But accurate calculation is difficult, since very often layouts and electronic circuits developed in the past are partly used, those newly developed during the current design will be used in other chip developments, and CAD may be partially used. Also, the cost depends on the designers' skills and required chip performance.

When a well-established processing technology is used instead of the most advanced one, the pseudo-full-custom design services with shorter development times and less cost are available from some firms, such as American Microsystems, Inc., who use symbolic layout and other CAD programs. Although comparison is difficult because of a variety of performance criteria and sizes and also a lack of reliable detailed data, it is estimated that the development cost for a gate array is usually $\frac{1}{3}$ to $\frac{1}{10}$ the cost for the pseudo-full-custom design approach with established processing techniques, and the development time is usually several weeks, while several months or more are required for the psuedo-full-custom design approach (*Electronics,* June 5, 1980, p. 141; July 3, 1980, p. 120; *Computer Design,* Sept. 1979, p. 174; *Business Week,* Apr. 21, 1980, p. 134I; July 27, 1981, p. 68N). The development cost for a gate array with reasonable packing density appears to be roughly in the range of $10,000 to $50,000 (*Electronics,* July 3, 1980, p. 122), though the number of gates on a chip is limited to about a few thousands. In addition to only a few connection masks to be custom made, CAD for placement and routing is very effective in reducing the development cost and time. For example, the connection mask preparation time for a 200-gate array is reduced from a week by hand to a few hours with CAD (*Electronics,* July 3, 1980, p. 123; Oct. 9, 1980, p. 144). A comparison example is shown in Table 9.5.1 for the design of a controller network which would require 50 off-the-shelf packages in low-power Schottky TTL, assuming 18 gates per package. A single CMOS gate array of 1,000 gates is sufficient to realize the controller. The development time from logic design to a prototype is about 14 weeks for the gate array. For the production volume range of 2,500 to 250,000 the gate array is most cost effective when packages are assembled in pc boards and cabinets (*Electronics,* July 3, 1980, p. 122). When the integration size of a gate array increases, the gate array is most cost effective in a wider range of production volumes, as illustrated in Fig. 9.5.2 (*Electronics,* June 5, 1980, p. 141).

Generally semiconductor manufacturers prefer full-custom design because of high production volumes with competitive pricing. But computer manufacturers who expect much lower production volumes prefer gate arrays for high speed (Armstrong, 1981; Cane, 1980) and PLAs for low speed (Logue et al., 1975), often mixed with microcomputers. Their CAD emphases are correspondingly different.

An approximate relationship between cost per package and production volume is illustrated in Fig. 9.5.3, though this may change depending on many factors, such as fabrication technology, logic families, system size, and

Table 9.5.1 Comparison of Different Design Approaches[a]

	SSI OR MSI OFF-THE-SHELF PACKAGES	CMOS GATE ARRAY	FULL-CUSTOM
Chip count	50	1 (1,000 gates)	1 (850 gates)
Gate count usage	95%	85%	100%
Board area	50 in.^{2}	2 in.^{2}	2 in.^{2}
Development costs	$10,000–$20,000	$20,000–$30,000	$50,000–$100,000
Component costs	$40–75	$20–90	$15–90
Testing and assembly	$20–50	$4–10	$4–12
Development time	10–16 weeks	9–16 weeks	9–12 months
Total cost per board for production of:			
250	$180	$200	—
2,500	$100	$ 90	$120
25,000	$ 80	$ 50	$ 60
250,000	$ 65	$ 30	$ 30
2.5 million	—	$ 25	$ 20

[a] Reprinted from *Electronics*, July 3, 1980, p. 122. Copyright © McGraw-Hill, Inc., 1980.

performance. It is roughly estimated that the semi-custom design is more economical than the full-custom design when the production volume is less than 50,000 (or 10,000 by another estimation), and the semi-custom design is less economical when the production volume is more than 300,000 (or 100,000 by another estimation). But, the comparison is difficult and has to be done carefully in each case, because each approach has variations and it makes a difference whether or not libraries of cells or macrocells are prepared from scratch. (Notice that design approaches are shown in thin-line curves for simplicity, but actually should be represented in very broad lines.) The cost per package for the off-the-shelf package design approach is fairly uniform over the entire range but increases for low production

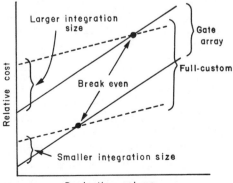

Fig. 9.5.2 Influence of integration size of gate array. (Reprinted from *Electronics*, June 5, 1980, p. 141. Copyright © McGraw-Hill, Inc., 1980. All rights reserved.)

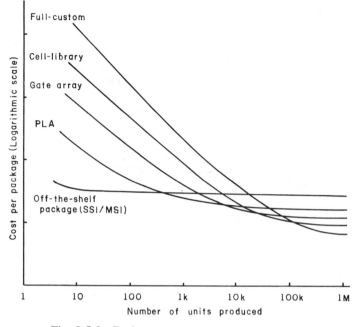

Fig. 9.5.3   Package cost versus production volume.

volumes, since the development cost becomes significant as initial investment in the overall package cost. The relationship shown in this figure will
change as the integration size increases, because the dependence on CAD
will inevitably increase.

References: Furlow, 1973; Exar Integrated Systems, 1979; Kroeger and
Tozun, 1980; Fischer, 1980; *EDN,* July 19, 1976, pp. 26–32; Nov. 5, 1977,
pp. 56–59; Jan. 20, 1978, pp. 28, 30; Apr. 20, 1980, p. 118; *Electronics,* Jan.
10, 1974, pp. 74–78; May 26, 1977, p. 130; Oct. 26, 1978, p. 54; *Electronics
Design,* Sept. 27, 1978, p. 38; *Computer Design,* Sept. 1977, pp. 148–154;
Sept. 1979, pp. 168–174; *Datamation,* Aug. 77, p. 150; American Microsystems, 1979; *Digital Design,* Sept. 1976, p. 11; Hartmann, 1981; Hartman
et al., 1981; Farina et al., 1981; Peters and Gold, 1981.

Custom–Vendor Relationship
When a customer asks a vendor to design and develop custom-design LSI/
VLSI packages, the customer can access the vendor with information at any
level, depending on the situation. For example, the customer can supply a
performance specification of the chip and the rest is done by the vendor.
Or the customer can supply logic network diagrams with or without partitioning, or electronic circuit diagrams. The customer can also supply layouts

and then the vendor immediately prepares masks and starts wafer fabrication. The customers must determine where they want to start and also which phases they need to check, by considering their capabilities as well as those of the vendor. It should be understood that any vendor usually avoids tying down capable designers in time-consuming aspects and prefers to use CAD programs as much as possible, unless orders of large production volumes can be expected after completion of the chips. (Vendors are also doing business and naturally have to squeeze out profits as much as possible.)

Many manufacturers who traditionally had not produced ICs in-house, such as watch manufacturers and computer manufacturers, have started LSI/VLSI production in-house at least as second sources to themselves. During the economy boom of 1978 and 1979, many manufacturers suffered financially because they could not obtain a sufficient supply of LSI/VLSI packages from semiconductor manufacturers, who were overloaded with orders, and consequently they could not manufacture their own products which required incorporation of these LSI/VLSI packages. Thus many of them are expanding their in-house LSI/VLSI production facilities. Despite the facility expansion, however, no single manufacturer can excel in every IC product of the different IC logic families. Consequently they still have to obtain supplies of many IC products from semiconductor manufacturers or nonsemiconductor manufacturers specialized in some IC products or design services. Because of the enormous complexity of LSI/VLSI design, unlike with other industrial products such as refrigerators and TV sets, customers and vendors need to establish trustworthy relationships. This is vitally important for the customers, because smooth design and development of good LSI/VLSI packages and also their uninterrupted supply are becoming a death or survival issue to them. Also, it is probably becoming very important for the customers to have their own designers, even though they do not own IC production facilities, for the following reasons. First, customers' engineers know better the market of their products, the details of their products, and what functions are to be added by LSI/VLSI packages, than the engineers of vendors. (For example, the engineers of a sewing machine manufacturer know the internal mechanism of a sewing machine better than the engineers of semiconductor manufacturers.) Second, customers would complete the design much more quickly because of direct pressure from the market, and also they can protect their proprietary information by designing or developing in-house as much as possible. If customers do every phase of design in-house and hand only masks to vendors, asking only for wafer fabrication, then trade secrets are probably fairly well protected. Third, if customers know LSI/VLSI technology better, custom-designed packages to be developed jointly with vendors will have better quality with shorter development times.

References: Walker, 1973; Phister, 1980; Robson, 1978; Hauer, 1978; *Electronic Design*, Jan. 18, 1979, p. 44.

9.6   Microcomputers

A **microcomputer** is a compact but complete general-purpose computer system, which consists of a microprocessor package (or packages) and other packages, such as memory packages and input/output control packages. If sufficiently many memory packages are added, a microcomputer can do any computation or tasks that any other computers, minicomputers, or mainframe computers can do, though its speed is generally slower. As the integration size increases, LSI/VLSI chips in general have limited usage and cannot be sold in high volumes. In this sense, off-the-shelf microprocessor chips are an exception.

9.6.1   Features and Applications

The major objective of a microcomputer is as follows. When the design of logic networks is too time-consuming or too expensive if the computational tasks or logic functions are to be realized with logic networks, we can realize them less expensively by programming with a microcomputer. A **microprocessor** package, which is a key part of the microcomputer, is an LSI/VLSI package usually designed for 4-, 8-, 12-, or 16-bit parallel processing, which is available from a semiconductor manufacturer as an off-the-shelf package.

The integration size of microprocessor chips continues to increase (Noyce and hoff, 1981). An early microprocessor chip 8008 of Intel contains about 3,000 *p*-channel MOSFETs (Shima et al., 1974). The first commercially successful microprocessor chip, the 8080 of Intel, contains about 4,500 *n*-channel MOSFETs (*Electronics*, Apr. 18, 1974, pp. 95–100). A microprocessor chip, the Z8000 of Zilog, contains 17,500 MOSFETs, and a more recent microprocessor chip, the 68000 of Motorola, contains about 35,000 MOSFETs (*Electronics*, Nov. 23, 1978, p. 48; Dec. 21, 1978, pp. 81–88).

Many microprocessor chips which cost a few hundred dollars when introduced, cost only a few dollars a few years later. Thus microprocessor chips by themselves are not very profitable for semiconductor manufacturers. They make profits by selling memory packages (very high production volumes), support packages (some special function packages have high prices due to uniqueness), and software (*EDN*, Nov. 5, 1980, pp. 97–99, 106).

Table 9.6.1 shows the microprocessor chip 8080A (an improved version of the 8080) and supporting packages, constituting the MCS-80 microcomputer system. The packages may be connected as shown in Fig. 9.6.1. Software support and other aids are partially shown in Table 9.6.2. In addition to static or dynamic RAMs and ROMs with fast access time, a magnetic bubble memory on the order of megabits with slow access time can be connected to the 8080A (Bryson et al., 1979).

An 8080-class microcomputer is estimated to replace 100 to 200 IC packages [or 30 to 50 TTL packages, depending on the situation (*Electronic Design*, June 7, 1976, p. 86)]. Thus a microcomputer of less than $50 can

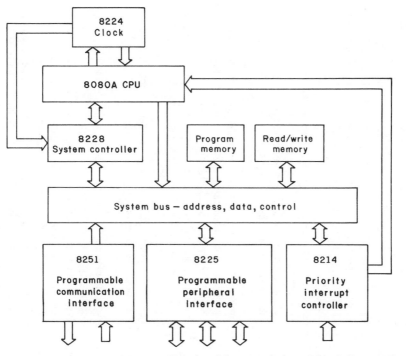

Fig. 9.6.1 **8080A microcomputer. (Reprinted by permission of Intel Corporation.)**

easily replace a few to several hundred dollars or more of manufacturing cost in a system with SSI and MSI off-the-shelf packages, which requires pc boards, cabinetry, power supplies, and others, as illustrated in Table 9.2.1. (Of course, more powerful microcomputers to be described later are more cost effective in complex tasks.) If logic design with a few hundred IC packages takes a few weeks, it can be reduced to a few days with a microcomputer mainly by programming, though a microcomputer is much slower to execute than logic networks in IC, probably by one or two orders of magnitude (or more, depending on the tasks to be performed). A microcomputer has great design flexibility. As seen in the case of the HP 2640 CRT terminal, which was designed based on a microprocessor (*Datamation,* Feb. 1976, p. 72), when new functions are found to be added or the current functions are found to be changed, we can do it in a short time simply by changing firmware stored in ROMs. This is much simpler, more economical, and quicker than logic gate network changes. Also microcomputer chips are much cheaper, with greater reliability, than a few hundred packages (Davidow, 1974).

Software Costs
The extremely low price of a microprocessor chip (a few dollars for 8-bit microprocessor packages and over \$10 for 16-bit microprocessors) is in a

Table 9.6.1 8080A Microprocessor and Typical Support Packages (Not All of Which Are Generally Used in One System)[a]

	PART NO.	MCS-80 SYSTEM COMPONENTS DESCRIPTION
CPU group	8080A	8-bit CPU, 2-μsec cycle	Built into the CPU group are the extra functions most designs require, as well as central logic and bus control:
	8224	Clock generator	· TTL and MOS crystal controlled clocks for system timing
	8228	System controller	· Auxiliary timing functions and single-level interrupt control
CPU options	8080A-1	1.3-μsec cycle	· High current sinking capability to keep memory and input/output interfaces simple, regardless of system size
	8080A-2	1.5-μsec cycle	
	M8080A	2-μsec cycle (−55 to +125°C)	
Input/Output	8212	8-bit input/output port	Implements virtually any 8-bit-input/output function with parallel latch/buffer and bus service request logic, Schottky bipolar for 15-mA output drive
	8251	Programmable communications interface	Operates under program control in virtually all serial data transmission protocols in use today, including IBM Bi-Sync
	8255	Programmable peripheral interface	Three 8-bit ports, software configurable for interface to printers, keyboards, displays, motor drives
Peripherals	8205	1 of 8 binary decoder	Expands memory and input/output address capability, Schottky bipolar for high speed
	8210	Dynamic RAM driver (8107B)	High-voltage clock driver and quad address driver, two enable inputs simplify address or data decoding
	8214	Priority interrupt control unit	Provides eight levels of interrupt control, cascades for simple expansion, current status register saves memory
	8216	Bidirectional bus driver	Noninverting 4-bit, Schottky bipolar driver, 50-mA outputs drive long bus lines and terminations
	8226	Bidirectional bus driver	Inverting version of 8216 bidirectional bus driver

	Part	Specification	Description
	8222	Dynamic RAM refresh controller (8107B)	Controls refresh of large, asynchronous dynamic RAM system, has adjustable control oscillator and internal address multiplexer
	8253	Programmable interval timer	Controls three active intervals with independent 16-bit counters at direct current to 3 MHz, counts binary or BCD
	8257	Programmable DMA controller	Provides four channels of priority DMA request logic for direct access of peripherals and memories
	8259	Programmable interrupt controller	Eight-level interrupt controller, priority algorithms can be varied with software, expandable to 64 levels
PROMs	8604	512 × 8, 100 nsec	All PROMs can be automatically programmed with the Intellec® MDS PROM programmer peripheral
	8702A	256 × 8 erasable, 1.3μsec	· High-speed programming—1 msec/bit for 8604 bipolar Schottky PROM, and 8704, 8708 reprogrammable MOS PROMs
	8704	512 × 8 erasable, 450 nsec	
	8708	1k × 8 erasable, 450 nsec	· All 2048 bits of 8702A reprogrammable MOS PROM can be programmed in 2 min
ROMs	8302	256 × 8, 1μsec	8302 2k ROM directly replaces 8702A PROM
	8308	1k × 8, 450 nsec	8308 8k ROM directly replaces 8704 or 8708 PROMs
	8316A	2k × 8, 850 nsec	High-density program storage with 8316A 16k ROM
RAMs	5101	256 × 4 static CMOS, 650 nsec	CPU interfaces directly with static RAMs:
	8101-2	256 × 4 static, 850 nsec	· 5101 CMOS RAM reduces standby power to 75 nW/bit
	8102A-6	1k × 1 static, 650 nsec	· Static RAMs are the low-cost, easy-to-use approach for small and medium systems
	8102A-4	1k × 1 static, 450 nsec	
	8111-2	256 × 4 static, common input/output, 850 nsec	Dynamic 4k RAM reduces cost and improves speed of large systems; system design simplified by use of Intel
	8107B	4k × 1 dynamic, 420 nsec	drivers and refresh controllers

a Reprinted by permission of Intel Corporation.

Table 9.6.2 8080A Design Aids[a]

Microcomputer development system	Intellec® MDS	With its ICE-80 in-circuit-emulator module, the Intellec® MDS supports programming, prototyping, and hardware/software debugging in the product's own environment. The mainframe is an 8080 system with expandable memory and input/output, DMA, interrupt logic, multiprocessor bus, clocks, and power supplies. Peripherals include diskette system, bipolar ROM simulator, universal PROM programmer, high-speed paper tape reader, and standard interfacing for a CRT console, teletypewriter, high-speed tape punch, and line printer
Comprehensive software packages	System monitor	Supports the system's comprehensive diagnostic aids, controls the system, and drives peripherals. Enables programs to be checked out in real time and supports simultaneous software/hardware debugging. Allows use of Intellec® MDS-800 hardware as prototyping resources, provides linkage to special peripherals, and loads developed programs into PROMs via PROM programmer.
	Macro assembler	You can compose 8080 programs in a symbolic assembly language, which the macro assembler translates to machine code. There is no need to rewrite similar program segments. Like MAC-80, this package provides full macro and conditional assembly capabilities.
	Text editor	A comprehensive tool for program entry and correction. Edits characters or lines of text. Commands include string search, substitution, insertion and deletion. The monitor provides input/output and other facilities required for easy entry and editing.
	DOS	Intel's new diskette operating system substantially reduces the time required to assemble, edit, and execute programs. Its comprehensive file management capabilities enable program and data files to be represented symbolically. Disk files can be created, edited, assembled, and executed easily and quickly through simple commands from the system console.

ICE-80   The in-circuit emulator ICE-80 provides a unique powerful tool for total hardware/ software system debug through the Intellec® MDS. ICE-80 allows all the resources of the MDS to be used directly in the prototype environment to run it, debug it and perform final production and field testing. It also allows software to be developed simultaneously with the prototype and to run on the prototype from the earliest possible time.

ROM-SIM   The ROM simulator is a high-speed random access memory which simulates Intel bipolar PROMs and ROMs. Its 130-nsec access time eliminates the necessity to program and use bipolar PROMs/ROMs when ultra-high-speed memory is required during prototype development.

Training   Intel regional training centers give courses in system design and programming, and also conduct weekly workshops that provide hands-on experience. On-site courses and seminars are also available.

Applications assistance   Call the nearest Intel sales office. Assistance is available in the field through Intel field applications engineers and field marketing engineers.

Production support   Intel delivers standard subsystems, standard boards, and custom boards, as well as system components. Standard products are stocked worldwide at Intel distributors.

MCS-80 system design kit   This kit contains all the components and software required to assemble and operate a basic 8080 system:
· CPU group (8080A, 8224, 8228)
· Two programmable input/output blocks (8251 and 8255)
· Two decoders (8205)
· PROM and RAM memory (8708, two 8111)
· Printed circuit board and connectors
· Clock crystal and other required components
· Control program (monitor stored in ROM)
· 8080 system user's and programming manual

407

Table 9.6.2. *(Continued)*

			Time Sharing Networks
Cross product software	PL/M™ cross compiler	Developed by Intel in 1973. PL/M is the only high-level program language for microcomputer system software design. It has significantly reduced programming costs. It produces code that can be stored in ROM. Should you ever need to use machine language coding, PL/M also provides the mechanism for linking to assembly language routines.	**United States** United Computing GE Tymshare **Europe** Tymshare Timesharing LTD Honeywell **Japan** Dentsu **Canada** GE **Australia** Honeywell
	MAC-80 cross assembler	This powerful macro assembler simplifies software design and eliminates the need to write redundant code. It has full macro capability, coupled with conditional assembly directives. The assembly language is fully compatible with the Intellec® resident assembler.	
	INTERP/80 simulator	INTERP/80 helps you quickly debug your programs. It enables computers to simulate program execution by the 8080A CPU. The package has complete debugging facilities, including all timing details, breakpoints, full file buffered input/output, and a host of commands that permit you to examine and modify program execution.	
	Availability	All three cross products are written in ANSI standard FORTRAN IV. All run on medium- or large-scale computers. 32-bit integer format. They can be purchased from Intel on magnetic tape or used via the computer time-sharing networks listed at the right.	
User's library		This large library contains hundreds of 8080 programs, such as floating-point math package, multiple precision arithmetic routines, a quick-sort program, a floating-point input/output conversion package, BCD to/from binary, binary to ASCII, and Gray to binary conversion. Members receive a manual documenting programs in the library, plus frequent updates. Memberships are free to users who submit accepted programs [more than 200 user programs are available (*Electronics*, Apr. 29, 1976, p. 118)]	

408

Documentation

· 8080 microcomputer systems user's manual*
· Intellec® MDS hardware reference manual
· Intellec® MDS operator's manual**
· 8080 assembly language programming manual**
· PL/M programming manual
· MAC-80 user's manual
· INTERP/80 user's manual

 * Includes MCS-80™ system design information and data sheets for system components
** Also documents resident software

a Reprinted by permission of Intel Corporation.

sense misleading. The price of a complete microcomputer system for complex tasks is not that low. If we want to design a complex system, many additional chips are needed in addition to the microprocessor chip, and the total system could cost hundreds or thousands of dollars. For complex tasks, software for the microcomputer system could cost more. It can cost from a few hundred dollars to thousands. For a $30 microprocessor, software in some cases has been known to cost a few hundred thousand dollars. Thus logic designers will soon be involved in cumbersome programming. Software cost has risen from 50% of the total system development cost in 1975 to 80% in 1979, and will exceed 90% as more complex tasks are done by microcomputers. As the computing capability of microprocessor chips increases with LSI/VLSI technology progress, such as Zilog's 16-bit microprocessor Z8000 and Motorola's 68000, which have a computing capability comparable to the minicomputers of the 1970s, software development costs increase sharply (Schindler, Jan. 4, 1979). Roughly speaking, while 4-bit microcomputers are used for simple tasks in high-volume production, such as control of home appliances, 8-bit microcomputers are chosen for programming tasks which require up to tens of thousands of bytes in memory, and 16-bit microcomputers are chosen for tasks that require more bytes, using higher level languages for more complex tasks. (*EDN,* Nov. 5, 1980, p. 96). In order to facilitate programming, manufacturers provide various forms of software support, such as cross assemblers (assemblers to be processed by computers other than the one under consideration), resident assemblers (assemblers to be processed by microprocessors themselves), debuggers, editors, interpreters, compilers, monitors, and simulators (*Electronic Design*, Dec. 20, 1975, pp. 20–26; Mar. 15, 1979, p. 104). Also many high-level languages such as BASIC, APL, PASCAL, COBOL, Language C, and FORTRAN are available, possibly with their compilers and interpreters stored in ROMs (*Electronics,* Apr. 13, 1978, pp. 119–124). Also PL/1 subset G is available, though PL/1's full set is available only with large computers (*Electronic,* Apr. 24, 1980, pp. 41, 102).

References: Ogdin, 1980; Schindler, 1978, Mar. 15, 1979, Mar. 19, 1981; Posa, Jan. 18, 1979; Wickham, 1979; Harakal, 1978; Hicks, 1978; Irvine, 1980; Wegner, 1980; Hemenway, 1980; Johnson, 1980; Masnick, 1978; Langer and Dugan, 1978; Hughes, 1978; Grossman, Nov. 8, 1978, *Electronics,* Aug. 7, 1975, pp. 107–111; Aug. 3, 1978, pp. 116, 117; Feb. 1, 1979, pp. 121–124; Feb. 15, 1979, pp. 41, 42; *EDN* Sept. 5, 1978, pp. 53, 54, 63; Nov. 5, 1980, pp. 277–338; *Electronic Design,* Mar. 15, 1980, pp. 225–240; Mar. 29, 1980, p. 32; *Spectrum* Jan. 1981, p. 41.

Unlike minicomputers and mainframe computers, for which uninterrupted supply of power sources is the major concern of their operations, microcomputers in many cases have their power supplies turned on or off frequently. Programs stored in RAMs are wiped out whenever the power supplies are turned off, so the programs are usually stored in ROMs. In this case designers test their new programs by storing them in PROMs. When the programs are completely debugged, if the product is manufactured in large quantities, the programs are stored in ROMs, because ROMs are much cheaper than PROMs.

In order to reduce programming efforts and also to improve the performance of microcomputer systems, many LSI/VLSI packages that replace some software by logic networks have been introduced, such as packages for floating-point calculation, as discussed in Chapter 10. Thus users of microcomputer systems require an increasingly good knowledge in both hardware and software, since they have to consider tradeoffs between hardware and software (Mihalik and Johnson, 1980; Huffman, 1980; Huston, 1978; Stakem, 1978; Ogdin, 1976; Bursky, Nov. 22, 1980).

In order to aid the development of microcomputer systems including software development, many microcomputer development systems, which are microcomputer systems themselves, are available (Yen, July 21, 1977, Zing, 1980; Garrow et al., 1975; *Mini-MicroSystems,* Aug. 1980).

Bit-Slice Microcomputers
If a microprocessor is implemented with MOS, it can be packed into a single chip. But the speed is too slow for some applications. For such applications, faster microprocessors with bipolar transistors are available. But because of heat, the entire microprocessor of bipolar transistors cannot be packed into a single chip, and it is divided into chips which process a few bits each. Such a microprocessor is called **bit-sliced**. This has the disadvantage of multiple chips, but it has the advantage of logic design flexibility. In other words, a microcomputer of an arbitrary word length can be easily constructed by cascading bit-slices. About a dozen bit-slice microprocessors, such as the 2900 by Advanced Micro Devices (Mick and Brick, 1980) and the Macrologic by Fairchild (*EDN,* Nov. 20, 1975, p. 59; Rallapalli and Verhofstadt, 1975), are available. Manufacturers provide supporting IC packages of various functions. The 2900 family, for example, consists of the following TTL packages: Am2901 (4-bit microprocessor slice) whose layout is shown in Fig. 9.6.2, Am2902 (high-speed look-ahead carry generator), Am2909 (microprogram sequencer), Am2915 (quad three-state bus transceiver with interface logic), Am29702/703 (inverting 64-bit RAM), Am29720/ 721 (256-bit RAM), Am29760/761 (256 × 4 bit PROM), Am29790/791 (FPLA), and others.

References: Myers, 1980; Adams, 1978; Adams and Smith, 1978; Harmon, 1979; *Electronics,* Oct. 23, 1980, p. 154; Chu, 1979, Aug. 2, 1979; Alexandridis, 1978; Blood, 1979, Mrazek and Bron, 1979.

One-Chip Microcomputers and Microcontrollers
The microcomputer discussed so far consists of more than a dozen IC packages in addition to a microprocessor chip (or chips in the case of slice microcomputers). For very simple tasks, the microcomputer in many pack-

Fig. 9.6.2 Layout of bit-slice microprocessor chip Am2901B of low-power Schottky TTL (538 equivalent gates or about 2000 bipolar transistors). (Courtesy of Advanced Micro Devices, Inc.)

ages is too expensive or too large. A much smaller microcomputer, which is packed in a single chip, is called **one-chip microcomputer**. In particular, a one-chip microcomputer, which is designed for controlling other products such as home appliances in real time (Walker, 1977), is called a **microcontroller**. More than 100 such microcomputers are commercially available. They are complete microcomputers which contain typically a ROM of 8k to 16k bits, a RAM of a few hundred bits, and a few dozen input/output ports, just to handle a limited number of control functions. Since all these must be packed into a single chip, the performance and flexibility are below those of many general-purpose multiple-chip microcomputers discussed so far. But it provides the cheapest solution to a wide spectrum of simple tasks in appliances, low-cost instruments such as digital thermometers, small games, toys, control of motors, floppy disk drives, gas pumps, scales, servo gears, and so on. One-chip microcomputers are also used as peripheral processors in large microprocessor-based systems such as point-of-sales systems, providing cheap and local processing, while taking the load off the central processor (*Electronics,* Feb. 19, 1976, p. 132; Mar. 4, 1976, p. 126).

References: Cushman, 1979, 1980; Bass et al., 1978; Bryce, 1980; Starbuck, 1978; Irwin, 1978; Peuto and Prosenko, 1978; Check, 1978; Bryant et al. 1980; Smith, 1980; Kohls, 1981; Millerick, 1980; Cragon, 1980.

Applications
There are currently about 200 microprocessor chips commercialy available. (Some magazines such as *EDN* and *Electronic Design* present annual survey articles about microprocessor chips.) Many support packages for input/output control and specialized functions are available (Bursky, Nov. 22, 1980).

The application range of microcomputers can be roughly classified by the number of bits. 4-bit microprocessor families handle small controller functions as microcontrollers. 8-bit families are used in most applications mentioned above, while 16-bit families and bit slices are used for data processing, entering into the application areas of minicomputers (*Electronics*, Apr. 15, 1976, p. 79; Bursky, Nov. 22, 1980; Brooks, 1976; Fattal, 1981; Mendelsohn, 1981).

In addition to the applications mentioned, microcomputers can be applied to designing desk calculators, point-of-sales terminals, automatic gasoline pumps, electronic weight scales, intelligent terminals, cash registers, teller machines, business machines for inventory control, game machines, automobiles, elevator controllers, process controllers, intelligent typewriters, TV sets (*Electronics,* Dec. 18, 1980, pp. 61, 62), blood cell counters, measurement instruments, printer controllers (*Electronic Design,* Jan. 8, 1981, p. 39), sensors, and minicomputers (*Proc. IEEE,* June 1976; Feb. 1978). Actually it is hard to predict how widely microcomputers will be used. For example, the first successful home computer TRS-80 by Radio Shack is based on the Z80 microprocessor chip. Traffic-light controllers can be built with 12 microprocessor family packages, while an equivalent TTL design requires 200 off-the-shelf SSI/MSI packages. A simple gas-pump meter needs one microprocessor and only 9 other packages. An electronic scale needs 8 chips (*Electronics,* Apr. 18, 1974, p. 82). The control of a motor which costs $10 by an analog feed-back loop can be done at $2 by a microcomputer (*Electronics,* Feb. 15, 1979, p. 41). Data-base management capabilities have been added to microcomputers, such as the highly successful home computer TRS-80 (*Electronics,* July 17, 1980, pp. 98, 99). Microcomputers are extensively used in manufacturing facilities of all kinds of industrial products, for example, as part of industrial robots (Allan, Jan. 1979). Microcomputers are used for energy saving of buildings (*Business Weeks,* Aug. 4, 1980, pp. 28D, H).

Microprocessors have been used in minicomputers and mainframe computers and also in digital communication networks. Many microprocessor chips have been custom-designed specifically tailored to minicomputers and mainframe computers (*Computer,* July 1978; Tjaden and Cohn, 1979; Wolfe, 1980; Boone et al., 1979; Appelt, 1979; Eldumiati et al., 1979; Roberson 1976, 1977; Roberts, Oct. 19, 1980; Clemens and Castleman, 1980; Warren 1981; *Electronics,* Oct. 23, 1980, pp. 196, 197, 201; Melear, 1981).

References on Microcomputers: Numerous books and papers. (Manufacturers of microcomputers also have many publications.) The following shows only samples (Noyce and Hoff, 1981; Bylinsky, 1975; *Electronics,* Apr. 15, 1976; Vacroux, 1975; Posa, Oct. 25, 1979; Russo, 1980; Odgin, 1978 (2 books), Dec. 1978; Wise et al., 1980; Winder, 1974; Shima et al., 1976; Smolin, 1978; Faiman et al., 1977).

9.6.2   Advanced or Specialized Microcomputers

Microcomputers have been improved in performance, and microcomputers specialized in analog signal processing have been developed.

Microcomputers with Improved Performance
More powerful microcomputers were introduced to open more applications (*Electronics,* Nov. 23, 1978, p. 48). In 1978 the 8086 was introduced by Intel (Morse et al., 1978; Alexy and Kop, 1978; Morse et al., 1980) and in 1979, the Z8000 was introduced by Zilog (Shima, 1978). The 16-bit microprocessor Z8000 has more computing capability than the minicomputer PDP 11/45 and slightly less than the PDP 11/70. Motorola's 16-bit microprocessor MC68000 designed with HMOS in 1980 is also powerful (Stritter and Gunter, 1979; Stritter and Tredennick, 1978; Daniels and Summers, 1978). Inside, the 68000 looks like a 32-bit computer: the arithmetic-and-logic unit handles 32-bit data, the registers hold 32-bit numbers, and the address bus directly accesses 16 Mbyte of memory (*Electronic Design,* Mar. 29, 1979, p. 93). Other manufacturers also developed 16-bit microcomputers (*Electronic Design,* Jan. 18, 1980, pp. 66–70; Jan. 18, 1980, pp. 66–70; Grappel and Hemenway, 1980). Intel announced the 32-bit microcomputer iAPX 432 consisting of three chips, whose computing capability is comparable to minicomputer VAX-11/780 and, when more processor chips are added, to mainframe IBM 370/158 (*Electronics,* Nov. 6, 1980, pp. 42, 44; Rattner and Lattin, 1981; Hemenway and Grappel, 1981).

These microprocessors and microcomputers can do much more complex tasks than those discussed so far, requiring extensive and expensive software development. Many support chips to control input/output handling or to do complex tasks which used to be done by software have been developed, as discussed in Chapter 10 (Grossman, Nov. 8, 1978; Peuto and Shustek, 1977; Freeman, 1980; Wiles and Lamb, 1981).

Microprocessors for Analog Signal Processing
A single-chip n-MOS microcomputer 2920 for processing analog signals was introduced by Intel in 1979. Analog signals are converted into digital signals and then digitally processed by this pipelining processor with software stored in EPROM. The results are converted back into analog signal outputs. The 2920 can perform functions, such as filtering, modulating, detecting, limiting, and mixing, which once required many components, such as inductances, capacitors, resistors, and operational amplifiers. With a few additional packages, the 2920 can be used to build such complex networks as modems, equalizers, tone sources, tone receivers, and process controllers (such as motor or servomotor drivers).

References: Hoff and Townsend, 1979; *Business Week,* Feb. 26, 1979, p. 88H; Hoff and Li, 1980.

9.6.3 Design of Microprocessors and Microcomputers

Off-the-shelf general-purpose microprocessor chips are usually started as full-custom design chips by semiconductor manufacturers with the most advanced new processing technology, and consequently their development is very time-consuming and expensive, though the development is much easier once the new processing technology has been established (i.e., once designers know how to use it with high reliability). Also microprocessor chips have been designed for in-house use by digital system manufacturers.

The architecture of a general-purpose microprocessor chip can be designed by conventional architecture design philosophy of general-purpose computers (see the many available textbooks on conventional architecture design), though new constraints, which we can see easily from the previous chapters and which are mostly due to the two-dimensional nature of the chip layout and the limited number of pins, must be taken into account. Many features of conventional architecture provided in large computers, such as virtual memory and pipelining, have been increasingly often incorporated into LSI/VLSI chips. In order to pack the processor into the chip most compactly, maintaining the best performance, each network in the processor must be allocated in an appropriate area in the chip by a chip plan. The registers, among which a number of signals in parallel are transmitted back and forth frequently, and transmission time is critical to chip performance, must be placed very close together. Otherwise performance is sacrificed, and large areas are wasted by the long connections among these networks, as discussed in Section 4.2. Placement and routing must be tried in many different ways until a best chip plan is found. In this sense, architecture design, which can yield a good chip plan, is very important. As an example, the chip plan of the 16-bit microprocessor chip 68000, together with its layout is shown in Fig. 9.6.3. [Circuits for testing microsequences occupy a 2% area and save greatly testing time during production (*Electronics,* Dec. 15, 1981, pp. 110, 112)].

Program instructions and data (which used to be stored in the same memory in the case of conventional computers with centralized memories) can be stored in separate ROMs if these ROMs can be placed in a chip. This is because memories can be located anywhere in the chip. By placing the ROMs in the appropriate places, the transmissson times to and from the ROMs can be improved, and the areas for the ROMs can be reduced, since the word length of each ROM can be chosen freely, unlike with conventional

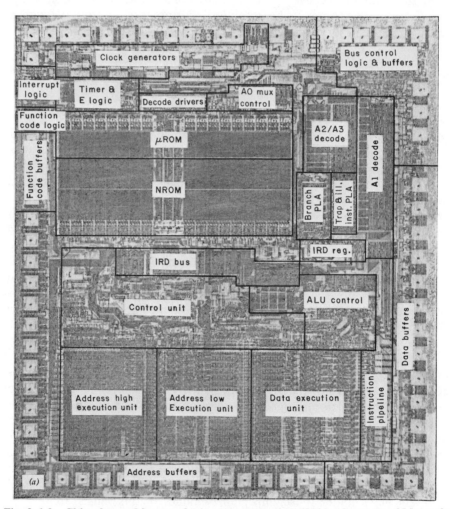

Fig. 9.6.3   Chip plan and layout of microprocessor chip 68000. (Courtesy of Motorola Semiconductor Products, Inc.) (*a*) Chip plan of 68000.

Fig. 9.6.3 *(Continued)* *(b)* **Layout of 68000 (HMOS).**

computers, where the data word length must be a multiple of the instruction word length because of sharing the same memory.

Unlike conventional architecture, which is usually designed on the basis of register-to-register transfer of signals, architecture design based on memory-to-memory transfers has been used increasingly often. This is because a logic gate occupies a chip area several times greater than each bit of static RAM, and registers can be realized with RAMs instead of logic gates, as discussed in Section 6.3. Thus we can pack more registers into the chip.

As the integration size increases, memories and control logic incorporated in a chip tend to occupy greater percentages of the entire area, in order to improve the computing capability. Control logic is increasingly realized with

PLA to save layout time, since control logic is most frequently changed and corrected.

References: Cragon, 1980; Scrupski, 1980; Hughes and Chappell, 1980; Tredennick, 1981; Price, 1980; Schindler, 1980; Mead and Conway, 1980; Posa, Oct. 25, 1979; Markowitz, 1981; Russo, 1980; Hayn et al., 1981; *Electronic Design,* May 14, 1981, pp. 131–140; *Electronics,* Mar. 10, 1981, pp. 39, 40; Apr. 21, 1981, pp. 40, 41; *Spectrum,* May 1981, pp. 47–50.

Microprocessor chips of greater integration sizes have been designed, as illustrated in Fig. 9.6.4. A most advanced microprocessor chip developed in the middle or late 1970s took typically several people for a few years, as mentioned already, though a precise estimation is difficult since the number of people involved does not remain constant during the entire period. As

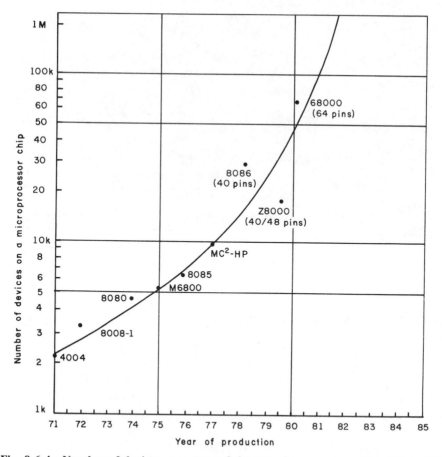

Fig. 9.6.4   Number of devices on a state-of-the-art microprocessor chip. (*Proc. 16th DA Conf.,* p. 548. Copyright © 1979, IEEE.)

the integration size increases, the man-years required increase, as shown in Fig. 9.1.2. Thus the development of a microprocessor chip is becoming expensive. The development of a complete microcomputer system is even more expensive, costing a few million dollars, since many support chips must be developed. It was claimed by Motorola that it had spent five million dollars and 60 man-years in developing the M6800 microcomputer family between early 1972 and August 1974 (*Barron's*, Feb. 2, 1976, p. 20). Also, the 35-odd LSI chips in Intel's 8086 family microcomputer chips introduced in 1978 required upwards of $250,000 apiece to design and develop, a top Intel manager estimates—not to mention a healthy investment in existing capital equipment (*Electronic Design,* Dec. 20, 1978, p. 45). [Also, according to *Business Week* (Apr. 21, 1980, p. 182), the 8086 family took five to ten million dollars and 2 years. In comparison, the development of a 16k dynamic RAM by MOSTEK took 2 years and three million dollars.] It is to be noticed that some support packages are becoming more complex than the cpu chips they are to be mated with. The Am9511 mathematics processor for cpu 8080A (or 9080A) is such an example, which provides a high-speed solution to many of the mathematical operations that the processor may be called upon to perform (*Electronic Design,* Mar. 29, 1979, p. 94; Mov. 22, 1980, pp. 123–138; *EDN,* Nov. 5, 1980, pp. 214–230). The current trend is that the man power for developing a microcomputer system doubles every $2\frac{2}{3}$ years as the integration size increases (*Spectrum,* Apr. 1979, pp. 30–37). But once it is manufactured in large quantities, the price of each chip can be as low as a few dollars, though a microprocessor chip of greater performance would cost more in the future.

The following example illustrates the difficulty of VLSI design. With present techniques and layout productivity, a microprocessor chip of 100,000 devices, which would be available by 1982, would take 60 man-years to lay out and another 60 man-years to debug. If one could reduce the number of devices in random-logic networks on a chip, much time could be reduced. The device ratio on the Intel microprocessor 8086, for example, is 4.4:1— 4.4 devices of all kinds, including those that can be generated by CAD, to every device that must be full-custom designed. If this ratio could be increased to between 10:1 and 20:1, then a 60-man-year effort could be reduced to 5 man-years (*Spectrum,* Jan. 1980, p. 34). In this sense, regular-structured networks seem to be preferred to logic gates because of their short programming and layout times, probably increasing the chip areas for them from the current 60% to 80% in the future. [For example, the 68000 shown in Fig. 9.6.3b contains memories (two dark square areas in Fig. 9.6.3b are ROMs of a total of about 30k bits) including a dozen PLAs, while the Z8000 in Fig. 4.7.1 contains no memories. It is to be noticed that microprocessor chips of greater integration sizes than the Z8000 (e.g., those shown in higher positions in Fig. 9.6.4) contain many memories. Accordingly, these chips contain a much greater number of devices than chips with totally random-logic networks, such as the Z8000. Also the number of active devices in these chips is usually much lower than the officially quoted number. For example, the 68000 theoretically can contain about 6,8000 devices, but the number of active devices is about half, because a number of intersections of lines in ROMs do not have MOSFETs.] Because memory access is mostly for intermediate computational results, more RAMs will be incorporated than ROMs, though the packing density of ROMs is higher by an order of magnitude. (When program instructions are stored in a separate chip, the processor chip does not need to contain ROMs inside.) For critical paths, which are vitally important to the speed of the processor chip, random-logic gate networks are usually used and painstakingly laid out compactly, though ROMs, PLAs, or previously designed networks whose layouts are stored in a cell library may be used for noncritical paths. CAD will be used more extensively with appropriate semi-custom design approaches, being mixed with the full-custom design approach.

References: Lattin, 1979; Moore, 1979; Allan, Apr. 1979.

It is, however, very important to notice that once a user's application of a microcomputer is fixed, custom design of the user's own microcomputer tailored to the application will improve the performance and reduce the cost further, if production in very large quantities is expected (*Electronic Design,* Apr. 26, 1974, pp. 90–95). When customers want to manufacture the merchandise in high production volumes (but this production volume is not high enough for a semiconductor vendor to justify the development of a new off-the-shelf microcomputer suitable for this customer), the customers should custom-design their own microcomputer, tailoring its architecture and instruction sets to their own needs (Ogdin, 1978 (2 books), Dec. 1978). If the production volume is reasonably high, custom-design microcomputers have lower cost and higher speed than an off-the-shelf microprocessor available from vendors, when an established processing technology can be used. But when the most advanced processing technology is to be adopted, custom-design microcomputers would not be cost effective unless the production volume is much greater.

After having introduced off-the-shelf microprocessor chips, semiconductor manufacturers constantly try to improve them in order to stay competitive. By the learning-curve process, yield is gradually improved, and with improved processing technology, the chip size is reduced with higher speed. Table 9.6.3 shows examples of such evolution (*Electronic Design,* May 24, 1979, p. 42). It is to be noticed that some chips contain a large number of MOSFETs for the chip size because of large memories on the chips.

References: Gold, 1976; Bursky, Mar. 15, 1980; Washburn, 1976; *Datamation,* June 1979, pp. 98–107; Shima, 1979.

Table 9.6.3 Evolution of Microprocessor Chips[a,b]

	NUMBER OF MOSFETs	1970–1972	1972–1974	1974–1976	1976–1978	1978–1980
Custom cpu	2,500	44k mil², 0.5MHz				
6800	5,000		46k mil², 1.0 MHz	31k mil², 1.5 MHz	22k mil², 2 MHz	15k mil², 2.5 MHz
6802	12,000 RAM is added to 6800			49k mil², 1.5 MHz	35k mil², 2 MHz	24k mil², 2.5 MHz
MC6809	15,000				40k mil², 2 MHz	30k mil², 3 MHz
6801	40,000 RAM, ROM, input/output are added to 6800				62k mil², 2 MHz	47k mil², 3 MHz
MC68000	68,000					63k mil², 4 MHz

[a] Reprinted from *Electronic Design*, May 24, 1979, p. 42. Copyright © Hayden Publishing Co., Inc., 1979.
[b] k means thousand.

9.7 Comparison of All Different Design Approaches

As discussed so far, we have a very wide spectrum of different design approaches, from full-custom design approaches to off-the-shelf microcomputers, as illustrated in Table 9.7.1. In other words, depending upon different criteria imposed by different design motivations, such as speed, power consumption, size, design time, ease of changes, and reliability, designers can use the following approaches:

1. Custom-design approaches
2. Off-the-shelf discrete components, SSI and MSI packages, along with memory packages
3. Off-the-shelf microcomputers along with LSI packages

The full-custom design approaches give us the highest performance and reliability or the smallest chip size, though they are most time-consuming. (Even in the case of microcomputers, the full-custom designed microcomputers have better performance and smaller sizes than off-the-shelf microcomputers, by being tailored to the users' specific needs.) This is one end of the wide spectrum of different design approaches. At the other end, the off-the-shelf microcomputers give us a design approach where the development time is shortest, by programming rather than by chip or logic design, and the design changes are the easiest. The off-the-shelf discrete, SSI, and MSI packages give us logic networks tailored to specific needs with less programming than the off-the-shelf microcomputers.

Custom design approaches, in particular the full-custom design approaches, are most economical for very high production volumes (on the order of a few hundred thousand), but the least economical for low production volumes.

When the production volume is low, the off-the-shelf discrete, SSI, and MSI packages give us the most economical approaches for simple tasks, but the off-the-shelf microcomputers are more economical for complex tasks, though performance is usually sacrificed. For more complex tasks, we had better use minicomputers (off-the-shelf but not custom-made).

In order to give a more concrete idea of the comparison, the microcontroller chip TMS 1000 of Texas Instruments, which has the largest production volume in industry [11 millions in the first quarter of 1980 according to *EDN,* (Nov. 5, 1980, p. 98)] is compared with other approaches in Tables 9.7.2 and 9.7.3. The TMS 1000 with p-MOS has 4-bit processing capability with a set of 43 instructions, a 256-bit RAM, and an 8k-bit ROM on the chip. As compared in Table 9.7.2, if a given task is programmed with the TMS 1000, the cost improvement is about 1000 times over the custom design of a digital system with discrete components and also about 20 times over the custom design of a digital system with an off-the-shelf microprocessor package added

Table 9.7.1 Comparison of Different Task-Realization Approaches

	FULL-CUSTOM	SEMI-CUSTOM	OFF-THE-SHELF DISCRETE, SSI, MSI	OFF-THE-SHELF MICROCOMPUTER	MINICOMPUTER (OFF-THE-SHELF)
Speed	Fastest	Fast	Medium	Slowest	Slow
Size	Smallest (chip size)	Small (chip size)	Large (many chips)	Medium (many chips)	Largest (cabinets)
Development time	Longest (layout)	Long (layout)	Medium (logic design)	Short (programming)	Shortest (programming)
Flexibility	Lowest	Low	Medium	High	Highest
Initial investment cost	Highest (layout)	High (layout)	Medium (logic design)	Low (programming)	Lowest (programming)
Unit cost					
High volume	Highest	High	Medium	Lowest	Not considered[a]
Low volume	Lowest	Low	Medium	Highest	Not considered[a]
Reliability	Highest	High	Low	Medium	Lowest

[a] A minicomputer is usually shared by many users.

Table 9.7.2   Comparison of Microcontroller and Design Approaches with Off-the-Shelf Components[a]

	NUMBER OF COMPONENTS OR PACKAGES	COSTS OF COMPONENTS AND ASSEMBLY[b]	RELATIVE COST
Discrete components	20,000–30,000	$6,000–9,000	1,000
SSI	350–500	$600–900	100
MSI	125–150	$250–450	50
Off-the-shelf microprocessor package with memory and other supporting packages	7–10	$100–200	20
Microcontroller	1	$5–10	1

[a] Reprinted from *Electronics*, June 5, 1980, pp. 138, 140. Copyright © McGraw-Hill, Inc., 1980. All rights reserved.
[b] Excluding backplanes, cables, cabinetry, etc.

with memory packages and other support packages (i.e., a hardware-oriented microcomputer, by selecting appropriate packages). Table 9.7.3 shows the advantage of the full-custom design of this programmable LSI chip. If 400 different chips, which are produced from the TMS 1000 by simply changing one connection mask, are custom-made from scratch, the total development cost would be 8 times greater, with about 4 times longer development time. No firm has the resource of experienced designers to custom-design so many chips during such a short period (Fischer, 1980).

The comparison in Table 9.7.1 greatly simplifies the entire situation, because each approach has variations, and consequently different approaches may overlap in some aspects, and also because the entire digital system or even each chip can be designed by a mixture of different approaches. Designers have to start the overall system design by determining which portions of the system require high speed (so ECL should be used), can tolerate slow speed (so IIL or CMOS networks can be used), or expect frequent design changes (so PLAs or ROMs should be used). Also there may be a possibility

Table 9.7.3   Comparison of Microcontroller and Custom-Design Approach[a]

	WITH TMS 1000	CUSTOM-DESIGNED SYSTEM
Number of applications	400	400
Design cost ($ millions)	10	80
Man-years	30	700
Number of days from specification to prototype per application	50–75	200–400

[a] Reprinted from *Electronics*, June 5, 1980, pp. 138, 140. Copyright © McGraw-Hill, Inc., 1980. All rights reserved.

of using parallel processing or pipelining to compensate for slow gate speeds. As the integration size increases, we have to mix more regular-structured networks (possibly by semi-custom design approaches, also with extensive use of CAD) with random-logic networks on a chip, because we need to finish the chip design within at most a few years in order not to miss the market. But when the integration size is reasonably small, we can finish the chip design within a reasonably short time, even if random-logic networks are used throughout the chip. Thus for different ranges of integration size we have to use an appropriate mixture of random-logic networks, regular-structured networks, and semi-custom design approaches.

As was seen already, tasks are realized by software (or firmware) in the case of microcontrollers, microcomputers, and minicomputers, whereas they are realized by logic design in the case of logic networks in off-the-shelf packages or custom-design approaches. But as illustrated in Fig. 9.7.1, microcontrollers, microcomputers, and minicomputers have different relative costs, depending on the complexity of the task to be realized (or processed). Microcomputers are less expensive to realize given tasks than microcontrollers, when the task complexity exceeds a certain threshold. Then minicomputers are less expensive when the task complexity further increases. When the task complexity is very low, logic networks in off-the-shelf discrete, SSI, or MSI packages are least expensive, in other words, logic design is less expensive than software. But when the complexity increases, logic design becomes very expensive. Actually in each case software and logic design can be mixed. In the case of microcomputers, for example, packages for special functions can be added to replace some software, such as floating-point arithmetic, making the microcomputers more hardware-oriented. Also in the case of logic networks, firmware, that is, software stored in ROMs, can be added as separate pckages or on chips.

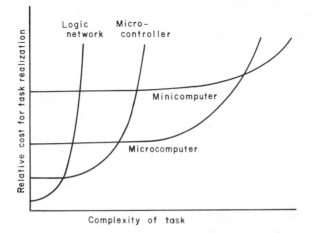

Fig. 9.7.1   Cost effective choice for task complexity.

Thus it is difficult to say where logic design in a conventional sense ends and where programming in a conventional sense starts. In every part of a task, designers must consider tradeoffs between programming and logic design, in other words, software–hardware tradeoffs.

In conclusion, as an enormous number of devices can be placed on a chip and the application of packages gets very diversified, methodologies by which designers should use ROMs, random-logic networks, programming, and minimum switching expressions for different portions of the chip are becoming essential for the success in the design and application of chips.

References: Lewis and Siena, 1973; *Spectrum,* Jan. 1976, p. 53; Schroeder, 1979; *Computer Design,* Jan. 1976, pp. 108, 109; Fischer, 1980; Roberts, *Proc. 17th DA Conf.,* 1980; Posa, Oct. 25, 1979; Kroeger and Tozun, 1980; Scherr, 1978; Mallach, 1981.

Problems

9.1 Design a TTL network for the function $\overline{x_1x_2x_3x_4x_5} \vee \overline{x_6x_7x_8x_9x_{10}x_{11}}$ in the gate array that consists of the cell of Fig. 9.4.2 only. (Show a layout.)

9.2 Suppose that we have a gate array of NAND gates with three inputs each and at most three connections are allowed to go through between any pair of adjacent (horizontally and vertically) gates (not over gates). Also suppose that all inputs of a gate are on the left side of each gate and the output is on the right side of the gate, as illustrated in Fig. P9.2. Place the gates of the NAND network shown in Fig. P3.5.2 and route the connections such that the entire network occupies the smallest area in the gate array.

9.3 Draw a layout of the CMOS network shown in Fig. 4.9.19c in the gate array of Fig. 9.4.22 such that the entire network occupies the smallest area.

9.4 Lay out the *n*-MOS network shown in Fig. 4.5.1b in the gate array of Fig. 9.4.25 such that the entire network occupies the smallest area.

9.5 Suppose that we have an *n*-MOS gate array where the *n*-MOS cell shown in Fig. P9.5a is arranged in matrix form.

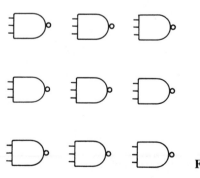

Fig. P9.2   Gate array.

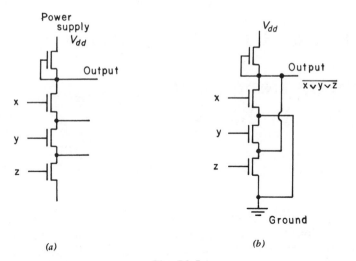

Fig. P9.5

(1) By using and/or connecting seven terminals differently, this cell can realize different functions at its output terminal. For example, the connection configuration shown in Fig. P9.5b realizes output function $\overline{x \vee y \vee z}$. Show all the different functions that this cell can realize and the corresponding connection configurations. (If there are functions that become identical by interchanging input variables, show only one of them. For example, $\overline{xy}$ needs to be shown among $\overline{xy}$, $\overline{xz}$, and $\overline{yz}$.) At least one of the terminals must be grounded.

(2) If this gate array is used for high speed, we have the high-ratio problem discussed in Section 4.3. If possible, determine the maximum channel width Q of each driver MOSFET, such that the high ratio mentioned in Section 4.3 is maintained for every connection configuration found in (1), assuming that the load MOSFET has channel width W and every MOSFET has channel length L.

9.6 Realize the CMOS network shown in Fig. 4.9.19c in the gate matrix exemplified with Fig. 9.4.29b such that the entire gate matrix occupies the smallest area.

9.7 Discuss what is the most economical design approach for a large high-speed computer of about 100,000 gates, assuming that 1,000 of them are manufactured. Then discuss major technical difficulties in design and manufacture. Also discuss the economy of this design approach. (Whenever necessary, introduce reasonable assumptions.)

9.8 Semiconductor manufacturers will be able to manufacture off-the-shelf general-purpose microprocessor packages of greater integration sizes with the same chip size by future technological progress. Discuss what factors will likely increase or decrease the cost per package.

CHAPTER **10**

System Design and Future Problems

System design aspects which are unique to LSI/VLSI chips are discussed in this chapter, partly summarizing the discussions in the previous chapters. Hardware–software tradeoffs and task-realization algorithms tailored to LSI/VLSI chip design are becoming important as specialized processor chips have been more frequently designed. Finally we discuss problems of LSI/VLSI system design, manufacturing, product planning, and social impacts in the future.

10.1 Problems in System Design

The logic design of digital systems, until the late 1960s, used to be straightforward, as illustrated in Fig. 10.1.1. After architecture design, ECL was chosen for high speed and TTL for low cost. So the selection of logic families was simple, and logic design was done with discrete components, SSI, and MSI packages. In this case, logic design, electronic circuit design, memory (magnetic core) design, development of software (if the system was a general-purpose computer, operating system), and development of peripheral equipment were done concurrently and almost independently, though there was some interplay.

Since a significant part of a digital system, including not only processors but also memories, software, and control logic of peripherals and communication, is integrated into a chip (totally or partially) due to LSI/VLSI technological progress, chip design now is enormously complex, as illustrated in Fig. 10.1.2. Architecture design is vitally important since appropriate architecture design determines speed, computing capability, power consumption, size, and cost of the entire system. Furthermore, if the architecture design is not appropriate, the performance, size, or other parameters of the entire system cannot be improved, even with best efforts in logic design, electronic circuit design, or layout. Also architecture design is becoming enormously complex. Unlike the case of Fig. 10.1.1, all design areas are integrated in a chip and must be done in a serial manner, rather than concurrently without much interaction in conventional design. For example,

428

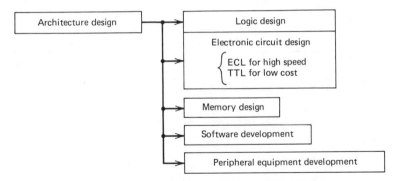

Fig. 10.1.1 **Simplicity of conventional system design.**

electronic circuit design and layout can be done only after the logic design is finished because the electronic circuits and layout shapes depend on the types of logic operations of gates. Logic design, electronic design, and lay-outs are highly interactive. Also for many logic networks, designers must consider hardware–software tradeoffs, that is, whether the tasks assigned to these logic networks (mixed with memories) can be done better with software stored in ROMs or RAMs. In other words, all design areas are integrated in each chip, but cannot be done concurrently without much interaction. Thus architecture design is to appropriately partition the entire digital system into chips, all design areas being together in each chip, such that each chip can be designed easily, with the entire system still having the best performance and economy. Conventional architecture design, to the contrary, is to simply divide the design of the entire system into separate design areas.

As the integration size increases, each chip, to which a specific task has been assigned by the above system architecture design, increasingly looks like a processor, rather than a logic network with simple tasks, such as a collection of registers, counters, and adders. (In other words, each chip is to perform the entire task rather than part of the task.) This is because the number of pins on each chip is limited and signal propagation on connections among chips is slow, requiring high power consumption. In designing each processor, the architecture design for the chip must be done prior to the logic design, considering a most appropriate chip plan. When the task as-signed to a processor chip is for a general purpose, we are designing a general-purpose microprocessor chip. Since this is discussed in Section 9.6.3, let us consider the case where the task assigned to a processor chip is special. We now must devise an algorithm which is most appropriate for LSI/VLSI realization of the task. (If the task is the calculation of trigono-metric functions, for example, the CORDIC algorithm is usually used, as explained later.) Then the chip architecture design follows. An appropriate design of this task-realization algorithm is more important in determining

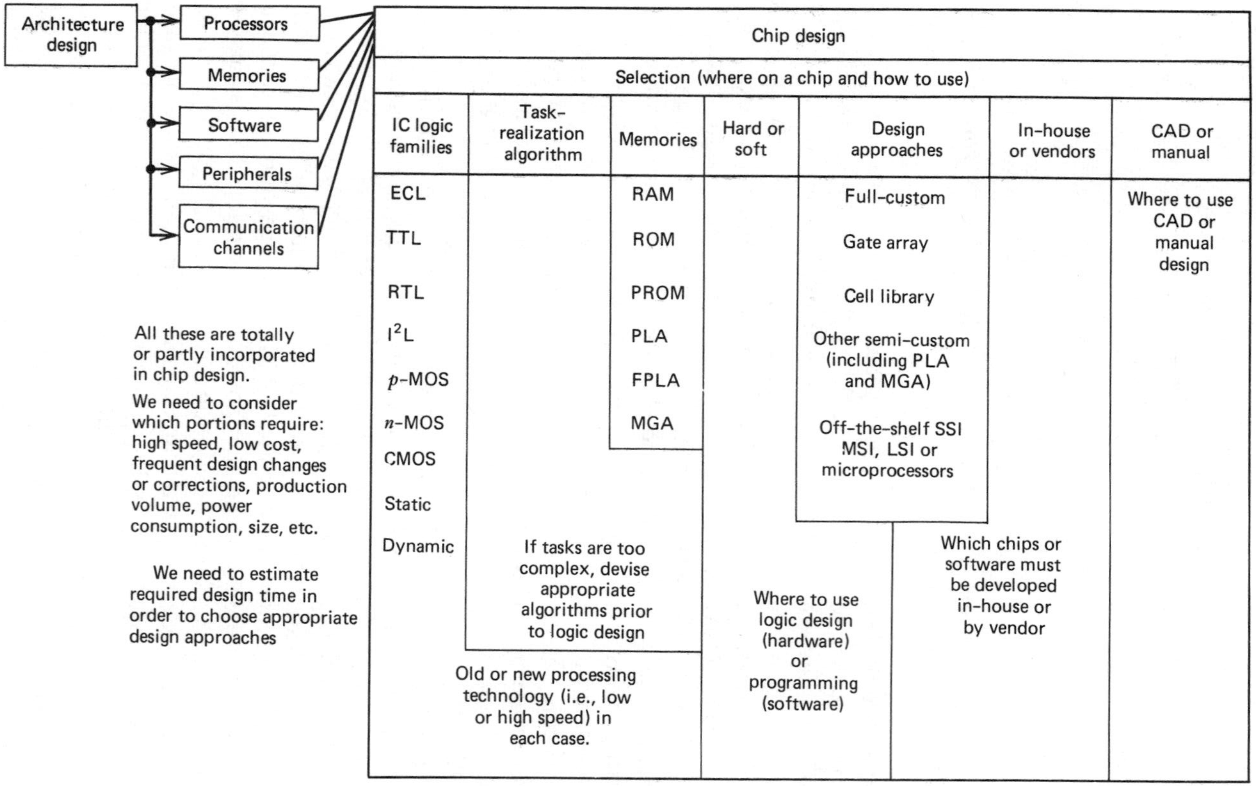

Architecture design → Processors, Memories, Software, Peripherals, Communication channels

Chip design

Selection (where on a chip and how to use)

IC logic families	Task-realization algorithm	Memories	Hard or soft	Design approaches	In-house or vendors	CAD or manual
ECL		RAM		Full-custom		Where to use CAD or manual design
TTL		ROM		Gate array		
RTL		PROM		Cell library		
I²L		PLA		Other semi-custom (including PLA and MGA)		
p-MOS		FPLA				
n-MOS		MGA		Off-the-shelf SSI MSI, LSI or microprocessors		
CMOS						
Static						
Dynamic	If tasks are too complex, devise appropriate algorithms prior to logic design		Where to use logic design (hardware) or programming (software)		Which chips or software must be developed in-house or by vendor	
Old or new processing technology (i.e., low or high speed) in each case						

All these are totally or partly incorporated in chip design.

We need to consider which portions require: high speed, low cost, frequent design changes or corrections, production volume, power consumption, size, etc.

We need to estimate required design time in order to choose appropriate design approaches

Fig. 10.1.2 Complexity of LSI/VLSI system design.

the chip size and performance than the subsequent chip architecture design. Some people collectively call both the task-realization design and the chip architecture design simply "algorithm design," or "dynamic architecture design." This task-realization algorithm design is a new design problem unique to LSI/VLSI technology, since complex tasks, which used to be realized in software in conventional computers with centralized magnetic core memories, become realizable in LSI/VLSI chips and also since the algorithms used for software processing with conventional computers may not be appropriate. For complex tasks we need to consider hardware–software tradeoffs as discussed in the next section.

The diagnosis of faulty logic networks by checking logic gates by software is becoming very difficult as the integration size of a chip increases. Fault-checking networks have often been incorporated on chip, in combination with the detection of faulty chips by functional diagnosis rather than gate-level diagnosis (*Electronic Design,* Jan. 8, 1981, pp. 35, 36; May 14, 1981, pp. 163–167; *Spectrum,* Jan. 1981, p. 43; *Electronics,* Dec. 4, 1980, pp. 137–141; Dec. 18, 1980, pp. 76, 78; Dec. 15, 1981, pp. 110, 112).

10.2 Hardware–Software Tradeoff

Hardware–software tradeoff considerations are becoming important, no matter whether the entire task is realized as a dedicated processor, or part of it is realized in memories, logic networks, or their mixture. As discussed already, if tasks are expected to be changed or not to be used often, we had better realize them in software and store them in RAMs, since software stored in ROMs or realized in logic networks is difficult to change, and the changes are expensive. If tasks are expected not to be changed but to be used often, we had better realize them in ROMs or logic networks. Firmware, that is, software stored in ROMs, is not only more compact, but also several times cheaper in even medium production volumes than are RAMs, though realization of ROMs is more time-consuming. Firmware is estimated generally to be 2 to 10 times faster in processing given tasks, and in some cases it can be 20 times faster than software (Long and Proske, 1979; *Electronics,* May 11, 1978, p. 14). But if we want to have a further improvement of speed, we have to realize the tasks in logic networks, in particular in full-custom design. Although realization of the tasks in logic networks is much more time-consuming and expensive than RAMs or ROMs, the execution of the task by logic networks is much faster, probably from 10 to 1000 times or more, depending on the nature of the tasks, and also logic networks are more compact and consequently cheaper in large production volumes than software in RAMs or firmware in ROMs. Increasingly often software has been realized by firmware and further by logic networks (i.e., by hardware).

Of course, the speed improvement is traded for programming or design time. Firmware is more time-consuming to program and to change. [Microprogramming is roughly 5 times or more time-consuming than ordinary programming. Also, the correction of a microprogram costs 20 times more at the test level than at the design level. Worse, it costs 20 times more to make the change at the field, that is, 400 times more than in original design (*Computer,* Nov. 1976, p. 68).] Logic network realization of a task is far more time-consuming to design and to change than if realized with software. This comparison, however, should not be confused with the case where a large number of different tasks are realized by software and hardware. For example, when software is to be run on a computer of different architecture, emulation by hardware is much less time-consuming to realize than reprogramming a large number of programs. Another example is the case of a certain type of tasks, where the design of a computer dedicated to these tasks is less time-consuming, since a general-purpose operating system need not be developed.

Another impetus for the hardware realization of tasks comes from the growing software development expenses for microcomputer applications. If microcomputers available from semiconductor manufacturers continue to be general-purpose computers, letting all tasks of users be realized in software, it has been estimated that the United States alone would require one million programmers by 1990, which is a prohibitive number. If software, including many portions of operating systems, is realized by hardware in LSI/VLSI chips, the users' software development expenses could possibly be reduced to roughly one fifth, according to A. S. Grove (*Computerworld,* Feb. 11, 1980, p. 71). Also the execution time is faster than the execution by software (*Electronics,* Feb. 28, 1980, pp. 89–95; *Forbes,* Mar. 3, 1980, pp. 61, 62; *Business Week,* Apr. 14, 1980, pp. 92–103).

Because of the tremendous cost reduction of logic gates by the progress of IC technology for high-volume production (roughly three orders of magnitude during the 1970s), LSI/VLSI chips, which realize special tasks, are commercially feasible. Let us find out why. Each model of a microprocessor chip is sold in tens of millions [a quarter to a half billion for a most popular model during its product life (*EDN,* Nov. 5, 1980, pp. 98, 99)], whereas each model of a mainframe computer is sold in hundreds to thousands (tens of thousands for a most popular model since the late 1970s as the price goes down). So LSI/VLSI chips for special tasks, which are to be hooked to a microprocessor as its **support chip** or **coprocessor chip** [examples of the support chips available are listed in *EDN,* (Nov. 5, 1980, pp. 214–230)], can be sold in large quantities, making the high initial investment a negligibly small percentage of the revenue generated by the chips. Also the manufacturing costs of such chips are low, on the order of tens or hundreds of dollars each. In contrast, if special tasks are realized with discrete components and SSI and MSI packages, such hardware realizations will not be sold in large quantities, since the mainframe computers to which these realizations should be hooked are sold only in hundreds to thousands. Thus the high initial investment will be a large percentage of the selling price. In addition, the manufacturing costs of these realizations are prohibitively high because a large number of IC packages are far more expensive by themselves and the selling prices will be even higher because of additional hardware and labor, such as pc boards, cabinets, cooling equipment, assembling, and wiring, as illustrated in Table 9.2.1. Thus hardware realization of software is commercially feasible only by LSI/VLSI technology.

Many tasks, such as those listed in Table 10.2.1 (Falk, 1974), have been considered for logic network realization. Many of them have been realized in software only (some in operating systems). Although some of them have been realized in logic networks in mainframe computers and, to a lesser extent, in minicomputers, such logic network realizations used to be very expensive, limiting their markets. However, logic network realizations in VLSI which are much lower priced can have very large markets. Examples of such tasks are arithmetic processors and floating-point processors. Both are single chips which are more complex than microprocessor chips and perform different types of operations specified by microprocessor chips. [A single chip for floating-point arithmetic can save over 500 instruction

Table 10.2.1 Tasks Which Are or May Be Suitable for Hardware Realization[a]

Memory allocation	Symbolic addressing
Memory reclamation	Variable field lengths
Virtual memory management	Variable data structures
Paging	Alphanumeric field manipulation
Segmentation; absent segment interrupt	Context switching
Memory and data protection	Emulation
Stack operations	Queues
Address generation	Links
Indexing	Compilation
Indirect addressing	Task dispatching
Storage protection	Next software instruction fetch
Multiple precision arithmetic	Interrupts; interrupt checking
Decimal multiply and divide	Trap catchers
Floating–point arithmetic	Peripheral data transfer
Sorting	Time–sharing supervision
Data manipulation algorithms	Text editing
	Control command instructions
Program linking and binding	Format checking
Program relocation	
Data relocation	Parity checks
Data structure	Error-control coding
Format checking	Automatic retry
Character string manipulation	Automatic diagnosis
Data-type conversion	

[a] From Falk, 1974. Copyright © IEEE, 1974.

steps (*EDP Industry Report,* Sept. 22, 1977, p. 6).] The arithmetic processor Am511A of Advanced Micro Devices, Inc., for example, performs fixed-point arithmetic operations in 16 or 32 bits and also floating-point arithmetic, trigonometric, and mathematical operations (square roots, logarithms, and exponentiation) in 32 bits. The floating-point processor Am9512 performs addition, subtraction, multiplication, and division in single (32 bits) or double (64 bits) precision. The Am9511 and Am9512 give a 5 to 50 times speed improvement over software realization of the tasks (*Electronics,* July 5, 1979, p. 100). As another example, the 8087 of Intel is a coprocessor chip to a microprocessor chip 8086 or 8088 and performs 2's complement integer operations in 16, 32, or 64 bits, floating-point operations in 32, 64, or 80 bits, and signed packed BCD integer operations in 80 bits. The operations include addition, subtraction, multiplication, division, binary-decimal conversion, square root, and transcendental functions, where transcendental functions such as trigonometric and logarithmic functions are computed based on a modification of the CORDIC algorithm to be described later. The 8087 chip, which is larger than 280 mil^2 based on HMOS, contains the equivalent of over 65,000 MOSFETs. (It is to be noticed that since the chip contains large RAMs and ROMs, it can contain a much greater number of MOSFETs than chips which contain only random-logic networks with or without small RAMs or ROMs.) The microcode utilizes over 30,000 bits of the ROMs, which is made feasible by use of a four-state ROM in HMOS (described in Section 6.4).

The burst-error processor chip AmZ8065 is a processor for error-correction based on Fire codes. Intel's microcontroller 8051 works as a Boolean processor also (*EDN,* July 20, 1980, p. 105; *Electronics,* May 5, 1981, pp. 121–134; *Electronic Design,* Nov. 22, 1980, p. 196). Its separate Boolean processor provides a direct method of converting the logic equations used in random-logic network design directly into software (Koehler, 1980).

Memory management chips handle virtual memories efficiently (*EDN,* April 15, 1981, pp. 60–62; Hu et al., 1980; Hu, 1981; Johnson, 1981). Operating systems have been partly realized in hardware (*Datamation,* Nov. 1976, p. 93; *Computer,* Dec. 1976, p. 36; McMinn et al., 1981). It is a complex task to realize the entirety of a large operating system in chips, because partitioning it into chips is complicated by the limit on the number of pins, and also because the signal transmission delay among chips can be a significant overhead in execution time (*Electronic Design,* Jan. 8, 1981, pp. 146, 147). Compilers for high-level languages such as PASCAL (*Electronic Design,* Nov. 24, 1979, p. 102; June 21, 1980, p. 23; Tanabe and Yamamoto, 1980), LISP (Ida and Goto, 1977; Steele and Sussman, 1980), and COBOL (Yamamoto et al., 1980) are realized in hardware (with many chips) (*Computer,* July 1981). Hardware realization of data-base management systems has been attempted (though many chips would be required) instead of software realization (*Datamation,* Jan. 1979). The relational type of data-base management systems would be efficient if large content-addressable memories were economically available (*Datamation,* Apr. 1976, p. 48).

References: Ogdin, 1976; Rice and Smith, 1971; Huffman, 1980; Bursky, May 10, 1980; Nov. 22, 1980, *Spectrum,* Jan. 1980, pp. 32–37; *Electronics,* July 5, 1979, pp. 99, 100; Schindler, May 24, 1979; Jan. 8, 1981; Langer and Dugan, 1978; *Electronics,* Nov. 24, 1977, p. 132; Stakem, 1978; Hughes, 1978; Conklin and Rodgers, 1978; *EDN,* Dec. 15, 1978, pp. 21, 22; Mar. 4, 1981, pp. 31–40; Stauffer, 1980; Gupta, 1980; Palmer et al., 1980; Bal et al., 1980; Eufinger, 1978; Sinha et al., 1980; Curtice, 1976; Stillman and Berra, 1977, Allison, 1977; Bremer, 1976; *Datamation,* June 1979, pp. 109–112; Smith, 1978, Rallapalli and Kroeger, 1980.

> **Remark 10.2.1:** A reliable cost comparison of software and logic gate realization is very difficult to find in the literature. But the following data may give some idea, though they are not from the same source. The cost per instruction is $10 to $20 in the case of programming with a microcomputer (*Computer,* Feb. 1977, p. 36; *EDN,* Mar. 20, 1980, p. 170; *Electronics,* June 22, 1978, p. 128), though this cost may vary depending on the microcomputer functions and languages. In the case of a logic network design with a dense, complex state-of-the-art technology, a cost of $100 per gate is unusually low (*Spectrum,* Mar. 1947, p. 46). Programmers and logic designers are in short supply. But as logic designers require more professional training, the shortage of logic designers appears to be more serious (*Electronics,* Dec. 20, 1979, p. 43). This may accelerate the increase of the cost per gate.
>
> *References:* Allison, 1977; Allison, 1980; Hughes, 1978; Heilmeier, 1979.□

For special complex tasks, significantly different algorithms may be appropriate for an improvement of performance and a reduction of chip size. The special architecture dedicated to these algorithms may be drastically different from general-purpose microcomputer architecture. Numerical analysis, for example, decomposes given mathematical functions into primitive steps, which are suitable for a conventional general-purpose computer with centralized memory, such that the entire computation can be done efficiently and accurately. Now with LSI/VLSI technology new numerical analysis algorithms need to be developed by tailoring these primitive steps to processing directly by logic gates and memories. Then we can develop architecture tailored to these algorithms, probably using unconventional logic networks with memories that are suitable to the primitive steps. Research into such algorithms is an important new area. In addition to those used in the support chips and coprocessor chips mentioned above, algorithms for sorting, fast Fourier transform, matrix/vector multiplication, speech synthesis, speech recognition, and many others have been explored. As an unconventional example, the CORDIC algorithm for the computation of elementary functions is explained in the following.

When the first scientific pocket calculator HP-35 was designed, designers at Hewlett-Packard did not adopt the conventional algorithms used in programming on mainframe computers. They considered the possibilities of using power series, polynomial expansions, continued fractions, and Chebyshev polynomials for the computation of elementary functions. But all these methods were too slow because of the number of multiplications and divisions required. They finally adopted an obscure algorithm called CORDIC to design their LSI chips. CORDIC (coordinate rotation digital computer) was devised by H. Briggs in 1624 and later by J. E. Volder in 1959. Although the computation is done by recursive operations by CORDIC, computation time is less than 1 sec for any function, while if a converging series, such as the Taylor series expansion, was adopted, the speed would have been several seconds or more (Osborne, 1974; Walther, 1971). We can realize algorithms for elementary functions with a processor that is designed based on CORDIC, consisting of shift registers, adders (which work also as subtractors), ROMs, and RAMs (as work memory space). In other words, by storing algorithm data in ROMs and using different control logic sequences for algorithms for different elementary functions, these elementary functions can be computed by this dedicated processor. When we calculate the trigonometric functions (which are elementary functions), for example, with conventional algorithms, using software on mainframe computers with centralized memory of magnetic cores, a large number of multiplications is involved. Also since the access time to magnetic core memories is slow (roughly 300 nsec or more), a reduction of the number of multiplications may not lead to significant improvement. But this dedicated processor, designed based on CORDIC, is very simple, as already mentioned, and the transfer of data among the shift registers, adders, and ROM can be done very quickly. The ROM is small and made of semiconductor devices, so its access time is very short. Furthermore no ordinary multiplication which is time-consuming is required. If a more specialized digital system, which contains more logic networks than the above simple configuration of shift registers, adders, and memories, is designed without using recursive computation such as CORDIC, such a digital system would be faster, but it would be more complex and also more difficult to design. The processor based on CORDIC, despite its simple architecture tailored to CORDIC, can efficiently calculate many elementary functions, such as sin, cos, tan, arcsin, arccos, arctan, sinh, cosh, tanh, arctanh, log, exp, and square root, by simply storing different sets of constants in ROMs. In this sense, CORDIC is a universal, powerful algorithm. CORDIC has been used in many calculators with some modifications (Volder, 1959; Egbert, 1977; Cochran, 1972; Perle, 1971; Haviland and Tuszynski, 1980).

References: Senzig, 1975; Whitney et al., 1972; Schmid, 1974; Hwang, 1979; de Lugish, 1970; Preparata and Vuillemin, 1981; *Electronic Design,* Jan. 22, 1981, pp. 40, 41; Kung, 1980; Preparata, 1978; Hayn, 1981; *Elec-*

tronics, Nov. 25, 1976, p. 77; May 24, 1979, p. 6; May 22, 1980, pp. 95–105; Oct. 23, 1980, pp. 194, 202; Schmidt, 1979; Foster and Kung, 1980; Thompson, 1980; Todd, 1978; *Computer Design,* Apr. 1979, pp. 104–116; July 1979, pp. 150–156; Kung and Leiserson, 1980; Majithia and Kitai, 1971; Cushman, 1981; Fischer, 1981; Henry, 1981.

Programmability of Hardware Realizations

Although speed is improved by the hardware realization of software, flexibility is lost. Firmware is much more time-consuming and expensive to change or to correct than software, and a logic network realization is far more time-consuming to change or to correct than firmware. In a sense, this is advantageous. This inflexibility forces programmers to accept standardized firmware or logic network realizations, reducing their reinventions of programs. Then by standardization, firmware or logic network realizations can be manufactured in high volume, pushing their prices down greatly. Also patent protection will be easier.

References: Schindler, May 24, 1979; Hartwig, 1979; Ashkenazi, 1981; *EDN,* Mar. 5, 1979, p. 31; Feb. 8, 1981, pp. 239, 240; *Electronic Design,* Jan. 8, 1981, pp. 134, 135; Conrad, 1981; Withington, 1980; Pressman, 1981.

If the hardware realization of software, however, is completely inflexible, it is inconvenient when we need to correct a design during development, or when we need to change interface specifications. But it is not so. When complex tasks are realized in firmware, firmware itself is more flexible than a logic network realization. When complex tasks are realized in logic networks on LSI/VLSI chips, the chips are usually processors which contain control logic in ROMs or PLAs. By changing ROMs (i.e., changing firmware) or PLAs, we can change control logic, making the logic network realization flexible (i.e., programmable) to some extent, though drastic changes are not possible. The loss of flexibility in applying chips of increasingly large integration size can be lessened to some extent by this programmability (Fischer, 1980; Scrupski, 1980).

10.3   Problems in the Future

When the integration size increases, we have to face a new formidable problem in developing LSI/VLSI chips. We have to integrate, without conflict, different scientific and engineering disciplines such as processing technology, software, algorithms, architecture, logic/electronic-circuit/layout designs, CAD, design verification, testing, and packaging. This problem is not really new, since we faced it in designing chips of smaller integration size. But when half a million or one million transistors can be packed on a single chip as the integration size increases, each of these disciplines becomes a complex difficult problem, and these disciplines as a whole become a much more complex problem with enormous difficulty because they interact in complex ways on a chip. If any aspect in these disciplines does not work, the entire chip does not work. In other words, the integration of many disciplines in a single chip in large quantities or qualities poses a new gigantic problem.

Despite these difficulties, we have to design, develop, manufacture, and market new products which incorporate LSI/VLSI chips and work reliably, in a reasonably short time, in order not to miss the market completely. Thus VLSI is essentially a system of compatible developments: chips, packages, CAD, tooling, fabrication automation, factory logistics, management of methodologies, software developments, and marketing strategies. In this section let us discuss related key future issues.

Large System Design

As the integration size increases, there are probably two different paths in future system design. One path is to pack cpu's, control logic, and memories in each chip as much as possible. This will provide the most cost effective systems in high-volume production. If a mainframe computer in a small number of VLSI chips is designed in this manner, such a digital system must be able to process already existing software (otherwise the market is too small), and IBM mainframe computers would be a natural target. (It is estimated that IBM software developed worldwide is worth about 200 billion dollars. This is why even IBM continues to design new computers that can run the old 370 programs. The next largest is perhaps 5 billion dollars of Digital Equipment Corporation's software (*Datamation,* Mar. 1979, p. 68). But the development of such digital systems could be plagued with the detection and correction of mistakes as the integration size and the computing capabilities increase.

The other path is to design a digital system as an aggregate of dedicated processor chips, probably mixed with general-purpose processor chips. Each chip is dedicated to advanced mathematical function computation, language compilers, data-base management, other key portions of conventional operating systems, sorting, input/output channel control, maintenance, application programs, and many others. The size of each chip is not too ambitiously large (not as large as in the case of the first path above), even if it may grow depending on technological progress. By designing dedicated processor chips appropriately, the system could have very high performance, even if it may not be as cost effective as in the case of the first path. Distributed data-processing networks can be realized using these processor chips. Computer systems based on nonconventional architecture, such as data-flow computers which are digital systems of parallel-processing with multiprocessors, can also be realized with dedicated processor chips.

As the cost of LSI/VLSI chips decreases, the prices of all computers from mainframes down to microcomputers decrease sharply, and the computer industry becomes more dependent on income from software. This is particularly true with mainframe manufacturers. This, however, does not mean that hardware is not important. If hardware has lower cost per performance (i.e., is cheaper and better), then the computers will have a greater market and the manufacturers' income from software for these computers will become proportionally greater. (The converse also can be true. In other words, if computers have comparable hardware price and performance, those with greater software support sell better.) In this case, what hardware is to software is something like what cameras are to film (i.e., if more cameras are sold, more film for them will also be sold). If some computers become dominant in the market, some other manufacturers may imitate these computers or supply software for these computers. But if these computers are sufficiently complex, this would not be easy. For example, if the operating systems of these computers are mixed with software, firmware, and logic gate networks (as IBM is doing gradually with the 4300 and 8100 computers), this would be difficult.

References: Faggin, 1978; Juliussen and Watson, 1978; Adams and Rolander, 1978; Durniak, 1979; Sugarman, 1980; Queyssac, 1979; Dennis, 1979; *Datamation,* Mar. 1979, pp. 68–71; *Computer,* Jan. 1980, pp. 4–7; *Computer Design,* June 1977, pp. 151–163; Oct. 1978, pp. 200–204; Mar. 1981, pp. 95–102; *Electronic Design,* May 14, 1981, pp. 51, 52; Fischer, 1980; *Computerworld,* Dec. 4, 1980, pp. 25–27; Hughes and Conrad, 1980; Alker, 1981; Schindler, May 4, 1981.

Chip Development versus Software Development

As the progress of IC technology lowers the manufacturing costs of hardware, the pricing of chips is becoming very similar to that of software. The initial investments for the development of large-scale chips and software are both expensive, but manufacturing costs are very low in the sense that the manufacturing cost of software is almost nothing, except for distribution cost, that is, the cost of a magnetic tape or a card deck, and the manufacturing cost of a chip is now very low. When the market, and consequently the production volume, are very large, both are very inexpensive, but when they are very small, both are very expensive. Thus manufacturers that have large customer bases are usually at an advantage. This tends to imply that manufacturers that reach the market first with new chips or products with new functions provided by chips occupy monopolistic positions, like the case of software developers. But there is an important difference in that the product life of software is practically unlimited if the software products are technology independent (in particular if the software is written on IBM mainframe computers, it can be run on new IBM computers), whereas the product life of chips is short because chips will be replaced by those based on new processing technology.

Because of the complex and time-consuming nature of software development, methodological programming approaches, called **software engineering,** which are psychologically appropriate for programmers to improve their programming productivity and to reduce errors, have been developed. (Structured programming is one such approach.) The selection of high-level languages, which are most appropriate for specific programming tasks, is also along this line. Since chip development is also becoming complex and time-consuming, and since the correction of mistakes or design changes is much more expensive than for software, serious attention to the psychological problems of designers, and consequently to methodologies in design approaches, are becoming vitally important in chip development also. The semi-custom design approaches discussed in Chapter 9 and the use of CAD discussed in Chapter 8 reduce such psychological problems. It appears that methodological considerations are becoming more important in chip development than technology, though technological progress is still the motive force in the progress of chip development techniques. In contrast, methodologies seem to be the almost sole consideration in software development because software is mostly technology independent (unless new software is required due to technological progress). Technology-independent methodologies are strongly desirable in chip development, but they appear to be much more difficult than in the case of software, because chip design contains electronic-circuit/solid-state technology, and technology-independent methodological approaches would lead to chips of poor quality. Of course the tradeoff between methodology and chip quality depends on how chips are to be used, like the efficiency difference between assembler language and higher-level languages.

Ironically the high technology of computers and ICs is becoming highly brain-labor intensive in both software and chip development, and the psychological and methodological management of programmers and chip designers is becoming a key problem for successful developments.

References: Brooks, 1975; Yourdon, 1979; Zelkowitz, 1978; Allison, 1980; Reifer and Trattner, 1977; Daly, 1977; Boehm, 1976; Belford et al., 1977; Kraft, 1977; *IBM Sys. J.,* 1980, pp. 414–477; 1981, pp. 119–267; *Electronic Design,* July 19, 1980, pp. 62–78; Aug. 16, 1980, pp. 45–48; *ACM Computing Surveys,* March 1981; Mannel and Johnson, 1981.

Future Manufacturing Problems

Because of the trends discussed so far, the structure of the manufacturing industry is gradually changing. Computer manufacturers, watch manufacturers, toy manufacturers, and possibly others are picking up LSI/VLSI technology know-how, because LSI/VLSI technology is becoming a vitally important basis of their products and consequently a survival-or-death problem. At the same time, semiconductor manufacturers are picking up system design capabilities and supporting software development capabilities. Yet manufacturing methods are drastically changing. Many components and software go into the inside of chips in unprecedented scale, with a number of new convenient functions which we could not imagine in the past. Assembly workers required for conventional manufacturing methods are being replaced mostly by chip designers. Each product (at least a major product line) requires many chip designers. So if a manufacturer has a number of different product lines, a large number of chip designers are needed, even if all the products in each line can be made by changing a few masks for a master product. Also these chip designers require advanced education or training in diversified disciplines. This will create an enormous shortage of chip designers (*Electronics*, Dec. 20, 1979, p. 43), even if chip design is facilitated by a corporate-wide CAD system.

Corporate-wide CAD systems are inevitable for essentially all manufacturers, requiring huge investments for many years. CRT terminals are distributed throughout a company so that designers can conveniently design LSI/VLSI chips by using a large number of CAD programs and a data base stored in central computers. The necessity of CAD systems is obvious if we recall that the initial investment for chip development is mostly salaries, as discussed in Section 2.3. So if the initial investment is reduced by shortening the design time by CAD, the selling prices of chips can be reduced greatly, even if chip size and performance are sacrificed (not as small or as good as time-consuming full-custom design chips). Of course CAD has other important advantages, such as the elimination of design mistakes. Thus unless a manufacturer develops an effective CAD system, it will lose a competitive edge over other manufacturers. CAD requires huge programming efforts. Also, extensive software management and maintenance are necessary in addition to the software that goes into the inside of the chips. This means that even manufacturers who have not been involved in extensive software development before will have to face the cumbersome management problem of extensive software development.

Market Changes

As the cost and size of LSI/VLSI chips decline sharply, a large number of new applications with these chips are surfacing in the market, as discussed in Section 1.1. Then, as illustrated in Table 10.3.1, the market shares of different types of computers in terms of shipped units are changing. Until the middle 1950s, computers meant mainframes, but this may change drastically during the 1980s. Notice, however, that in terms of percentage, the share of mainframes is shrinking, but in terms of units or dollars the market is expanding as the cost/performance of mainframes is improving. Since the production volume increases as mainframe prices decreases, mainframe

Table 10.3.1   Unit Shipments Classified by Computer Types[a]

COMPUTER TYPE	1956	1966	1976	1986
Mainframes ($100,000 and up)	100%	70%	16%	1%
Minicomputers ($15,000 to $100,000)	—	30%	75%	31%
Microcomputers (less than $15,000)	—	—	9%	68%

[a] From *Computerworld*, June 22, 1981, p. 65. Courtesy of Gartner Group.

manufacturers must change their ways of production from handcrafting individual computers of low production and high price into assembly-line-type production of high-volume, low-priced computers. In other words, even mainframes come to bear the nature of consumer electronic products, while manufacturers of consumer electronic products, toys, and watches are becoming computer manufacturers in their own ways. Also mainframe development is more time-consuming, because the design and correction are more time-consuming for semi-custom LSI chips than for off-the-shelf packages (*Datamation*, 1980, pp. 50, 51; *Business Week*, Oct. 20, 1980, pp. 99, 100; June 1, 1981, p. 110). Even though the price of a mainframe is about one million dollars, its production volume is on the order of a few thousands, even for IBM. [If the price is lower, the production volume is higher (*EDP Industry Report*, June 26, 1981).] A most popular microcomputer chip probably will sell in billions (*EDN*, Nov. 5, 1980, p. 98). Even if each chip costs a few dollars or less, new products with it can cost much more than those without it. The price difference, that is, the value added by a chip, could be on the order of tens or hundreds of dollars. (For example, the price of a food blender could double, at least initially, if a microcomputer is incorporated. Also Atari's video game console of about $150 has reportedly been sold in millions, and each owner will buy game cartridges of $20 to 30 each in tens of millions every year (Bernstein, 1981). Accordingly, the number of cartridges sold for each game is far greater than the number of copies sold for each software package of mainframe computers.) Thus the total market value of a microcomputer can be much greater than that of a mainframe, though the market values of software to be additionally sold must be compared carefully.

Chips are becoming enormously complex, and chip designers are making strenuous design efforts. But all these are being paid off by great simplicity at the user level and almost endless new products, as mentioned in Chapter 1. There are many possibilities for incorporating digital function capabilities into products. For example, even tiny wrist watches can have time in different time zones, an alarm, an engagement calendar, computers, and musical melodies. Also many products could not explore large markets without microprocessor chips. Video cassette recorders could not be popular without microcomputers for recording TV programs at specified times. Furthermore they will replace the photographic movie market when inexpensive solid-state cameras with LSI/VLSI chips are available, replacing movie film by magnetic tapes. Personal computers for small business and professionals have the fastest growing computer market, prompted by a number of computer retail stores opened nationwide (*Business Week*, Dec. 1, 1980, pp. 55, 56; Sept. 28, 1981, pp. 76–82). Video game consoles are sold in millions with "Space Invader" game craze. Home computers will be a big market, as a number of computer-aided instruction programs become available (*Business Week*, July 27, 1981, pp. 66, 68, 68E–H). In the future every residential home will have a video room with a projector TV screen, where a family gathers for watching videotex for shopping or news (*Business Week*, June 29, 1981, pp. 74–83; Fedida and Malik, 1979; Sigel et al., 1980), learning practically everything by computer-aided instruction with personal computers and video disks (e.g., using courseware developed for the computer-aided instruction project PLATO at the University of Illinois), or enjoying "vusic" (visual music) by video disks. At the same time, many manufacturers are becoming consumer electronics type digital system manufacturers, changing the nature of their products, manufacturing methods, and marketing methods. When they plan new products with IC chips, they have to determine marketing strategies at the same time. Otherwise, they cannot determine appropriate design approaches among those discussed in the previous chapters (in particular, in Chapter 9), because design approaches are highly dependent on production volumes, model changes, and many other factors.

When many products incorporate complex functions by digital systems on chips, we have a difficult input/output interface problem. If there are too many key buttons, users cannot remember which key buttons to push in what sequence. Microwave ovens are such examples. Also as the integration size of chips increases, complete testing is too time-consuming and too expensive for manufacturers. So we will have a number of products that may malfunction at unexpected times. The repair of such products is becoming very expensive. Thus product definition or planning with good human engineering and self-diagnosis capability is a key to successful new products, with appropriate marketing strategy. IBM's word processor Displaywriter (*Business Week*, Nov. 24, 1980, pp. 104, 106) and Xerox's intelligent terminal Star (*Datamation*, July 1981, pp. 36–49) are good examples of such products.

LSI/VLSI is the basis for the second industrial revolution, but before that age we have to solve a number of difficult problems which are not only technological but also managerial, methodological, psychological, and even sociological. The nature of changes is changing. Everything is becoming more complex, and interrelationship is even more complex. The understanding of each subject or area is important. But finding a good balance in all kinds of tradeoffs among many things (such as cost, performance, market size, maintainability, repairability, upgradability, and marketing channels) is probably more essential to good system design.

References: Toffler, 1980; Orme, 1980; Wise et al., 1980; Evans, 1979; Barron and Curnow, 1979; Lecht, 1977; *EDN*, Nov. 20, 1980, pp. 154–156; *Digital Design*, Mar. 1979, p. 96; April 1979, p. 80; July 1979, p. 80.

APPENDIX

Theoretical Background for Procedure 4.2.1 for the Design of Networks with a Minimum Number of Negative Gates

Let us discuss why we can design MOS networks with a minimum number of negative gates by Procedure 4.2.1. [For algebraic terminology, see Muroga (1979).]

First, let us prove why the feed-forward network in Fig. 4.2.2 can express any loopless network. Consider an arbitrary loopless network. There is at least one gate to which no other gates supply inputs. Thus the leftmost gate in Fig. 4.2.2 has been chosen. The remaining gates constitute a loopless network, and there is at least one gate to which no other gates except the chosen gate supply inputs. Thus the second gate in Fig. 4.2.2 has been chosen. The continuation of this procedure will lead to the network in Fig. 4.2.2.

As can be easily verified using DeMorgan's theorem, any negative function may be rewritten as a disjunctive form with complemented literals only. For example, $\overline{x_1 x_2 \vee x_1 x_3}$ may be rewritten as $\overline{x}_1 \vee \overline{x}_2 \overline{x}_3$. Let us show that if a negative function f has such a disjunctive form, an irredundant disjunctive form of f also has such a property.

Theorem A1: Any negative function f has the unique irredundant disjunctive form. A function f is a negative function if and only if the irredundant disjunctive form consists of complemented literals only.

Proof: When we apply the iterated consensus method on a disjunctive form for f, there is no consensus, so only subsuming terms are deleted. Since the complete sum (i.e., the disjunction of all prime implicants) which results has no consensus, this is the unique irredundant disjunctive form and it consists of complemented literals only. The converse is trivial. *Q.E.D.*

When a function f is expressed in the form of a lattice like Fig. 4.2.3, vectors (x_1, x_2, x_3) such that $f = 0$ are clustered in an upper part if f is a negative function. An example for negative function $f = \overline{x_1 x_2 \vee x_1 x_3}$ is illustrated in Fig. A1. This property is formally stated in Theorem A2.

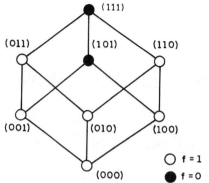

$\bigcirc$ f = 1

$\bullet$ f = 0   **Fig. A1   Lattice for function** $f = \overline{x_1 x_2} \vee \overline{x_1 x_3}$

Theorem A2:   A necessary and sufficient condition for a function f to be a negative function is

$$f(A) \le f(B)$$

for every pair of vectors A and B, such that $A > B$. [$A > B$ means that A has 1 in every position where B has 1. For example, $(1011) > (1001)$.]

Proof: Let f be a negative function and be represented as

$$f = P_1 \vee P_2 \vee \cdots \vee P_k$$

where P_j is a product of complemented literals for $j = 1, \ldots, k$. Consider a pair A and B such that $A > B$. Suppose that $f(A) = 0$. Then $f(A) \le f(B)$ holds, no matter whether $f(B) = 1$ or 0. Suppose that $f(A) = 1$. Then there must be $P_g = 1$ for A, where $1 \le g \le k$. Since P_g is a product of complemented literals and $A > B$ holds, $P_g = 1$ holds for B. That is, $f(A) = f(B) = 1$. Thus when f is a negative function, $f(A) \le f(B)$ holds for every pair $A > B$.

Conversely, assume that $f(A) \le f(B)$ whenever $A > B$. We need to show that f has a disjunctive form consisting of complemented literals only.

Among all vectors C's, such that $f(C) = 1$, let $C_1, C_2, \ldots, C_k$ be vectors for each of which there is no other C such that $C > C_i$ and $f(C_i) = 1$ for $i = 1, 2, \ldots, k$. Then corresponding to each C_i, form the product P_i of complemented literals of variables only which have 0 in C_i (e.g., if $C_i = (1010)$, $P_i = \overline{x_2} \overline{x_4}$). Then a disjunctive form which consists of these products P_i's,

$$P_1 \vee P_2 \vee \cdots \vee P_k$$

expresses f, because, by the assumption, the form assumes the value 1 only for each C_i and all the C's such that $C < C_i$. Because the disjunctive form consists of complemented literals only, f is a negative function. *QED*

Write the lattice of all vectors of n variables and label all nodes. When A and B differ in only one digit and $A > B$ holds, the directed edge from A to B is denoted with $\overrightarrow{AB}$. Edge $\overrightarrow{AB}$ is called an **inverse edge** if $L(A) > L(B)$ holds between two binary numbers $L(A)$ and $L(B)$, labeled on A and B.

(Note that $>$ here means "greater than" in the comparison of two binary numbers, whereas $>$ means in comparison of two vectors.)

As can be easily seen, when the value of a negative function f is labeled on each node of the lattice, f has no inverse edge.

Theorem A3: For a feed-forward network of R negative gates for f in Fig. 4.2.2, the $(u_1, \ldots, u_R)$'s labeled on the lattice of all vectors of $x_1, \ldots, x_n$ have no inverse edge. Conversely if the $(u_1, \ldots, u_R)$'s have no inverse edge, a feed-forward network of R negative gates whose outputs are $u_1, \ldots, u_R$ can be realized, where $u_R = f$.

Proof: Let us prove the first statement by mathematical induction. Since $g_1(x_1, \ldots, x_n)$ is a negative function,

$$u_1(A) = g_1(A) \leq g_1(B) = u_1(B)$$

holds for any edge $\vec{AB}$ in the lattice because of Theorem A2. Thus (u_1) has no inverse edge.

Assume that $(u_1, \ldots, u_i)$ has no inverse edge, that is, $(u_1, \ldots, i_i)_A \leq (u_1, \ldots, u_i)_B$ holds for every edge $\vec{AB}$, where $(u_1, \ldots, u_i)_A$ and $(u_1, \ldots, u_i)_B$ denote the binary numbers $(u_1, \ldots, u_i)$ labeled on A and B, respectively. If $(u_1, \ldots, u_i)_A < (u_1, \ldots, u_i)_B$ holds, $(u_1, \ldots, u_{i+1})_A < (u_1, \ldots, u_{i+1})_B$ holds, regardless of the values of $u_{i+1}(A)$ and $u_{i+1}(B)$. If $(u_1, \ldots, u_i)_A = (u_1, \ldots, u_i)_B$ holds, $(A, u_1(A), \ldots, u_i(A)) > (B, u_1(B), \ldots, u_i(B))$ because $A > B$ and $u_k(A) = u_k(B)$ for $k = 1, 2, \ldots, i$ hold. Consequently, $u_{i+1}(A) = g_{i+1}(A, u_1(A), \ldots, u_i(A)) \leq g_{i+1}(B, u_1(B) \ldots, u_i(B)) = u_{i+1}(B)$ holds because g_{i+1} is a negative function. Thus $(u_1, \ldots, u_{i+1})_A \leq (u_1, \ldots, u_{i+1})_B$ holds. Therefore $(u_1, \ldots, u_{i+1})$'s, and consequently $(u_1, \ldots, u_R)$'s by induction, have no inverse edge.

Conversely, assume that $(u_1, \ldots, u_R)$'s have no inverse edge, that is, $(u_1, \ldots, u_R)_A \not> (u_1, \ldots, u_R)_B$ holds for every edge $\vec{AB}$. This leads to $(u_1, \ldots, u_R)_A \leq (u_1, \ldots, u_R)_B$ for every pair of A and B such that $A > B$.

Then $u_1(A) \leq u_1(B)$ holds for every pair of A and B such that $A > B$. So u_1 can be realized by a negative gate by Theorem A2.

Assume that $u_1, \ldots, u_i$ are realized as negative gates $g_1, \ldots, g_i$ in the configuration of a feed-forward network. By the assumption that the $(u_1, \ldots, u_R)$'s have no inverse edge, $(u_1, \ldots, u_i)_A \not> (u_1, \ldots, u_i)_B$ holds for every pair of A and B such that $A > B$. In other words, only one of the following cases holds:

(i) $(u_1(A), \ldots, u_i(A)) < (u_1(B), \ldots, u_i(B))$

(ii) $(u_1(A), \ldots, u_i(A)) = (u_1(B), \ldots, u_i(B))$

Only when case (ii), that is, $u_k(A) = u_k(B)$ for $k = 1, \ldots, i$ holds, $(A, u_1(A), \ldots, u_i(A)) > (B, u_1(B), \ldots, u_i(B))$ can hold. [Case (i) need not be considered because $(A, u_1(A), \ldots, u_i(A))$ and $(B, u_1(B), \ldots, u_i(B))$ become incomparable, and this is not involved in the condition stated in Theorem A2. Notice that in Theorem A2, only a pair of vectors such that one vector is greater than the other are considered.] Furthermore since the $(u_1, \ldots, u_R)$'s are assumed to have no inverse edge, $u_{i+1}(A) \leq u_{i+1}(B)$ holds. Thus $g_{i+1}(A, u_1(A),$

. . . , $u_i(A)) \leq g_{i+1}(B, u_1(B), \ldots, u_i(B))$ holds when $(A, u_1(A), \ldots, u_i(A))$ $> (B, u_1(B), \ldots, u_i(B))$ holds. Therefore by Theorem A2, u_{i+1} can be realized with a negative gate g_{i+1} which has inputs $x_1, \ldots, x_n, u_1, \ldots, u_i$. By induction, $u_1, \ldots, u_R$ can be realized with R negative gates in a feed-forward network whose output is f. *QED.*

Procedure 4.2.1 follows from Theorem A3 and yields the minimum value of R because the bit length R of $L(00. . .0)$ is minimized. It is important to notice that if the lattice is labeled, without inverse edge, by not necessarily minimizing R, a feed-forward network in which no cell has undesirable complexity can be obtained due to Theorem A3.

References

Some publication names are abbreviated as follows: **IEEE** for the Institute of Electrical and Electronics Engineers, **T(E)C** for *Transactions on (Electronic) Computers*, **TED** for *Transactions on Electron Devices*, **TCHMT** for *Transactions on Components, Hybrids, and Manufacturing Technology*, **TCT** for *Transactions on Circuit Theory*, **IRE** for *The Institute for Radio Engineers*, **ISSCC** for *International Solid-State Circuits Conference (IEEE)*, **JSSC** for *Journal of Solid-State Circuits*, **DA** for *Design Automation*, **SJCC** for *Spring Joint Computer Conference*, **IEDM** for *International Electron Devices Meeting*, **NCC** for *National Computer Conference*, **ICCC** for *International Conference on Circuits and Computers*, **AFIPS** for *American Federation of Information Processing Societies*, **IFIP** for *International Federation of Information Processing*, and **IBM JRD** for *IBM Journal of Research and Development*.

Adams, G., and T. Rolander (Intel), "Design motivations for multiple processor microcomputer systems," *Computer*, March 1978, pp. 81–89.

Adams, P. M., "Microprogrammable microprocessor survey," *SIGmicro newsletter*, Part I, March 1978, pp. 23–49; Part II, June 1978, pp. 7–38.

Adams, W. T., and S. M. Smith, "How bit-slice families compare," *Electronics*, Part 1, "Evaluating processor elements," Aug. 3, 1978, pp. 90–98; Part 2, "Sizing up the microcontrollers," Aug. 17, 1978, pp. 96–102.

Adler, D., "Amorphous-semiconductor devices," *Scientific American*, May 1977, pp. 36–48.

Agajanian, A. H. (IBM), *Semiconducting Devices: A Bibliography of Fabrication Technology, Properties, and Applications*, Plenum Press, 1976, 944 pp.

Agajanian, A. H. (IBM), *Computer Technology: Logic, Memory, and Microprocessors—A Bibliography*, Plenum Press, 1978, 346 pp.

Agajanian, A. H. (IBM), *MOSFET Technologies: A Comprehensive Bibliography*, Plenum Press, 1980.

Agrawal, D. P., and K. K. Agrawal, "Efficient sorting with CCDs and magnetic bubble memories," *IEEE TC*, Feb. 1981, pp. 153–157.

Aho, A. V., J. E. Hopcroft, and J. D., Ullman, *The Design and Analysis of Computer Algorithms*, Addison-Wesley, 1974, 470 pp.

Ahrens, P., K. Skoge, and D. Vetter (Milton Bradley), "Speech chip timeshares a 2-pole section to create a 12-pole filter," *Electronics*, Mar. 10, 1981, pp. 177–180 (toys with synthesized speech).

Ahrons, R. W., and P. D. Gardner, "Interaction of technology and performance in complementary symmetry MOS integrated circuits," *IEEE JSSC*, Feb. 1970, pp. 24–29.

Akazawa, Y., et al. (NTT and NEC), "A high speed 1,600-gate bipolar LSI processor," *ISSCC 78*, pp. 208, 209.

Akers, S. B. (GE), "A modification of Lee's path connection algorithm," *IEEE TEC*, Feb. 1967, pp. 97, 98.

Akers, S. B. (GE), "Routing," in *Design Automation of Digital Systems*, Vol. 1, edited by M. A. Breuer, Prentice-Hall, 1972, pp. 283–333, Chap. 6.

Akers, S. B. (GE), "Logic system for fault test generation," *IEEE TC*, June 1976, pp. 620–630.

Akers, S. B. (GE), "Test generation techniques," *Computer*, Mar. 1980, pp. 9–15.

Alexandridis, N. A., "Bit-sliced microprocessor architecture," *Computer,* June 1978, pp. 56–80.

Alexy, G., and H. Kop (Intel), "Get minicomputer features at ten times the 8080 speed with the 8086," *Electronic Design,* Sept. 27, 1978, pp. 60–66.

Alfke, P., and C. Alford, "Low-power Schottky TTL," *Progress, Fairchild J. of Semiconductor,* Sept./Oct. 1975, pp. 3–9.

Alford, C., B. Bechdolt, D. Ferris, S. Goodspeed, and P. Griffith, "FAST, an elegant use of power," *Progress, Fairchild J. Semiconductor,* Mar./Apr. 1979, pp. 4–10.

Alker, P. L. (Convergent Technologies), "Distributed intelligence vs. centralized logic," *MINI-MICRO Systems,* May 1981, pp. 103–115. (Cost analysis.)

Alkhateeb, D., S. Kawasaki, and S. Muroga, "An improvement of a branch-and-bound algorithm for designing NOR optimal networks," Rep. UIUCDCS-R-80-1033, Dept. of Computer Science, University of Illinois, Sept. 1980, 35 pp.

Allan, R., "Designer's bookshelf," *Spectrum,* Oct. 1975, pp. 54, 55.

Allan, R., "The microcomputer invades the production line," *Spectrum,* Jan. 1979, pp. 53–57.

Allan, R., "VLSI: Scoping its future," *Spectrum,* Apr., 1979, pp. 30–35.

Allen, C. A. (Master Logic), "Digital CMOS gate arrays," *Wescon,* 1979, 13/5, 3 pp.

Allen, R. A. (TRW), "Digital CCD arithmetic technology," *Compcon 77,* Spring, pp. 342–344.

Allison, A., "Follow three simple rules to improve software productivity," *EDN,* Mar. 20, 1980, pp. 167–171. (Software cost is analyzed. An extensive list of literature on software productivity is included.)

Allison, A., and R. J. Frankenberg, "Cost advantage of PROM vs. ROM open to debate," *EDN,* Jan. 5, 1976, p. 99.

Allison, D. F., and L. K. Russell, U.S. patent no. 3,600,642, Aug. 17, 1971. (MOS structure with precisely controlled channel length and method.)

Allison, D. R., "A design philosophy for microcomputer architecture," *Computer,* Feb. 1977, pp. 35–41. (Hardware realization of software.)

Alsing, C. J., K. D. Holberger, C. J. Holland, E. J. Rasala, and S. J. Wallach (Data General), "Minicomputer fills mainframe's shoes," *Electronics,* May 22, 1980, pp. 130–137. (PALs are used extensively in 32-bit Eclipse MV/8000 minicomputer, p. 133.)

Altman, L. "Special report: CMOS enlarges its territory," *Electronics,* Mar. 15, 1975, pp. 77–88.

Amazeen, B. E., and M. P. Timko (Analog Devices Semiconductors), "Linear compatible I^2L," *Wescon, 1979,* 16/1, 5 pp.

Amelio, G. F., "Charge-coupled device," *Scientific American,* Feb. 1974, pp. 22–31.

American Microsystems, Inc., *Six Steps to Success with Custom LSI,* 1979, 18 pp.

Amey, D. I. (Univac), and J. W. Balde (Western Electric), "Four-chip hybrid carrier holds down system costs," *Electronics,* Jan. 17, 1980, pp. 113–117.

AMI staff, *MOS Integrated Circuits,* Van Nostrand, 1972, 474 pp.

Anacker, W. (IBM), "Computing at 4 degrees Kelvin," *Spectrum,* May 1979, pp. 26–37. (Josephson junction computer.)

Anderson, R. B., "Cut binary to BCD conversion costs," *Electronic Design,* Oct. 11, 1969, pp. 104–110.

Antipov, I. (IBM), "Proposed process modifications for dynamic bipolar memory to reduce emitter–base leakage current," *IEEE JSSC,* Aug. 1980, pp. 714–719.

Antoniadis, D. A., S. E. Hansen, and R. W. Dutton, "SUPREM II—A program for IC process modeling and simulation," Tech. Rep. SEL 78-020, Stanford Electronics Lab., June 1978.

Appelt, D. R. (Texas Instruments), "Making it compatible and better: Designing a new high-end computer," *Electronics,* Oct. 11, 1979, pp. 131–136. (990/12.)

Appleton, D. S. (Borg Warner), "A strategy for manufacturing automation," *Datamation,* Oct. 1977, pp. 64–70.

Armstrong, R. A. (Digital Equipment Co.), "Applying CAD to gate arrays speeds 32-bit minicomputer design," *Electronics,* Jan. 13, 1981, pp. 167–173.

Ashida, M. et al. (Fujitsu), "A 3000-gate CMOS masterslice LSI," *Proc. 11th Conf. Solid State Dev.*, Tokyo, Japan, 1979, pp. 203–206.

Ashkenazi, D. A., "Hardware comes to the aid of modular high-level language," *Electronics*, Apr. 21, 1981, pp. 175–177.

Asija, S. P. (Univac), "Four-phase logic is practical," *Electronic Design*, Dec. 20, 1977, pp. 160–163.

Athanas, T. G. (RCA), "Development of COS/MOS technology," *Solid State Technology*, June 1974, pp. 54–59.

Ayres, R. (Xerox), "Silicon compilation—A hierarchical use of PLAs," *Proc. 16th DA Conf.*, 1979, pp. 314–326.

Bal, S., E. Burdick, R. Barth, and D. Bodine (National Semiconductor), "System capabilities get a boost from a high-powered dedicated slave," *Electronic Design*, Mar. 1, 1980, pp. 77–82. (Floating-point processor chip NS16081.)

Balde, J. W. (Western Electric), and D. I. Amey (Univac), "New chip carrier package concepts," *Computer*, Dec. 1977, pp. 58–68.

Balde, J. W. (Western Electric), and D. I. Amey (Univac), "New chip carrier concepts will impact LSI-based designs," *EDN*, Sept. 20, 1978, pp. 119–126.

Balph, T. (Motorola), "Use ECL 10,000 layout rules to help solve pc-board interconnect problems," *Electronic Design*, Aug. 17, 1972, pp. 72–76.

Balph, T., W. Blood and J. Prioste (Motorola), "A card program for high speed logic element interconnections," *Computer Design*, May 1975, pp. 135–139. (ECL interconnections.)

Bandler, J. W., and M. R. M. Rizk, "Optimization on electrical circuits," Mathematical Programming Study 11, Oct. 1979, pp. 1–64.

Barbe, D. F., Ed., *Charge-Coupled Devices*, Springer, 1980, 180 pp.

Barbe, D. F., Ed., *Very Large Scale Integration (VLSI): Fundamentals and Applications*, Srpinger, 1980, 279 pp.

Barna, A., and C. A. Liechti (Hewlett-Packard), "Optimization of GaAs MESFET logic gates with subnanosecond propagation delays," *IEEE JSSC*, Aug. 1979, pp. 708–715.

Barna, A., and D. I. Porat, *Integrated Circuits in Digital Electronics*, John Wiley, 1973, 483 pp.

Barnes, D., "ROMs and PROMs are moving to greater densities, compatibility," *Electronic Design*, July 5, 1978, pp. 40–42, 44.

Barnes, D., "More speed, greater density keep PROMs in the game against EPROMs," *Electronic Design*, July 19, 1978, pp. 42–45.

Barnes, D., "CMOS performance, cost makes digital just part of its story," *Electronic Design*, Nov. 8, 1978, pp. 52–56.

Barraclough, W., A. C. L. Chiang, and W. Sohl (Microdata), "Techniques for testing the microcomputer family," *Proc. IEEE*, June 1976, pp. 943–950.

Barron, I., and R. Curnow, *The Future with Microelectronics*, Frances Printer Ltd., 1979, 243 pp.

Bass, C., J. Estrin, and B. L. Peuto, "Rich instructions, nine addressing modes make coding easy—Introducing the Z8, part 2," *Electronics*, Aug. 31, 1978, pp. 134–137.

Bassak, G., "Microelectronics takes to the road in a big way: A special report," *Electronics*, Nov. 20, 1980, pp. 113–122. (IC for automobiles.)

Batni, R. P., and C. R. Kime, "A module-level testing approach for combinational networks," *IEEE TC*, June 1976, pp. 594–604.

Baugh, C. R., C. S. Chandersekaran, R. S. Swee, and S. Muroga, "Optimal networks of NOR-OR gates for functions of three variables," *IEEE TC*, Feb. 1972, pp. 153–160.

Baugh, C. R., and B. A. Wooley (Bell Labs), "A two's complement parallel array multiplication algorithm," *IEEE TC*, Dec. 1973, pp. 1045–1047.

Beall, R. J. (Amdahl), "Packaging for a super computer," *IEEE Intercon 74*, 18/3, 9 pp.

Bechade, R., and W. K. Hoffman (IBM), "Generalized 2 bit slice ALU," *ICCC 80*, pp. 1094–1098.

Bechdolt, R., D. Ferris, and P. Griffith (Fairchild), "Oxide isolation builds a better Schottky TTL," *Electronics,* Mar. 1, 1979, pp. 111–116.

Belford, P. C., J. D. Donahoo, and W. J. Heard (Computer Sciences Corp.), "An evaluation of the effectiveness of software engineering techniques," *Compcon 77,* Fall, pp. 259–267.

Bell, C. G. (DEC), "Computers—Past, present and future," *Research/Development,* May 1976, pp. 22–26.

Bening, L. (CDC), "Developments in computer simulation of gate level physical logic—A tutorial," *Proc. 16th DA Conf.,* 1979, pp. 561–567.

Bentley, A. W. (Cubic), "FPLA arbiter concept adapts to application needs," *Computer Design,* June 1981, pp. 149–155.

Berger, H. H., and K. Kelwig (IBM), "An investigation of the intrinsic delay (speed limit) in MTL/I^{2}L," *IEEE TED,* Apr. 1979, pp. 405–415.

Berger, H. H., and S. K. Wiedmann (IBM), "Merger-transistor logic (MTL)—A low cost bipolar logic concept," *IEEE JSSC,* Oct. 1972, pp. 340–346.

Berger, H. H., and S. K. Wiedmann (IBM), "The bipolar LSI breakthrough," *Electronics,* Part 1, Sept. 4, 1975, pp. 89–95; Part 2, Oct. 12, 1975, pp. 99–103.

Berglund, N. C. (IBM), "Level-sensitive scan design tests chips, boards, system," *Electronics,* Mar. 15, 1979, pp. 108–110. (Test of logic of IBM System/38.)

Bernhard, R., "The 64-kb RAM teaches a VLSI lession," *Spectrum,* June 1981, pp. 38–42.

Bernstein, P. W., "Atari and the video-game explosion," *Fortune,* July 27, 1981, pp. 40–46.

Beynon, J. D. E., and D. R. Lamb, *Charge-Coupled Devices and Their Applications,* McGraw-Hill, 1980, 275 pp.

Besant, C. B., *Computer-Aided Design and Manufacture,* John Wiley, 1981, 170 pp.

Bhattacharyya, A., et al., "1/N circuit and device technology," *IBM JRD,* May 1980, pp. 378–389.

Biancomano, V., "Logic-simulator programs set pace of computer-aided design," *Electronics,* Oct. 13, 1977, pp. 98–101.

Bingham, D. (Intersil), "CMOS: Higher speeds, more drive and analog capability expand its horizons," *Electronic Design,* Nov. 8, 1978, pp. 74–82.

Birkner, J. (Monolithic Memories), "Microprogramming random logic," *Compcon 78,* Spring, pp. 75–80.

Birkner, J. (Monolitic Memories), "Reduce random-logic complexity by using arrays of fuse-programmable circuits," *Electronic Design,* Aug. 16, 1978, pp. 98–105.

Birkner, J. (Monolithic Memories), "PALs: Programmable logic functions help minimize hardware," *Wescon,* 1979, 18/4, 6 pp.

Blaauw, G. A., *Digital System Implementation,* Prentice-Hall, 1976, 384 pp.

Blattner, D. J., "Choosing the right programs for computer-aided design," *Electronics,* Apr. 29, 1976, pp. 102–105.

Bloch, E. (IBM), "VLSI for the 1980s," *Circuits Manufacturing,* Aug. 1979, pp. 16, 18, 22, 24, 26. (Summary of the speech at the IEEE/EIA's 26th Electronic Components Conf., May 1979.)

Block, E., and D. Galage (IBM), "Component progress: Its effect on high-speed computer architecture and machine organization," *Computer,* Apr. 1978, pp. 64–76.

Blodgett, A. J., Jr. (IBM), "A multilayer ceramic multichip module," *IEEE TCHMT,* Dec. 1980, pp. 634–637.

Blood, W. R., Jr. (Motorola), "M10800 MECL LSI circuits are designed for high-performance microprogrammed processors," *1979 Wescon,* 2/4, 12 pp.

Blood, W. R., Jr., (Motorola) "Computer aided design: The engineering interface for Macrocell array circuits," *Compcon 81,* Spring, pp. 182–185.

Blood, W. R., Jr. (Motorola)," CAD toolbox holds all gear for the design of custom logic arrays," *Electronics,* Sep. 8, 1981, pp. 140–142.

Blood, W. R., Jr., et al., *MECL System Design Handbook,* 2nd ed., Motorola, 1972, 237 pp.

Blumberg, R. J., and S. Brenner (IBM), "A 1500 gate, random logic, large-scale integrated (LSI) masterslice," *IEEE JSSC,* Oct. 1979, pp. 818–822. Also *ISSCC 79,* pp. 60–61.

Bobeck, A. H., P. I. Bonyhard, and J. E. Geusic, "Magnetic bubbles—An emerging new memory technology," *Proc. IEEE*, Aug. 1975, pp. 1176–1195.

Bobeck, A. H., and H. E. D. Scovil (Bell Labs), "Magnetic bubbles," *Scientific American*, June 1971, pp. 78–90.

Boehm, B. W. (TRW), "Software engineering," *IEEE TC*, Dec. 1976, pp. 1226–1241. (Survey.)

Bollinger, D., and R. Fink (Veeco Instruments), "A new production technique: Ion milling—Part II," *Solid State Technology*, Dec. 1980, pp. 97–103.

Bongiovanni, G., and F. Luccio, "Maintaining sorted files in a magnetic bubble memory," *IEEE TC*, Oct. 1980, pp. 855–863.

Boone, L. A., G. A. Champine, and B. A. Borgerson (Univac), "The microarchitecture of UNIVAC's 1100/60," *Datamation*, July 1979, pp. 173–178.

Bossen, D. C., and M. Y. Hsiao, "A system solution to the memory soft error problem," *IBM JRD*, May 1980, pp. 390–397.

Bowers, J. C., et al., "A survey of computer-aided-design and analysis program," Electrical and Electronics Department, University of South Florida, Tampa, FL (Air Force Aero Propulsion Lab., Wright-Patterson Air Force Base), Apr. 1976, 230 pp.

Boysel, L. L., and J. P. Murphy, "Four-phase LSI logic offers new approach to computer design," *Computer Design*, Apr. 1970, pp. 141–146.

Braeckelmann, W., et al. (Siemens), "A masterslice LSI for subnanosecond random logic," *IEEE JSSC*, Oct. 1979, pp. 829–832.

Branin, F. H., Jr. (IBM), "The analysis and design of power distribution nets on LSI chips," *ICCC 80*, pp. 785–790.

Branin, F. H., Jr., G. R. Hogsett, R. L. Lunde, and L. E. Kugel (IBM), "ECAP II," *Spectrum*, June 1971, pp. 14–25.

Branin, F. H., Jr., et al. (IBM), "ECAP II—A new electronic circuit analysis program," *IEEE JSSC*, Aug. 1971, pp. 146–166.

Bratt, B., "A CRT display system using NMOS memories," Motorola Application Note AN-706, 1973.

Bredeson, J. G., "A cellular array for integer and fractional BCD–binary conversion," *Computer Design*, May 1974, pp. 104–108.

Bremer, J. W. (Honeywell Information System), "Hardware technology in the year 2001," *Computer*, Dec. 1976, pp. 31–36. (More of the hard core of the OS will be put into hardware, partly for security reasons.)

Breuer, M. A., "General survey of design automation of digital computer," *Proc. IEEE*, Dec. 1966, pp. 1708–1721.

Breuer, M. A., Ed., *Design Automation of Digital Systems*, vol. 1, Prentice-Hall, 1972, 420 pp.

Breuer, M. A., Ed., *Digital System Design Automation: Languages, Simulation and Data Base*, Computer Science Press, 1975, 417 pp.

Breuer, M. A., "Recent developments in automated design and analysis of digital systems, *Computer*, May/June 1977, pp. 23–35.

Breuer, M. A., and A. D. Friedman, *Diagnosis and Reliable Design of Digital Systems*, Computer Science Press, 1976, 308 pp.

Breuer, M. A., and A. C. Parker, "Digital system simulation: Current status and future trends or Darwin's theory of simulation," *Proc. 18th DA Conf.*, 1981, pp. 269–275.

Breuer, M. A., A. D. Friedman, and A. Iosupovicz, "A survey of the state of the art of design automation," *Computer*, Oct. 1981, pp. 58–75.

Bricaud, P., and J. Campbell (Signetics), "Multiple output PLA minimization: EMIN," *Wescon, 1978*, 33/3, 9 pp.

Brockman, D. M. (Boeing), "Special PROM mode effects binary-to-BCD converter," *Electronics*, Mar. 31, 1977, pp. 105–107.

Brodsky, M. (Zilog), "Hardening RAMs against soft errors," *Electronics*, Apr. 24, 1980, pp. 6, 117–122.

Brooks, F. P., Jr., *Mythical Man-Month*, Addison-Wesley, 1975, 195 pp.

Brooks, F. P., Jr., "An overview of microcomputer architecture and software," *Proc. Euromicro 1976,* North Holland, pp. 1–6.

Brown, P. M., Jr. (Exar), "Complex LSI design using I²L gate arrays," *Wescon, 1980,* 30/2, 6 pp.

Brown, R. F., et al. (IBM Zürich), "Model for a 15 ns 16k RAM with Josephson junctions," *IEEE JSSC,* Aug. 1979, pp. 690–699.

Brown, W. L., et al. (Bell Labs), "Ion beam lithography," *Solid State Technology,* Aug. 1981, pp. 60–67.

Browne, B. T., and J. J. H. Miller, Eds., *Numerical Analysis of Semiconductor Devices,* Boole Press, Dublin, 1979.

Bruederle, S., and P. Smith (Signetics), "Designing with I²L," *Wescon,* 1975, 4 pp.

Brunner, R. H., E. J. Holden, J. C. Luber, D. T. Nozer, and N. G. Wu (IBM), "Automated semiconductor line speeds custom chip production," *Electronics,* Jan. 27, 1981, pp. 121–127. (IBM's QTAT—quick turn around time.)

Bryant, J. D., et al. (Texas Instruments), "TMS 2100 series single chip microcomputers," *Electro 80,* 29/2, 8 pp.

Bryce, H. (Motorola), "The MC68120 8-bit intelligent peripheral controller enhances any system as a slave processor," *Electro 80,* 33/1, 6 pp.

Bryson, D., D. Clover, and D. Lee (Intel Magnetics), "Megabit bubble-memory chip gets support from LSI family," *Electronics,* Apr. 26, 1979, pp. 105–111.

Buchwald, S. (Data General), "Microcode increases minicomputer processing capability," *Computer Design,* Oct. 1977, pp. 91–99. (Comparison of hard-wired control and microprogrammed control.)

Bucklen, W., et al. (TRW/LSI Products), "Single-chip digital multipliers form basic DSP building blocks," *EDN,* Apr. 1, 1981, pp. 153–163. (Multiplier survey.)

Buie, J. (TRW), "Improved tripke diffusion means densest ICs yet," *Electronics,* Aug. 7, 1975, pp. 101–106.

Buie, J. L., and E. E. Swartzlander, Jr. (TRW), "High density bipolar logic technology," *Compcon 77,* Spring, pp. 338–341.

Burck, C., "Getting to know the smart phone," *Fortune,* Feb. 25, 1980, pp. 134–146.

Burke, R. E., and J. G. van Bosse (Automatic Electric Lab.), "NAND-AND circuits," *IEEE TC,* Feb. 1965, pp. 63–65. (Comparison of TTL networks and NAND networks.)

Burns, J. R., "Switching response of complementary-symmetry MOS transistor logic circuits," *RCA Rev.,* vol. 25, 1964, p. 627.

Bursky, D., "Overview of programmable logic and memory devices," *Wescon, 1979,* 7/0, 5 pp.

Bursky, D., "μC boards pack the CPUs and the support for every system—but how much is enough?" *Electronic Design,* Mar. 15, 1980, pp. 85–89.

Bursky, D., "Coprocessor implements floating-point math," *Electronic Design,* May 10, 1980, pp. 35–37.

Bursky, D., "Memory and μP ICs pack more, consume less," *Electronic Design,* June 7, 1980, pp. 75–87.

Bursky, D., "Microprocessors—4 to 32-bit—Push back performance limits," *Electronic Design,* Nov. 22, 1980, pp. 109–115. (IBM 370 data flow on chips.)

Bursky, D., "PALs pursue a logical goal—easier, quicker more testable arrays," *Electronic Design,* July 5, 1980, pp. 33–34. (PALs used in Data General's Eclipse.)

Bursky, D., "Support circuits—The 'power' behind powerful processors," *Electronic Design,* Nov. 22, 1980, pp. 123–130, 132, 134, 136, 138, 140.

Bursky, D., "UVEPROMs and EEPROMs crash speed and density limits," *Electronic Design,* Nov. 22, 1980, pp. 55–58, 60, 62, 64, 66.

Bursky, D., "Processing advances shrink lines and sharpen device geometries to enhance performance," *Electronic Design,* June 11, 1981, pp. 78–85.

Burstein, S., T. Mariner, and J. Wunner (General Instruments), "Software-controlled sound IC gives 'command performance'," *Electronic Design,* Mar. 29, 1979, pp. 110–115. (Music and sound effects—a combo on a chip.)

Business Week, "The next big leap in electronics," Oct. 24, 1977, pp. 94B–94L.

Business Week, "How 'silicon spies' get away with copying," Apr. 21, 1980, pp. 180, 182, 187, 188.

Business Week, "Voice mail arrives in the office," June 9, 1980, pp. 80–84.

Bylinsky, G., "Here comes the second computer revolution," *Fortune,* Nov. 1975, pp. 134–139, 182, 184.

Bylinsky, G., "Industry's new frontier in space," *Fortune,* Jan. 29, 1979, pp. 76–79, 82, 84.

Byrd, J. S., and J. D. Jenkins (Dupont), "Random addressing with PROMs save time and money," *EDN,* Sept. 5, 1974, pp. 69–71.

Calebotta, S. (National Semiconductor), "Use three-state logic with confidence," *Electronic Design,* July 6, 1972, pp. 70–72.

Cane, D. (DEC), "Semicustom technology derives minicomputer architecture," *Computer Design,* Dec. 1980, pp. 103–108.

Cange, T. P., J. Kocsis, H. J. Sigg, and G. D. Vendelin (Signetics), "Double-diffused MOS transistor achieves microwave gain," *Electronics,* Feb. 5, 1971, pp. 99–104.

Cantor, D., "Logarithmic and exponential function evaluation," *IEEE TEC,* Apr. 1962, pp. 155–164.

Capece, R. P., "Tackling the very large-scale problems of VLSI: A special report," *Electronics,* Nov. 23, 1978, pp. 111–125.

Capece, R. P., "The race heats up in fast static RAMs," *Electronics,* Apr. 26, 1979, pp. 125–135.

Capece, R. P., "Older processes revamped as new arrivals extend performance limits," *Electronics,* Sept. 13, 1979, pp. 109–115.

Capece, R. P., "Memories," *Electronics,* Oct. 25, 1979, pp. 124–134.

Capell, A., D. Knoblock, L. Mather, and L. Lopp (Hewlett-Packard), "Process refinements bring C-MOS on sapphire into commercial use," *Electronics,* May 26, 1977, pp. 99–105.

Carmody, P. (IBM), "An interactive graphics system for large scale integration design," *Proc. Int. Conf. Interactive Techniques in Computer Aided Design,* Sept. 1978, pp. 281–294.

Carr, W. N., and J. P. Mize, *MOS/LSI Design and Application,* McGraw-Hill, 1972, 331 pp.

Carroll, J. A. (Dynamics Measurements Corp.), "Solving mass-produced ROM programming problems with base registers," *Computer Design,* Aug. 1977, pp. 99–105. (Software in ROM.)

Cavinaugh, G. (Texas Instruments), "Using Schottky to upgrade performance of your TTL systems," *EDN,* Sept. 20, 1973, pp. 38–42.

Cavlan, N. (Signetics), "Field-PLAs simplify logic designs," *Electronic Design,* Sept. 1, 1975, pp. 84–90.

Cavlan, N., "Signetics' field programmable logic arrays," 1976, 51 pp.

Cavlan, N., (Signetics), "Design flexibility with programmable logic", *Wescon,* 1979, 7/2, 10 pp. (FDLA and FPRP.)

Cavlan, N., and R. Cline (Signetics), "FPLA applications—Exploring design problems and solutions," *EDN,* Apr. 5, 1976, pp. 63–69.

Cavlan, N., and S. J. Durham (Signetics), "Field-programmable arrays: Powerful alternatives to random logic," *Electronics,* July 5, 1979, pp. 109–114.

Cavlan, N., and S. J. Durham (Signetics), "Field-programmable logic: Sequencers and arrays transform truth tables into working systems," *Electronics,* July 19, 1979, pp. 132–139.

Cayton, B. (General Instrument), "Designing with nitride-type EAROMs," *Electronics,* Sept. 15, 1977, pp. 107–113.

Chan, J. Y., et al. (Fairchild), "A 100 ns 5 V only 64k × 1 MOS dynamic RAM," *IEEE JSSC,* Oct. 1980, pp. 839–846.

Chang, H., T. C. Chen, and C. Tung (IBM), "The realization of symmetric switching functions using magnetic bubble technology," *NCC 1973,* pp. 413–420.

Chang, H. Y., E. G. Manning, and G. Metze, *Fault Diagnosis of Digital Systems,* John Wiley, 1970, 159 pp.

Chang, H. Y., et al., "LAMP: Logic analyzer for maintenance planning," 5 papers in *Bell Sys. Tech. J.,* 1974, pp. 1431–1535.

Chatterjee, P. K., G. W., Taylor, R. L. Easley, H. S. Fu, and A. F. Tasch, Jr. (Texas Instruments), "A survey of high-density dynamic RAM cell concepts," *IEEE TED,* June 1979, pp. 827–839.

Chawla, B. R., H. K. Gummel, and P. Kozak (Bell Labs), "MOTIS—An MOS timing simulator," *IEEE Trans. Commun. Systems,* Dec. 1975, pp. 901–910.

Check, W. (Intel), "Single chip microcomputers minimizing system cost through higher integration," *Wescon, 1978,* 27/6, 8 pp.

Chen, J. Z., W. B. Chin, T. S. Jen, and J. Hutt, "A high-density bipolar logic masterslice for small systems," *IBM JRD,* May 1981, pp. 142–151. (STL gate arrays.)

Chester, M., "Magnetic bubble memory update," *Computer Design,* May 1980, pp. 232–240.

Chiba, T., et al. (Sharp), "SHARPS: a hierarchical layout system for VLSI," *Proc. 18th DA Conf.,* 1981, pp. 820–827.

Chin, W. J., Chen, and T. S. Jen (IBM), "High density cost performance bipolar masterslice for low end systems," *ICCC 80,* pp. 197–202. (STL gate array.)

Cho, Y. E., A. J. Korenjak, and D. E. Stockton (RCA), "Floss: An approach to automated layout for high-volume designs," *Proc. 14th DA conf.,* 1977, pp. 138–141.

Chow, W. C. (Amdahl), "Packaging tradeoffs for high performance computer," *Compcon 79,* Spring, pp. 290–294.

Chrones, C. (Testing Technology), "Parametric testers evaluate wafer processing," *EDN,* Apr. 15, 1981, pp. 117–122.

Chu, P. (Fairchild), "ECL accelerates to new system speeds with high-density byte-slice parts," *Electronics,* Aug. 2, 1979, pp. 120–125. (Fairchild bit slice F100220.)

Chu, P. (Fairchild), "Design philosophy and architecture of an 8 bit microprogrammable ECL bit-slice family," *Wescon,* 1979, 2/1, 6 pp.

Chua, L. O., and P. M. Lin, *Computer-Aided Analysis of Electronic Circuits: Algorithms and Computational Techniques,* Prentice-Hall, 1975, 737 pp.

Chung, K. M. (Wang Lab.), F. Luccio (Università di Pisa), and C. K. Wong (IBM), "A tree storage scheme for magnetic bubble memories," *IEEE TC,* Oct. 1980, pp. 864–874.

Chung, K. M., F. Luccio, and C. K. Wong, "On the complexity of permuting records in magnetic bubble memory systems," *IBM JRD,* Jan. 1980, pp. 75–84.

Claiborne, J. (Daisy Systems), "Automated work station fills CAD gap," *Electronic Design,* Jan. 21, 1982, pp. 193–201.

Clare, C. R., *Designing Logic Systems Using State Machines,* McGraw-Hill, 1973, Chap. 5.

Clark, B. T., and Y. M. Hill, "IBM multichip multilayer ceramic modules for LSI chips— Design for performance and density," *IEEE TCHMT,* Mar. 1980, pp. 89–93. (IBM 4300 package.)

Clemens, D. P., and G. Castleman (HP), "Distributed computer network takes charge in IC facility," *Electronics,* June 5, 1980, pp. 151–155.

Clemens, J. T., R. H. Doklan, and J. J. Nolen, "An *n*-channel Si-gate integrated technology," *IEDM,* Dec. 1975, pp. 299–302.

Clements, M. R. (Amdahl), "And now, fifth-generation supercomputers," *Computerworld,* Dec. 31, 1979–January 7, 1980.

Cline, R. C. (Signetics), "A single-chip sequential logic element," *ISSCC 78,* pp. 204, 205.

Cochran, D. S., "Algorithms and accuracy in the HP-35," Hewlett-Packard J., June 1972, pp. 10, 11.

Code, B., "Can you count on your calculator?" *Electronics,* Nov. 25, 1976, pp. 77, 78. (Numerical analysis errors in calculators.)

Coe, J. E., and W. G. Oldham (Intel), "Enter the 16,384-bit RAM," *Electronics,* Feb. 19, 1976, pp. 116–121.

Cole, B., "Depletion mode shrinks cpu chips," *Electronics,* May 13, 1976, pp. 65, 66.

Coleman, V. (AMD), "Microcoded arithmetic speeds up division," *Electronic Design,* July 19,

1980, pp. 163–169. (Microcode with 4-bit slice microprocessor Am2903 speeds up arithmetic calculation.)

Conklin, P. F., and D. P. Rodgers (DEC), "Advanced minicomputer designed by team evaluation of hardware/software tradeoffs," *Computer Design*, Apr. 1978, pp. 129–137. (VAX-11/780, considering custom LSI versus off-the-shelf packages.)

Conleur, J. F. (General Electric), "BIDEC—A binary-to-decimal or decimal-to-binary converter," *IRE TEC*, Dec. 1958, pp. 313–316.

Conrad, M. (Texas Instruments), "'Component' software slashes system-development costs," *Electronic Design*, Jan. 22, 1981, pp. 95–101.

Cook, P. W., D. L. Critchlow, and L. M. Terman, "Comparison of MOSFET logic circuits," *IEEE JSSC*, Oct. 1973, pp. 348–355.

Cook, P. W., et al. (IBM), "One micron MOSFET PLAs," *ISSCC 79*, pp. 62, 63, 278, 279.

Cook, P. W., C. W. Ho, and S. E. Schuster (IBM), "A study in the use of PLA-based macros," *IEEE JSSC*, Oct. 1979, pp. 833–840.

Cooper, J. A., J. A. Copeland, and R. H. Krambeck (Bell Labs), "A CMOS microprocessor for telecommunications applications," *ISSCC 77*, pp. 138, 139. (Pseudo *n*-MOS is shown.)

Cooperman, M. (GTE), "High speed current mode logic for LSI," *IEEE Trans. Commun. Systems*, July 1980, pp. 626–635.

Cox, A., and A. Davis (Interdesign), "Array spectrum from low power CMOS to 20 MHz bipolar," *Wescon, 1980*, 30/4, 10 pp.

Cox, D. T., W. T. Devine, and G. J. Kelly (IBM), "High density logic array," U.S. patent no. 3,987,287, Oct. 19, 1976.

Cragon, H. G. (Texas Instruments), "The elements of single-chip microcomputer architecture," *Computer*, Oct. 1980, pp. 27–41. (Comparison of single-chip microcomputer architecture and conventional computer architecture.)

Cranswick, A., R. Heatley, and C. Lingham (Plessey), "CAPRICE: Circuit analysis programme for I. C. evaluation," *ICCC 80*, pp. 578–581.

Crippen, R., and H. Hingarh (Fairchild), "4000 I³L gates in array," *Progress, Fairchild J. of Semiconductor*, Mar./Apr. 1980, vol. 8, no. 2, pp. 4–11.

Crocker, N. R., R. W. McGuffin, and A. Micklethwaite (ICL), "Automatic ECL LSI design," *Proc. 14th DA conf.*, 1977, pp. 158–167.

Culliney, J. N., "Topics in MOSFET network design," Ph.D. dissertation, Dept. of Computer Science, University of Illinois, 1977.

Cunningham, J. A. (National Semiconductor), "Using the learning curve as a management tool," *Spectrum*, June 1980, pp. 45–48.

Curtice, R. M. (A. D. Little), "The outlook for data base management," *Datamation*, Apr. 1976, pp. 46–49.

Cushman, R., "Leapfrog ahead with standard-family MSI/LSI," *EDN*, Apr. 5, 1973, pp. 30–38.

Cushman, R. H., "Are single-chip microcomputers the universal logic of the 1980s?" *EDN*, Jan. 5, 1979, pp. 83–89.

Cushman, R. H. "1-chip µCs, high-level languages combine for fast prototyping," *EDN*, Aug. 5, 1980, pp. 89–96.

Cushman, R. H., "Digitization is on the way for FFT designs," *EDN*, Aug. 5, 1981, pp. 99–106.

Cutler, R. B., "Algebraic derivation of minimal sums for functions of a large number of variables," Ph.D. dissertation, Dept. of Computer Science, University of Illinois, 1980. (Computational efficiency improvement of minimization by new algorithms.)

Dais, J. L., J. S. Erich, and D. Jaffe (Bell Labs), "Face-down TAB for hybrids," *IEEE TCHMT*, Dec. 1980, pp. 623–634.

Daly, E. B. (GTE), "Management of software development," *IEEE Trans. Software Eng.*, May 1977, pp. 229–242.

Daniels, R. G., and G. J. Summers (Motorola), "16 bit MUP's: Past, present and future—The 6800 approach," *Wescon, 1978*, 20/2, 8 pp.

Danielsson, P. D., "A note on wired gates," *IEEE TC*, Sept. 1970, pp. 849, 850.

Dao, T. T. (Fairchild), "Recent multi-valued circuits," *Compcon 81*, Spring, pp. 194–204.

Darringer, J. A., and W. H. Joyner, Jr. (IBM), "A new look at logic synthesis," *Proc. 17th DA Conf.*, 1980, pp. 543–549.

Darringer, J. A., et al. (IBM), "Experiments in logic synthesis," *ICCC 80*, pp. 234–237, 237A.

Datamation, "Current status of digital system design languages," Aug. 1973, pp. 112–114.

Davidow, W. (Intel), "How microprocessors boost profits," *Electronics*, July 11, 1974, pp. 105–108; Jan. 23, 1975, p. 92.

Davidson, A. (IBM), "A Josephson latch," *IEEE JSSC*, Oct. 1978, pp. 583–590.

Davidson, E. E., and R. D. Lane (IBM), "Diode damp line reflections without overloading logic," *Electronics*, Feb. 19, 1976, pp. 123–127.

Davis, G. R. (Signetics), "ISL gate arrays: Fast, easy, certain," *Electro 81*, 16/3, 8 pp.

Decker, R. W., "Computer aided design and manufacturing at GM," *Datamation*, May 1978, pp. 159–165.

DeFalco, J. A., "Comparison and uses of TTL circuits," *Computer Design*, Feb. 1972, pp. 63–68. (Survey.)

de Lugish, B. G., "A class of algorithms for automatic evaluation of certain elementary functions in a binary computer," Rep. 399, Dept. of Computer Science, University of Illinois, 1970, 182 pp.

Dennard, R. H., F. H. Gaensslen, L. Kuhn, and H. N. Yu, "Design of micron MOS switching devices," *IEDM Digest Tech. Papers*, 1972, pp. 168–170.

Dennard, R. H., et al., "Design of ion implanted MOSFETs with very small physical dimensions," *IEEE JSSC*, Oct. 1974, pp. 256–268.

Dennis, J. B., "The varieties of data flow computers," *1st Int. Conf. on Distributed Computer Systems*, Huntsville, AL, Oct. 1979, pp. 430–439.

Deo, N., *Graph Theory with Applications to Engineering and Computer Science*, Prentice-Hall, 1974, 489 pp.

Des Roches, G. (Hughes), "EEPROM eclipses other reprogrammable memories," *Electronic Design*, Nov. 22, 1980, pp. 247–250.

DeTroye, N. C., "Integrated injection logic—Present and future," *IEEE JSSC*, Oct. 1974, pp. 206–211.

Deutsch, D. N., and P. Glick (Bell Labs), "An over-the-cell router," *Proc. 17th DA Conf.*, 1980, pp. 32–39.

Devitt, D., and J. George (Solid-State Scientific), "Beam tape plus automated handling cuts IC manufacturing costs," *Electronics*, July 6, 1978, pp. 116–119.

Dhaka, V. A., J. E. Muschinske, and W. K. Owens, "Subnanosecond emitter-coupled logic gate circuit using Isoplanar II," *IEEE JSSC*, Oct. 1973, pp. 368–372.

DiPietro, D. M. (Hewlett-Packard), "A 5-GHz f_T monolithic IC process for high-speed digital circuits," *ISSCC 75*, pp. 118, 119, 222.

Dietmeyer, D. L., *Logic Design of Digital Systems*, Allyn and Bacon, 1971, 800 pp.

DiLorenzo, J., and K. Khandelwal, Ed., *GaAs FET Principles and Technology*, Artech House, 1981, 350 pp.

Dingwall, A. G. F., and R. E. Stricker (RCA), "C^2L: A new high-speed high-density bulk CMOS technology," *IEEE JSSC*, Aug. 1977, pp. 344–349.

Director, S. W., Ed., *Computer Aided Circuit Design, Simulation and Optimization*, Dowden, Hutchinson and Ros, 1974, 400 pp.

Dirilten, H., "On the mathematical models characterizing faulty four-phase MOS logic arrays," *IEEE TC*, Mar. 1972, pp. 301–305.

Doane, D. A. (RCA), "Optical lithography in the 1-μm limit," *Solid State Technology*, Aug. 1980, pp. 104–114.

Dobriner, R., "Ion implantation: From a speciality to a standard method for new ICs," *Electronic Design*, May 24, 1974, pp. 36–40.

Dobson, P. "The next microchip revolution," *New Scientist*, Mar. 19, 1981, pp. 747–750.

Donaldson, J. (Teradyne), "Partial RAMs can fill today's memory boards," *Electronics*, Jan. 17, 1980, pp. 131–134.

Donath, W. E., "Equivalence of memory to random logic," *IBM JRD*, Sept. 1974, pp. 401–407.

Donath, W. E., "Wire length distribution for placements of computer logic," *IBM JRD*, May 1981, pp. 152–155.

Donlan, T. G., "A CAD/CAM world?" *Barron's*, Dec. 22, 1980, pp. 4–6, 20.

Donlan, T. G., "Mobile phone drive," *Barron's*, May 11, 1981, pp. 4, 5, 24–26.

Durniak, A., "VLSI shakes the foundations of computer architecture," *Electronics*, May 24, 1979, pp. 111–133. (Use of memories in mainframe computers. PALs used in Computer Automation's Naked Mini 4/10S.)

Dussine, R., "Evolution of ROM in computers," *Honewell Computer J.*, vol. 5, no. 2, 1971, pp. 79–87.

Dutton, R. W., and D. A. Antoniadis, "Process simulation for device design and control," *ISSCC 79*, pp. 244, 245.

Dworsky, L. N. (Motorola), *Modern Transmission Line Theory and Applications*, John Wiley, 1979, 236 pp.

Eaton, S. S. (RCA), "Sapphire brings out the best in C-MOS," *Electronics*, June 12, 1975, pp. 115–120.

Ebbinghaus, K., and G. Kreilkamp (IBM), "An interactive graphic tool for logic-design," *Proc. Int. Conf. Interactive Techniques in Computer Aided Design*, Sept. 1978, pp. 307–316.

Eberlein, D. D. (Cray Research), "Custom MSI for very high speed computers," *Compcon 79*, Spring, pp. 295–298.

Eden, R. C., B. M. Welch, and R. Zucca (Rockwell), "Planar GaAs IC technology: Applications for digital LSI," *IEEE JSSC*, Aug. 1978, pp. 419–426.

Eden, R. C., B. M. Welch, R. Zucca, and S. I. Long (Rockwell), "The prospects for ultrahigh-speed VLSI GaAs digital logic," *IEEE JSSC*, Apr. 1979, pp. 221–239.

Egbert, W. E., "Personal calculator algorithms," Hewlett-Packard J., Part I, "Square roots, "May 1977, pp. 22–24; Part II, "Trigonometric functions," June 1977, pp. 17–20; Part III, "Inverse trigonometric functions," Nov. 1977, pp. 22, 23; Part IV, "Logarithmic functions," Apr. 1978, pp. 29–32.

Eichelberger, E. B., and E. Lindbloom, "A heuristic test-pattern generator for programmable logic arrays," *IBM JRD*, Jan. 1980, pp. 15–22.

Eichelberger, E. B., and T. W. Williams (IBM), "A logic design structure for LSI testing," *Proc. 14th DA conf.*, 1977, pp. 462–468. (LSSD.)

Eichelberger, E. B., et al. (IBM), "A logic design structure for testing internal arrays," *Proc. 3rd USA-Japan Comp. Conf.*, San Francisco, Oct. 1978, pp. 266–272. (LSSD.)

Eidson, J. C. (Hewlett-Packard), "Fast electron-beam lithography," *Spectrum*, July 1981, pp. 24–28.

Eimbinder, J., Ed., *Semiconductor Memories*, John Wiley, 1971, 214 pp.

Eldumiati, I. I., et al. (Bell Labs), "MAC-4: A single-chip microcomputer," *Compcon 79*, Fall, pp. 13–17.

Ellenberger, D. J., and Y. W. Ng (Bell Labs), "AIDE—A tool for computer and architechture design," *Proc. 18th DA Conf.*, 1981, pp. 796–803. See *Electronics*, Apr. 7, 1981, pp. 44, 46.

El-Mansy, Y. A. (Intel), "On scaling MOS devices for VLSI," *ICCC 80*, pp. 457–460. Also *Electronics*, Oct. 23, 1980, pp. 44, 46.

El-Ziq, Y. M. (Univac), "Logic design automation of MOS combinational networks with fan-in, fan-out contraints," *Proc. 15th DA Conf.*, 1978, pp. 240–249.

El-Ziq, Y. M., and S. Y. C. Su, "Computer-aided logic design of two-level MOS combinatorial networks with statistical results," *IEEE TC*, Oct. 1978, pp. 911–923.

Eschenfelder, A. H., *Magnetic bubble technology*, second ed., Springer, 1981, 345 pp.

Eufinger, R. J. (Rockwell), "Integrated peripherals into processing systems," *Computer Design*, Dec. 1978, pp. 77–83.

Evangelisti, C. J., G. Goertzel, and H. Ofek (IBM), "Designing with LCD; language for computer design," *Proc. 14th DA Conf.*, 1977, pp. 369–376.

Evans, C., *The Micro Millenium*, Viking Press, 1979, 255 pp.

Evans, S. A. (Texas Instruments), "Scaling I²L for VLSI," *IEEE TED,* Apr. 1979, pp. 396–405.

Evans, S. A., et al (Texas Instruments), "A 1-μm bipolar VLSI technology," *IEEE JSSC,* Aug. 1980, pp. 438–444.

Exar Integrated Systems, Inc., *Exar Linear and Digital Semi-Custom IC Design Programs,* June 1979, 37 pp.

Faggin, F. (Zilog), "How VLSI impacts computer architecture," *Spectrum,* May 1978, pp. 28–31.

Faggin, F., and T. Klein, "Silicon-gate ICs produce a faster generation of MOS devices," in *Large and Medium Scale Integration,* edited by S. Weber, McGraw-Hill, 1974, pp. 24–30. Also *Electronics,* Sept. 29, 1969, pp. 88–94.

Faggin, F., T. Klein, and L. Vadasz (Intel), "Insulated gate field effect transistor circuits with silicon gates," *IEDM,* Oct. 1968.

Faiman, M., A. C. Weaver, and R. W. Catlin, "MUMS—A reconfigurable microprocessor architecture," *Computer,* Jan. 1977, pp. 11–17.

Fairchild, "Micromosaic logic design," in *Fairchild OPTIMOS,* 1972, pp. 235–280, Appendix D.

Fairchild, *The ECL Handbook,* 1974.

Fairchild, *CMOS Data Book,* 1977. (Isoplanar CMOS.)

Fairchild, *ECL Data Book,* 1977.

Fairchild Corp., *Semiconductor and Integrated Circuit Fabrication Techniques,* Prentice-Hall, 1979, 209 pp.

Falconer, W. E., and C. Skrzypczak (ATT), "The Bell system—On its way to a digital network," *Bell Labs. Rec.,* May/June 1981, pp. 138–145.

Falk, H., "Hard–soft tradeoffs," *Spectrum,* Feb. 1974, pp. 34–39.

Falk, H., "Design for production," *Spectrum,* Oct. 1975, pp. 48–53.

Falk, H., "Computers: All pervasive," *Spectrum,* Jan. 1977, pp. 38–42.

Falk, H., "'Chipping in' to digital telephones," *Spectrum,* Feb. 1977, pp. 42–46.

Fan, S. P., et al., "MOTIS-C: A new circuit simulator for MOS LSI circuits," *IEEE Proc. Intnt'l Symp. on Circuits and Systems,* 1977, p. 700.

Faris, S. M. (IBM), "Loop decoder for Josephson memory arrays," *IEEE JSSC,* Aug. 1979, pp. 699–707.

Farina, D. E., J. R. Duffy, and T. L. Kellgren (Alphatron), "Cell-library system accomodates any degree of design expertise," *Electronics,* Nov. 30, 1981, pp. 126–128.

Farr, T. M., Jr. (Xerox), "Ready-only memory controls universal counter," *EDN,* May 5, 1976, p. 114.

Faster, J. C. (Bell Labs), "The evolution of an integrated data base," *Proc. 12th DA Conf.,* 1975, pp. 394–398.

Fattal, J. (Motorola), "Innovations in microcomputer architectures," *Electro 81,* 1/0, 4 pp.

Fedida, S., and R. Malik, *Viewdata Revolution,* John Wiley, 1979, 186 pp.

Feller, A., R. Noto, and A. M. Smith (RCA), "Standard cell approach for generating custom CMOS/SOS devices using a fully automatic layout program," *ICCC 80,* pp. 311–314.

Femling, D. (National Semiconductor), "Enhancement of modular design capability by use of tri-state logic," *Computer Design,* June 1971, pp. 59–64.

Fette, B., "Principle of dynamic MOS," *Motorola Monitor,* Oct. 1971, pp. 21–25.

Fette, B. A. (Motorola), "Dynamic MOS—A logical choice," *EDN,* Nov. 15, 1971, pp. CH6-CH14.

Fetterolf, J. (AMP), "Transmission-line methods speed 1 ns data along," *Electronic Design,* June 21, 1980, pp. 95–99.

Feuer, M., K. H. Khokhani, and D. Metha (IBM), "Computer-aided design wires 5,000 a circuit chip," *Electronics,* Oct. 9, 1980, pp. 144–145. (Placement.)

Fiduk, K. W., Jr., D. W. Hightower, and F. A. Coady (Texas Instruments), "A nonmodular placement algorithm with applications," *ICCC 80,* pp. 83–86.

Fischer, J. L. (Texas Instruments), "Programmable components: The shape of VLSI to come," *Electronics*, June 5, 1980, pp. 138–142.

Fischer, T. (ITT), "Digital VLSI breeds next-generation TV receivers, *Electronics*, Aug. 11, 1981, pp. 6, 97–103.

Fisher, M., and A. Young (RCA), "CMOS technology can do it all in digital and linear systems," *EDN*, June 20, 1979, pp. 107–114.

Fleisher, H., and L. I. Maissel, "An introduction to array logic," *IBM JRD*, Mar. 1975, pp. 98–109.

Foltz, J., and F. Musa, "Waveform analysis of C/MOS logic," *Electronics*, June 19, 1972, pp. 85–89.

Ford, D. C. (Motorola), and F. Hrobak and L. White (Texas Instruments), "128 vs. 256 cycle refresh," *Electronic Products Mag.*, Mar. 1981, pp. 38–43.

Foss, R. C., and R. Harland (Mosaid), "Standards for dynamic MOS RAMs," *Electronic Design*, Aug. 16, 1977, pp. 66–70.

Foss, R. C., and R. Harland (Mosaid), and J. A. Roberts (Computing Devices), "Check list for 4,096-bit RAMs flags potential problems in memory design," *Electronics*, Sept. 2, 1976, pp. 103–107.

Foster, M. J., and H. T. Kung, "The design of special-purpose VLSI chips," *Computer*, Jan. 1980, pp. 26–40.

Fox, P. E., and W. J. Nestork, "Design of logic circuit technology for IBM System/370 Model 145 and 155," *IBM JRD*, Sept. 1971, pp. 384–390.

Franson, P., "Don't overlook semicustom ICs for your next design project," *EDN*, Feb. 5, 1977, pp. 72–76.

Franson, P., "Logic family update—SSI/MSI still thrive in the world of LSI," *EDN*, Feb. 20, 1977, pp. 79–85.

Franson, P., "Advent of standard chip software sure to ease system implementation," *EDN*, Oct. 5, 1977, pp. 21, 22. (Software in ROMs.)

Franson, P., "EPROMs will soon overtake PROMs," *EDN*, June 5, 1978, pp. 22–30. (Overtake in demand.)

Freeman, R. (Zilog), "Architectural concepts for microprocessor peripheral families," *Wescon, 1980*, 31/5, 3 pp.

Freese, D. R. (Universal Instruments), "Streamlining insertion of components in pc boards," *Electronics*, Sept. 5, 1974, pp. 117–123.

Frei, A. H., W. K. Hoffman, and K. Shepard (IBM), "Minimum area parity circuit building block," *ICCC 80*, pp. 680–684. (Building block based on the circuit of Fig. 4.5.3.)

Friedman, A. D., and P. R. Menon, *Fault Detection in Digital Circuits*, Prentice-Hall, 1971, 220 pp.

Friedman, T. D., and S. C. Yang, "Quality of designs from an automatic logic generator— ALERT," *Proc. 7th DA Workshop*, 1970, pp. 71–82.

Fukushima, T. (Fujitsu), "The bipolar PROM: An overview and history," *Wescon, 1979*, 7/1, 5 pp.

Funatsu, S., et al. (NEC), Test generation system in Japan," *Proc. 12th DA Conf.*, 1975, pp. 114–122.

Furlow, W., "Today's hardest design decision: An overview of custom CMOS/LSI," *EDN*, June 5, 1973, pp. 42–47.

Gaensslen, F. H. (IBM), "MOS devices and integrated circuits at liquid nitrogen temperature," *ICCC 80*, pp. 450–452.

Gaensslen, F. H., et al. (IBM), "Very small MOSFETs for low-temperature operation," *IEEE TED*, Mar. 1977, pp. 218–229.

Gajski, D. D., "Parallel compressors," *IEEE TC*, May 1980, pp. 393–398.

Galling, W., and R. Ball, "How Omega and Tissot got ticking again," *Fortune*, Jan. 14, 1980, pp. 68–70.

Galloway, D., B. Hartman, and D. Wooten (INMOS), "64k dynamic RAM speeds well beyond the pack," *Electronic Design*, Mar. 19, 1981, pp. 221–225.

Garrett, L. S., "Integrated-circuit digital I—Requirements and features of a logic family; RTL, DTL, and HTL devices," *Spectrum, Oct. 1970, pp. 46–58.*

Garrett, L. (Motorola), "CMOS may help majority logic win designer's vote," *Electronics,* Jan. 19, 1973, pp. 107–112.

Garrow, R., et al. (Intel), "Microcomputer-development system achieves hardware-software harmony," *Electronics,* May 29, 1975, pp. 95–102.

Garry, P., "I²L: it's getting faster and smaller," *Spectrum,* June 1977, pp. 30–33.

Gaskill, J. "A simplified approach to testing CMOS inverters and gates," *EDN,* Aug. 20, 1973, pp. 64–67.

Gaur, S. P. (IBM), "Performance limitations of silicon bipolar transistors," *IEEE TED,* Apr. 1979, pp. 415–421.

George, J., and B. Schmidt, "A strong commitment to complementary MOS," *Motorola Monitor,* May 1972, pp. 16–21.

Gerson, S. (Mostek), "Package piggybacks standard E-PROM to emulate one-chip microcomputer," *Electronics,* Jan. 31, 1980, pp. 89–92.

Getreu, E. E. (Tektronix), *Modeling the Bipolar Transistor,* Elsevier, 1978, 261 pp. Book review: *IEEE TED,* Sept. 1979, p. 1379.

Geyer, J. M. (GE), "Correction routing algorithm for printed circuit boards," *IEEE TCT,* Jan. 1971, pp. 95–100.

Gheevala, T. R. (IBM), "A 30-ps Josephson current injection logic," *IEEE JSSC,* Oct. 1979, pp. 787–793.

Gibson, D., and S. Nance (AMI), "SLIC—Symbolic layout of integrated circuits," *Proc. 13th DA Conf.,* 1976, pp. 434–440.

Giles, M., et al. (National Semiconductors), "Two-chip radio link pilots toys and models," *Electronics,* June 5, 1980, pp. 145–149. (Radio control toys.)

Glaser, A. B., and G. E. Subak-Sharpe, *Integrated Circuit Engineering,* Addison-Wesley, 1977, 811 pp.

Glasser, L. A., and P. Penfield, Jr. (MIT), "An interactive PLA generator as an archetype for a new VLSI design methodology," *ICCC 80,* pp. 608–611.

Glock, H. (Siemens), "Design considerations for high speed ECL RAMs," *ICCC 80,* pp. 16–18.

Goetz, M. A., "What's good for IBM," *Datamation,* Apr. 1975, pp. 103, 104.

Gold, J., "To use custom LSI or μP's? That is often the question," *Electronic Design,* July 19, 1976, pp. 26, 28, 30, 32.

Golden, R. L., P. A. Latus, and P. Lowy, "Design automation and the programmable logic array macro," *IBM JRD,* Jan. 1980, pp. 23–31.

Goldsmith, T. M. (Fujitsu), "A family of CMOS arrays with TTL performance," *Electro 81,* 16/4, 6 pp.

Goodman, T. (Fairchild), "F300 advanced gate array offering power/delay tradeoffs," *1980 Electro Professional Program,* 22/1, 5 pp.

Goodman, T., and W. Owens (Fairchild), "Flexible subanosecond gate array family," *Wescon, 1980,* 30/1, 5 pp.

Gosling, W., W. G. Townsend, and J. Watson, *Field Effect Electronics,* Wiley-Interscience, 1971, 364 pp.

Gossen, R. N., Jr. (Texas Instruments), "The 64-kbit RAM: A prelude to VLSI," *Spectrum,* Mar. 1979, pp. 42–45.

Goto, T., and N. Manabe (NEC), "How Japanese manufacturers achieve high IC reliability," *Electronics,* Mar. 13, 1980, pp. 140–147.

Grappel, R., and J. Hemenway, "Compare the newest 16-bit μPs to evaluate their potential," *EDN,* Sept. 5, 1980, pp. 197–201.

Grason, J., and A. W. Nagle (Bell Labs), "Digital test generation and design for testability," *Proc. 17th DA Conf.,* 1980, pp. 175–189.

Grasselli, A., and U. Montanari, "On the minimization of read-only memories in microprogrammed digital computers," *IEEE TC,* Nov. 1970, pp. 1111–1114.

Gray, J. (Monolithic Memories), "Power switch ROMs and PROMs quickly," *Electronic Design*, Apr. 26, 1977, pp. 102–104.

Greenbaum, J. R., et al. (GE), "Digital-IC models for CAD," *Electronics*, Part 1, "TTL NAND gates," Dec. 6, 1973, pp. 121–125: Part 2, "TTL flip-flops," Dec. 20, 1973, pp. 107–112: Part 3, "Monostable multivibrator," Jan. 24, 1974, pp. 98–101: Part 4, "AND-OR-INVERT gate," Feb. 7, 1974, pp. 124–126: Part 5, "Shift register," Mar. 7, 1974, pp. 125–128.

Greenfield, J. A., and R. W. Dutton, "Nonplanar VLSI device analysis using the solution of Poisson's equation," *IEEE TED*, Aug. 1980, pp. 1520–1532 (GEMINI).

Greenwood, R. (Plessey), "MNOS devices provide flexibility in nonvolatile logic," *Computer Design*, Nov. 1980, pp. 232–237.

Greer, D. L. (GE), "An assoiciative matrix," *IEEE JSSC*, Oct. 1976, pp. 679–691. Also *ISSCC 76*, pp. 18, 19, 222.

Groover, M. P., *Automation, Production Systems and Computer-Aided Manufacturing*, Prentice-Hall, 1980, 601 pp.

Grossman, M., "Second-generation 16-bit µCs are here—but so are plenty of software problems," *Electronic Design*, Nov. 8, 1978, pp. 66–77. (Comparison of Z8000, MC68000, and PDP-11/45.)

Grossman, M. "Focus on relays: Do solid-state relays and electromechanicals really compete? *Electronic Design*, Dec. 20, 1978, pp. 119–129.

Grossman, R. M., "Developments in microlithography spur VLSI circuit progress," *EDN*, Feb. 5, 1979, pp. 55–62.

Grossman, M., "High-density packaging goes denser with QUIPs and square-chip carriers," *Electronic Design*, Mar. 15, 1979, pp. 36–38, 42, 43, 46, 48.

Grossman, M., "Processing packs up for the new frontier: VLSI," *Electronic Design*, Mar. 29, 1979, pp. 82–86.

Grossman, M., "Autodesign and safe chip removal make multilayered modules feasible," *Electronic Design*, Mar. 15, 1980, pp. 42, 44, 46.

Grossman, M., "E-beams, new processes write a powerful legacy," *Electronic Design*, June 7, 1980, pp. 65–72.

Grossman, M. "Challenge: Interconnections for rising data speeds," *Electronic Design*, June 21, 1980, pp. 63–68.

Gunn, J. F., D. J. Lynes, and R. L. Pritchett (Bell Labs), "A bipolar 16k ROM utilizing Schottky diode cells," *Computer*, July 1977, pp. 14–17.

Gupta, A., and J. W. Lathrop, "Yield analysis of large integration circuit chips," *IEEE JSSC*, Oct. 1972, pp. 389–395.

Gupta, B. K., "Arithmetic processor chips enhance microprocessor system performance," *Computer Design*, July 1980, pp. 85–94.

Gutman, F. (California Microwave), "8080 routine emulates PLA hardware," *EDN*, Oct. 5, 1979, pp. 81, 82.

Hachtel, G. D., L. A. Sangiovanni-Vincentelli (IBM), and A. R. Newton (University of California), "Some results in optimal PLA folding," *ICCC 80*, pp. 1023–1027.

Hage, C. J., and J. C. Krueger (Signetics), "An efficient symbolic layout and design rule checking system for gate arrays," *ICCC 80*, pp. 1040–1043.

Hagiwara, T., et al. (Hitachi), "A 16 kb electrically erasable programmable ROM," *ISSCC 79*, pp. 50, 51.

Halbo, L., "Status and prospects for I²L," *ICCC 80*, pp. 210–214.

Hamilton, D. J., and W. G. Howard, *Basic Integrated Circuit Engineering*, McGraw-Hill, 1975, 587 pp.

Hamilton, P., and J. Lyman, "The 1980 achievement award: Developers of projection mask aligner provided basic IC fabrication tool," *Electronics*, Oct. 23, 1980, pp. 105–108.

Hanan, M., and J. M. Kurtzberg (IBM), "Placement techniques," in *Design Automation of Digital Systems*, vol. 1, edited by M. A. Breuer, Prentice-Hall, 1972, pp. 213–282, Chap. 5.

Hanan, M., P. K. Wolff, Sr., and B. J. Agule (IBM), "Some experimental results on placement techniques," *Proc. 13th DA Conf.,* 1976, pp. 214–224.

Harakal, J. P. (Intel), "Software components: Key to lowered system development cost," *Wescon,* 1978, 30/4, 5 pp.

Harmon, W. J., Jr. (AMD), "A high-performance 16-bit bipolar microprocessor—The Am29116," *Wescon,* 1979, 2/3, 13 pp.

Harrison, R. A., R. W. Wolzwarth, and P. R. Motz (General Motors), and R. G. Daniels, J. S. Thomas, and W. H. Wiemann (Motorola), "Logic fault verification of LSI: How it benefits the user," *Wescon,* 1980, 34/1, 4 pp.

Hart, C. M., A. Slob, and H. E. J. Wulms (Philips), "Bipolar LSI takes a new direction with integrated injection logic," *Electronics,* Oct. 3, 1974, pp. 111–118. (Interface circuits are shown.)

Hart, K., and A. Slob, "Integrated injection logic: A new approach to LSI," *IEEE JSSC,* Oct. 1972, pp. 346–351.

Hart, P. A. H., T. Van't Hof, and F. M. Klaasen (Philips), "Device down scaling and expected circuit performance," *IEEE JSSC,* Apr. 1979, pp. 343–351.

Hartmann, R. (IC Cost Consultants), "Gate arrays—an overview," *Electro 81,* 16/1, 5 pp.

Hartmann, R. (IC Cost Consultants), and R. Walker (LSI Corp. Logic), "LSI gate arrays outpace standard logic," *Electronic Design,* Mar 15, 1981, pp. 107–112 (correction Aug. 6, 1981, pp. 13, 16).

Hartwig, G. C. (Data Communications), "It's the user's turn to pull the software strings," *Data Communications,* June 1979, pp. 49, 50, 53–55. (Standardization of software by hardware realization.)

Hashimoto, A., and J. Stevens, "Wire routing by optimizing channel assignment within large apertures," *Proc. 8th DA Workshop,* 1971, pp. 155–169.

Hassler, E. (Texas Instruments), "Data base design automation—A tutorial," *Proc. 11th DA Workshop,* 1974, pp. 1–13.

Hatfield, W. B., D. R. Simon, and W. H. Dunn (Singer), "Top-of-the-line sewing machines embroider from memory," *Electronics,* Mar. 1, 1979, pp. 133–136.

Hauer, D. (Micro Power Systems), "Custom LSI," *Wescon,* 1978, 32/5, 2 pp.

Haviland, G. L., and A. A. Tuszynski, "A CORDIC arithmetic processor chip," *IEEE TC,* Feb. 1980, pp. 68–79.

Hayes, J. (Synertek), "MOS scaling," *Computer,* Jan. 1980, pp. 8–13.

Hayes, J. P., and E. J. McCluskey, "Testability consideratins in microprocessor-based design," *Computer,* Mar. 1980, pp. 17–26.

Hayn, J. (Texas Instruments), "LPC speech-synthesis chips mate easily with micros," *Electronic Design,* July 23, 1981, pp. 161–168.

Hayn, J. K. McDonough, and J. Bellay (Texas Instruments), "Strip architecture fits microcompter into less silicon," *Electronics,* Jan. 27, 1981, pp. 107–111.

Hebenstreit, E., and K. Horninger (Siemens), "High-speed programmable logic arrays in ESFI SOS technology," *IEEE JSSC,* June 1976, pp. 370–374.

Heilmeier, G. H. (Texas Instruments), "Needed: A 'miracle slice' for VLSI fabrication," *Spectrum,* Mar. 1979, pp. 45–47.

Heller, W. R. (IBM), "Contrasts in physical design between LSI and VLSI," *Proc. 18th DA Conf.,* 1981, pp. 676–683.

Heller, W. R., W. F. Mikhail, and W. E. Donath, (IBM), "Prediction of wiring space requirements for LSI," *J. Design Automation and Fault-Tolerant Computing,* June 1978, pp. 117–144. Also *Proc. 14th DA Conf.,* 1977, pp. 32–42.

Hemel, A. (Philco Ford), "Slash ROM sizes with equivalent functions," *Electronic Design,* Feb. 15, 1969, pp. 66–74.

Hemel, A. (Communication Products), "Making small ROMs do math quickly, cheaply and easily," *Electronics,* May 11, 1970, pp. 104–111.

Hemel, A. (Memorex), "Square root extraction with read-only memories," *Computer Design,* Apr. 1972, pp. 100–102.

Hemel, A. (Monolithic Memories), "The PLA: A 'different kind' of ROM," *Electronic Design,* Jan. 5, 1976, pp. 78–84.

Hemenway, J., "Microcomputer operating system comes of age," *Mini-Micro Systems,* Oct. 1980, pp. 97, 98, 103–112, 114, 119.

Hemenway, J., and R. Grappel, "Understand the newest processor to avoid future shock," *EDN,* Apr. 29, 1981, pp. 129–136. (Comparison of iAPX 432 with other microcomputers.)

Henisch, H. K., "Amorphous-semiconductor switching," *Scientific American,* Nov. 1969, pp. 30–41.

Henkels, W. H., and J. H. Greiner (IBM), "Experimental single flux quantum NDRO Josephson memory cell," *IEEE JSSC,* Oct. 1979, pp. 794–796.

Henkels, W. H., and H. H. Zappe (IBM), "An experimental 64-bit decoded Josephson NDRO random access memory," *IEEE JSSC,* Oct. 1978, pp. 591–600.

Henle, R. A., and J. T. Ho (IBM), "Structured logic," *AFIPS,* 1969, pp. 61–68.

Henry, P. S., "Fast decryption algorithm for the Knapsack cryptographic system," *Bell Sys. Tech. J.,* May/June 1981, pp. 767–773.

Herrick, K. C. (ESI Electronics), "A PROM-based detector detects and corrects errors," *EDN,* June 20, 1980, p. 169–172.

Hewlett, W. H., Jr. (Texas Instruments), "Oxide-isolated integrated Schottky logic," *IEEE JSSC,* Oct. 1980, pp. 800–802.

Hibberd, R. G., *Integrated Circuits,* McGraw-Hill, 1969, 177 pp.

Hicks, S. M. (Forth), "FORTH: A cost saving approach to software development," *Wescon, 1978,* 30/2, 5 pp.

Hightower, D. W., "A solution to line-routing problems on the continuous plane," *Proc. 6th DA Workshop,* 1969, pp. 1–24.

Hightower, D. W. (Bell Labs), "The interconnection problem—A tutorial," *Computer,* Apr. 1974, pp. 18–32. Also *Proc. 10th DA Workshop,* 1973, pp. 1–21.

Hightower, D. W., and F. G. Alexander (Texas Instruments), "A mature IIL/STL gate array layout system," *Compcon 80,* Spring, pp. 149–155.

Hightower, D., and R. Horton (Texas Instruments), "Computer-aided logic arrays," *Wescon, 1980,* 30/3, 7 pp. (Schottky transistor logic gate arrays.)

Hightower, D., and M. Roberts (Texas Instruments), "Automated logic arrays and the customer interface," *Compcon 81,* Spring, pp. 186–190.

Hingarh, H. K. (Fairchild), "A high performance 400-gate I^{3}L array," *Wescon,* 1980, 30/5, 6 pp. (I^{3}L gate array F9480.)

Hitchcock, R. (IBM), "Cellular wiring and the cellular modeling technique," *Proc. 6th DA Workshop,* 1969, pp. 25–41.

Hitchcock, R. B. (IBM), "Partitioning of logic gates—A theoretical analysis of pin reduction," *Proc. 7th DA Workshop,* 1970, pp. 54–63.

Hittinger, W. C., "Metal-oxide-semiconductor technology," *Scientific American,* Aug. 1973, pp. 48–57.

Hively, J. W. (Fairchild), "Subnanosecond ECL gate array," *Wescon,* 1978, 3/3, 11 pp. (F200 gate array.)

Hnatek, E. R. (DCA Reliability Lab.), "4-Kilobit memories present a challenge to testing," *Computer Design,* May 1975, pp. 117–125.

Hnatek, E. R., "User's tests, not data sheets, assure IC performance," *Electronics,* Nov. 27, 1975, pp. 108–113.

Hnatek, E. R., *A User's Handbook of Semiconductor Memories,* John Wiley, 1977, 652 pp.

Hnatek, E. R. (Monolithic Memories), "Semiconductor memory update," *Computer Design,* Part 1, "ROMs," Dec. 1979, pp. 67–77; Part 2, "RAMs," Jan. 1980, pp. 119–131; Part 3, "High density technologies," Feb. 1980, pp. 147–159.

Ho, C. W., "Time-domain sensitivity computation for networks containing transmission lines," *IEEE TCT,* Jan. 1971, pp. 114–122.

Hochman, H. T., "The art of building LSI's," *Spectrum,* Sept. 1969, pp. 29–36.

Hodges, D. A., Ed., *Semiconductor Memories,* IEEE Press, 1972, 287 pp.

Hodges, D. A., and D. O. Pederson, "The here and now of computer-aided circuit design," *ISSCC 74,* pp. 38, 39, 222.

Hoel, J. H., "Some variations of Lee's algorithm," *IEEE TC,* Jan. 1976, pp. 19–24.

Hoeneisen, B., and C. A. Mead, "Fundamental limitations in microelectronics," *Solid-State Electronics,* Aug. 1972, pp. 819–829; 981–987.

Hoff, M. E., and W. Li (Intel), "Software makes a big talker out of the 2920 microcomputer," *Electronics,* Jan. 31, 1980, pp. 102–107. (Speech synthesis by an analog microprocessor.)

Hoff, M. E., and M. Townsend (Intel), "Single-chip *n*-MOS microcomputer processes signal in real time," *Electronics,* Mar. 1, 1979, pp. 105–110.

Hoffman, G. (Mostek), "Applications of ion-implanted depletion-mode MOS devices," *Wescon, 1971.* Also Application note from Mostek.

Hoffmann, K. (Siemens), "Continuously charged-coupled random-access memory (C^3RAM)," *ISSCC 76,* pp. 130–131.

Hoffpauir, P. (National Semiconductors), "LS^2 family squeezes more speed from existing TTL sockets," *Electronics,* Feb. 28, 1980, pp. 149–153.

Hong, S. J., R. G. Cain, and D. L. Ostapko, "Mini: A heuristic approach for logic minimization," *IBM JRD,* Sept. 1974, pp. 443–458.

Hong, S. J., R. Nair, and E. Shapiro, "A physical design machine," in *VLSI 81,* ed. by J. P. Gray, 1981, pp. 257–266.

Horiuchi, M., and H. Katto (Hitachi), "A low voltage, high-speed alterable *n*-channel nonvolatile memory," *IEDM,* 1978, pp. 336–339.

Horninger, K. (Siemens), "A high-speed ESFI SOS programmable logic array with an MNOS version," *IEEE JSSC,* Oct. 1975, pp. 331–336.

Horowitz, E., and S. Sahni, *Fundamentals of Computer Algorithms,* Computer Science Press, 1978, 626 pp.

Horton, R., D. Thomas, and R. Rozeboom (Texas Instruments), "Design automation speeds through customization of logic arrays," *Electronics,* July 14, 1981, pp. 132–137. (CAD of STL gate arrays.)

Hosoya, T., S. Muramoto, and S. Matsuo (NTT), "A self-aligning contact process for MOS LSI," *IEEE TED,* Jan. 1981, pp. 77–82.

Houston, T. W., et al. (Texas Instruments), "Silicon MESFET circuit performance for VLSI," *ISSCC 79,* pp. 80, 81. (Silicon MESFET with a short channel of about 1 μm with comparable speed and lower power consumption than GaAs MESFET.)

Howes, M. J., and D. V. Morgan, Eds., *Charge-Coupled Devices and Systems,* John Wiley, 1979, 300 pp. (Physics, fabrication, signal processing, memories, and imaging are discussed.)

Hu, K. C., "Logic design methods for irredundant MOS networks," Ph.D. dissertation, Dept. of Computer Science, University of Illinois, Aug. 1978, 317 pp.

Hu, K. C. (Zilog), "Z8010 MMUs help microprocessors in handling large memory systems," *Electro 81,* 6a/3, 8 pp.

Hu, K. C., H. Yonezawa, and B. Peuto (Zilog), "Memory-management units help 16-bit μPs to handle large memory systems," *Electronic Design,* Apr. 26, 1980, pp. 128–135.

Huffman, D. (Mostek), "Polysilicon-load RAMs plug into mainframes or microprocessors," *Electronics,* Sept. 27, 1979, pp. 131–139.

Huffman, G. "As μc-system complexity grows, support chips assume host tasks," *EDN,* Nov. 5, 1980, pp. 214–230.

Hughes, G. P., and R. C. Fink (General Instruments), "X-ray lithography breaks the VLSI cost barrier," *Electronics,* Nov. 9, 1978, pp. 99–106.

Hughes, J., and P. Chappell (Texas Instruments), "Memory-to-memory for future controllers," *Electronic Design,* Nov. 8, 1980, pp. 131–135.

Hughes, J., and M. Conrad, (Texas Instruments), "Microfunctions distribute VLSI advantages," *Electronic Design,* Dec. 20, 1980, pp. 81–88.

Hughes, P. (National Semiconductors), "Factoring in software costs avoids red ink in microprocessor projects," *Electronics*, June 22, 1978, pp. 126–130. (Good cost analysis of software and firmware.)

Hurst, S. L., "Custom LSI design: The universal-logic-module approach," *ICCC 80*, pp. 1116–1119.

Husson, S. S., *Microprogramming—Principles and Practices*, Prentice-Hall, 1970, 614 pp.

Huston, W. D. (Motorola), "Processor architecture can cut software cost and risk," *Wescon, 1978*, 30/3, 8 pp.

Hwang, K., *Computer Arithmetic*, John Wiley, 1979, 423 pp.

IEEE, "Diagnosis of decoded PLAs," *Digest of Semiconductors Test Symp.*, Cherry Hill, NJ (Catalog no. 77 CH 126–76), Oct. 1977, pp. 88–101.

Ibaraki, T., "Gate interconnection minimization of switching networks using negative gates," *IEEE TC*, 1971, pp. 698–706.

Ibaraki, T., and S. Muroga, "Synthesis of networks with a minimum number of negative gates", Rep. 309, Dept. of Computer Science, University of Illinois, Feb. 1969. Also *IEEE TC*, Jan. 1971, pp. 48–59.

Ida, T., and E. Goto, "Performance of a parallel hash hardware with key deletion," *Proc. IFIP Congress*, 1977, pp. 643–647.

Ilardi, F. A., *Computer Circuit Analysis*, Prentice-Hall, 1976, 406 pp.

Intel, *Memory Design Handbook*, 1977.

Intel, *The Semiconductor Memory Book*, John Wiley, 1978, 524 pp.

Ipri, A. C. (RCA), "Sub-micron polysilicon gate CMOS/SOS technology," *IEDM*, 1978, pp. 46–49.

Ipri, A. C. (RCA), "Impact of design rule reduction on size, yield, and cost of integrated circuits," *Solid State Technology*, Feb. 1979, pp. 85–89, 91.

Irvine, C. A. (Softech Microsystems), "PASCAL: A software engineering language for microcomputers," *Electro 80*, 31/1, 7 pp.

Irwin, J. (National Semiconductors), "The whole system/the microcontroller," *Wescon, 1978*, 27/2, 6 pp.

Ishikawa, K., *Guide to Quality Control*, Asian Productivity Organization, Tokyo, 1976, 226 pp.

Iwamatsu, S., and K. Asanami (VLSI Tech. Res. Assoc.), "Deep UV projection system," *Solid State Technology*, May 1980, pp. 81–85.

Jaeger, R. (Signetics), "ECL 10K interconnects economically," *Electronic Design*, Sept. 27, 1974, pp. 90–94.

Jecman, R. M., C. H. Hui, A. V. Ebel, V. Kynett, and R. J. Smith (Intel), "HMOS II static RAMs overtake bipolar competition," *Electronics*, Sept. 13, 1979, pp. 124–128.

JEDEC, "Terms and definitions applicable to microcomputers and memory integrated circuits," Publ. 100, Electronic Industries Association, Washington, DC, 1979, 25 pp.

Jen, T. S., and N. Nan (IBM), "Gate array experiences in IBM," *1980 Electro Professional Program*, 22/3, 8 pp.

Jenné, F. B. (AMI), "Grooves add new dimension to V-MOS structure and performance," *Electronics*, Aug. 18, 1977, pp. 100–106.

Jensen, R. W., and L. P. McNamee, *Handbook of Circuit Analysis, Languages and Techniques*, Prentice-Hall, 1976. (Survey of ECA, ASTAP, etc.)

Johnson, D. W. (CDC), "Go from flowchart to hardware," *Electronic Design*, Sept. 1, 1976, pp. 90–95.

Johnson, R. C., "Microsystems and software," *Electronics*, Oct. 23, 1980, pp. 150–163. (Microcomputer software survey.)

Johnson, R. C., "Special report on memory management: microsystems exploit mainframe methods," *Electronics*, Aug. 11, 1981, pp. 119–127.

Johnson, W. S., G. L. Kuhn, A. L. Renninger, and G. Perlegos (Intel), "16-k EE-PROM relies on tunneling for byte-erasable program storage," *Electronics*, Feb. 28, 1980, pp. 113–117.

Johnson, W. S., et al. (Intel), "A 16 kb electrically erasable nonvolatile memory," *ISSCC 80*, pp. 152–153, 271. (FLOTOX.)

Juliussen, J. E., and J. Watson (Texas Instruments), "Problems of the 80's: Computer system organization," *Proc. Conf. on Computing in the 1980s, IEEE,* 1978, pp. 14–23.

Juran, J., *Quality Control Handbook,* 3rd ed., McGraw-Hill, 1974, 1600 pp.

Jurgen, R. K., "All electronic dashboards coming," *Spectrum,* June 1981, pp. 34–37.

Kaestner, O., "Implementing branch instructions with polynomial counters," *Computer Design,* Jan. 1975, pp. 69–75.

Kambayashi, Y., "Logic design of programmable logic arrays," *IEEE TC,* Sept. 1979, pp. 609–617.

Kambayashi, Y., and S. Muroga, "Properties of wired logic," to be published.

Kamikawai, R., et al. (Hitachi), "Placement and routing program for master-slice LSIs," *Proc. 13th DA Conf.,* 1976, pp. 245–250.

Kamuro, S., et al., "64 DSA ROM," *IEEE JSSC,* Apr. 1980, pp. 253–254.

Kang, S., and W. M. van Cleemput, "Automatic PLA synthesis from a DDL-P description," *Proc. 18th DA Conf.,* 1981, pp. 391–397.

Kani, K., "Design automation techniques," in *LSI Technology,* edited by M. Watanabe, Institute of Electronics Communication Engineers of Japan, 1979, pp. 222–257, Chap. 6 (in Japanese).

Kani, K., H. Kawanishi, and A. Kishimoto (NEC), "ROBIN: a building block LSI routing program," *IEEE Proc. Intnt'l Symp. on Circuits and Systems,* Apr. 1976, pp. 658–661.

Kaplan, C., "Computer-aided design," *Spectrum,* Oct. 1975, pp. 40–47.

Karp, J., and E. de Atley, "Use four-phase MOS IC logic," *Electronic Design,* Apr. 1, 1967, pp. 62–66.

Karstad, K. (RCA), "CMOS for general purpose logic design," *Computer Design,* May 1973, pp. 99–106.

Katto, H., et al. (Hitachi), "64-k RAM rebuffs external noise," *Electronics,* July 31, 1980, pp. 103–106.

Katz, K. (Bell Labs), "Gate array chips and computer aided design methods for custom LSI/VLSI," *ICCC 80,* pp. 207–209.

Kawasaki, T., Y. Kambayashi, and S. Muroga, "Optimal networks with NOR-OR gates and Wired-OR logic," to be published.

Kawato, N., et al. (Fujitsu), "Design and verification of large scale computer by using DDL," *Proc. 16th DA Conf.,* 1979, pp. 360–366.

Kenney, G. C., et al. (Philips), "An optical disk replaces 25 mag tapes," *Spectrum,* Feb. 1979, pp. 33–38.

Kernighan, B. W., and S. Lin, "An efficient heuristic procedure for partitioning graphs," *Bell Sys. Tech. J.,* 1970, pp. 291–307.

Kerkhoff, H. G., M. L. Tervoert, and J. A. Stemerdink, "The design and application of a CCD four-valued full adder circuit," *Compcon 81,* Spring, pp. 96–99.

Keyes, R. W., "Physical problems and limits in computer logic," *Spectrum,* May 1969, pp. 36–45.

Keyes, R. W., "Physical limits in digital electronics," *Proc. IEEE,* May 1975, pp. 740–767.

Keyes, R. W., "Power dissipation in planar digital circuits," *IEEE JSSC,* June 1975, p. 181.

Keyes, R. W., "Physical limits on computer devices," *Compcon 78,* Spring, pp. 294–296.

Keyes, R. W., "The evolution of digital electronics toward VLSI," *IEEE JSSC,* Apr. 1979, pp. 193–201.

Khokhani, K. H., and A. M. Patel (IBM), "The chip layout problem: A placement procedure for LSI," *Proc. 14th DA Conf.,* 1977, pp. 291–297. (CAD procedures for the use of IBM gate arrays.)

Khokhani, K. H., et al. (IBM), "Placement of variable size circuits on LSI masterslices," *Proc. 18th DA Conf.,* 1981, pp. 426–434. (APLACE.)

Kidder, T., *The Soul of a New Machine,* Atlantic/Little, Brown, 1981, 293 pp. (Design story of Data General's supermini Eagle.)

King, G., "Silicon-on sapphire ICs: technology forecast," *Circuits Manufacturing,* Sept. 1975, p. 48.

King, M. C. (Perkin-Elmer), "Future developments for 1:1 projection photolithography," *IEEE TED,* Apr. 1979, pp. 711–716.

King, M. C. (Perkin-Elmer), "Emerging technology of 1:1 optical projection photography," *Compcon 80,* Spring, pp. 294–297.

King, R. (Motorola), "Using MECL 10K Macrocell Array as a basic building block for standard product development," *1980 Electro Professional Program,* 22/2, 7 pp.

Kirk, W. J., Jr., L. S. Carter, and M. L. Waddell (Bendix), "Eliminate static damage to circuits," *Electronic Design,* Mar. 29, 1976, pp. 80–85.

Kitamura, T., S. Tashiro, and M. Inagaki (NEC), "Diagnosis techniques and methodologies for digital systems," Chapt. 1 in *Advances in Information Systems Science,* vol. 7, Plenum Press, 1978, pp. 1–87.

Klein, R., W. H. Owen, R. T. Simko, and W. E. Tchon (Xicor), "5-volt-only, nonvolatile RAM owes it all to polysilicon," *Electronics,* Oct. 11, 1979, pp. 111–116. (Survey of products.)

Knepper, R. W. (IBM), "Dynamic depletion mode: An E/D MOSFET circuit method for improved performance," *IEEE JSSC,* Oct. 1978, pp. 542–549.

Knuth, D. E., *The Art of Computer Programming,* Addison-Wesley, vol. 1, 1968; vol. 2, 1969; vol. 3, 1973.

Kobylar, A. W., R. L. Lindsay, and S. G. Pitroda, "ROMs cut cost, response time of *m*-out-of-*n* detectors," *Electronics,* Feb. 15, 1973, pp. 112–114.

Kodres, U. R., "Partitioning and card selection," in *Design Automation of Digital Systems,* vol. 1, edited by M. A. Breuer, Prentice-Hall 1972, pp. 173–212, Chap. 4.

Koehler, R. (Intel), "Microcontroller doubles as Boolean processor," *Electronic Design,* May 24, 1980, pp. 57–62. (Microcontroller 8051.)

Kohls, R. J. (Motorola), "A broad family of single-chip microcomputer," *Electro 81,* 1/5, 9 pp.

Konian, R. R., and J. L. Walsh (IBM), "Low voltage inverter (LVI) bipolar logic circuit logic circuit for gate-arrays," *Compcon 81,* Spring, pp. 191–192.

Koo, J. T. (National Semiconductors), "VLSI design and testing," *ICCC 80,* pp. 245–247.

Korenjak, A. J., and A. H. Teger (RCA), "An integrated CAD data base system," *Proc. 12th DA Workshop,* 1975, pp. 399–406.

Kosonocky, W. F., and D. J. Sauer (RCA), "The ABCs of CCDs," *Electronic Design,* Apr. 12, 1975, pp. 58–63.

Kostopoulos, G. K. (Singer), "Computing the square root of binary numbers," *Computer Design,* Aug. 1972, pp. 53–57. (Not based on ROMs.)

Kozak, P., H. K. Gummel, and B. R. Chawla (Bell Labs), "Operational features of an MOS timing simulator," *Proc. 12th DA Conf.,* 1975, pp. 95–101.

Kozawa, T., et al. (Hitachi), "Advanced LILAC—An automated layout generation system for MOS/LSI," *Proc. 11th DA Workshop,* 1974, pp. 26–46.

Kraft, P., *Programmers and Managers: The Routinization of Computer Programming in the United States,* Springer, 1977, 118 pp.

Krautner, G. N. (Exar), "Semi-custom LSI design with I²L gate arrays," *Wescon, 1978,* 32/1, 6 pp.

Kroeger, J., and B. Threewitt (Signetics), "Review the basics of MOS logic," *Electronic Design,* Mar. 15, 1974, pp. 98–105.

Kroeger, J., and B. Threewitt (Signetics), "Heed the limitations of MOS I/O circuitry," *Electronic Design,* May 10, 1974, pp. 83–88.

Kroeger, J. H., and O. N. Tozun (International Microcircuits Inc.), "CAD pits semicustom chips against standard slices," *Electronics,* July 3, 1980, pp. 119–123.

Krohn, H. E. (CDC), "Design verification of large scale scientific computers," *Proc. 14th DA Conf.,* 1977, pp. 354–361.

Kung, H. T., "The structure of parallel algorithms," in *Advances in Computers,* vol. 19, edited by M. C. Yovits, Academic Press, 1980, pp. 65–112.

Kung, H. T., and C. Leiserson, "Algorithms for VLSI processor arrays," Sec. 8.3 in *Introduction to VLSI Systems,* edited by C. Mead and L. Conway, Addison-Wesley, 1980, pp. 271–292.

Kuo, F. F., and W. G. Magnuson, Eds., *Computer-Oriented Circuit Design,* Prentice-Hall, 1969, 560 pp.

Kurtzberg, J. M. (Burroughs), "On approximation methods for the assignment problem," *J. ACM,* Oct. 1962, pp. 419–439.

Kvamme, F. (National Semiconductor), "Standard read-only memories simplify complex logic design," *Electronics,* Jan. 5, 1970, pp. 88–95.

LaBrie, P. J., and W. D. Miller (Semiconductor Circuits, Inc.), "Lick power-supply heat problems with good thermal management," *EDN,* Apr. 20, 1980, pp. 191–196.

Lai, H. C., "A study of current logic design problems," Ph.D. dissertation, Dept. of Computer Science, University of Illinois, Jan. 1976.

Lai, H. C., and S. Muroga, "Minimum parallel binary adders with NOR (NAND) gates," *IEEE TC,* Sept. 1979, pp. 648–659.

Lake, D. W. (Rockwell), "Logic simulation in digital systems," *Computer Design,* May 1970, pp. 77–83. (Simulation cost.)

Lam, H. W., R. R. Shah, A. F. Tasch, Jr., and L. Crosthwait (Texas Instruments), "Device and circuit fabrication and physical characterization of pulsed-laser-annealed polysilicon on SiO_2 and $Si_3 N_4$, *IEDM,* Dec. 1979.

Lancaster, D., *TTL Cookbook,* Howard W. Sams & Co., 1974, 335 pp.

Lancaster, D., *CMOS Cookbook,* Howard W. Sams & Co., 1977, 414 pp.

Landman, B. S., and R. L. Russo (IBM), "On a pin versus block relationship for partitions of logic graphs," *IEEE TC,* Dec. 1971, pp. 1469–1479.

Langenberg, D. N., D. J. Scalapino, and B. N. Taylor, "The Josephson effects," *Scientific American,* May 1966, pp. 30–39.

Langer, R., and T. Dugan (National Semiconductor), "Say it in a high-level language with 64-k read-only memories," *Electronics,* Apr. 13, 1978, pp. 119–124. Reader's comments, June 22, 1978, p. 6. (Interpreters and compilers for BASIC, FORTRAN, PL/M etc., to be stored in ROMs.)

Larson, R. P. (Rockwell), "Symbolic layout system speeds mask design for ICs," *Electronics,* July 20, 1978, pp. 125–128.

Lassen, C. L. (Exacta Circuits, Ltd.), "Wanted: A new interconnection technology," *Electronics,* Sept. 27, 1979, pp. 113–121.

Lattin, W. (Intel), "LSI computer system's technology for the 1980s," *Proc. Conf. on Computing in the 1980s, IEEE,* 1978, pp. 232–236.

Lattin, W. (Intel), "VLSI design methodology—The problem of the 80's for microprocessor design," *Proc. 16th DA Conf.,* 1979, pp. 548, 549.

Lau, S. Y. (Signetics), "High performance 1200-gate LSI array," *Wescon, 1979,* 13/4, 8 pp.

Laub, L. J. (Exxon Enterprises), "Optical mass storage technology," *Compcon 81,* Spring, pp. 470–472.

Lauther, U. (Siemens), "The Siemens CALCOS systems for computer aided design of cell based IC layout," *ICCC 80,* pp. 768–771.

Leblond, A., G. Serrero, and A. Verdillon, "Automatic layout of symbolic MD-MOS circuits," *ICCC 80,* pp. 772–776.

Lecht, C. P. (Advanced Computer Techniques), *The Wave of Change: A Techno-Economic Analysis of the Data Processing Industry,* McGraw-Hill, 1977, 194 pp.

Lee, C. Y., "An algorithm for path connections and its applications," *IRE TEC,* Sept. 1961, pp. 346–365.

Lee, F. S., et al. (Rockwell), "High speed LSI GaAs integrated circuits," *ICCC 80,* pp. 697–700.

Lee, S. Y., T. C. Chen, H. Chang, and C. Tung (IBM), "Text editing with magnetic bubbles," *Compcon 74,* pp. 69–72.

Leonard, D. N. (Texas Instruments), "MOS content-addressable memories," Chap. 8 in *Semiconductor Memories,* edited by J. Eimbinder, John Wiley, 1971, 214 pp.

Leonard, M. G., "Promise of RAM market pushes chip technology frontiers," *High Technology,* June 1980, pp. 57–63.

Leung, K. C., C. Michel, and P. Le Beux, "Logical systems design using PLAs and Petri nets—programmable hardwired systems," *Proc. IFIP Congress*, 1977, pp. 607–611.

Lepselter, M. P. (Bell Labs), "X-ray lithography breaks the submicrometer barrier," *Spectrum*, May 1981, pp. 26–29.

Leventhal, L. A., "Cut your processor's computation time by storing information in tables," *Electronic Design*, Aug. 16, 1977, pp. 82–89.

Lewin, D., *Computer-Aided Design of Digital Systems*, Crane Russak, 1977, 313 pp.

Lewis, D. R., and W. R. Siena, "Microprocessor or random logic," *Electronic Design*, Sept. 1, 1973, pp. 106–110. Also p. 71.

Liguori, F., Ed., *Automatic Test Equipment: Hardware, Software and Management*, IEEE Press, 1974, 253 pp.

Lin, C. M. (DEC), "A 4µm NMOS NAND structure PLA," *IEEE JSSC*, Apr. 1981, pp. 103–107.

Lindsay, B. W., and B. T. Preas (Sandia), "Design rule checking and analysis of IC mask designs," *Proc. 13th DA Conf.*, 1976, pp. 301–308.

Linford, J. R. (Motorola), "Binary-to-BCD conversion with complex IC functions," *Computer Design*, Sept. 1970, pp. 53–61. (Speeds of different approaches are roughly compared.)

Liu, C. L., *Introduction to Combinatorial Mathematics*, McGraw-Hill, 1968, 393 pp.

Liu, C. L., *Elements of Discrete Mathematics*, McGraw-Hill, 1977, 294 pp.

Liu, S. S., et al. (Intel), "HMOS III technology", *ISSCC 82*, pp. 234, 235, 326.

Liu, T. K., "Synthesis of logic networks with negative functions (with special emphasis on MOS logic networks)," Ph.D. thesis, Dept. of Computer Science, University of Illinois, 1972.

Liu, T. K., "Synthesis of multilevel feed-forward MOS networks," *IEEE TC*, June 1977, pp. 581–588.

Liu, T. K., "Synthesis of feed-forward MOS network with cells of similar complexities," *IEEE TC*, Aug. 1977, pp. 826–831.

Liu, T. K., K. Hohulin, L. E. Shiau, and S. Muroga, "Optimal one-bit full adder with different types of gates," *IEEE TC*, Jan. 1974, pp. 63–70.

Lo, T. C., R. E. Scheuerlein, and R. Tamlyn, "A 64k FET dynamic random access memory: Design considerations and description," *IBM JRD*, May 1980, pp. 318–327.

Logue, J. C., et al., "Hardware implementation of a small systems in programmable logic arrays," *IBM JRD*, March 1975, pp. 110–119.

Logue, J. C., et al., "Techniques for improving engineering productivity of VLSI designs," *IBM JRD*, May 1981, pp. 107–115.

Lohstroh, J., and A. Slob (Philips), "Punch-through cell for dense bipolar ROMs," *ISSCC 78*, pp. 20–21.

Long, J., and T. Proske (Hewlett-Packard), "Microprogramming boosts computer power— especially if you can do it yourself," *Electronic Design*, Mar. 15, 1979, pp. 84–88.

Loosemore, K. J. (Compeda), "Gaelic '80—The integrated system approach to CACD," *ICCC 80*, pp. 525–528.

Lopez, A. D., and H. F. S. Law (Bell Labs), "A dense gate matrix layout method for MOS VLSI," *IEEE JSSC*, Aug. 1980, pp. 736–740.

Losleben, P., and K. Thompson, (Dept. of Defense), "Topological analysis for VLSI circuits," *Proc. 16th DA Conf.*, 1979, pp. 461–473.

Lowe, L. "Projection aligners stepping out," *Electronics*, Mar. 24, 1981, pp. 104, 106.

Luckey, J. D. (Harris), "CMOS devices with other logic families," *Computer Design*, Dec. 1981, pp. 179–181.

Ludwig, J. A. (IBM), "A 50k bit Schottky cell bipolar read-only memory," *IEEE JSSC*, Oct. 1980, pp. 816–820.

Luecke, G., J. P. Mize, and W. C. Carr, *Semiconductor Memory Design and Application*, McGraw-Hill, 1973, 320 pp.

Luscher, P. E., W. S. Knodle, and Y. Chai (Varian), "Automated molecular beams grow thin semiconductor films," *Electronics,* Aug. 28, 1980, pp. 160–168.

Lyman, J. "Film carriers win productivity size," *Electronics,* Oct. 16, 1975, pp. 122–125.

Lyman, J., "Film carriers star in high-volume IC production," *Electronics,* Dec. 25, 1975, pp. 61–68.

Lyman, J., "Growing pin count is forcing LSI package changes," *Electronics,* Mar. 17, 1977, pp. 81–91.

Lyman, J. "Demands of LSI are turning chip makers toward automation, production innovations," *Electronics,* July 21, 1977, pp. 81–92.

Lyman, J. "Chip carriers are making inroads," *Electronics,* Nov. 24, 1977, pp. 86–89.

Lyman, J., "Packaging technology responds to the demand for higher densities," *Electronics,* Sept. 28, 1978, pp. 117–125.

Lyman, J., "Lithography chases the incredible shrinking line," *Electronics,* Apr. 12, 1979, pp. 105–116.

Lyman, J., "Scaling the barriers to VLSI's fine lines," *Electronics,* June 19, 1980, pp. 115–126.

Lyman, J., "Tape automated bonding meets VLSI challenge," *Electronics,* Dec. 18, 1980, pp. 100–105.

Lyman, J., and A. Rosenblatt, "The drive for quality and reliability," *Electronics,* May 19, 1981, pp. 125–148.

MacDougall, M. H. (CDC), "Computer system simulation: An introduction," *ACM Comp. Rev.,* Sept. 1970, pp. 191–209.

Maddox, R. L. (Rockwell), "Reverse CMOS processing," *Solid State Technology,* Feb. 1981, pp. 128–131, 140.

Mahoney, G. W., et al. (IBM), "Program called ASTAP makes fast work of analyzing large-scale circuits," *Electronics,* Apr. 18, 1974, pp. 109–113.

Majitha, J. C., and R. Kitai, "A cellular array for the nonrestoring extraction of square roots," *IEEE TC,* Dec. 1971, pp. 1617, 1618.

Maley, G. A. (IBM), *Manual of Logic Circuits,* Prentice-Hall, 1970, 256 pp., Chap. 4.

Mallach, E. G. (Honeywell Inf.), "Is the 16-bit minicomputer dead?" *Mini-Micro Systems,* May 1981, pp. 119–124.

Manabe, K., et al. (Toshiba), "A C^2MOS 16-bit parallel microprocessor," *ISSCC 76,* pp. 14, 15.

Mannel, T., and R. C. Johnson, "Buying software gets systems to market sooner: A special report," *Electronics,* Apr. 21, 1981, pp. 163–170.

Marino, D., "New algorithms for the approximate evaluation in hardware of binary logarithms and elementary functions," *IEEE TC,* Dec. 1972, pp. 1416–1421.

Marion, L., "Mainframe builders making more ICs," *Electronics,* May 22, 1980, pp. 106, 107. (Number of IC-related employees in mainframe manufacturers.)

Markkula, A. C., Jr. (Intel), "N-channel RAMs," *Wescon,* 1973, 16/1, pp. 1–5.

Markle, D. A. (Perkin-Elmer), "New projection printer," *Solid State Technology,* June 1974, pp. 50–53.

Markowiz, R. J. (Intel), "Software impact on microcomputer architecture: A case study," *Compcon 81,* Spring, pp. 40–48.

Marlowe, F. J., and J. P. Hasili, "Generating pulses with CMOS flip-flops," *Electronics,* June 7, 1973, p. 120.

Marshall, M., "Through the memory cells—Further explorations of ICs in testingland," *EDN,* Feb. 20, 1976, pp. 77–85. (The intricate mechanisms of memory testing.)

Marshall, M., "Microcomputers adding data-base management," *Electronics,* July 17, 1980, pp. 98, 99.

Marshall, M., and L. Waller, "VLSI pushes super-CAD techniques," *Electronics,* July 31, 1980, pp. 73–80.

Masessa, A. J., and R. G. Mohr (Bell Labs), "Design and developement of a 68-lead nonhermetic loaded chip carrier," *IEEE TCHMT,* Sept. 1980, pp. 424–430.

Masnick, B. (Hazeltine), "ROM-based software development," *Wescon, 1978*, 30/1, 6 pp.

Masuda, H., et al. (Hitachi), "A 5 v-only 64k dynamic RAM based on high s/n design," *IEEE JSSC*, Oct. 1980, pp. 846–854.

Matick, R. (IBM), *Computer Storage Systems and Technology*, John Wiley, 1977, 667 pp.

Matisoo, J. (IBM), "The superconducting computer," *Scientific American*, May 1980, pp. 50–65.

Mattera, L., "Component reliability," *Electronics*, Part 1, Oct. 2, 1975, pp. 91–98; Part 2, Oct. 30, 1975, pp. 87–94; Dec. 11, 1975, pp. 6, 8; Dec. 25, 1975, pp. 83–85.

Mattison, R. L. (GTE), "Design automation of MOS artwork," *Computer*, Jan. 1974, pp. 21–27.

Maul, L. (Motorola), "ECL 10,000 layout and loading rules," *EDN*, Part I, July 20, 1973, pp. 32–36; Part II, Aug. 5, 1973, pp. 38–41.

May, P., and F. C. Schiereck (Philips), "High-speed static programmable logic array in LOC-MOS," *IEEE JSSC*, June 1976, pp. 365–369.

McCluskey, E. J., and J. F. Wakerly, "A circuit for detecting and analyzing temporary failures," *Compcon 81*, Spring, pp. 317–321.

McDermott, J., "E-beam and x-ray technology—A must for the high-density ICs of the 1980s," *Electronic Design*, June 21, 1978, pp. 40–44.

McDermott, J. "Thermoelectric coolers tackle jobs heat sinks can't," *EDN*, May 20, 1980, pp. 111–117.

McDermott, J., "Hand-held personal calculators," *EDN*, June 20, 1980, pp. 116–130.

McDermott, J., "Fitness and health-care products incorporate advanced electronics," *EDN*, Sept. 20, 1980, pp. 69–75, 77.

McDowell, J. J. (Monolithic Memories), "Improve ROM systems with PROMs," *Electronic Design*, July 5, 1974, pp. 92–96.

McDowell, J. J. (Monolithic Memories), "Sequence detector uses programmable ROMs," *EDN*, July 20, 1974, pp. 100, 102.

McKenny, V. (Mostek), "Depletion-mode devices hike speed of MOS random access memory," *Electronics*, Feb. 15, 1971, pp. 80–85.

McMinn, C., R. Markowitz, J. Wharton, and W. Grundmann (Intel), "Silicon operating system standardizes software," *Electronics* Sep. 8, 1981, pp. 135–139. (Intel 80130.)

McMullen, J. (R. V. Weatherford Co.), "Programming the PROM: A make or buy decision," *EDN*, Mar. 5, 1974, pp. 59–62.

Mead, C., and L. Conway, Eds., *Introduction to VLSI Systems*, Addison-Wesley, 1980, 396 pp.

Mei, K. C. (Hewlett-Packard), "Multiple-output logic circuits," U.S. patent no. 3,965,367, June 22, 1976.

Melear, C. (Motorola), "Enhanced microcomputer help host in DP systems," *Electronic Design*, May 14, 1981, pp. 185–191.

Melving, D. (Intel), "LSI helps telephones go digital," *Spectrum*, June 1980, pp. 30–33.

Mendelsohn, A., "Forum on 8 and 16-bit microprocessors," *Electronic Products Mag.*, June 15, 1981, pp. 47–53.

Mennoue, A., and R. L. Russo (IBM), "An example computer logic graph and its partitions and mappings," *IEEE TC*, Nov. 1974, pp. 1198–1204. (Partitioning in design automation.)

Metze, G., "Minimum square rooting," *IEEE TEC*, Apr. 1965, pp. 181–185.

Mick, J. R. (Advanced Micro Devices), "Using Schottky 3-stage outputs in bus-organized systems," *EDN*, Dec. 5, 1974, pp. 35–39.

Mick, J. R., and J. Brick (Advanced Micro Devices), *Bit-Slice Microprocessor Design*, McGraw-Hill, 1980, 398 pp.

Mihalik, M., and H. Johnson (Tektronix), "μP-based product design starts with μP selection," *Electronic Design*, Sept. 1, 1980, pp. 119–125.

Mikami, K., and K. Tabuchi (Mitsubishi Electric), "A computer program for optimal routing of printed circuit conductors," *Proc. IFIP Congress*, 1968, pp. 1475–1478.

Mikkelson, J. M., et al. (Hewett-Packard), "An nMOS VLSI process for fabrication of a 32b cpu chip," *ISSCC 81*, pp. 106, 107.

Miles, G. (Intersil), "For sequential control, FPLA make sense," *Electronic Design*, Oct. 25, 1976, pp. 164–169.

Miles, T. E. (Signetics), "Schottky TTL vs ECL for high speed logic," *Computer Design*, Oct. 1972, pp. 79–86.

Millerick, J. (National Semiconductors), "COP 420C, CMOS controller with enhanced hardware and software capabilities," *Electro 80*, 29/1, 6 pp.

Millman, J., and C. C. Halkias, *Integrated Electronics*, McGraw-Hill, 1972, 911 pp.

Minato, O., et al. (Hitachi), "A high-speed low-power hi-CMOS 4k static RAM," *IEEE TED*, June 1979, pp. 882–885.

Minato, O., et al. (Hitachi), "2k × 8 bit hi-CMOS static RAMs," *IEEE TED*, Aug. 1980, pp. 1591–1595.

Minnick, R. C., P. T. Bailey, R. M. Sandfort, and W. L. Semon, "Magnetic bubble computer systems," *AFIPS Conf. Proc.*, Part 2, Fall 1972. (Magnetic bubble logic.)

Mitchell, C., and P. Holland (National Semiconductors), "The use of high speed PROMs to generate Boolean functions," *Wescon, 1978*, 3/4, 7 pp.

Mizell, L. C., "Thermal analysis of microcircuits," *Microelectronics*, Sept. 1969, pp. 12–18.

Mohammadi, F., "Silicides for interconnection technology," *Solid State Technology*, Jan. 1981, pp. 65–72, 92.

Mokhoff, N., "Office automation: A challenge," *Spectrum*, Oct. 1979, pp. 66–69.

Mokhoff, N., "Four targets of VLSI," *Spectrum*, June 1980, pp. 34–36.

Mokhoff, N., "Beating an electronic opponent is a challenge," *Spectrum*, Nov. 1980, pp. 26–30. (Electronic games.)

Monolithic Memories, *PAL Handbook*, 1978.

Monolithic Memories, "Use of bipolar ROM and P.ROMs," Application Note 110, early 1970s.

Moon, D., "The polynomial counter design technique with applications in four-phase logic," *Computer Design*, Nov. 1969, pp. 135–143.

Moore, B. E., "What level of LSI is best for you," *Electronics*, Feb. 16, 1970, pp. 126–130.

Moore, G. (Intel), "VLSI: Some fundamental challenges," *Spectrum*, Apr. 1979, pp. 30–37.

Moorehead, F. F., Jr., and B. L. Crowder, "Ion implantation," *Scientific American*, Apr. 1973, pp. 64–71.

Mori, H., et al. (NEC), "BRAIN: An advance iterative layout design system for printed wiring boards," *ICCC 80*, pp. 754–757.

Morris, R. L., and J. R. Miller, Eds., *Designing with TTL Integrated Circuits*, McGraw-Hill, 1971, 322 pages.

Morse, S. P., W. B. Pohlman, and B. W. Ravenel (Intel), "The Intel 8086 microprocessor: A 16-bit evolution of the 8080," *Computer*, June 1978, pp. 18–27.

Morse, S. P., B. W. Ravenel, S. Mazor, and W. B. Pohlman (Intel), "Intel microprocessors—8008 to 8086," *Computer*, Oct. 1980, pp. 42–60.

Moses, R. J. (GTE), "Distributed processing in manufacturing at GTE Automatic Electric," *Proc. 15th DA Conf.*, 1978, pp. 26–33.

Motorola, "Polycell LSI—Complex, custom designed MOS integrated circuits," 2nd ed., 1971, 167 pp. Supplement 1 (Shift register cells), 1971, 26 pp.

Motorola, "LSI with 112 gate TTL array," 1972.

Motorola Inc., *McMOS Handbook*, 1974.

Motorola, "Motorola Megalogic—The modular approach to bipolar LSI," 1975, 16 pp.

Mrazek, D. (National Semiconductor), "Save ROMs in high resolution dot-matrix displays and printers," *EDN*, May 5, 1973, pp. 56–64. (Intermediate coding.)

Mrazek, D., and M. Bron (National Semiconductor), "Boost bit-slice microsequence speed: Choose from eight different circuits," *Electronic Design*, May 24, 1979, pp. 118–123. (Comparison of different architectures.)

Mrazek, D., and M. Morris (National Semiconductor), "How to design with programmable logic arrays," National Semiconductor, AN-89, Aug. 1973, Also in *Memory Databook,* National Semiconductor, 1977.

Mrazek, D., and M. Morris (National Semiconductor), "PLAs replace ROMs for logic designs," *Electronic Design,* Oct. 25, 1973, pp. 66–70.

Muehldorf, E. I. (TRW), "High speed integrated circuit characterization and test strategy," *Solid State Technology,* Sept. 1980, pp. 93–98.

Muehldorf, E. I. (TRW), and A. Savkar (IBM), "LSI logic testing," *IEEE TC,* Jan. 1981, pp. 1–17.

Muller, H. H., W. K. Owens, and P. J. Verhofstadt (Fairchild), "Fully compensated emitter-coupled logic: Eliminating the drawbacks of conventional ECL," *IEEE JSSC,* Oct. 1973, pp. 362–367.

Muroga, S., "Logical design of optimal digital networks by integer programming," in *Advances in Information Systems Science,* vol. 3, edited by J. T. Tou, Plenum Press, 1970, pp. 283–348, Chap. 5.

Muroga, S., *Threshold Logic and Its Applications,* John Wiley, 1971, Chap. 14.

Muroga, S., *Logic Design and Switching Theory,* John Wiley, 1979, 617 pp.

Muroga, S., and H. C. Lai, "Minimization of logic networks under a generalized cost function," *IEEE TC,* Sept. 1976, pp. 893–907.

Murphy, B. T. (Bell Labs), "Cost-size optima of monolithic integrated circuits," *Proc. IEEE,* Dec. 1964, pp. 1537–1545.

Murphy, B. T., and R. Edwards (Bell Labs), "A CMOS 32b single chip microprocessor," *ISCC 81,* pp. 230, 231, 276.

Musgrave, G., Ed., *Computer-Aided Design of Digital Electronic Circuits and Systems,* North-Holland, 1979, 325 pp.

Myers, G. J. (IBM), *Digital System Design with LSI Bit-Slice Logic,* John Wiley, 1980, 338 pp.

Nagel, L. W., "SPICE-2: A computer program to simulate semiconductor circuits," Ph.D. dissertation, Dept. of Electrical Engineering, University of California, Berkeley, May 1975.

Nagel, L. W., and D. O. Pederson, "SPICE, simulation programs with integrated circuits analysis," Memorandum ERL-M382, Dept. of Electrical Engineering, University of California, 1973.

Nair, R., S. M. Thatte, and J. A. Abraham, "Efficient algorithms for testing semiconductor random-access memory," *IEEE TC,* June 1978, pp. 572–576.

Nakamura, K., "Synthesis of gate-minimum multi-output two-level negative gate networks," *IEEE TC,* Oct. 1979, pp. 768–772.

Nakamura, K., N. Tokura, and T. Kasami, "Minimal negative gate networks," *IEEE TC,* Jan. 1972, pp. 5–11.

Nakashiba, H., et al. (NEC-Tohiba), "An advanced PSA technology for high-speed bipolar LSI," *IEEE TED,* Aug. 1980, pp. 1390–1394.

Nakaya, M., et al. (Mitsubishi Electric), "High-speed MOS gate array," *IEEE JSSC,* May 1980, pp. 1665–1670.

Nance, S. (AMI), "SLIC layout saves time," *Spectrum,* Oct. 1978, p. 58.

Nash, D., and H. Willman (Raytheon), "Software engineering applied to computer-aided (CAD) software development," *Proc. 18th DA Conf.,* 1981, pp. 530–539.

National Semiconductor, "MM 4232 sine look-up table specification sheet," Mar. 1972.

National Semiconductor, "The application of ROMs," AN-61, June 1972.

National Semiconductor, "DM 7575/DM 8575, DM 7576/DM 8576 programmable logic array," 1973.

National Semiconductor, "IMP-4A/521D MOS/LSI control and read only memory unit (CROM) for 4-bit microprocessors," preliminary data, Sept. 1974.

National Semiconductor, "PAL—Programmable Array Logic," 1979, 8 pp.

Nelson, Bob (National Semiconductor), "Power-switching circuits cut PROM power consumption," *EDN,* May 20, 1980, pp. 127–130.

Nelson, C. T. (National Semiconductor), "Supermatched bipolar transistors improve dc and ac designs," *EDN,* Jan. 5, 1980, pp. 115–120.

Nevins, J. L., and D. E. Whitney, "Computer-controlled assembly," *Scientific American*, Feb. 1978, pp. 62–74.

Nguyen-Huu, A. (Teradyne), "Advances in CMOS device technology," *Computer Design*, Jan. 1975, pp. 87–92.

Nichols, J. L. (Fairchild), "A logical next step for read-only memories," *Electronics*, June 12, 1967, pp. 111–113.

Nicolino, S. J., Jr. (Intel Magnetics), "LSI devices support bubble memory system design," *Computer Design*, Nov. 1980, pp. 159–167.

Nishizawa, J., T. Nonaka, Y. Mochida, and T. Ohmi, "Approaches to high performance SITL," *IEEE JSSC*, Oct. 1979, pp. 873–875.

Noorlag, D. J. W., L. M. Terman, and A. G. Konheim (IBM), "The effect of alpha-particle-induced soft errors on memory systems with error correction," *IEEE JSSC*, June 1979, pp. 319–325.

Noto, R., F. Borgini, and B. Suskind (RCA), "Automated universal array," *ICCC 80*, pp. 315–318, (CMOS/SOS gate array.)

Noyce, R. N. (Fairchild), "Making integrated electronics technology work," *Spectrum*, May 1968, pp. 63–66.

Noyce, R. N. (Intel), "Large-scale integration: What is yet to come?" *Science*, Mar. 18, 1977, pp. 1102–1106.

Noyce, R. N., and M. E. Hoff, Jr. (Intel), "A history of microprocessor development at Intel," *IEEE Micro*, Feb. 1981, pp. 8–21.

Nuzillart, G., C. Arnodo, and J-P. Puron (Thompson-CSF), "A subnanosecond integrated switching circuit with MESFETs for LSI," *IEEE JSSC*, June 1976, pp. 385–394.

Oakes, M. F. (Racal-Redac), "The complete VLSI design system," *Proc. 16th DA Conf.*, June 1979, pp. 452–460.

Oberman, R. M. M., *Digital circuits for binary arithmetic*, Halstead, 1979, 340 pp.

O'Donovan, P. L., and L. F. Lind, "Heuristic algorithm for reducing large product-of-sums logic expressions," *IEEE Trans. Systems, Man, Cybernetics*, Oct. 1976, pp. 711–714.

Offerdahl, R. E. (Control Data Corp.), "High utilization of a master slice LSI array," *Wescon, 1978*, 3/5, 10 pp.

Ogdin, C. A., "Designers must know when to make hardware/software trade-offs," *EDN*, Nov. 20, 1976, pp. 289–294. (Hardware–software tradeoffs for UART.)

Ogdin, C. A., "Using EAROMs with your microcomputer requires some special tricks," *EDN*, Nov. 20, 1978, pp. 237–247.

Ogdin, C. A., "The microprocessor popularity race," *Mini-Micro Systems*, Dec. 1978, pp. 58–66.

Ogdin, C. A., *Microcomputer Design*, Prentice-Hall, 1978, 208 pp.

Ogdin, C. A., *Software Design for Microcomputers*, Prentice-Hall, 1978, 200 pp.

Ogdin, C. A., (Software Technique), "Language selection criteria," *Electro 80*, 31/4, 4 pp.

Ohta, K., et al. (NEC), "A high-speed logic LSI using diffusion self-aligned enhancement depletion MOST," *ISSCC 75*, pp. 124, 125, 224.

Ohzone, T. (Matsushita Electric), "64-k static RAM surrounds *n*-MOS cells with C-MOS circuits," *Electronics*, Nov. 6, 1980, pp. 145–148.

Ohzone, T., et al. (Matsushita Electric), "An 8k × 8 bit static MOS RAM fabricated by *n*MOS/ *n*-well CMOS technology," *IEEE JSSC*, Oct. 1980, pp. 854–861.

Okada, K., et al. (NEC), "A polysilicon self-aligned process for bipolar LSIs," *ICCC 80*, pp. 689–692.

Olson, H. T. (GTE), "A user's experience in computer aided manufacturing or 'CAM' at GTE-Automatic Electric," *Proc. 15th DA Conf.*, 1978, pp. 23–25.

O'Neil, W. D. (Exar Integrated Systems), "Combined analog/digital LSI design using I^{2}L gate arrays," *Wescon, 1979*, 13/2, 6 pp.

O'Neill, L. A., et al. (Bell Labs), "Designers workbench—Efficient and economical design aids," *Proc. 16th DA Conf.*, 1979, pp. 185–199.

Orme, M., *Micros: A Pervasive Force* (A study of the impact of microelectronics on business and society 1946–1990), John Wiley, 1980, 214 pp.

Osborne, T. (Hewlett-Packard), "Tradeoffs in electrical calculator design," *Spectrum*, Feb. 1974, pp. 36, 37.

Otsuka, W. (Monsanto), "Microprocessor makes alphanumeric display smart," *Electronics*, Dec. 7, 1978, pp. 137–141.

Otsuka, W. (General Instrument Optoelectronics), "PROMs decode ASCII characters for segmented LED displays," *EDN*, Aug. 5, 1979, pp. 105–110.

Owens, W. K. (Fairchild), "Gate arrays versus custom LSI," *Compcon 81*, Spring, pp. 174–177.

Paillotin, J. F., "Optimization of the PLA area," *Proc. 18th DA Conf.*, 1981, pp. 406–410.

Paivanas, J. A., and J. K. Hassan (IBM), "A new air film technique for low contact handling of silicon wafers," *Solid State Technology*, Apr. 1980, pp. 148–155.

Palmer, J., R. Nave, C. Wymore, R. Koehler, and C. McMinn (Intel), "Making mainframe mathematics accessible to microcomputers," *Electronics*, May 8, 1980, pp. 114–121. (8087.)

Panish, M. G., and A. Y. Cho (Bell Labs), "Molecular beam epitaxy," *Spectrum*, Apr. 1980, pp. 18–23.

Pape, U., "Implementation and efficiency of Moore-algorithms for the shortest route problem," *Mathematical Programming*, Oct. 1974, pp. 212–222.

Pashley, R., et al., "H-MOS scales traditional devices to higher performance level," *Electronics*, Aug. 18, 1977, pp. 94–99. Correction, Feb. 2, 1978, pp. 6, 8.

Patil, S. S., and A. T. Welch, "A programmable logic approach for VLSI," *IEEE TC*, Sept. 1979, pp. 594–601.

Pearman, R. A., *Power Electronics: Solid State Motor Control*, Prentice-Hall, 1980, 368 pp.

Pederson, R. A. (Bell Labs), "Integrated injection logic: A bipolar LSI technique," *Computer*, Feb. 1976, pp. 24–29.

Peltier, A. W. (Motorola), "A new approach to bipolar ISI: C^3L," *ISSCC 75*, pp. 168–196.

Penoyer, R. F., et al. (IBM), "An 18k bipolar dynamic random access memory," *IEEE JSSC*, Oct. 1980, pp. 861–865.

Percival, R., "The application of ROMs," National Semiconductor, AN-61, June 1972.

Percival, R. (National Semiconductor), "ROMs are versatile in digital systems," *Electronic Design*, June 8, 1972, pp. 66–71.

Perle, M. D. (Zwicker Electric), "CORDIC reduces trigonometric function look-up," *Computer Design*, June 1971, pp. 72, 74, 76, 78. (Comparison of computer time and accuracy between Taylor series expansion and CORDIC for cos x.)

Perrin, J. P., M. Denouette, and E. Daclin, *Switching Machines*, vol. 2, Springer, 1972, 421 pp.

Persky, G., D. N. Deutsch, and D. G. Schweikert (Bell Labs), "LTX—A system for the directed automatic design of LSI circuits," *Proc. 13th DA Conf.*, 1976, pp. 399–407.

Persky, G., D. N. Deutsch, and D. G. Schweikert (Bell Labs), "LTX—A minicomputer-based system for automated LSI layout," *J. Design Automation and Fault-Tolerant Computing*, May 1977, pp. 217–255.

Peters, T., and M. Gold, "Logic arrays supplant discrete packages," *Electronic Design*, Dec. 24, 1981, pp. 63–69.

Petrale, J. J. (Loral Corp.), "PROM controller makes fast work of serial jobs," *Electronics*, Apr. 12, 1979, pp. 134–138.

Peuto, B. L., and G. J. Prosenko (Zilog), "One-chip microcomputer excels in I/O—and memory-intensive users—Introducing the Z8, part 1," *Electronics*, Aug. 31, 1978, pp. 129–133.

Peuto, B. L. (Zilog), and L. J. Shustek (Stanford), "Current issues in the architecture of microprocessors," *Computer*, Feb. 1977, pp. 20–25. (Comparison of microprocessors and mainframes.)

Phadnis, M. G., and D. G. Joshi, "BCD/binary conversion with single IC cell," *Computer Design*, Dec. 1978, pp. 94–103.

Phelps, M., "Control MOS/LSI yield by design," *EDN/EEE*, Oct. 15, 1971, pp. 21–27.

Phister, M., Jr., "Technology and economics: Integrated circuit manufacturing costs," *Computer Design*, Oct. 1979, pp. 34, 38, 39, 42.

Phister, M., Jr., "Technology and economics: Integrated circuit geometry and optimum chip size," *Computer Design*, Nov. 1979, pp. 30, 32, 37, 38.

Phister, M., Jr., "Technology and economics: Manufacturing cost for a microprogrammed system," *Computer Design*, Dec. 1979, pp. 26, 30–32.

Phister, M., Jr. (Systems Consulting), "Technology and economics: The engineering audit," *Computer Design*, Jan. 1980, pp. 42, 47, 50, 54.

Pierce, J. A., E. Malitz, and M. Wilson, "Software eases a hard task-wire routing ECL," *Electronic Design*, June 21, 1980, pp. 113–115.

Pitts, R. C., "Gate arrays—cost-slashing replacements for SSI, MSI," *Electronic Design*, Dec. 24, 1981, pp. 127–133.

Plessey Semiconductors, *MNOS IC Handbook*, 1981, 52 pp. (Nonvolatile logic applications.)

Pogge, R. D. (Naval Weapon Center), "Lookup tables provide quick logarithmic calculations," *EDN*, Aug. 5, 1977, pp. 87–91.

Pomeranz, J., R. Nijhuis, and C. Vicary (IBM), "Customized metal layers vary standard gate-array chip," *Electronics*, Mar. 15, 1979, pp. 105–108. (704 Schottky TTL gate array.)

Posa, J. G., "Programming microcomputer systems with high-level languages," *Electronics*, Jan. 18, 1979, pp. 105–112.

Posa, J. G., "Microprocessors and microcomputers," *Electronics*, Oct. 25, 1979, pp. 144–157. (Comparison of microprocessors and microcomputers.)

Posa, J. G., "Gate arrays—A special report," *Electronics*, Sept. 25, 1980, pp. 145–158. Readers' comments Dec. 18, 1980, p. 8; Jan. 13, 1981, p. 8; Feb. 24, 1981, p. 8.

Powell, R. E. (E-Systems), "Justification and financial analysis for CAD," *Proc. 17th DA Conf.*, 1980, pp. 564–571.

Powers, G. L. (Fairchild), "Comment on NOR-NAND synthesis," *IEEE TC*, Aug. 1965, p. 648.

Preas, B. T., and C. W. Gwyn (Sandia), "Methods for hierarchical automatic layout of custom LSI circuits masks," *Proc. 15th DA Conf.*, June 1978, pp. 155–169.

Preparata, F. P., "Universal logic modules as an approach to functional standardization," *Proc. 1st Texas Symp. Computer Systems*, 1972, 1-3, 5 pp.

Preparata, F. P., "New parallel-sorting schemes," *IEEE TC*, July 1978, pp. 669–673.

Preparata, F. P., and J. Vuillemin, "The cube-connected cycles: A versatile network for parallel computation," *Commun. ACM*, May 1981, pp. 300–309.

Preparata, F. P., and R. T. Yeh, *Introduction to Discrete Structures*, Addison-Wesley, 1973, 354 pp.

Pressman, D., "Use five methods of protection to safeguard computer software," *EDN* Aug. 5, 1981, pp. 182–184.

Price, J. E. (Amdahl), "LSI chip architecture for large high performance computers," *ICCC 80*, pp. 497–502.

Priel, U., "Take a look inside the TTL IC," *Electronic Design*, Apr. 15, 1971, pp. 68–71.

Priel, U., and P. Holland (National Semiconductor), "Application of a high speed programmable logic array," *Computer Design*, Dec. 1973, pp. 94–96.

Prioste, J., *MECL 10,000 Macrocell Array Design Manual*, Motorola, 1978, 44 pp.

Prioste, J. (Motorola), "Macrocell approach customizes fast VLSI," *Electronic Design*, June 7, 1980, pp. 159–166.

Prioste, J., R. Rao, and W. R. Blood, Jr. (Motorola), "Functional array eases custom ECL design," *Electronics*, Feb. 5, 1979, pp. 113–117.

PRO-LOG Corp., *PROM User's Guide*, 1977, 47 pp.

Puckett, H. E., and W. W. Lattin (Motorola), "Why *N*-channel MOS?" *IEEE Intercon 1973*, 23/1, pp. 1–5.

Queisser, H.-J., Ed., *X-Ray Optics: Applications to Solids*, Springer, 1977, 227 pp.

Queyssac, D. (Motorola), "Projecting VLSI's impact on microprocessors," *Spectrum*, May 1979, pp. 38–41.

Quinn, P. M., J. M. Early, W. B. Sander, and T. A. Longo (Fairchild), "A 16k × 1 I³L dynamic RAM," *ISSCC 78*, pp. 154–155.

RCA, *COS/MOS Integrated Circuits Manual*, 1972.

RCA, *RCA COS/MOS Custom IC Layout Guide*, 1973, 39 pp.

Rallapalli, K., and J. Kroeger (AMD), "Chips make fast math a map for microprocessors," *Electronics*, Apr. 24, 1980, pp. 153–157. (Am9511A and Am9512.)

Rallapalli, K., and P. Verhofstadt (Fairchild), "Macrologic-versatile-functional blocks for high performance digital systems," *NCC*, 1975, pp. 67–73.

Ramamoorthy, C. V., J. R. Goodman, and K. H. Kim, "Some properties of iterative square-rooting methods using high-speed multiplication," *IEEE TC*, Aug. 1972, pp. 837–847.

Ramsay, F. R. (Ferranti), "Automation of design for uncommitted logic array," *Proc. 17th DA Conf.*, 1980, pp. 100–107.

Ramsey, F. R. (Ferranti), "A remote design station for customer uncommitted logic array design," *Proc. 18th DA Conf.*, 1981, pp. 498–504.

Ranada, D., "High-performance digital logic," *EDN*, Feb. 5, 1979, pp. 83–89.

Raphael, H. A. (Singer Business Machines), "Fast BCD/binary conversions," *Electronic Design*, Oct. 25, 1973, pp. 84–89. (Speeds of different approaches are roughly compared.)

Raskin, J., and T. Whitney (Apple Computer), "Perspectives on personal computers," *Computer*, Jan. 1981, pp. 62–73.

Rattner, J., and W. W. Lattin (Intel), "ADA determines architecture of 32-bit microprocessor," *Electronics*, Feb. 24, 1981, pp. 6, 119–126. (iAPX 432.)

Reddy, R. (Nitron), "MNOS devices remember after the lights go out," *EDN*, Aug. 20, 1978, pp. 99–104.

Redfern, T. P., and J. Jorgensen (National Semiconductor), "New CMOS voltage translation and buffering techniques," *EDN*, Dec. 20, 1974, pp. 24, 25.

Reece, D. A., "CMOS full adder converts BCD to binary," *EDN*, Oct. 5, 1980, p. 161.

Reed, Z. J., and R. Ligget (Motorola), "Upgraded logic family boosts system capability," *Electronic Design*, Aug. 16, 1980, pp. 103–107. (MECL 10KH.)

Reifer, D. J., and S. Trattner (Aerospace Corp.), "A glossary of software tools and technique," *Computer*, July 1977, pp. 52–60.

Reingold, E. M., J. Nievergelt, and N. Deo, *Combinatorial Algorithms: Theory and Practice*, Prentice-Hall, 1977, 433 pp.

Reyling, G. (National Semiconductor), "PLAs enhance digital processor speed and cut component count," *Electronics*, Aug. 8, 1974, pp. 109–114.

Reynolds, J. A. (Micro Mask), "An overview of E-beam mask-making," *Solid State Technology*, Aug. 1979, pp. 87–94.

Rice, D. (Fairchild), "Isoplanar-S scales down for new heights in performance," *Electronics*, Dec. 6, 1979, pp. 137–141.

Rice, R., and W. R. Smith (Fairchild), "SYMBOL—A major departure from classic software dominated von Neumann computing systems," *1971 SJCC*, pp. 575–587.

Richman, P., *MOS Field-Effect Transistors and Integrated Circuits*, John Wiley, 1973, 259 pp.

Rideout, V. L. (IBM), "One-device cells for dynamic random-access memories: A tutorial," *IEEE TED*, June 1979, pp. 839–852.

Riley, W. B., Ed., *Electronic Computer Memory Technology*, McGraw-Hill, 1971, 270 pp.

Rivard, J. G. (Ford Motor), "Microcomputers hit the road," *Spectrum*, Nov. 1980, pp. 44–47. (Engine control by microcomputers.)

Roberson, D. A. (IBM), "A microprocessor-based portable computer: The IBM 5100," *Proc. IEEE*, June 1976, pp. 994–999.

Roberson, D. A. (IBM), "An architecture for a small interactive computer—The IBM 5100," *Compcon 77*, Spring, pp. 136–139.

Roberts, M. B. (Texas Instruments), "VLSI—A challenge for system designers," *Proc. 17th DA Conf.*, 1980, p. 345.

Roberts, M. (ECS Microsystems), "Multiprocessing networks vs mainframes," *Mini-Micro Systems,* Oct. 1980, pp. 121–128. (Hypercube.)

Robson, W. (Maruman), "Procurement approaches to custom LSI," *Wescon, 1978,* 32/3, 2 pp.

Rockwell, "Programmable logic array," Publ. 2519-D19, 1973.

Rodgers, T. J., and J. D. Meindl, "VMOS: High-speed TTL compatible MOS logic," *IEEE JSSC,* Oct. 1974, pp. 239–249.

Rodgers, T. J., et al. (AMI), "VMOS ROM," *IEEE JSSC,* Oct. 1976, pp. 614–622.

Rodgers, T. J., et al. (AMI), "VMOS memory technology," *IEEE JSSC,* Oct. 1977, pp. 515–524.

Rodriguez, E. T. (Theta-J), "Model semiconductor thermal designs even with scanty vendor data," *Electronic Design,* Feb. 15, 1979, pp. 102–105.

Rodriguez, E. T. (Theta-J), "Fast DMOS optocouplers beat SSRs in speed and reed relays in performance," *Electronic Design,* Mar. 29, 1980, pp. 60–63.

Rohner, P., *Fluid Power Logic Circuit Design,* John Wiley, 1979, 226 pp.

Ronen, R. S., and F. B. Micheletti (Rockwell), "Recent SOS technology, advances and applications," *Solid State Technology,* Aug. 1975, pp. 39–46.

Rosenberg, L. M. (RCA), "The evolution of design automation to meet the challenges of VLSI," *Proc. 17th DA Conf.,* 1980, pp. 3–11.

Rosenberg, L., and C. Benbassat (RCA), "CRITIC, An integrated circuit design rule checking program," *Proc. 11th DA Workshop,* 1974, pp. 14–18.

Roth, J. P. (IBM), "Programmed logic array optimization," *IEEE TC,* Feb. 1978, pp. 174–176.

Roth, J. P. (IBM), *Computer Logic, Testing, and Verification,* Computer Science Press, 1980, 176 pp.

Roy, J. C., "A silicon-on-sapphire integrated video controller," *Hewlett-Packard J.,* Mar. 1981, pp. 16–19.

Rozeboom, R. W. (Texas Instruments), "Current problems related to LSI functional testing," *Proc. 13th DA Conf.,* 1976, pp. 203, 204.

Rozeboom, R. W., and J. J. Crowley (Texas Instruments), "An implementation of computer aided test generation techniques," *Proc. 13th DA Conf.,* 1976, pp. 194–202.

Ruehli, A. E. (IBM), "Survey of computer-aided electrical analysis of integrated circuit interconnections," *IBM JRD,* Nov. 1979, pp. 626–639.

Ruehli, A. E. (IBM), "Survey of analysis, simulation and modeling for large scale logic circuits," *Proc. 18th DA Conf.,* 1981, pp. 125–129.

Rung, R. D. (Hewlett-Packard), "Determining IC layout rules for cost minimization," *IEEE JSSC,* Feb. 1981, pp. 35–43.

Russo, P. M. (RCA), "VLSI impact on microprocessor evolution, usage, and system design," *IEEE JSSC,* Aug. 1980, pp. 397–406.

Russo, R. L., "Partitioning for LSI: On the tradeoff between logic performance and circuit-to-pin ratio," IBM Research RC 3191, Dec. 21, 1970, 8 pp.

Ryan, J., *Electronic Assembly,* Prentice-Hall, 1980, 166 pp.

Sakai, T., et al. (NTT), "Elevated electrode integrated circuits," *IEEE TED,* Apr. 1979, pp. 379–385.

Salkin, H. M., *Integer Programming,* Addison-Wesley, 1975, 537 pp.

Sander, W. B., J. M. Early, and T. A. Longo (Fairchild), "A 4096 × 1 (I^{3}L) bipolar RAM," *ISSCC 76,* pp. 182, 183.

Sandfort, R. M. (Monsanto), and E. R. Burke, "Logic functions for magnetic bubble devices," *IEEE Trans. on Magnetics,* Sept. 1971, pp. 358–361.

Santoni, A., "Computer's multilayer ceramic modules permit greater circuit-packaging density," *EDN,* June 20, 1980, pp. 79–80. (Comparison of IBM 370 and 4331/4341 packagings.)

Sasaki, T. et al. (NEC), "MIXS: A mixed level simulator for large digital system logic verification," *Proc. 17th DA Conf.,* 1980, pp. 626–633.

Sasao, T., "An application of multiple-valued logic to a design of programmable logic arrays," *Proc. 8th Int. Symp. on Multiple-Valued Logic,* 1978, pp. 65–72.

Sasao, T., and H. Terada, "Multiple-value logic and the design of programmable logic arrays with decoders," *Proc. 9th Int. Symp. on Multiple-Valued Logic,* 1979, pp. 27–37.

Sasao, T., and H. Terada, "On the complexity of shallow logic functions and the estimation of programmable logic array size," *Proc. 10th Int. Symp. on Multiple-Valued Logic,* 1980, pp. 65–73.

Sato, F., et al. (NEC-Toshiba), "A 400ps bipolar 18 bit RALU using advanced PSA," *IEEE JSSC,* Oct. 1980, pp. 802–808.

Sato, K., et al. (Mitsubishi Electric), "MIRAGE—A simple model routing program for the hierarchical layout design of IC masks," *Proc. 15th DA Conf.,* 1978, pp. 155–169.

Scavuzzo, R. J. (Bell Labs), "Digital logic and circuit design for improved power-delay product in LSI," *ICCC 80,* pp. 693–696.

Schamis, R. S., "Reduce system noise with CMOS circuits," *Electronic Design,* Dec. 6, 1973, pp. 112–115.

Scherr, A. L., "Distributed data processing," *IBM System J.,* vol. 17, no. 4, 1978, pp. 324–343.

Schilling, D. L., and C. Belove, *Electronic Circuits Discrete and Integrated,* 2nd ed., McGraw-Hill, 1979, 811 pp.

Schindler, M. J., "Computer-aided design beats trial-and-error—just find the software," *Electronic Design,* June 7, 1977, pp. 44–49.

Schindler, M. J., "Fit your μc with the right software package," *Electronic Design,* July 5, 1978, pp. 64–72.

Schindler, M. J., "Computers, big and small, still spreading as software grows," *Electronic Design,* Jan. 4, 1979, pp. 88–92.

Schindler, M. J., "Focus on software: while problems abound, so do solutions—if you can find them," *Electronic Design,* Mar. 15, 1979, pp. 103–110. (Software problems for microcomputers.)

Schindler, M. J., "Software's main challenger is a 'hybrid' that can be mass-produced," *Electronic Design,* May 24, 1979, pp. 78–80, 82, 83. (Advantages of firmware compared with those of software.)

Schindler, M. J., "System performance hinges on cpu architecture," *Electronic Design,* May 10, 1980, pp. 115–122.

Schindler, M. J., "Software," *Electronic Design,* Jan. 8, 1981, pp. 190–199.

Schindler, M. J., "Microcomputer operating systems branch into mainframe territory," *Electronic Design,* Mar. 19, 1981, pp. 179–182, 186–219.

Schindler, M. J., "New architectures keep pace with throughput needs," *Electronic Design,* May 4, 1981, pp. 97–102, 104, 106.

Schmid, H., "BCD: Logic of many uses," *Electronic Design,* June 21, 1973, pp. 90–95.

Schmid, H. (GE), *Decimal Computation,* John Wiley, 1974, 266 pp.

Schmidt, L. (Hewlett-Packard), "Implementing a digital filter in custom LSI," *Computer Design,* "Chip area considerations," June 1979, pp. 184, 185, 188; "Reducing multiplier area," July 1979, pp. 180, 182, 185.

Schmookler, M. S., "Design of large ALUs using multiple PLA macros," *IBM JRD,* Jan. 1980, pp. 2–14.

Schnable, G. L., L. J. Gollace, and H. L. Pujol (RCA), "Reliability of CMOS integrated circuits," *Computer,* Oct. 1978, pp. 6–17.

Schroeder, K. (RCA), "Software: Micros vs. minis," *Digital Design,* April 1979, pp. 20–22, 24, 26.

Scrupski, S. E., "TI looks to VLSI with memory-to-memory architectures," *Electronic Design,* Oct. 11, 1980, pp. 33, 34.

Seeds, R. B., "Yield and cost analysis of bipolar LSI," *IEDM,* Oct. 1967.

Seely, J. H., and R. C. Chu (IBM), *Heat Transfer in Microelectronic Equipment—A Practical Guide,* Marcel Dekker, 1972, 350 pp.

Seidman, A. H., "A new world of glassy semiconductor," in *1982 Yearbook of Science and Technology,* Encyclopaedia Britannica, 1981, pp. 164–181.

Selberherr, S., W. Fichtner, and H. W. Pötzl, "MINIMOS—A program package to facilitate MOS device design and analysis," in *Numerical Analysis of Semiconductor Devices,* edited by B. T. Browne and J. J. H. Miller, Boole Press, Dublin, 1979, pp. 275–279.

Seliger, R. L., and P. A. Sullivan (Hughes), "Ion beam promise practical systems for submicrometer wafer lithography," *Electronics,* Mar. 27, 1980, pp. 46, 48, 50, 142–146.

Senzig, D. (Hewlett-Packard), "Calculator algorithms," *Compcon 75,* Spring, pp. 139–141.

Sequin, C. H., and M. Tompsett, *Charge Transfer Devices,* Academic Press, 1975, 309 pp.

Shackil, A. F., "Design case history: Singer's electronic sewing machine," *Spectrum,* Feb. 1981, pp. 40–43. (Details of Singer Athena 2000).

Shackil, A. F., "Microprocessors and the M. D.," *Spectrum,* Apr. 1981, pp. 45–49. (Smart medical equipment.)

Shah, P. (Texas Instruments), "A unified device—process optimization—A key to VLSI design," *ICCC 80,* pp. 446–449.

Shanks, R. R. (Energy Conv.), "Amorphous semiconductors for electrically alterable memory applications," *Computer Design,* May 1974, pp. 94–100.

Shanks, R. R., and C. Davis (Burroughs), "A 1024-bit nonvolatile 15 ns bipolar read-write memory," *ISSCC 78,* pp. 112, 113.

Shapiro, G. (Analogic), "Exploit LSI memory components, today, instead of waiting for arithmetic devices," *Wescon,* 1979, 18/5, 4 pp. (sin and cos in ROMs.)

Sherwood, W., "PLATO-PLA translator/optimizer," *Proc. Design Automation and Microprocessors,* 1977, pp. 28–35.

Shi, S. Y. (NCR), "Shortcut to logarithms combines table, lookup and computation," *Computer Design,* May 1976, pp. 184, 186.

Shichman, H., and D. A. Hodges, "Modeling and simulation of insulated-gate field-effect transistor switching circuits," *IEEE JSSC,* 1968, pp. 285–289.

Shima, M. (Zilog), "Two versions of 16-bit chip span microprocessor, minicomputer needs," *Electronics,* Dec. 21, 1978, pp. 81–88. (Z8000.)

Shima, M. (Intel), "Demystifying microprocessor design," *Spectrum,* July 1979, pp. 22–30.

Shima, M., and F. Faggin (Intel), "In switch to *n*-MOS microprocessor gets on 2μs cycle time," *Electronics,* Apr. 18, 1974, pp. 95–100.

Shima, M., F. Faggin, and S. Mazor (Intel), "An *n*-channel 8-bit single chip microprocessor," *ISSCC 74,* pp. 56, 57, 229. (8008 and 8080.)

Shima, M., F. Faggin, and R. Ungermann (Zilog), "Z-80 chip sets heralds third microprocessor generation," *Electronics,* Aug. 19, 1976, pp. 89–93.

Shinozaki, S., and Y. Nishi (Toshiba), "Integrated injection logic," *J. Institute of Electronics Communication Engineers of Japan,* Jan. 1978, pp. 46–54 (in Japanese).

Shinozaki, T., "Computer program for designing optimal networks with MOS gates," Rep. UIUCDCS-R-72-502, Dept. Computer Science, University of Illinois, Apr. 1972, 118 pp.

Shirakawa, I., et al., "A layout system for the random logic portion of MOS LSI," *Proc. 17th DA Conf.,* 1980, pp. 92–99. Also *IEEE TC,* Aug. 1981, pp. 572–581.

Sigel, E., et al., *Videotext: The Coming Revolution in home/office information retrieval,* Knowledge Industry Publications, 1980, 154 pp.

Sigg, H., and S. Lai (Signetics), "D-MOS for fast low power digital circuits," *Compcon 77,* Spring, pp. 100–102.

Sinha, B. P., et al., "Hardware implementation of assembler using associative memories," *ICCC 80,* pp. 1–3, 3A.

Skokan, Z. E. (Hewlett-Packard), "Emitter function logic—Logic family for LSI," *IEEE JSSC,* Oct. 1973, 356–361.

Slater, S. P., and A. M. Cox (Ferranti), "A bipolar gate array family with a wide application-performance coverage," *Wescon,* 1979, 13/3, 7 pp.

Smith, D. E. (Rockwell), "Multi-function single chip microcomputers," *Electro 80,* 29/3, 10 pp.

Smith, F., and R. Yu (National Semiconductor), "Single-supply 16-k, 64-k RAMs simplify upgrading," *Electronic Design*, May 24, 1980, pp. 85–88.

Smith, J. E. (Burroughs), Ed., *Integrated Injection Logic*, John Wiley, 1980, 421 pp.

Smith, P. W., and W. J. Tomlinson (Bell Labs), "Bistable optical devices promise subpicosecond switching," *Spectrum*, June 1981, pp. 26–33.

Smith, R. J., *Circuits, Devices, and Systems*, 3rd ed., John Wiley, 1976, 767 pp.

Smith, R. J., *Electronics: circuits and devices*, 2nd ed., John Wiley, 1980, 494 pp.

Smith, S. (Integrated Computer Systems), "External arithmetic processors," *Computer Design*, Dec. 1978, pp. 144–149. (Survey of multiplier chips.)

Smith, S., and E. R. Garen (Integrated Computer Systems), "CMOS on sapphire," *Computer Design*, Sept. 1978, pp. 194, 196, 198.

Smith, W., and S. B. Crook (Texas Instruments), "Phonemes, allophones, and LPC team to synthesize speech," *Electronic Design*, June 25, 1981, pp. 121–127.

Smolin, M. (National Semiconductor), "Microprocessor trends," *Wescon, 1978*, 20/4, 4 pp.

Solomon, P. M. (IBM), "The performance of GaAs logic gates in LSI," *IEDM*, 1978, pp. 201–204.

Solomon, P. M., and D. D. Tang (IBM), "Bipolar circuit scaling," *ISSCC 79*, pp. 86, 87.

Soukup, J. (Bell Northern Res.), "Fast maze router," *Proc. 15th. DA Conf.*, 1978, pp. 100–102.

Soukup, J. (Bell Northern Res.), "Global router," *Proc. 16th DA Conf.*, 1979, pp. 481–484.

Southard, R. K. (AMP), "Interconnection system approaches for minimizing data transmission problem," *Computer Design*, Mar. 1981, pp. 107–116.

Spence, H. W. (Texas Instruments), "Computer aids for semiconductor product design," *1979 Wescon Professional Program*, 26/3, 5 pp. Also *Electronic Design*, Sept. 1, 1979, pp. 45, 46.

Spencer, R. F., Jr., "Complex gates in digital systems design," *IEEE Computer Group News*, Sept. 1969, pp. 47–56. Correction January/February, 1970, p. 41.

Stakem, P. H. (OAO Corp.), "Weigh hardware and software options to optimize your μP design solutions," *Electronic Design*, Sept. 13, 1978, pp. 106-110. (Hardware–software tradeoff for UART.)

Starbuck, D. E. (Rockwell), "Single-chip microcomputer experience pays off in success formula," *Wescon, 1978*, 27/1, 10 pp.

Stark, M. (Intel), "Two bits per cell ROM," *Compcon 81*, Spring, pp. 209–212.

Stauffer, M. K. (AMD), "Math processor chips boost μc computing power," *EDN*, Aug. 20, 1980, pp. 113–120.

Stebnisky, W. W., and A. Feller (RCA), "State-of-the-art CMOS/SOS technology for next generation computers," *ICCC 80*, pp. 329–332.

Steele, G. L., Jr., and G. J. Sussman, "Design of a LISP-based microprocessor," *Commun. ACM*, Nov. 1980, pp. 628–645.

Stehlin, R. A. (Texas Instruments), "Two Schottky TTL families," *Computer Design*, July 1980, pp. 154, 156, 158.

Steinberg, D. S. (Litton), *Cooling Techniques for Electronic Equipment*, John Wiley, 1980, 400 pp.

Stenzel, W. J., W. J. Kubitz, and G. H. Garcia, "A compact high-speed parallel multiplication scheme," *IEEE TC*, Oct. 1977, pp. 948–957.

Stephan, G. A. (Interdesign), "Semicustom integrated circuits—A rapid approach to reducing circuit size," *Wescon, 1978*, 32/2, 4 pp.

Stephenson, K. (Harris), "Avoid CMOS noise sensitivity problems," *Electronic Design*, Dec. 6, 1975, pp. 80–84.

Stetson, R. J. (Storage Technology), "Select a character/function decoder," *Electronic Design*, July 5, 1977, pp. 80–83.

Stewart, R. G. (RCA), "High density CMOS ROM arrays," *ISSCC 77*, pp. 20, 21. Also *IEEE JSSC*, Oct. 1977, pp. 502–506.

Stewart-Warner Corp., *SWAP Design Manual*, 1976.

Stiglianese, M. (Cincinnati Electronics), "Interface CMOS logic with switches using standards IC," *Electronic Design*, Aug. 16, 1974, pp. 80–83.

Stillman, J., and P. B. Berra, "A comparison of hardware and software associative memory in the context of computer graphics," *Commun. ACM*, May 1977, pp. 331–339.

Stinehelfer, J., et al. (Fairchild), "Large ECL bipolar RAMs," *Compcon 81*, Spring, pp. 120–124.

Streetman, B. G., *Solid State Electronic Devices*, 2nd ed., Prentice-Hall, 1980, 461 pp.

Stritter, E., and T. Gunter (Motorola), "A microprocessor architecture for a changing world: The Motorola 68000," *Computer*, Feb. 1979, pp. 43–52.

Stritter, E., and N. Tredennick (Motorola), "Microprogrammed implementation of a single chip microprocessor," *SIGmicronewsletter*, Dec. 1978, pp. 11–19.

Su, S. C., "Low-temperature silicon processing techniques for VLSI fabrication," *Solid State Technology*, Mar. 1981, pp. 72–82.

Sud, R., and K. C. Hardee (Inmos), "16-k static RAM takes new route to high speed," *Electronics*, Sept. 11, 1980, pp. 117–123.

Sugarman, R., "'Superpower' computers," *Spectrum*, Apr. 1980, pp. 28–34.

Sullivan, R. M. (Honeywell), "Challenges of the 80's packaging LSI chips," *Proc. Conf. Computing in the 1980s*, 1978, pp. 251–253.

Sumney, L. W., "VLSI with a vengeance," *Spectrum*, Apr. 1980, pp. 24–27. (VHSIC program.)

Suran, J. J., "A perspective on integrated electronics," *Spectrum*, Jan. 1970, pp. 67–79.

Suzuki, Y., K. Odagawa, and T. Abe (Toshiba), "Clocked CMOS calculator circuitry," *IEEE JSSC*, Dec. 1973, pp. 462–469.

Swartzlander, E. E., Jr., Ed., *Computer Arithmetic*, Dowden, Hutchinson and Ross, 1980, 379 pp.

Szygenda, S. A., and E. W. Thompson, "Digital logic simulation in a time-based, table-driven environment," *Computer*, Mar. 1975, Part 1, "Design verification," pp. 24–36; Part 2, "Parallel fault simulation," pp. 38–49.

Tanabe, N., H. Nakamura, and K. Kawakita (NEC), "MOSTAP: An MOS circuit simulator for LSI circuits," *IEEE 1980 Int. Symp. on Circuits Systems*, pp. 1035–1038.

Tanabe, N., and M. Yamamota (NEC), "Single chip PASCAL processor: Its architecture and performance evaluation." *Compcon 80*, Fall, pp. 395–399.

Tanaka, C., et al. (Mitsubishi Electric), "An integrated computer aided design system for gate array masterslices," *Proc. 18th DA Conf.*, 1981, pp. 812–819.

Tang, C. K. (IBM), "A storage cell reduction technique for ROS design," *AFIPS*, 1971, pp. 163–170.

Tang, D. D., et al. (IBM), "Subnanosecond self-aligned I²L/MTL circuits," *IEEE JSSC*, Aug. 1980, pp. 444–449.

Tanimoto, M., J. Murota, Y. Ohmori, and N. Ieda (NTT), "A novel MOS PROM using a highly resistive poly-Si resistor," *IEEE TED*, Mar. 1980, pp. 517–520.

Tarbox, E. F. (Sylvania Electric Products), "Converting from binary to binary coded decimal numbers," in *Semiconductor Memories*, edited by J. Eimbinder, John Wiley, 1971, pp. 193–203.

Tarui, Y., Y. Hayashi, and T. Sekigawa, "Diffusion self-aligned enhance-depletion MOS-IC (DSA-ED-MOS-IC)" *Proc. 2nd Conf. Solid State Devices*, Tokyo, Japan, 1970, pp. 193–198.

Tarui, Y., Y. Hayashi, H. Teshima, and T. Sekigawa, "Transistor Schottky-barrier-diode integrated logic circuit," *IEEE JSSC*, Feb. 1969, pp. 3–12.

Tasch, A. F., Jr., P. K. Chatterjee, H-S. Fu, and T. C. Holloway (Texas Instruments), "The Hi-C RAM cell concept," *IEEE TED*, Jan. 1978, pp. 33–41.

Taub, H., and D. Schilling, *Digital Integrated Electronics*, Mc-Graw-Hill, 1977, 650 pp.

Teja, E. R., "Voice input and output," *EDN*, Nov. 20, 1979, pp. 159–167.

Texas Instruments, "MOS programmable logic arrays," Application Bulletin CA-158.

Texas Instruments, "Types SN 74S 330, SN 74S 331 expandable 12-input, 50-term field-programmable logic arrays," data sheet (Schottky TTL FPLA with response time 35 ns), Feb., 1979.

Theis, D. J., "An overview of memory technologies," *Datamation*, Jan. 1978, pp. 113–131.

Thompson, C. D., "Fourier transforms in VLSI," *ICCC 80*, pp. 1046–1051.

Thurber, K. J. (Honeywell), "Universal logic modules implemented using LSI memory techniques," *AFIPS*, 1971, pp. 177–194.

Titus, J. A., et al., "Interfacing fundamentals: Lookup tables," *Computer Design*, Feb. 1979, pp. 130–132, 134.

Tjaden, G., and N. Cohn (Univac), "Some considerations in the design of mainframe processors with microprocessor technology," *Computer*, Aug. 1979, pp. 68–74.

Tobey, A. C. (GCA Corp.), "Wafer stepper steps up yield and resolution in IC lithography," *Electronics*, Aug. 16, 1979, pp. 109–112.

Todd, S., "Algorithm and hardware for a merge sort using multiple processors," *IBM JRD*, Sept. 1978, pp. 509–517.

Toffler, A., *The Third Wave*, Morrow, 1980, 517 pp.

Tomisawa, O., et al. (Mitsubishi Electric), "A 920 gate DSA MOS masterslice," *IEEE JSSC*, Oct. 1978, pp. 536–541. Also *ISSCC 78*, pp. 64, 65, 267.

Torimaru, Y., K. Miyano, and H. Takeuchi (Sharp), "DSA 4k static RAM," *IEEE JSSC*, Oct. 1978, pp. 647–650.

Torrero, E. A., "Focus on CMOS," *Electronic Design*, Mar. 15, 1974, pp. 86–95.

Toyabe, T. and S. Asai (Hitachi), "Analytical models of threshold voltage and breakdown voltage of short-channel MOSFETs derived from two-dimensional analysis," *IEEE JSSC*, Apr. 1979, pp. 375–383. (CADDET.)

Travis, T. (Aerospace Corp.), "Patching a program into a ROM," *Electronic Design*, Sept. 1, 1976, pp. 98–101.

Tredennick, N. (IBM), "How to flowchart for hardware," *Computer*, Dec. 1981, pp. 87–102. (Flowcharting for the control logic of 68000.)

Tsantes, J. "Programmable-memory choices expand design options," *EDN*, Jan. 5, 1980, pp. 80–98. (Are EAROMs losing out to EEPROMs?)

Tsantes, J., "Silicon-on-sapphire technology advances toward VLSI visibility," *EDN*, Sept. 20, 1980, pp. 55–60, 62, 64.

Tsantes, J., "Leadless chip carriers revolutionize IC packaging," *EDN*, May 27, 1981, pp. 49–58, 66–68, 70, 72, 74.

Tsukiyama, S., E. S. Kuh, and I. Shirakawa, "On the layering problem of multilayer PWB wiring," *Proc. 18th DA Conf.*, 1981, pp. 738–745.

Turner, T. (Mostek), "IC failure-rate calculations evaluate realistically," *EDN*, Apr. 15, 1981, pp. 111–114.

Twaddell, W., "Uncommitted IC logic," *EDN*, Apr. 19, 1980, pp. 89–98. (List of characteristics of PLAs, FPLAs, PROMs, and gate arrays available from different firms.)

Twaddell, W., "ICs and semiconductors," *EDN*, July 20, 1980, pp. 74, 83, 84, 86, 88, 90, 92, 94. (Survey of multipliers and floating-point processor chips.)

Twaddell, W., "Semiconductor RAMs," *EDN*, Sept. 20, 1980, pp. 118–130, 132.

Twaddell, W., "EEPROMs gain in density and speed, threaten to displace UVEPROMs," *EDN*, Jan. 21, 1981, pp. 37–46.

Twaddell, W., "CMOS ICs," *EDN*, June 24, 1981, pp. 89–100.

Uchiumi, K., and T. Makimoto (Hitachi), "16-k EE-PROM keeps MNOS in the running," *Electronics*, Feb. 24, 1981, pp. 154–156.

Ueda, M., Y. Sugiyama, and K. Wada (NTT), "An automatic layout system for masterslice LSI: MARC," *IEEE JSSC*, Oct. 1978, pp. 716–721.

Vacca, A. A., "The case for emitter-coupled logic," *Electronics*, Apr. 26, 1971, pp. 48–52.

Vacca, A. A. (Control Data), "Considerations for high performance LSI applications," *Compcon 79*, Spring, pp. 278–284.

Vacroux, A. G., "Microcomputers," *Scientific American*, May 1975, pp. 32–40.

van Cleemput, W. M., *Computer Aided Design of Digital Systems: A Bibliography,* Computer Science Press, vol. I, "1960–1974," 1976; 374 pp.; vol. II, "1975–1976," 1976, 278 pp.; vol. III, "1976–1977," 1977.

van Cleemput, W. M., Ed., *Computer-Aided Design Tools for Digital Systems,* 2nd ed., IEEE, 1979.

van Cleemput, W. M., "Computer hardware description languages and their applications," *Proc. 16th DA Conf.,* 1979, pp. 554–560.

van Cleemput, W. M., et al., "HIDRIS: A hierarchical system for the design and realization of integrated systems," *Proc. 3rd USA–Japan Comp. Conf.,* 1978, pp. 394–400.

Van Tuyl, R., and C. A. Liechti (Hewlett-Packard), "Gallium arsenide spawns speed," *Spectrum,* Mar. 1977, pp. 40–47.

Van Tuyl, R. L., C. A. Liechti, R. E. Lee, and E. Gowen (Hewlett-Packard), "GaAs MESFET logic with 4-GHz clock rate," *IEEE JSSC,* Oct. 1977, pp. 485–496.

Varsos, S. G., H. V. Taylor, and W. E. Bentley (Lockheed Electronics), "It's time to convert to complex BCD," *Electronic Design,* Jan. 18, 1978, pp. 86–91.

Volkelis, W. V., and R. A. Henle (IBM), "Performance vs. circuit packing density," *Compcon 79,* Spring, pp. 285–289.

Vodovoz, E., "Programmable logic arrays: A dormant giant awakening," Designer's roundtable, *EDN,* Mar. 5, 1975, pp. 29–35. (Comparison of PLAs with ROMs and microcomputers.)

Volder, J. E. (General Dynamics), "The CORDIC trigonometric computing technique," *IRE TEC,* Sept. 1959, pp. 330–334.

Voyer, J. L. (Martin Marietta), "Approximate logs easily," *Electronic Design,* Sept. 26, 1974, pp. 176–182.

Wagner, F. V. (Informatics), "Is decentralization inevitable?" *Datamation,* Nov. 1976, pp. 86–87, 91, 93, 97. (Emulation is easier than writing new operating system.)

Walker, G. M., "The HP-35: A tale of teamwork with vendors," *Electronics,* Feb. 1, 1973, pp. 102–106.

Walker, G. M., "LSI controls gaining in home appliances," *Electronics,* Apr. 14, 1977, pp. 91–99. (Use of microcomputers in microwave ovens, air-conditioners, and dryers.)

Walker, R. (Fairchild), "Converting TTL logic to ECL," *Computer Design,* Oct. 1973, pp. 102, 104, 108.

Walker, R. (Fairchild), "CMOS specifications: Don't take them for granted," *Electronics,* Jan. 9, 1975, pp. 103–107.

Wallace, R. K., and A. J. Learn (Intel), "Simple process propels bipolar PROMs to 16-k density and beyond," *Electronics,* Mar. 27, 1980, pp. 147–150. (Cleanly burned fuses don't grow back.)

Wallmark, J. T., and J. G. Carlstedt, *Field Effect Transistors in Integrated Circuits,* John Wiley, 1974, 153 pp.

Walther, J. S. (Hewlett-Packard), "Unified algorithm for elementary functions," *SJCC 1971,* pp. 379–385.

Wang, A., "Calculating apparatus," U.S. patent no. 3,402,285, Sept. 17, 1968.

Warner, R. M., Jr., "Comparing MOS and bipolar integrated circuits," *Spectrum,* June 1967, pp. 50–58.

Warner, R. M., Jr., "Applying a composite model to the IC yield problem," *IEEE JSSC,* June 1974, pp. 86–95.

Warner, R. M., Jr., "I-squared L: a happy merger," *Spectrum,* May 1976, pp. 42–47.

Warren, C., "High-performance buses clear a path for future MCs," *EDN,* June 24, 1981, pp. 157–187.

Washburn, J. (Computer Automation), "Low power computers: A make or buy decision," *Computer Design,* Nov. 1976, pp. 120–122, 124. (Economy of microprocessors.)

Watanabe, M., Ed., *LSI Technology,* Denki-tsushin-gakkai, Tokyo, 1979, 327 pp. (in Japanese).

Watkins, B. G., "A low-power multiphase circuit technique," *IEEE JSSC,* Dec. 1967, pp. 213–220.

Watson, D., "Thermal design," in *Reliability and Maintainability of Electronic Systems,* edited by J. E. Arsenault and J. A. Roberts, Computer Science Press, 1980, pp. 162–187, Chap. 9.

Waxman, R. (IBM), "VLSI—A design challenge," *Proc. 16th DA Conf.*, 1979, pp. 546, 547.

Weber, E. V., and H. S. Yourke (IBM), "Scanning electron-beam system turns out IC wafers fast," *Electronics*, Nov. 10, 1977, pp. 96–101.

Weeks, W. T., et al. (IBM), "Algorithm for ASTAP—A network analysis program," *IEEE TCT*, Nov. 1973, pp. 628–634.

Wegner, P., "The ADA language and environment," *Electro 80*, 31/3, 7 pp.

Weinberger, A., "Large scale integration of MOS complex logic: A layout method," *IEEE JSSC*, Dec. 1967, pp. 182–190.

Weinberger, A., "High-speed programmable logic array adders," *IBM JRD*, Mar. 1979, pp. 163–178. (Look-ahead adder in decoded PLA.)

Weindling, M. N., "A method for best placement of units on a plane," *Proc. DA Workshop*, 1964.

Weisbecker, J. (RCA), "A practical low-cost, home/school microprocessor system," *Computer*, Aug. 1974, pp. 20–30.

Weiss, D. G. (Motorola), "Computer architecture in LSI—Lessons and prospects," *Compcon 81*, Spring, pp. 48–52.

Welch, B. M. (Rockwell), "Advances in GaAs LSI/VLSI processing technology," *Solid State Technology*, Feb. 1980, pp. 95–101.

Werbizky, G. G., P. Winkler, and F. W. Haining (IBM), "Making 100,000 circuits fit where at most 6,000 fit before," *Electronics*, Aug. 2, 1979, pp. 109–114. Correction, Aug. 16, 1979, p. 6.

Werbizky, G. G., P. E. Winkler, and F. W. Haining (IBM), "Packaging technology for the IBM 4300 processors," *Compcon 80*, Spring, pp. 304–311.

White, R. M., "Disk-storage technology," *Scientific American*, Aug. 1980, pp. 138–148.

Whitney, T. M., F. Rodé, and C. C. Tung, "The 'powerful pocketful': An electronic calculator challenges the slide rule," *Hewlett-Packard J.*, June 1972, pp. 2–9.

Wickham, K. (Texas Instruments), "Pascal is a 'natural'", *Spectrum*, Mar. 1979, pp. 35–41.

Wiemann, W. (Motorola), "CAD system for VLSI," *Proc. 16th DA Conf.*, 1979, p. 550.

Wiggins, R., and L. Brantigham (Texas Instruments), "Three-chip system synthesizes human speech," *Electronics*, Aug. 31, 1978, pp. 109–116.

Wilcox, P., and H. Rombeek (Bell-Northern Res.), "F-LOGIC—An interactive fault and logic simulator for digital circuits," *Proc. 13th DA Conf.*, 1976, pp. 68–73. Also *Electronics*, Sept. 13, 1979, p. 152.

Wilenken, D. (Intersil), "Reduce CMOS-multiplexer troubles through proper device selection," *EDN*, Jan. 20, 1979, pp. 75–79.

Wiles, M. F., and S. Lamb (Motorola), "Special-purpose processor makes short work of host's I/O chores," *Electronics*, May 19, 1981, pp. 165–168.

Williams, J. D. (Hewlett-Packard), "STICKS—A graphical compiler for high level LSI design," *NCC*, 1978, pp. 289–295.

Williams, M. J. Y., and J. B. Angell, "Enhancing testability of large-scale integrated circuits via test points and additional logic," *IEEE TC*, Jan. 1973, pp. 46–60.

Williams, T., "Computer terminals and printers get even smarter," *Electronic Design*, May 14, 1981, pp. 113–124, 126.

Williams, W. T. (IBM), and K. P. Parker (Hewlett-Packard), "Testing logic networks and designing for testability," *Computer*, Oct. 1979, pp. 9–21.

Wilson, D. R., and P. R. Schroeder (Mostek), "A 100 ns 150 mw 64k bit ROM," *ISSCC 78*, pp. 152, 153, 273.

Wimmer, K. (Tektronix), "Field-programmable patches simplify fireware maintenance," *EDN*, Apr. 29, 1981, pp. 139–142, 144.

Winder, R. O. (RCA), "COSMAC—A COM/MOS microprocessor," *ISSCC 74*, pp. 64, 65.

Winder, R. O. (RCA), "The COS/MOS microprocessor," *Wescon*, 1974, 15/4, 6 pp.

Winter, K., and A. Wagner-Korne (National Semiconductor), "All-CMOS components build a μc system that checks power loss, erases regulation," *Electronic Design,* Apr. 12, 1980, pp. 141–146. (P²CMOS.)

Wise, K. D., K. Chen, and R. E. Yokely, *Microcomputers—A Technology Forecast and Assessment to the Year 2000,* John Wiley, 1980, 251 pp.

Withington, F. G. (A. D. Little), "The golden age of packaged software," *Datamation,* Dec. 1980, pp. 131, 132, 134.

Wolfe, C. F. (Univac), "Bit-slice processors come to mainframe design," *Electronics,* Feb. 28, 1980, pp. 118–123.

Wolfendale, E., Ed., *MOS Integrated Circuit Design,* Halsted–John Wiley, 1973, 120 pp.

Wollesen, D. L. (AMI), "C-MOS LSI: Comparing second-generation approaches," *Electronics,* Sept. 13, 1979, pp. 116–123.

Wollesen, D. L. (AMI), "CMOS LSI—The computer component process of the 80's," *Computer,* Feb. 1980, pp. 59–67.

Wong, B. W., and W. D. Jackson, "A high-performance bipolar LSI counter chip using EFL and I²L circuits," *Hewlett-Packard J.,* Jan. 1979, pp. 12–17.

Wood, R. A., "High-speed dynamic programmable logic array chip," *IBM JRD,* July 1975, pp. 379–383.

Wood, R. A. (IBM), "A high density programmable logic array chip," *IEEE TC,* Sept. 1979, pp. 602–608.

Wood, R. A., Y. N. Hsieh, C. A. Price, and P. P. Wang (IBM), "An electrically alterable PLA for fast turnaround-time VLSI development hardware," *IEEE JSSC,* Oct. 1981, pp. 570–577.

Woods, M. H. (Intel), "An E-PROM's integrity starts with its cell structure," *Electronics,* Aug. 14, 1980, pp. 132–136.

Woods, R. (Electronic Arrays), "Evaluate UV EPROM data retention," *Electronic Design,* Mar. 29, 1978, pp. 82–84.

Wu, L. C. (Amdahl), "VLSI and mainframe computers," *Compcon 78,* Spring, pp. 26–29.

Wyland, D. C. (Monolithic Memories), "Using PROMs as logic elements," *Computer Design,* Sept. 1974, pp. 98–100. (Use of PROM in output interface of Data General's Nova.)

Wyland, D. C. (Monolithic Memories), "Shift registers can be designed using RAMs and counter chips," *EDN,* Jan. 5, 1974, pp. 64–67.

Xylander, M. P. (IBM), "Low-power bipolar technique begets low-power LSI logic," *Electronics,* July 31, 1972, pp. 80–82. (TTL for LSI.)

Yamada, A. (NEC), "Design automation statistics in Japan," *Proc. 18th DA Conf.,* 1981, pp. 43–50.

Yamamoto, K., "Design of irredundant MOS networks: A program manual for the design algorithm DIM," Rep. UIUCDCS-R-76-784, Dept. of Computer Science, University of Illinois, Feb. 1976, 120 pp.

Yamamoto, M., et al., "Design of a COBOL oriented high level language machine," *NEC Research and Development,* Apr. 1980, pp. 29–38.

Yano, S., et al. (NEC and NTT), "A masterslice approach to low energy high speed LSIs," *ICCC 80,* pp. 203–206.

Yasutoshi, Y., and K. Kani (NEC), "A survey of computer aided electronic circuit design," *J. Institute of Electronics Communication Engineers of Japan,* July 1978, pp. 724–730 (in Japanese).

Yeh, C. C., "Design of irredundant multiple-level MOS networks for multiple-output and incompletely specified function," Rep. UIUCDCS-R-77-896, Dept. of Computer Science, University of Illinois, Sept. 1977, 125 pp.

Yen, M. Y. (Intel), "Fast emulator debugs 8085-based microcomputers in real time," *Electronics,* July 21, 1977, pp. 108–112.

Yen, T. T. (Stathan Instruments, Inc.), "CMOS flip-flop can do more than logic tasks," *Electronics,* Mar. 20, 1975, pp. 123–126.

Yen, Y. T., "Transient analysis of four-phase MOS switching circuits," *IEEE JSSC,* Mar. 1968, pp. 1–5.

Yen, Y. T., "A mathematical model characterizing four-phase MOS circuits," *IEEE TC*, Sept. 1968, pp. 822–826.

Yen, Y. T., "A method of automatic fault-detection test generation for four-phase MOS LSI circuits," *SJCC*, 1969, pp. 215–220.

Yen, Y. T., "Computer-aided test generation for four-phase MOS LSI circuits," *IEEE TC*, Oct. 1969, pp. 890–894.

Yenni, D. R., Jr., and J. R. Huntsman (3M), "Guarding ICs against static discharge," *Electronics*, July 17, 1980, pp. 115–121.

Yoshida, K., et al. (Toshiba), "A layout checking system for large scale integrated circuits," *Proc. 14th DA Conf.*, 1977, pp. 322–330.

Young, M. H., "The minimal covering problem and automated design of two-level AND/OR optimal networks," Ph.D. dissertation, Dept. of Computer Science, University of Illinois, 1978, (Computational efficiency improvement of minimization for symmetric functions by group theory), 186 pp.

Young, M. H., "Program manual of programs for minimal covering problems: ILLOD-MINIC-B, ILLOD-MINIC-BP, ILLOD-MINIC-BC, ILLOD-MINIC-BA, ILLOD-MINIC-BG," Rep. UIUCDCS-R-78-924, Dept. of Computer Science, University of Illinois, 1978, 129 pp.

Young, M. H. (STC-Microtechnology), "PRIM: A placement and routing implementation system for master-slice LSI chip design," *Compcon 81*, Spring, pp. 417–420.

Young, M. H., and R. B. Cutler, "Program manual for the programs ILLOD-MINSUM-CBS, ILLOD-MINSUM-CBSA, ILLOD-MINSUM-CBG, and ILLOD-MINSUM-CBGM to derive minimal sums for irredundant disjunctive forms for switching functions," Rep. UIUCDCS-R-78-926, Dept. of Computer Science, University of Illinois, 1978, 43 pp.

Young, S. (Mostek), "Memories have hit a density ceiling but new processes will push through," *Electronic Design*, Oct. 25, 1978, pp. 42–49.

Yourdon, E. N., Ed., *Classics in Software Engineering*, Yourdon Press, 1979, 424 pp.

Zelkowitz, M. V., "Perspectives on software engineering," *ACM Computing Review*, June 1978, pp. 197–216.

Zeskind, D., "Use of automotive electronics grows, but applications shift their focus," *EDN*, Jan. 7, 1981, pp. 61, 62, 64–66, 68.

Zimmerman, T. A., R. A. Allen, and R. W. Jacobs (TRW), "Digital charge-coupled logic (DCCL)," *IEEE JSSC*, Oct. 1977, pp. 473–485.

Zimmerman, T. A. (TRW), and D. F. Barbe (Naval Res.), "A new role for charge-coupled devices: Digital signal processing," *Electronics*, Mar. 31, 1977, pp. 97–103. (Logic operations by CCD).

Zimmermann, G., "Computer aided design of control structures for digital computers," *ICCC 80*, pp. 103–106.

Zing, C. P. (Intel), "Development system puts two processors on speaking terms," *Electronics*, July 31, 1980, pp. 93–97.

Index

Pages in *italics* refer to key terms.